Lecture Notes in Computer Science 6612

Commenced Publication in 1973
Founding and Former Series Editors:
Gerhard Goos, Juris Hartmanis, and Jan van Leeuwen

Xiaoyong Du   Wenfei Fan   Jianmin Wang
Zhiyong Peng   Mohamed A. Sharaf (Eds.)

Web Technologies and Applications

13th Asia-Pacific Web Conference, APWeb 2011
Beijing, China, April 18-20, 2011
Proceedings

 Springer

Volume Editors

Xiaoyong Du
Renmin University of China, School of Information, Beijing 100872, China
E-mail: duyong@ruc.edu.cn

Wenfei Fan
University of Edinburgh, LFCS, School of Informatics
10 Crichton Street, Edinburgh, EH8 9AB, UK
E-mail: wenfei@inf.ed.ac.uk

Jianmin Wang
Tsinghua University, School of Software, Beijing 100084, China
E-mail: jimwang@tsinghua.edu.cn

Zhiyong Peng
Wuhan University, Computer School
Luojiashan Road, Wuhan, Hubei 430072, China
E-mail: peng@whu.edu.cn

Mohamed A. Sharaf
The University of Queensland
School of Information Technology and Electrical Engineering
St. Lucia, QLD 4072, Australia
E-mail: m.sharaf@uq.edu.au

ISSN 0302-9743   e-ISSN 1611-3349
ISBN 978-3-642-20290-2   e-ISBN 978-3-642-20291-9
DOI 10.1007/978-3-642-20291-9

Springer Heidelberg Dordrecht London New York

Library of Congress Control Number: 2011924095

CR Subject Classification (1998): H.2-5, C.2, J.1, K.4

LNCS Sublibrary: SL 3 – Information Systems and Application, incl. Internet/Web
and HCI

Typesetting: Camera-ready by author, data conversion by Scientific Publishing Services, Chennai, India

Printed on acid-free paper

Springer is part of Springer Science+Business Media (www.springer.com)

Preface

This volume contains the proceedings of the 13th Asia-Pacific Conference on Web Technology (APWeb), held April 18–20, 2011, in Beijing, China. APWeb is a leading international conference on research, development and applications of Web technologies, database systems, information management and software engineering, with a focus on the Asia-Pacific region.

Acceptance into the conference proceedings was competitive. The conference received 104 submissions overall, from 13 countries and regions. Among these submissions the Program Committee selected 26 regular papers and 10 short papers. The acceptance rate for regular papers is 25%, while the overall acceptance rate is 34.6% when short papers are included. The submissions ranged over a variety of topics, from traditional topics such as Web search, databases, data mining, to emerging emphasis on Web mining, Web data management and service-oriented computing. The submissions were divided into nine tracks based on their contents. Each submission was reviewed by at least three members of the Program Committee or external reviewers. Needless to say, to accept only 36 papers we were forced to reject many high-quality submissions. We would like to thank all the authors for submitting their best work to APWeb 2011, accepted or not.

In addition to the research papers, this volume also includes four demo papers, which were selected by a separate committee chaired by Gao Hong and Egemen Tanin. It also includes papers accompanying three keynote talks, given by Philip S. Yu, Deyi Li and Christian S. Jensen.

This volume also contains APWeb workshop papers, from the Second International Workshop on XML Data Management (XMLDM 2011) organized by Jiaheng Lu from Renmin University of China, and the Second International Workshop on Unstructured Data Management (USD 2011) organized by Tenjiao Wang from Peking University, China. These two workshops were treated as special sessions of the APWeb conference. The details on the workshop organization can be found in the messages of the workshop organizers.

APWeb 2011 was co-organized and supported by Renmin University and Tsinghua University, China. It was also supported financially by the National Natural Science Foundation of China, and the Database Society of China Computer Federation.

The conference would not have been a success without help from so many people. The Program Committee that selected the research papers consisted of 118 members. Each of them reviewed a pile of papers, went through extensive discussions, and helped us put together a diverse and exciting research program. In particular, we would like to extend our special thanks to the Track Chairs: Diego Calvanese, Floris Geerts, Zackary Ives, Anastasios Kementsietsidis, Xuemin Lin, Jianyong Wang, Cong Yu, Jeffrey Yu, Xiaofang Zhou, for leading and monitoring the discussions. We would also like to thank the APWeb conference Steering Committee, the APWeb Organizing Committee, as well as the local organizers and volunteers in Beijing, for their effort and time to help make the conference a success. We thank Qian Yi for maintaining the conference website, and the Microsoft Conference Management Tool (CMT) team for their support and help. Finally, we are extremely indebted to Yueguo Chen for overseeing the entire reviewing process, and to Tieyun Qian for handling the editing affairs of the proceedings.

March 2011

Xiaoyong Du
Wenfei Fan
Jianmin Wang
Zhiyong Peng
Mohamed A. Sharaf

In Memoriam of Prof. Shixuan Sa
(December 27, 1922 – July 11, 2010)

We dedicate this book to our respected Prof. Shixuan Sa, who passed away on July 11, 2010 after a long battle against Parkinson disease.

Professor Sa was a founder of database education and research in China. He wrote the first textbook on databases in China, with his former student Prof. Shan Wang. It is this book that introduced many Chinese scholars to database research.

Professor Sa was also a leader in the Chinese database community. The NDBC Sa Shixuan Best Student Paper Award has been established by the National Database Conference of China (NDBC) in memoriam of his contributions to the database society in China. In fact, the best student paper award was initiated and promoted by Prof. Sa for NDBC in 1984.

Shixuan Sa, Professor of Computer Science at Renmin University of China, was born in Fuzhou city, Fujian Province on December 27, 1922. He studied mathematics at Xiamen University and graduated in 1945. Prior to his move to Renmin University in 1950, he was a lecturer of mathematics at Sun Yat-Sen University and at Huabei University. He was a founder and the first Chairman of the Department of Economic Information Management at Renmin University, which is the first established academic program on information systems in China.

Many of us remember Prof. Sa as a friend, a mentor, a teacher, a leader of the Chinese database community, and a brilliant and wonderful man, a man with great vision, an open mind and a warm heart. Shixuan Sa inspired and encouraged generations of database researchers and practitioners, and was a pioneer in establishing international collaborations. His contributions to the research community, especially in database development and education, are unprecedented. Above all, he was a man admired for his humor, his modesty, and his devotion to his students, family, and friends. In the database world Prof. Sa will be remembered as the man who helped to produce the database community in China.

We would like to express our deep condolences to the family of Shixuan Sa.

Shan Wang
Jianzhong Li
Xiaoyong Du
Wenfei Fan
Jianmin Wang
Zhiyong Peng

Conference Organization

Honorary General Chair

Jiaguang Sun   Tsinghua University, China

Conference Co-chairs

Jianzhong Li   Harbin Institute of Technology, China
Philip Yu   University of Illinois at Chicago, USA
Ramamohanarao (Rao)
   Kokagiri   The University of Melbourne, Australia

Program Committee Co-chairs

Xiaoyong Du   Renmin University of China
Wenfei Fan   The University of Edinburgh, UK
Jianmin Wang   Tsinghua University, China

Local Organization Co-chairs

Xiao Zhang   Renmin University of China
Chaokun Wang   Tsinghua University, China

Workshop Co-chairs

Zhanhuai Li   Northwest Polytechnical University, China
Wookshin Han   Kyungpook National University, Korea

Tutorial Co-chairs

Ling Feng   Tsinghua University, China
Anthony K.H. Tung   National University of Singapore

Panel Co-chairs

Ge Yu   Northeast University of China
Ee-Peng Lim   Singapore Management University, Singapore

Demo Co-chairs

Hong Gao   Harbin Institute of Technology, China
Egemen Tanin   University of Melbourne, Australia

Publication Co-chairs

Zhiyong Peng   Wuhan University, China
Mohamed Sharaf   University of Queensland, Australia

Publicity Co-chairs

Shuigeng Zhou   Fudan University, China
Takeo Kunishima   Okayama Prefectural University, Japan

Treasurer Chair

Tengjiao Wang   Peking University, China

Paper Award Co-chairs

Aoying Zhou   East China Normal University, China
Katsumi Tanaka   Kyoto University, Japan

Program Committee

Program Committee Co-chairs

Xiaoyong Du	Renmin University of China
Wenfei Fan	The University of Edinburgh, UK
Jianmin Wang	Tsinghua University, China

Track Chairs

Diego Calvanese	Free University of Bozen-Bolzano
Floris Geerts	University of Edinburgh
Zackary Ives	University of Pennsylvania
Anastasios Kementsietsidis	IBM T.J. Watson Research Center at Hawthorne, USA
Xuemin Lin	University of New South Wales
Jianyong Wang	Tsinghua University
Cong Yu	Google Research
Jeffrey Yu	Chinese University of Hong Kong
Xiaofang Zhou	University of Queensland

Program Committee Members

Toshiyuki Amagasa	University of Tsukuba, Japan
Denilson Barbosa	University of Alberta, Canada
Pablo Barcelo	Universidad de Chile, Chile
Geert Jan Bex	Hasselt University, Belgium
Athman Bouguettaya	CSIRO, Australia
Zhipeng Cai	Mississippi State University, USA
Chee Yong Chan	National University of Singapore
Jae-Woo Chang	Chonbuk National University, Korea
Wenguang Chen	Peking University, China
Haiming Chen	Chinese Academy of Sciences, China
Hong Chen	Renmin University of China
Hanxiong Chen	University of Tsukuba, Japan
Jinchuan Chen	Renmin University of China
Jin Chen	Michigan State University, USA
Lei Chen	Hong Kong University of Science and Technology
Reynold Cheng	The University of Hong Kong
David Cheung	The University of Hong Kong
Richard Connor	University of Strathclyde, UK
Bin Cui	Beijing University, China

Zhiyong Peng	Wuhan University, China
Alex Poulovassilis	University of London, UK
KeunHo Ryu	Chungbuk National University, Korea
Marc Scholl	Universitaet Konstanz, Germany
Mohamed Sharaf	University of Queensland, Australia
Shuming Shi	MSRA, China
Jianwen Su	U.C. Santa Barbara, USA
Kazutoshi Sumiya	University of Hyogo, Japan
Katsumi Tanaka	Kyoto University, Japan
David Taniar	Monash University, Australia
Alex Thomo	University of Victoria, Canada
Anthony Tung	National University of Singapore
Stijn Vansummeren	Free University of Brussels, Belgium
Chaokun Wang	Tsinghua University, China
Daling Wang	Northeastern University, China
Tengjiao Wang	Peking University, China
Hua Wang	University of Southern Queensland, Australia
Hongzhi Wang	Harbin Institute of Technology, China
Wei Wang	University of New South Wales, Australia
Wei Wang	Fudan University, China
Xiaoling Wang	East China Normal University
Ji-Rong Wen	MSRA, China
Jef Wijsen	University of Mons-Hainaut, Belgium
Peter Wood	University of London, UK
Yuqing Wu	Louisianna State University, USA
Yingyuan Xiao	Tianjing University of Technology, China
Jianliang Xu	Hong Kong Baptist University
Linhao Xu	IBM Research China
Xiaochun Yang	Northeastern University, China
Jian Yin	Sun Yat-Sen University, China
Haruo Yokota	Tokyo Institute of Technology, Japan
Jae Soo Yoo	Chungbuk National University, Korea
Hwanjo Yu	POSTECH, Korea
Lei Yu	State University of New York at Binghamton, USA
Philip Yu	University of Illinois at Chicago, USA
Ge Yu	Northeast University, Northeastern
Ming Zhang	Beijing University, China
Rui Zhang	The University of Melbourne, Australia
Yanchun Zhang	Victoria University, Australia
Ying Zhang	University of New South Wales, Australia
Wenjie Zhang	University of New South Wales, Australia
Xiao Zhang	Renmin University of China
Zhenjie Zhang	ADSC Singapore
Shuigeng Zhou	Fudan University, China
Xiangmin Zhou	CSIRO, Australia
Xuan Zhou	CSIRO, Australia

External Reviewers

Md. Hijbul Alam
Nikolaus Augsten
Sukhyun Ahn
Bhuvan Bamba
Jiefeng Cheng
Hyunsouk Cho
Gabriel Pui Cheong Fung
Christian Grn
Guangyan Huang
Reza Hemayati
Hai Huang
Yu Jiang
Neila Ben Lakhal
Sau Dan Lee
Chunbin Lin
Brahim Medjahed
Kenta Oku

Jiajie Peng
Jianzhong Qi
Abdelmounaam Rezgui
Byung-Gul Ryu
Lee Ryong
Feng Shi
Gao Shen
Yousuke Watanabe
Zhilin Wu
Xike Xie
Ling Xu
Hao Yan
Jaesoo Yoo
Jianwei Zhang
Linlin Zhang
Yuanchun Zhao
Lixiao Zheng

Table of Contents

Keynotes

Session 1: Classiffication and Clustering

Session 2: Spatial and Temporal Databases

Session 3: Personalization and Recommendation

Session 4: Data Analysis and Application

Session 5: Web Mining

Session 6: Web Search and Information Retrieval

Session 7: Complex and Social Networks

Session 8: Secure and Semantic Web

Special Session 1: Demo Papers

Special Session 2: The Second International Workshop on Unstructured Data Management

Special Session 3: The Second International Workshop on XML Data Management

Information Networks Mining and Analysis

Philip S. Yu

University of Illinois at Chicago
Chicago, Illinois
psyu@cs.uic.edu

Abstract. With the ubiquity of information networks and their broad applications, there have been numerous studies on the construction, online analytical processing, and mining of information networks in multiple disciplines, including social network analysis, World-Wide Web, database systems, data mining, machine learning, and networked communication and information systems. Moreover, with a great demand of research in this direction, there is a need to understand methods for analysis of information networks from multiple disciplines. In this talk, we will present various issues and solutions on scalable mining and analysis of information networks. These include data integration, data cleaning and data validation in information networks, summarization, OLAP and multidimensional analysis in information networks. Finally, we illustrate how to apply network analysis technique to solve classical frequent item-set mining in a more efficient top-down fashion.

More specifically, on the data integration, data cleaning and data validation, we discuss two problems about correctness of information on the web. The first one is *Veracity*, i.e., *conformity to truth*, which addresses how to find true facts from a large amount of conflicting information on many subjects provided by various web sites. A general framework for the Veracity problem will be presented, which utilizes the relationships between web sites and their information, i.e., a web site is trustworthy if it provides many pieces of true information, and a piece of information is likely to be true if it is provided by many trustworthy web sites. The second problem is *object distinction*, i.e., how to distinguish different people or objects sharing identical names. This is a nontrivial task, especially when only very limited information is associated with each person or object. A general object distinction methodology is presented, which combines two complementary measures for relational similarity: set resemblance of neighbor tuples and random walk probability, and analyzes subtle linkages effectively.

OLAP (On-Line Analytical Processing) is an important notion in data analysis. There exists a similar need to deploy graph analysis from different perspectives and with multiple granularities. However, traditional OLAP technology cannot handle such demands because it does not consider the links among individual data tuples. Here, we examine a novel graph OLAP framework, which presents a multi-dimensional and multi-level view over graphs. We also look into different semantics of OLAP operations, and discuss two major types of graph OLAP: informational OLAP and topological OLAP.

We next examine summarization of massive graphs. We will use the connectivity problem of determining the minimum-cut between any pair of nodes in the network to illustrate the need for graph summarization. The problem is well

X. Du et al. (Eds.): APWeb 2011, LNCS 6612, pp. 1–2, 2011.

solved in the classical literature for memory resident graphs. However, large graphs may often be disk resident, and such graphs cannot be efficiently processed for connectivity queries. We will discuss edge and node sampling based approaches to create compressed representations of the underlying disk resident graphs. Since the compressed representations can be held in main memory, they can be used to derive efficient approximations for the connectivity problem.

Finally, we examine how to apply information network analysis technique to perform frequent item-set mining. We note that almost all state-of-the-art algorithms are based on the bottom-up approach growing patterns step by step, hence cannot mine very long patterns in a large database. Here we focus on mining top-k long maximal frequent patterns because long patterns are in general more interesting ones. Different from traditional bottom-up strategies, the network approach works in a top-down manner. The approach pulls large maximal cliques from a pattern graph constructed after some fast initial processing, and directly uses such large-sized maximal cliques as promising candidates for long frequent patterns. A separate refinement stage is then applied to further transform these candidates into true maximal patterns.

iMiner: From Passive Searching to Active Pushing

Deyi Li

Institute of Electronic System Engineering, Beijing 100039, China

The success of a search engine in cloud computing environment relies on the numbers of users and their click-through. If we take the previous search key words as tags of users to study and differentiate the user interaction behaviors, the search engine is able to actively push related and useful information to users based on their previous actions instead of passively waiting for users' queries. However the user searching behavior is affected by lots of factors, and it is quite complex and uncertain. The log files provided by a search engine have recorded all the information of the user interaction process on their servers or browsers, such as key words, click-through rate, time stamp, time on page, IP address, browser type and system stats, even the user location etc, which are all important information to understand and categorize users' searching behavior. Is there any statistical property almost independent to search key words? How to push recommendation based on the queried key words? And how to extract user behavior models of searching actions in order to recommend the information to meet users' real needs more timely and precisely?

In order to answer these questions, we don't think there is only one correct or optimal solution. We take the do-the-best strategy with uncertainty. Statistics play a great rule here. From the statistical point of view there are two very important distributions: the Gauss distribution and the power-law distribution. We suggested and have already studied a cloud model to bridge the gap between the two by using 3 mathematical characteristics: Expectation, Entropy and Hyper entropy. The collective intelligence may appear under certain conditions. The cloud model can well describe the preferential attachment in users' habits, which means that a continuous or periodic habit may reappear again and again, and even the user interest is changed from time to time.

We are developing an iMiner system at present, a mining platform that can actively push useful information to users, rather than a traditional search engine which only passively waits for users' key words. To support this kind of user-oriented pushing service, iMiner models users' behaviors according to their identities, status, interests, and previous querying records, preference of results, temporal search patterns, and interest shifts. It is made possibly with the modules of huge log data pre-processing, users' actions modeling, query modeling and query matching etc. We try to implement all of the mining services on a virtual cloud mining service center. The supported services in cloud computing are user-oriented to meet various groups which may be small but very active and frequently click-through the pages. The supported services may even be sensitive to the users' environment such as the present location and time.

X. Du et al. (Eds.): APWeb 2011, LNCS 6612, p. 3, 2011.
© Springer-Verlag Berlin Heidelberg 2011

On the Querying for Places on the Mobile Web

Christian S. Jensen

Department of Computer Science, Aarhus University
Denmark
csj@cs.au.dk

The web is undergoing a fundamental transformation: it is becoming mobile and is acquiring a spatial dimension. Thus, the web is increasingly being used from mobile devices, notably smartphones, that can be geo-positioned using GPS or technologies that exploit wireless communication networks. In addition, web content is being geo-tagged. This transformation calls for new, spatio-textual query functionality. The research community is hard at work enabling efficient support for such functionality.

We refer to points of interest with a web presence as spatial web objects or simply as *places*. Such places thus have a location as well as a textual description. Spatio-textual queries return places that are near an argument location, typically the location of a user, and are relevant to a text argument, typically a search engine query. The talk covers recent results by the speaker and his colleagues aimed at providing efficient support for the spatio-textual querying of places. Specifically, the talk touches on questions such as the following:

– How to use user-generated content, specifically large collections of GPS records, for identifying semantically meaningful places?
– How to assign importance to places using GPS records and logs of queries for driving directions?
– How to retrieve the k most relevant places according to both text relevancy and spatial proximity with respect to a query?
– How to support continuous queries that maintain up-to-date results as the users who issue queries move?
– How to take into account the presence of nearby places that are relevant to a query when determining the relevance of a place to a query?

References

1. Cao, X., Cong, G., Jensen, C.S.: Retrieving Top-K Prestige-Based Relevant Spatial Web Objects. PVLDB 3(1), 373–384 (2010)
2. Cao, X., Cong, G., Jensen, C.S.: Mining Significant Semantic Locations From GPS Data. PVLDB 3(1), 1009–1020 (2010)
3. Cong, G., Jensen, C.S., Wu, D.: Efficient Retrieval of the Top-k Most Relevant Spatial Web Objects. PVLDB 2(1), 337–348 (2009)
4. Venetis, P., Gonzales, H., Jensen, C.S., Halevy, A.: Hyper-Local, Directions-Based Ranking of Places. PVLDB 4(1) (2011) (to appear)
5. Wu, D., Yiu, M.L., Jensen, C.S., Cong, G.: Efficient Continuously Moving Top-K Spatial Keyword Query Processing. In: ICDE (2011) (to appear)

X. Du et al. (Eds.): APWeb 2011, LNCS 6612, p. 4, 2011.

Music Review Classification Enhanced by Semantic Information[*]

Wei Zheng[1], Chaokun Wang[1,2,3], Rui Li[1], Xiaoping Ou[1], and Weijun Chen[4]

[1] School of Software, Tsinghua University, Beijing 100084, China
[2] Tsinghua National Laboratory for Information Science and Technology
[3] Key Laboratory for Information System Security, Ministry of Education
[4] Department of Computer Science and Technology, Tsinghua University
{zhengw04,lirui09,ouxp09}@mails.thu.edu.cn, {chaokun,cwj}@tsinghua.edu.cn

Abstract. In this paper, we put forward the semantic-based music review classification problem by illustrating application scenarios. In this problem, we aim to classify music reviews into categories according to their semantic meanings. A solution called the SEMEC model is proposed. In SEMEC, besides the music reviews, related semantic music information is also utilized. A heuristic SVM classification algorithm is proposed to build the classifiers. SEMEC makes use of the concept of *entropy* and the *probability model* to combine the results from different classifiers. Extensive experimental results show that SEMEC is effective and efficient.

Keywords: Music Review, Semantic Classification, Heuristic Algorithm.

1 Motivation

A *music review* is an evaluation on a recording, a singing, a composition, a performance of a live music concert, a musical theater show, lyrics of a music piece, even everything music related. With the rapid growth of the Internet, music review plays an important role in multimedia systems. People contribute abundant music reviews which host vast quantities of valuable and useful information. Music review related data mining works are attracting more and more attentions in the research area of user interaction, music information retrieval, multimedia system, etc. Music reviews are usually associated with music objects, such as music pieces, vocal concerts, albums, etc. The mining of music review is quite different from other text mining problems due to (1) Music objects usually have ambiguous artistic conceptions in different people's minds, especially for the sophisticated symphonies, orchestral music or piano music. The different comprehension on a music object results in quite discretional contents in music

[*] The work is supported by the National Natural Science Foundation of China (No. 60803016), the National HeGaoJi Key Project (No. 2010ZX01042-002-002-01) and Tsinghua National Laboratory for Information Science and Technology (TNLIST) Cross-discipline Foundation.

X. Du et al. (Eds.): APWeb 2011, LNCS 6612, pp. 5–16, 2011.

reviews. (2) Compared to other kinds of review, music review usually holds a much larger range of contents, such as the evaluation of the musical performance, music arrangement, lyrics composing, album designation, even to daily lives of the singers. (3) Even within a single review, there may be comments from different aspects simultaneously. As a result, the mining and analysis of music reviews has become a difficult task via existing text mining techniques. Let us consider the following scenarios.

Categorical Information Retrieval. In traditional information retrieval (IR) tasks, results are ranked according to their relationships to the request. However, even if the result seems to be relevant to the request, the ranked result may also be out of users' interests. For example, in the purpose of finding a melancholy melody, a user only focuses on the music reviews on melody, rather than the reviews that have nothing to do with melody. Meanwhile, if you want to find out why an artist performs better in melancholy music than the brighter one, the reviews involving background information about the artist may be a good resource. In categorical IR, when a query "melancholy" arrives, all retrieved results will be classified in categories, such as artist, melody, lyrics, album. Users can select the category they are interested in to get further information.

Music Data Mining. Multimedia systems, such as music social web sites and music related forums, hold abundant contents which can be utilized as the resource of data mining tasks. The music review classification methods can be applied in music-data-related mining tasks. Different kinds of music reviews usually hold disparate meanings and are useful in different tasks. For example, for content based music data mining tasks, those reviews related with melody are more valuable than those on lyrics. But if we want to study the writing style of a lyrics writer, those reviews on lyrics may be an important resource.

In this paper, we address the above application scenes by introducing the *SEmantic-based Music rEview Classification* model (SEMEC). The main contributions of this work are as follows.

• We put forward the semantic-based music review classification problem and its typical applications and propose the SEMEC model as a solution. Besides the music reviews, the semantic music information is also utilized to classify the reviews.
• In order to build the classifier, the *Heuristic SVM Classification* algorithm is proposed. It can be widely utilized in semantic-based review mining tasks. Based on the *confusion matrix*, we evaluate the *diversity* of a classifier, by which SEMEC combines multi-classifiers.
• Extensive experimental results on actual data set show that SEMEC is an effective and efficient solution to the problem. Meanwhile, the combined classifier can be more effective than anyone of the constituent classifiers.

The rest of this paper is organized as follows. Section 2 reviews the background of music review mining tasks and some related research works. In Section 3, we introduce the SEMEC model in detail. In Section 4, the experimental results are reported. At last, a brief conclusion and the future works are mentioned.

2   Related Work

In recent years, research works on review mining are attracting more and more attentions. Many exciting ideas have been proposed in this area.

Dave proposed a method for automatically distinguishing between positive and negative product reviews [5]. Their classifier drew on information retrieval techniques for feature extraction and scoring. But the results for various metrics and heuristics varied according to the testing situation. Considering that some features usually delivered positive evaluations on the product while others did the contrary, Hu and Liu proposed a method aiming at mining relationships between user opinions and product features [8]. Besides the positive/negative judgement, Kim proposed an opinion summarization approach which extracted advantages and disadvantages of a product from related reviews [12]. Another interesting research topic was differentiating subjective reviews from objective ones via objective classification, which was useful in opinion mining tasks [18]. Apart from product review, other kinds of review, such as movie review and book review, also attracted attentions of research communities. Chaovalit introduced a movie review mining method to classify movie reviews into positive and negative ones [3]. Review mining techniques were also applied in tag extraction and automatic rating tasks, e.g., Zhang adopted the *random walk algorithm* for keyword (tag) generation on book reviews [20].

Turnbull detailed surveyed five popular music tag generation approaches, among which music review was considered as an important resource for tag harvesting [16]. Based on *part-of-speech (POS) tagging* tools and *frequent pattern mining* methods, a one-to-five star rating based music evaluation approach was proposed by Downie [9]. Downie classified music objects into different genres by analyzing related music reviews [6]. There were also a lot of exciting research works on music tags or ratings [7,2,16]. However, this paper mainly concerns the music reviews and aims to classify music reviews according to their semantic meanings.

3   The SEMEC Model

Since we did not find any existing research on this problem, a detailed survey of the existing music social web sites was carried out (See Section 4.1 for detail). We found that most music reviews could be classified into five classes, i.e. *Comments on Album, Artist, Melody, Lyrics* and *External Character*. SEMEC classifies each music review into one of these classes. The following subsections describe the workflow of SEMEC in detail, which is shown in Fig. 1.

3.1   Semantic Preprocessing

A music spider is designed to build a *semantic music dictionary* containing different kinds of semantic music phrases, such as the names of artists, music composers and lyrics writers, titles of albums and songs, user contributed music tags and the lyrics. Given a collection of music reviews, the part-of-speech (POS)

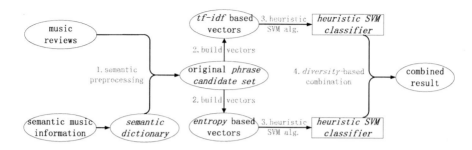

Fig. 1. The workflow of the SEMEC model

tagger is applied to generate the original *phrase candidate set*. We define the *phrase candidate set* as a set consisting of keywords and phrases which will be utilized in discriminating different kinds of semantic documents. And then, we define the *Semantic-based Classification* problem as

Definition 1 (Semantic-based Classification Problem). *Given a semantic dictionary* $D = \{d_1, d_2, \ldots, d_n\}$, *an original phrase candidate set* $R(R \subseteq D)$ *and a document set* S, *the vector collection generated on* R *and* S *is denoted as* $g(R, S)$. *Given a classifier* c *and its evaluation function* $f_c(g(R, S))$, *it is aimed to find a phrase candidate set* RP, *such that*

$$RP = argmax\{f_c(g(R, S)) : R \subseteq D\}. \tag{1}$$

In traditional text classification problems, we build vectors directly on the input documents and apply attribute reduction or attribute association analysis, such as Quick Reduction [10], to get the inputs to the classifier(s). No additional semantic information is utilized. But in this paper, the input vectors are built on both the music review collection and related semantic music phrases. The phrase candidate set RP is utilized as a differentiation measurement, which means the phrases in RP indicate the differences between different classes.

3.2   Feature Selection and Classifier Building

As $g(R, S)$ in $Equ.(1)$, two kinds of methods are utilized to build the vectors, i.e. the *tf-idf* based and the *entropy* based vectors. In the former case, the classical *tf-idf* schema [15] is utilized. For the latter, (1) we first define the *Phrase Entropy* of a phrase p_i on class c_j as (Let N be the number of classes)

$$Entropy_j(p_i) = -\frac{n_{ij}}{n_i} \log \frac{n_{ij}}{n_i} \quad (j = 1, 2, \ldots, N), \tag{2}$$

where n_{ij} is the number of music reviews both belonging to c_j and including p_i, n_i denotes the number of reviews including p_i. (2) We define the *Review Entropy* of r_k (r_k is a music review) on class c_j as

$$Entropy(r_k)[j] = \sum_{p_i \in r_k} Entropy_j(p_i) \quad (j = 1, 2, \ldots, N), \tag{3}$$

Algorithm 1. Heuristic SVM Classification Algorithm

Input: Music Review set S, Semantic Dictionary D
Output: SVM classifier C, Phrase candidate set RP

```
 1  begin
 2      P ← GetOriginalPhraseCandindateSet(D, S)// Build original phrase candidate set
 3      repeat // Remove high-frequency phrases until no improvement
 4          P ←RemoveMostFrequentPhrase(P)
 5          ErrRate, C ←BuildAndTestSVM(S, P)// Test the classifier
 6      until ErrRate is not changed;
 7      repeat // Remove low-frequency phrases until no improvement
 8          P ←RemoveLeastFrequentPhrase(P)
 9          ErrRate, C ←BuildAndTestSVM(S, P)// Test the classifier
10      until ErrRate is not changed;
11      repeat // Remove each phrase heuristically
12          foreach phrase p in P do
13              ErrRate′, C′ ←BuildAndTestSVM(S, RP \ {p})
14          p ← the phrase conducting lowest ErrRate when removed
15          RP ← RP \ {p}
16          C ←TheClassifierCauseLowestErrRate()
17          // C is the classifier conducting the lowest error rate
18      until ErrRate is not changed;
19      return C, RP
20  end
```

---

where $Entropy(r_k)[j]$ is the jth element of the vector $Entropy(r_k)$. (3) This vector is normalized as an input to the classifier(s).

Support Vector Machine (SVM) is widely used in review analysis [19,11,4]. Meanwhile, the heuristic methods are applied to some SVM related problems [17,1]. But our work is quite different, in which we apply the SVM heuristically to find an appropriate phrase candidate set. The original phrase candidate set does not only contain the phrases in music reviews, but also additional semantic phrases. The existence of some phrases takes a negative impact on the classification results. For example, the high-frequency phrases are likely to be the common adjective phrases and appearing in many kinds of reviews, such as the phrase of "amazing", which can be utilized to describe both the performance of an artist and the melody of a music piece. This kind of phrase is not suitable to be utilized to build the input vectors to the classifier(s). Meanwhile, the low-frequency phrases should also be removed due to their negative effects on content analysis [15].

Therefore, the *Heuristic SVM Classification* algorithm is proposed to build a proper phrase candidate set, shown as Algorithm 1. At a certain iteration, there may be more than one phrase which conducts an improvement on the classification result (evaluated by F-measure) if it is removed from the phrase candidate set. The one causing the most significant improvement is removed (Line 14–15 of Algorithm 1). Heuristically, we remove phrases until no improvement can be achieved. And then, the final phrase candidate set is obtained, on which the vectors will be built as the input to the SVM classifier.

For example, there is a data set consisting of 5 reviews, denoted as I_1, I_2, I_3, I_4 and I_5, the actual classes of which are c_2, c_2, c_3, c_1 and c_3. The semantic dictionary is $D = \{a, b, c, d\}$. We aim to find the phrase candidate set $RP(RP \subseteq D)$,

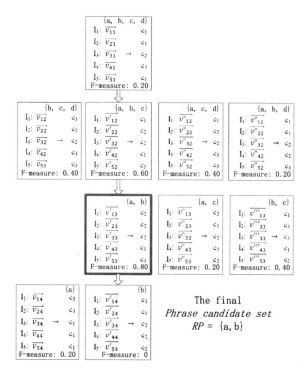

Fig. 2. An example of the Heuristic SVM Classification Algorithm. There are 4 iterations in this example. In the algorithm, only the subset conducting the highest increase of precision will be retained at each iteration. Therefore, the classifier is only rebuilt 3 times at the third iteration.

on which the vectors are generated as the input of the SVM classifiers. The whole processing is shown in Fig. 2 step by step.

3.3 Diversity-Based Classifier Combination

As mentioned before, the *tf-idf* based vectors and the *entropy* based ones, are utilized to build two SVM classifiers separately. Combination on multi-classifiers has been widely researched in these years and a lot of works have been proposed, such as the voting method, ranking model, neural network model [13,14].

In our work, we make use of the *confusion matrix M* to combine the results. The confusion matrix is an $N \times N$ matrix, where N denotes the total number of classes (in this work N equals to 5), and M_{ij} denotes the number of reviews which belong to c_i and have been classified to c_j. When a classifier f is built, a 5-fold cross validation is carried out and the confusion matrix M is calculated. We evaluate the effectiveness of f on each class by defining the concept of *diversity*. The *diversity* of the classifier f on the class c_j is defined as

$$Diversity_f(c_j) = -N \cdot \frac{1}{N} \log \frac{1}{N} + \sum_{i=1}^{N} \frac{M_{ij}}{\sum_i M_{ij}} \log \frac{M_{ij}}{\sum_i M_{ij}} \quad (j = 1, 2, \ldots, N). \quad (4)$$

It characterizes the *Information Gain* on each class when the classifier f is utilized to build the model.

Additionally, the conditional probabilities are calculated on M, such as

$$P_f(Actual = c_j | Predict = c_i) = \frac{M_{ij}}{\sum_{i=1}^{N} M_{ij}} \quad (i, j = 1, 2, \ldots, N). \tag{5}$$

Suppose there are k classifiers, denoted as $f_1, f_2, \ldots, f_k$. When an unclassified review r is fed, from each classifier we get a prediction, denoted as $c_{f_1}, c_{f_2}, \ldots, c_{f_k}$. We define the *rating scores* on each class c_j as

$$R_j = \sum_{l=1}^{k} P(Actual = c_j | Predict = c_{f_j}) \cdot \frac{Diversity_{f_l}(c_j)}{\sum_l Diversity_{f_l}(c_j)} \quad (j = 1, 2, \ldots, N). \tag{6}$$

The predicted probability that the review r belongs to class c_j will be

$$P_j = \frac{R_j}{\sum_{i=1}^{k} R_i} \quad (j = 1, 2, \ldots, N). \tag{7}$$

4   Experiments

4.1   Data Set and Experimental Models

Due to lack of benchmarks, we collected music reviews written by both the review specialists and the music enthusiasts from "Expert Music Reviews", "Meta critic", "QQ Music", "Epinions" and "DouBan"[1] . We invited 120 teachers and students from Dept. of Computer Science, Academic of Arts and Design, School of Art and Education Center, School of Software of Tsinghua University to make user studies. With anonymous questionnaires, the participants classified the reviews manually. With a detailed analysis of all the questionnaires, we found an interesting phenomenon. That was, even on a same review, different users might have different opinions. For example, here is a segment from a review.

"... *Tanya keeps up her performance style all along in the new album. But compared with her last album, there is still a gap. Tanya expert herself to accomplish every song and her sincerity and love on music touched us. ...*"

Three people marked this review as "comment on artist" while another two classified it into the class of "comment on album". The similar phenomenon could be found on many reviews.

As shown in Table 1, the top five classes containing 1034 reviews were selected to form the experimental data set. Each experiment in this paper was carried out with a 5-fold cross validation. The experiments were carried out on a PC with Core 2 2.4GHz CPU and 4GB memory. The *linear kernel* was selected as the kernel function of SVM.

---

[1] http://www.crediblemusicreviews.com, http://www.metacritic.com/music, http://ent.qq.com/, http://www.epinions.com,http://www.douban.com

Table 1. Statistical Info. of Experimental Data

class	size
comments on album	227
comments on artist	216
comments on melody	205
comments on lyrics	178
comments on external character	208
other comments	63
Total	1097

The following models were utilized to make comparisons.

Baseline–Basic Model. Since we did not find any existing research work on this problem, we took the *tf-idf* based SVM classification model as the baseline, in which the classical *tf-idf* model were utilized to build the input vectors of SVM classifiers. In the basic model, we did not make use of any additional semantic information, nor the heuristic algorithm. Top 10% and bottom 10% frequent phrases were removed as noisy phrases.

Naïve Model. Additional semantic information was utilized in the naïve model. Compared to SEMEC, the naïve model built the classifiers with different methods. Let the phrase candidate set at some iteration be P_u. In SEMEC, we evaluated the classifier if a phrase was removed from P_u. We denote the phrase conducting the highest increase of F-measure as p_x. The updated phrase candidate set utilized in the next iteration would be $P_{u+1} = P_u \backslash p_x$. However, in the naïve model, if a phrase p_y was removed and the F-measure of the classifier increased, the phrase candidate set $P_{u+1} = P_u \backslash p_y$ would be utilized in the next iteration. As a result, in the naïve model, there might be more than one phrase candidate set at each iteration. In the worst case the time complexity would be $O(|P|^2)$ ($|P|$ was the size of the original phrase candidate set).

SEMEC Model. The SEMEC model including semantic preprocessing, heuristic SVM classification, diversity-based classifier combination, was utilized. Both the *tf-idf* based vectors and the *document entropy* based vectors were utilized to build a classifier, respectively. Meanwhile, the SEMEC model was an approximate model to the naïve one.

Table 2. The semantic dictionary

The Type of phrases	Size
Names(artist, composer, lyrics writer)	278
Titles(album and music)	420
User contributed tags	1617
Music related and common Phrases	227,961
Total	230,276

Table 3. The size of the original *phrase candidate set*

The size of the data set	The size of the *phrase candidate set*
600	10,581
700	13,783
800	14,532
900	14,920
1034	15,873

4.2   Semantic Preprocessing

In preprocessing, a semantic dictionary, shown as Table 2, was generated, on which the original phrase candidate set would be built. We randomly selected 600, 700, 800, 900 reviews from the data set and carried out the semantic preprocessing. As we enlarged the data set, the size of the original phrase candidate grew slowly, shown as Table 3.

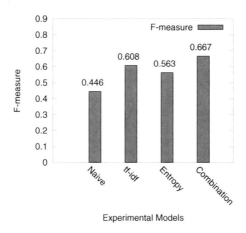

Fig. 3. Comparisons on F-measure[2]

4.3   Comparison between SEMEC and Other Models

SEMEC and the Basic Model. We took experiments on SEMEC and the basic model to make comparisons. At each step in the SEMEC model, the output was evaluated in F-measure. As Fig. 3, experimental results showed that the effectiveness (F-measure) of SEMEC was going forward. In the basic model, the F-measure was only 0.446 due to lack of additional semantic information. Also, it fully embodied the difficulty of classifying music reviews. By applying the heuristic classification algorithm with additional semantic music information, the values of F-measure were improved to 0.608 and 0.563, with the tf-ldf and entropy based SVM classifier, respectively. With diversity based classifier combination, the F-measure value of SEMEC was improved to 0.667, higher than any one of the two classifiers. Since a classifier might be suitable for the prediction on some special classes but unconvincingly on others. In our combination method, if a classifier provided a good prediction on some classes, it would be assigned a higher weight on these classes. Therefore, it was possible for the combined

---

[2] "Naive" meant the F-measure on the naïve model. "tf-idf" and "Entropy" meant the F-measures on the outputs of SEMEC after the *tf-idf* based or *entropy* based heuristic SVM classifiers, respectively. "Combination" meant the F-measure on SEMEC after classifier combination.

Table 4. The final *confusion matrix* of the SEMEC model[3]

	Album	Artist	Melody	Lyrics	Ex.Char
Album	**142**	22	45	11	7
Artist	36	**130**	15	19	16
Melody	19	25	**141**	13	7
Lyrics	8	10	23	**125**	12
Ex.Char	6	35	5	10	**152**

Table 5. Comparisons on the size of the final *phrase candidate set*

Size of the data set	Size of the *phrase candidate set* by SEMEC	Size of the *phrase candidate set* by Naive Model
600	120	155
700	126	206
800	166	183
900	224	248
1034	301	208

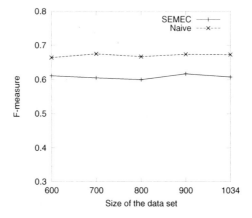

Fig. 4. Comparisons between SEMEC and the naïve model

classifier to achieve a better prediction result than some, even all the individual classifiers. The confusion matrix of SEMEC was shown in Table 4.

SEMEC and the Naïve Model. Experiments were carried out to make comparisons between the naïve and the SEMEC model. As an approximate model to the naïve one, SEMEC sacrificed less than 0.062 in F-measure (Fig. 4). But SEMEC was much more efficient than the naïve one (Fig. 5). On the data set consisting of 1034 reviews, SEMEC took less than 16 minutes to build the classifier, while the naïve model took more than 48 hours. Compared to SEMEC, the size of the phrase candidate set built by the naïve model was not strictly increasing as the size of data set grew (Table 5). It was because the naïve model had tested much more phrase combinations than SEMEC in building the final phrase candidate set.

A typical output of SEMEC illustrating the probability distribution of each review was shown in Fig. 6. In conclusion, the SEMEC model was quite a stable and reliable solution to the semantic-based music review classification problem.

---

[3] "Album" meant "Comments on Album", and so forth.

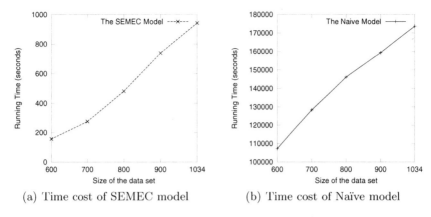

(a) Time cost of SEMEC model   (b) Time cost of Naïve model

Fig. 5. Time costs of SEMEC and the naïve model on building model

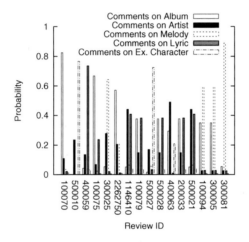

Fig. 6. A sample output of SEMEC

5   Conclusions

In this paper, we proposed the sematic-based music review classification problem and a solution called the SEMEC model. SEMEC makes use of additional semantic music information as well as the music reviews as the input of two heuristic SVM classifiers. It also combines the different classifiers by defining the concept of *diversity* on confusion matrices. In future works, we will try to further reduce the time consumption of SEMEC.

References

1. Anguita, D., Ghio, A., Ridella, S., Sterpi, D.: K-fold cross validation for error rate estimate in support vector machines. In: ICDM, pp. 291–297 (2009)
2. Bischoff, K., Firan, C.S., Paiu, R.: Deriving music theme annotations from user tags. In: WWW, pp. 1193–1194 (2009)

3. Chaovalit, P., Zhou, L.: Movie review mining: a comparison between supervised and unsupervised classification approaches. In: HICSS, pp. 112c–112c (2005)
4. Cheung, K., Kwok, J., Law, M., Tsui, K.: Mining customer product ratings for personalized marketing. Decision Support Systems 35(2), 231–243 (2003)
5. Dave, K., Lawrence, S., Pennock, D.M.: Mining the peanut gallery: opinion extraction and semantic classification of product reviews. In: WWW, pp. 519–528 (2003)
6. Downie, J.S., Hu, X.: Review mining for music digital libraries: phase ii. In: JCDL, pp. 196–197 (2006)
7. Eck, D., Lamere, P., Bertin-Mahieux, T., Green, S.: Automatic generation of social tags for music recommendation. In: NIPS, pp. 385–392 (2007)
8. Hu, M., Liu, B.: Mining opinion features in customer reviews. In: AAAI, pp. 755–760 (2004)
9. Hu, X., Downie, J.S., West, K., Ehmann, A.F.: Mining music reviews: Promising preliminary results. In: ISMIR, pp. 536–539 (2005)
10. Jensen, R., Shen, Q.: Semantics-preserving dimensionality reduction: Rough and fuzzy-rough-based approaches. IEEE Trans. Knowl. Data Eng. 16(12), 1457–1471
11. Kennedy, A., Inkpen, D.: Sentiment classification of movie reviews using contextual valence shifters. Computational Intelligence 22(2), 110–125 (2006)
12. Kim, W.Y., Ryu, J.S., Kim, K.I., Kim, U.M.: A method for opinion mining of product reviews using association rules. In: Proceedings of the 2nd International Conference on Interaction Sciences, pp. 270–274 (2009)
13. Kotsiantis, S.B.: Supervised machine learning: A review of classification techniques. In: Emerging artificial intelligence applications in computer engineering: real word AI systems with applications in eHealth, HCI, information retrieval and pervasive technologies, pp. 3–24 (2007)
14. Ou, G., Murphey, Y.L.: Multi-class pattern classification using neural networks. Pattern Recognition 40(1), 4–18 (2007)
15. Salton, G., Buckley, C.: Term-weighting approaches in automatic text retrieval. Inf. Process. Manage. 24(5), 513–523 (1988)
16. Turnbull, D., Barrington, L., Lanckriet, G.: Five approaches to collection tags for music. In: ISMIR, pp. 225–230 (2008)
17. Wang, W., Xu, Z.: A heuristic training for support vector regression. Neurocomputing 61, 259–275 (2004)
18. Wiebe, J.: Learning subjective adjectives from corpora. In: AAAI, pp. 735–740 (2000)
19. Ye, Q., Zhang, Z., Law, R.: Sentiment classification of online reviews to travel destinations by supervised machine learning approaches. Expert Syst. Appl. 36(3), 6527–6535 (2009)
20. Zhang, L., Wu, J., Zhuang, Y., Zhang, Y., Yang, C.: Review-oriented metadata enrichment: a case study. In: JCDL, pp. 173–182 (2009)

Role Discovery for Graph Clustering

Bin-Hui Chou and Einoshin Suzuki

Department of Informatics, ISEE, Kyushu University, Japan
{chou,suzuki}@i.kyushu-u.ac.jp

Abstract. Graph clustering is an important task of discovering the underlying structure in a network. Well-known methods such as the normalized cut and modularity-based methods are developed in the past decades. These methods may be called non-overlapping because they assume that a vertex belongs to one community. On the other hand, overlapping methods such as CPM, which assume that a vertex may belong to more than one community, have been drawing attention as the assumption fits the reality. We believe that existing overlapping methods are overly simple for a vertex located at the border of a community. That is, they lack careful consideration on the edges that link the vertex to its neighbors belonging to different communities. Thus, we propose a new graph clustering method, named *RoClust*, which uses three different kinds of roles, each of which represents a different kind of vertices that connect communities. Experimental results show that our method outperforms state-of-the-art methods of graph clustering.

1 Introduction

Graph clustering is the task of clustering vertices in a graph into clusters, namely communities, with many edges connecting vertices of the same community and relatively few edges connecting vertices of different communities [5,14]. It has been drawing much attention in years because it allows us to understand the underlying structure of a graph, of which examples include social networks and the World Wide Web.

Well-known methods such as the normalized cut [3,15] and modularity-based methods [2,10,11] are proposed. In these methods, each vertex belongs to one single community and is considered equally important. Recently, researchers started to propose overlapping methods for which a vertex may belong to more than one community because it is natural to think that communities may overlap in most real networks [13]. To discover overlapping communities, overlapping methods including the Clique Percolation Method (CPM) [13] and split betweenness [7] are proposed. In these methods, a vertex can belong to one or more communities and, vertices that belong to multiple communities are considered overlapping vertices.

Consider the example of in Fig. 1, which is a subgraph of selecting words CHEMICAL, CHAIN and PHYSICS from the USF word association dataset[1] [9].

---

[1] Words A and B are connected if one writes A when he is asked to write the first word that comes to mind given B.

X. Du et al. (Eds.): APWeb 2011, LNCS 6612, pp. 17–28, 2011.

Assume that the graph in Fig. 1 consists of three corresponding word communities, CHEMICAL, CHAIN and PHYSICS. Word SCIENCE is an overlapping vertex between communities because it connects many words in two communities.

Existing overlapping methods discover words CHEMISTRY and SCIENCE as overlapping vertices between the communities of CHEMICAL and PHYSICS. Existing methods often attribute REACTION into the community of CHAIN and not that of CHEMICAL because REACTION is relatively less densely-connected when compared to vertices in the community of CHEMICAL.

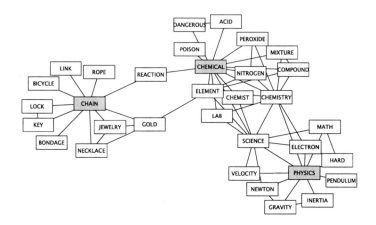

Fig. 1. A subgraph of selecting CHEMICAL, CHAIN and PHYSICS and words of their corresponding neighbors from the USF word association dataset

However, CHEMISTRY should belong to the community of CHEMICAL, which conforms to its semantics, because it connects relatively more vertices in the community of CHEMICAL. REACTION should be a peculiar vertex that does not belong to any community because each of communities connects it with only one edge. From these observations, we see that existing overlapping methods are overly simple for a vertex located at the border of a community. That is, they lack careful consideration on the edges that link the vertex with its neighbors belonging to different communities.

To overcome the shortcomings of the existing methods, we propose a new graph clustering algorithm, named *RoClust*, in this paper. In [1], three roles, bridges, gateways and hubs, are proposed to identify vertices that connect communities. Each of them represents a different kind of relationships between its neighbors. In Fig. 1, words REACTION, CHEMISTRY and SCIENCE correspond to bridges, gateways and hubs, respectively. These roles are shown to identify important vertices at the border of communities [1]. In this paper, we discover borders of communities by using bridges, gateways and hubs, and then uncover the whole community structure by propagating memberships obtained from the roles to vertices located in the central community. Experimental results show that our method outperforms state-of-the-art methods in terms of both non-overlapping and overlapping community detection.

2 Related Works

Modularity Q [12] is a quality function to measure how good a partition is. Modularity-based methods [2,10] aim to discover the community structure by maximizing Q, which is defined as $Q = \sum_i \left[\frac{l_i}{L} - \left(\frac{d_i}{2L} \right)^2 \right]$ where l_i, L and d_i represent the number of edges between vertices in community i, the number of edges in the graph, and the sum of degrees of vertices in community i, respectively. Modularity measures the fraction of all edges within each community minus the expected value of the same quantity in a graph with the same community division but random connections between vertices.

The normalized cut is based on the idea of minimizing the number of edges between communities. The normalized cut [15] $Ncut$ is defined as $Ncut = \frac{cut(A,B)}{assoc(A,V)} + \frac{cut(A,B)}{assoc(B,V)}$, where $cut(A,B)$ and $assoc(A,V)$ represent the number of edges to be removed to separate vertices in community A from those in community B, and the sum of edges that connect vertices in community A to all vertices in the graph, respectively.

Besides modularity-based methods and the normalized cut, there are also many methods proposed such as GN algorithms [6,12], and density-based clustering [16]. Please refer to [5,14] for more details. These methods may be called non-overlapping because they assume that a vertex belongs to one community. Accordingly, they cannot discover the overlapping community structure.

The Clique Percolation Method (CPM) [13] finds overlapping communities based on the observation that a community consists of several fully connected subgraphs, i.e., cliques, that tend to share many of their vertices. A community is defined as the largest connected subgraph obtained by the union of all adjacent k-cliques. Two k-cliques are adjacent if they share $k-1$ vertices. Gregory [7] extends the GN algorithm by introducing a new measure named split betweenness. A vertex v is split into v_1 and v_2 and becomes an overlapping vertex if the edge betweenness value of (v_1, v_2) is larger than any adjacent edge of v.

Zhang et al. [17] present a method of combining the modularity function and fuzzy c-means clustering. By using fuzzy c-means clustering, a vertex is assigned different degrees of community memberships. In [4], the overlapping community structure can be discovered by defining clusters as sets of edges rather than vertices. [8] proposes a method that performs a local exploration of the network, searching for the community of each vertex. During the procedure, a vertex can be visited several times even if it is assigned to a community. By this way, the overlapping community structure can be discovered.

As discussed in the introduction, existing overlapping methods are overly simple for a vertex located at the border of a community. They lack careful consideration on the edges that link the vertex with its neighbors belonging to different communities.

3   Problem Definition

We tackle a problem of clustering vertices of a single graph into communities in this paper. Intuitively, vertices within a community connect each other with many edges while vertices between communities connect each other with comparatively few edges. Our problem is formalized as follows.

The input is an undirected and simple graph $G = \langle V, E \rangle$, where $V = \{v_1, v_2, \ldots, v_n\}$ is a non-empty finite set of vertices and E is a set of edges, where an edge is a binary relation of an unordered pair of distinct elements of V. The output is $C = \{C_0, C_1, \ldots, C_k\}$ where C_0 represents the set of vertices that does not belong to any community and $\cup C_i = V$ where $0 \le i \le k$. C_i, where $1 \le i \le k$, satisfies the following conditions: $C_i \neq \emptyset$ and $C_i \neq C_j$. Note that $C_i \cap C_j = \emptyset$ is not included in the definitions, which indicates that vertices can belong to more than one community.

4   Proposed Method

4.1   Bridges, Gateways, and Hubs

In this paper, we propose a new algorithm for graph clustering, named *RoClust*, which uses three roles, bridges, gateways and hubs, to discover communities. In [1], we propose bridges, gateways and hubs, each of which represents a different kind of important vertices in connecting communities.

A bridge is a vertex that connects communities, each of which links the bridge with a single edge. A gateway is a vertex, in whose neighborhood most of neighbors belong to the same community while there exists at least one neighbor that belongs to a different community. A hub is an overlapping vertex to which vertices belonging to several communities are linked. The definitions of bridges, gateways and hubs are given as follows [1],

$$Bridge(v) \Leftrightarrow \forall x, y \in N(v) - \{v\}, x \neq y : CIN(x, y) \wedge \neg loner(x) \wedge \neg loner(y)$$
$$Gateway(v) \Leftrightarrow (1)\exists x, y \in N(v) - \{v\}, x \neq y : SC(x, y)$$
$$(2)\exists z \in N(v) - \{v\}, \forall u \in N(v) - \{v, z\} : \neg loner(z) \wedge CIN(z, u)$$
$$Hub(v) \Leftrightarrow \exists w, x, y, z \in N(v) - \{v\}, w \neq x, y \neq z :$$
$$SC(w, x) \wedge SC(y, z) \wedge CIN(w, y) \wedge CIN(x, z)$$

where node v's neighborhood $N(v) = \{u \in V \mid (v, u) \in E\} \cup \{v\}$, and $loner(v) \Leftrightarrow |N(v)| = 2$. Nodes v and w are connected via an intermediate node $CIN(v, w) \Leftrightarrow |N(v) \cap N(w)| = 1$; nodes v and w are strongly connected $SC(v, w) \Leftrightarrow |N(v) \cap N(w)| \ge 2$.

The main intuition behind $CIN(v, w)$ and $SC(v, w)$ lies in the shared neighborhood. A pair of vertices that have two or more common vertices between their neighborhoods are much more likely to be in the same community than a pair of vertices that have only one common vertex. In other words, $CIN(v, w)$ implies that vertices v and w belong to different communities and $SC(v, w)$ implies that

v and w belong to the same community. In this paper, we extend the definitions of $SC(v,w)$ and $CIN(v,w)$ by using the structural similarity [16]. The structural similarity $\sigma(v,w)$ is defined as $\sigma(v,w) = \frac{|N(v) \cap N(w)|}{\sqrt{|N(v)||N(w)|}}$, which is a measure to evaluate the similarity between neighborhoods of two vertices v and w. A high value for $\sigma(v,w)$ implies that v and w share many common vertices between their neighborhoods. To avoid poor discovery of bridges, gateways and hubs in complex graphs, $CIN(v,w)$ and $SC(v,w)$ are reformulated as $CIN'(v,w)$ and $SC'(v,w)$ by introducing the structural similarity,

$$SC'(v,w) \Leftrightarrow \sigma(v,w) \geq \beta$$
$$CIN'(v,w) \Leftrightarrow \sigma(v,w) < \alpha$$

where α $(0 < \alpha < 0.5)$ and β $(0.5 \leq \beta < 1)$ represent two thresholds that control the number of discovered vertices with roles. The larger the value of α is, the larger the numbers of bridges, gateways and hubs are. A small value for β has the same effect. Compared to $CIN(v,w)$, where nodes v and w share only one common neighbor, by using $CIN'(x,y)$ we can discover more vertices of the roles because nodes x and y whose $\sigma(x,y)$ is smaller than the threshold may share more than one common neighbor.

4.2 Clustering Algorithm

We adopt the approach of hierarchical clustering, specifically an agglomerative method, which iteratively combines two communities into one until one community remains. The advantage of hierarchical clustering is that it returns a hierarchical structure of communities, which is often observed in real world networks [8].

Algorithm 1. Overview of RoClust

Input   : $G = \langle V, E \rangle$, α, β
Output: Hierarchical clustering result on G
1 $Q.clear(); l = 1;$
   // Step 0. Discover bridges, gateways and hubs
2 $B \leftarrow bridge(G, \alpha);$ $W \leftarrow gateway\_hub(G, \alpha, \beta);$
3 $J \leftarrow gateway(G, \alpha, \beta);$ $H \leftarrow hub(G, \alpha, \beta);$
   // Step 1. Divide G into m communities
4 $DivideByBridge(B, l);$ $DivideByGH(W, l);$
5 $DivideByGateway(J, l);$ $DivideByHub(H, l);$
6 $MembershipPropagate(Q);$
   // Step 2. Merge communities
7 Compute the distance between communities;
8 **while** # of communities $\neq 1$ **do**
9 $\qquad$ Find communities C_{ip} and C_{jp} that have the minimum distance;
10 $\qquad$ Merge C_{ip} and C_{jp} into one community;
11 $\qquad$ Update the distance between communities;

Algorithm 1 shows an overview of our algorithm. We first discover vertices of roles, and then perform the clustering algorithm, which contains two main steps, the dividing step and the merging step. Note that a vertex may be at the same time a gateway and a hub according to their definitions so we classify vertices of roles into four types as shown in lines 2 and 3.

Dividing Step. In existing hierarchical graph clustering, a vertex is initially regarded as a community so there are n communities before combinations of communities. Because bridges, gateways and hubs are vertices between communities, information of relationships between their neighbors are important for the clustering. As a result, in this paper, to import the community information obtained from the role discovery into the clustering, we initially divide the graph G into m ($m < n$) communities. Our clustering algorithm starts with m communities instead of n communities before the merging step.

In the dividing step, we obtain m communities by propagating labels of communities from borders of communities to center parts of communities. That is, we start to decide labels of communities for vertices from bridges, gateways, hubs and their neighborhoods. In Algorithm 1, $DivideByBridge()$, $DivideByGH()$, $DivideByGateway()$ and $DivideByHub()$ are procedures to decide community memberships for neighborhoods of bridges, vertices that are at the same time gateways and hubs, gateways, and hubs, respectively. In the procedures, l represents a label for community memberships. Once community memberships of vertices of the roles and their neighbors are decided, we further propagate the memberships to other vertices that do not have memberships, which is implemented in the procedure of $MembershipPropagate(Q)$, where Q is a queue that stores an ordered list of vertices waiting for propagation.

It is simple to determine community labels for neighbors of a bridge because every two of its neighbors have the CIN relation so we simply assign a different community label l to each of them. We mark the bridge as a peculiar vertex.

However, it becomes challenging to determine community labels for neighbors of a gateway and those of a hub because some neighbors may share the SC relation, some may share the CIN relation, and some may share neither of them.

Procedure DivideByGateway(J, l)

1 **foreach** $j \in J$ **do**
2   Remove j from the subgraph g formed by $N(j)$;
3   **foreach** v *of the largest connected component graph of g* **do**
4         $label(v, l); Q.push(v)$;
5      $label(j, l)$;
6   **foreach** $v \in N(j)$ **do**
7   **if** $loner(v)$ **then**
8            $label(v, l)$;
9      $l = l + 1$;

Procedure DivideByGH(W, l)

---

1   $smallest = |V|$;
2   **foreach** $w \in W$ **do**
3 | Remove w from the subgraph g composed of $N(w)$;
4 | Get $CG = \{cg \subset g | cg$ is a component graph$\}$;
5 | **foreach** $cg \in CG$ **do**
6 | | **if** $|cg| \neq 1$ **then**
7 | | | **if** $|cg| < smallest$ **then**
8 | | | | $lb = l$;
9 | | | **foreach** v of cg **do**
10 | | | | $label(v,l)$; $Q.push(v)$;
11 | | | $label(w,l)$; $l = l + 1$;
12 | **foreach** $cg \in CG$ **do**
13 | | **if** $|cg| = 1$ **then**
14 | | | **foreach** v of cg **do**
15 | | | | **if** $loner(v)$ **then**
16 | | | | | $label(v, lb)$;
17 | | | | **else**
18 | | | | | $label(v,l)$; $l = l + 1$;

The SC and CIN relations are considered important clues for the community structure. Suppose that nodes x and y that are neighbors of a gateway have the SC relation, which implies that nodes x and y share many common neighbors besides the gateway. Accordingly, nodes x and y remain connected even without adjacent edges to the gateway because of other shared common neighbors. On the contrary, neighbors that have the CIN relation are very likely to become separated when adjacent edges to the gateway are absent. From the above observations, we remove a gateway or a hub from the subgraph composed of its neighborhoods to determine community memberships for the neighborhoods.

If a loner which is a neighbor of a gateway forms a community by itself, it may not be merged to the community of the gateway. To avoid this counter-intuitive situation, we assign the membership that most neighbors of the gateway have to the loner. Similarly, a loner which is a hub's neighbor is given the membership of vertices in the smallest component graph of the hub.

The pseudo codes for $DivideByGH()$ and $DivideByGateway()$ are shown in Procedures 3 and 2, respectively. $DivideByHub()$ performs the same procedures in Procedure 3 but without lines 17 and 18.

The procedure $MembershipPropagate()$ propagates the memberships from vertices that are near bridges, gateways or hubs to vertices that are far from them by using the queue Q until every vertex has its membership. During the propagation, the membership of a vertex is determined from its neighbors. Roughly speaking, a vertex v that does not have any membership, i.e., $getLabel(v) =$ NULL in Procedure 4, obtains the community membership that most of its neighbors have. This strategy fits our intuition because a vertex belongs to the community that most of its neighbors belong to. The queue Q keeps being updated during the propagation.

Procedure `MembershipPropagate(`Q`)`

---

1 $U = \{u \in V \mid getLabel(u) = $ NULL $\}$;
2 **while** $|U| > 0$ **do**
3      $x = Q.front()$; $Q.pop(x)$;
4   **foreach** $v \in N(x)$ **do**
5   **if** $getLabel(v) = NULL$ **then**
6            $lb \leftarrow$ the community membership that most of v's neighbors have;
7   **if** $lb \neq NULL$ **then**
8               $label(v, lb)$;
9   **else**
10              $Q.push(x)$;

Merging Step. In the merging step, two communities with the minimum distance are iteratively merged until one community remains. To evaluate the distance, we define a distance function $d(C_i, C_j)$ between communities C_i and C_j as

$$d(C_i, C_j) = D_{C_i}\sqrt{|C_i|} + D_{C_j}\sqrt{|C_j|} + g_{C_iC_j}\sqrt{|C_i||C_j|}$$

where D_C, $|C|$, and $g_{C_iC_j}$ represent the diameter of a community C, the number of vertices of a community, and the length of the shortest path from community C_i to community C_j, respectively. The main idea behind the distance function is that, we consider merging candidate communities from small to large ones because a candidate of a small size is less likely to form a community. Since a candidate of a high density resembles a community more than that of a low density, we consider the diameter in the distance function. Also, it fits our intuition to merge two candidate communities located near instead of two candidate communities located distant into one community, so $g_{C_iC_j}$ is employed.

Complexity Analysis of RoClust. The complexity of our clustering algorithm is dominated by the merging step. For the distance function, we can compute the diameter of a community in $O(|V||E|)$ time and obtain the length of the shortest path between communities in $O(s^2)$ time, where s represents the size of a community, by calculating the length of the shortest path between each pair of vertices beforehand. As a result, line 7 in Algorithm 1 can be carried out in $O\left(m^2(|V||E| + s^2)\right)$ time, where m represents the number of communities after we divide the graph in the dividing step. In each loop of merging, we can find the pair of communities with the minimum distance in $O(m^2)$ time, merge them in $O(|V|)$ time, and update the distance in $O\left(m(|V||E| + s^2)\right)$ time.

5 Experimental Results

We compare our method with four existing methods, a modularity-based method, the normalized cut, CPM and CONGA. We implemented the greedy optimization of modularity [2]. We use a software named Graclus [3] that computes the normalized cut in experiments. For CPM and CONGA, we use the software provided by their authors. Note that we cannot assign the number of communities in a modularity-based method and CPM while in the normalized cut and CONGA, we assign the number of communities.

5.1 USF Word Association

The USF word association dataset contains 7207 vertices and 31784 edges. In the experiments, we extracted subgraphs from the dataset by carefully selecting words for experiments. One of the reasons of the word selection is that we can evaluate the clustering result based on selected words. As shown in the motivating example where we select CHEMICAL, CHAIN and PHYSICS, the subgraph is made of the three words and their neighbors. The neighbors of each word have more edges with each other than with neighbors of another word, resulting in the subgraph of three corresponding communities. The other reason is for clarity due to the space limitation.

Fig. 2 shows the results of experiments where we select COUNTRY and SEA from the dataset. In Fig. 2a, our method successfully discovers two communities. Word LAND is an overlapping vertex between communities. Words AIR, WOODS and COTTAGE do not belong to any community and do not have much relation to SEA and COUNTRY, which is confirmed by their semantics.

Fig. 2b and Fig. 2c show the clustering results of the normalized cut and the modularity-based method, respectively. Both methods cannot discover overlapping vertices so they assign the overlapping vertex the membership of one community. Furthermore, the peculiar vertices are also uncovered and are partitioned into the white community.

Fig. 2d and Fig. 2e show the clustering results of CPM and CONGA, respectively. Because CPM aims to obtain the union of cliques, one of the communities is so large that most of vertices are included. In CONGA, vertices that should be in the white community are clustered in the gray community, which is counterintuitive. The reason is probably that most of the vertices between the two communities connect communities with two edges, each of which connects one community. This results in the outcome that the split betweenness, which is an important measure in CONGA to discover overlapping vertices, fails to discover borders of communities.

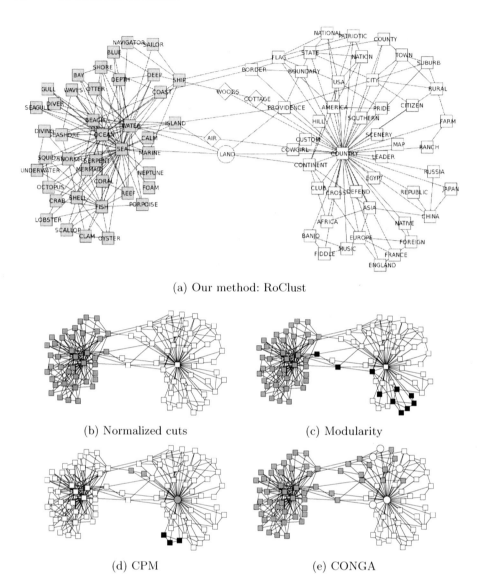

(a) Our method: RoClust

(b) Normalized cuts   (c) Modularity

(d) CPM   (e) CONGA

Fig. 2. Results of selecting COUNTRY and SEA from the USF word association dataset (Different shades represent different communities. Circular vertices are overlapping vertices, and diamond vertices are vertices that do not belong to any community).

5.2 DBLP

DBLP[2] is a database that provides bibliographic information on major computer science journals and proceedings. We use coauthorship networks[3] of KDD and

---

[2] http://www.informatik.uni-trier.de/~ley/db
[3] Two authors are connected if they have coauthored at least one paper.

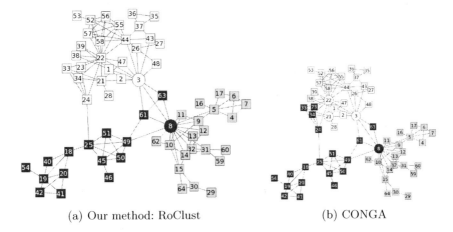

(a) Our method: RoClust   (b) CONGA

Fig. 3. Results of a subgraph of the coauthorship network of IJCAI (Different shades represent different communities. Circular vertices represent overlapping vertices).

IJCAI from 2005 to 2009 in experiments. We extracted connected subgraphs each of which has more than 50 vertices in our experiments. We focus on comparing our method with the overlapping methods in the experiment because we think that there exist overlapping vertices in the graphs.

Our method, RoClust, has competitive results with those of CONGA in most experiments. However, Fig. 3 shows a typical example of superior results, which is a subgraph from the IJCAI coauthorship network. In our method, three communities are successfully divided, and between communities, overlapping vertices are also discovered. However, the black and white communities are confounded in CONGA. The result of nodes 22, 23, 24, 33, and 34 in the black community is counter-intuitive, which can be confirmed from the topology. We found that this kind of results occur in CONGA when there are more than one path between communities in the graph. Specifically speaking, node 22 has two kinds of paths to reach the black community: one is via nodes 3, 61; the other is via node 24, which results in the split of node 22 according to the algorithm of CONGA. Note that the split of node 22 indicates there are two communities around node 22.

CPM has an unfavorable result, which is similar to the result in Fig. 2d because CPM aims to obtain the union of cliques to form a community, resulting in one community with half of all vertices and several communities with few vertices. Due to the space limitation, we do not show the figure.

6   Conclusions

In this paper, we propose a new algorithm for graph clustering, named Ro-Clust. Our method uses three roles, bridges, gateways and hubs, each of which represents a different kind of vertices that connect communities. Our method overcomes the shortcomings of existing methods, which are overly simple for a vertex located at a border of communities. Experimental results show that our method outperforms the state-of-the-art methods of graph clustering. In the experiments, we also found that our method may not behave appropriately in graphs in which many vertices located at borders of communities connect with each other densely, which will be our future work to focus on.

Acknowledgments. This work was partially supported by the grant-in-aid for scientific research on fundamental research (B) 21300053 from the Japanese Ministry of Education, Culture, Sports, Science and Technology, and the Strategic International Cooperative Program funded by the Japan Science and Technology Agency (JST).

References

1. Chou, B.-H., Suzuki, E.: Discovering community-oriented roles of nodes in a social network. In: Bach Pedersen, T., Mohania, M.K., Tjoa, A.M. (eds.) DAWAK 2010. LNCS, vol. 6263, pp. 52–64. Springer, Heidelberg (2010)
2. Clauset, A., Newman, M.E.J., Moore, C.: Finding Community Structure in Very Large Networks. Phys. Rev. E 70(066111), 1–6 (2004)
3. Dhillon, I., Guan, Y., Kulis, B.: Weighted Graph Cuts without Eigenvectors A Multilevel Approach. IEEE Trans. PAMI 29(11), 1944–1957 (2007)
4. Evans, T.S., Lambiotte, R.: Line Graphs, Link Partitions, and Overlapping Communities. Phys. Rev. E 80(1) (2009)
5. Fortunato, S.: Community Detection in Graphs. Phys. Rep. 486, 75–174 (2010)
6. Girvan, M., Newman, M.E.J.: Community Structure in Social and Biological Networks. PNAS 99(12), 7821–7826 (2002)
7. Gregory, S.: An Algorithm to Find Overlapping Community Structure in Networks. In: Kok, J.N., Koronacki, J., Lopez de Mantaras, R., Matwin, S., Mladenič, D., Skowron, A. (eds.) PKDD 2007. LNCS (LNAI), vol. 4702, pp. 91–102. Springer, Heidelberg (2007)
8. Lancichinetti, A., Fortunato, S., Kertész, J.: Detecting the Overlapping and Hierarchical Community Structure in Complex Networks. New J. Phys. 11(3) (2009)
9. Nelson, D., McEvoy, C., Schreiber, T.: USF Word Association (1998), http://w3.usf.edu/FreeAssociation/
10. Newman, M.E.J.: Fast Algorithm for Detecting Community Structure in Networks. Phys. Rev. E 69(066133) (2004)
11. Newman, M.E.J.: Modularity and Community Structure in Networks. PNAS 103(23), 8577–8582 (2006)
12. Newman, M.E.J., Girvan, M.: Finding and Evaluating Community Structure in Networks. Phys. Rev. E 69(026113) (2004)
13. Palla, G., Derényi, I., Farkas, I., Vicsek, T.: Uncovering the Overlapping Community Structure of Complex Networks in Nature and Society. Nature 435(7043), 814–818 (2005)
14. Schaeffer, S.E.: Graph Clustering. Computer Science Review 1(1), 27–64 (2007)
15. Shi, J., Malik, J.: Normalized Cuts and Image Segmentation. IEEE Trans. PAMI 22(8), 888–905 (2000)
16. Xu, X., Yuruk, N., Feng, Z., Schweiger, T.: SCAN: A Structural Clustering Algorithm for Networks. In: Proc. KDD, pp. 824–833 (2007)
17. Zhang, S., Wang, R., Zhang, X.: Identification of Overlapping Community Structure in Complex Networks Using Fuzzy c-means Clustering. Physica A 374(1), 483–490 (2007)

Aggregate Distance Based Clustering
Using Fibonacci Series-FIBCLUS

Rakesh Rawat, Richi Nayak, Yuefeng Li, and Slah Alsaleh

Faculty of Science and Technology, Queensland University of University
Brisbane Australia
{r.rawat,r.nayak,y2.li,s.alsaleh}@qut.edu.au

Abstract. This paper proposes an innovative instance similarity based evaluation metric that reduces the search map for clustering to be performed. An aggregate global score is calculated for each instance using the novel idea of Fibonacci series. The use of Fibonacci numbers is able to separate the instances effectively and, in hence, the intra-cluster similarity is increased and the inter-cluster similarity is decreased during clustering. The proposed FIBCLUS algorithm is able to handle datasets with numerical, categorical and a mix of both types of attributes. Results obtained with FIBCLUS are compared with the results of existing algorithms such as k-means, x-means expected maximization and hierarchical algorithms that are widely used to cluster numeric, categorical and mix data types. Empirical analysis shows that FIBCLUS is able to produce better clustering solutions in terms of entropy, purity and F-score in comparison to the above described existing algorithms.

Keywords: Clustering numeric, categorical and mix datasets, Fibonacci series and golden ratio, similarity evaluation.

1 Introduction

Evaluation of similarity of attributes between instances is the core of any clustering method. The better a similarity function the better the clustering results would be. If the dataset contains numeric attributes, distance measures such as Euclidean, Manhattan and cosine, are effective to evaluate the similarity between objects [1],[2],[3]. However when the dataset contains categorical (finite and unordered) attributes or a mix of numeric and categorical attributes then such distance measures may not give good clustering results [3]. Comparison of a categorical attribute in two objects would either yield 1 for similar values and 0 indicating that two instances are dissimilar. Such similarity measures are defined as overlap measure [4], and mostly suffer from the problem of clustering dissimilar instances together when the number of attributes matched is same, but attributes that are matched are different [5]. Data driven similarity measures are becoming a focus of research [5]. Datasets containing a mix of numerical and categorical attributes have become increasingly common in modern real-world applications.

In this paper, we present a novel algorithm called as FIBCLUS (Fibonacci based Clustering) that introduces effective similarity measures for numeric, categorical and

X. Du et al. (Eds.): APWeb 2011, LNCS 6612, pp. 29–40, 2011.

a mix of both these types of attributes. Due to the mapping of all attributes of an instance to a global aggregate score, this method reduces the complexity inherent in the clustering process. Moreover, due to the use of Fibonacci numbers to separate the attribute values, this method enables higher intra-cluster similarity and lower inter-cluster similarity and, in hence, better clustering. Experiments with the proposed method are conducted using a total of 9 datasets, containing a mix of numeric, categorical and combinational attributes. The quality of clusters obtained is thoroughly analyzed. Empirical analysis shows that there was an average improvement of 14.6% in the purity values, 28.5% in the entropy values and about 8% in the F-score values of clusters obtained with FIBCLUS method on all the datasets in comparison to clustering solutions obtained using the existing methods such as k-means, x-means expected maximization and hierarchical algorithms .

The contributions of this paper can be summarized as: 1) A novel clustering similarity metrics that utilises Fibonacci series to find similarities between numerical, categorical and a mix of both the data types; 2) A global score representation method for these types of attributes; and 3) Enhancing existing clustering algorithms by using FIBCLUS as a similarity metrics.

2 Problem Statement

When pure categorical datasets or mixed datasets consisting of both the categorical and numerical attributes are to be clustered, the problem is how to measure the similarity between the instances represented by categorical attributes. A similarity measure, overlap, between two categorical instances X_i and X_j can be defined as follows:

$$S(X_i, X_j) \equiv \sum_{k=1}^{m} \delta(x_{ik}, x_{jk}), \text{ where } \delta(x_{ik}, x_{jk}) = \begin{cases} 1, & x_{ik} = x_{jk} \\ 0, & \mathbf{x}_{ik} \neq x_{jk} \end{cases}. \tag{1}$$

Such similarity measures may result in weak intra similarity when calculating the similarity between categorical attributes [2]. Other similarity measures for categorical attributes such as Eskin, Goodall, IOF, OF, Lin, Burnaby [5] are based on the overlap similarity measure and inherit the same problems. Moreover, in modern real-world applications, data with various instances containing a mix of both categorical and numerical attributes are common. A problem arises when assignment of an instance to a particular cluster is not easy. This problem is shown by the example in deck of cards problem.

Consider two datasets, one containing a single deck of 52 cards and another consisting of two decks of cards. Each deck of cards is identified by the distinct cover design it has. Clustering deck of cards may be a trivial problem, but it represents perfect clustering and the major shortcomings of clustering methods, which is when assignment of an instance to a cluster becomes difficult. As the number of deck increases, the number of clusters and the complexity inherent within the clustering process increases. As the number of deck increases from 1..n the number of perfect clusters increases to $4n$ where n is the number of decks. The ideal clustering results are shown in Table 2 for the deck of cards dataset problem. The corresponding clustering results obtained by different algorithms such as expectation minimization (denoted as EM), K means (KM) and extended K means (XM) are shown in Table 3.

Table 1. Data description for deck of cards clustering problem

SN	Attribute Name	Attribute type	Value Range	Description
1	Card No	Numeric/discrete	1-13	1-13 of all cards
2	Colour	Categorical	2	Red or Black
3	Category	Categorical	4	Hearts, Diamonds, Spade, Clubs
4	Deck Id	Numeric/Binary	1,2	1-1st Deck,2-2nd Deck

Table 2. Deck of cards cluster accuracy measure criteria (D1=deck1,D2=deck2)

2 Clusters	4 Clusters	8 Clusters
1-13, Red	1-13,Red , Hearts	1-13,Red , Hearts, D1
1-13, Black	1-13,Black , Spade	1-13,Red , Hearts, D2
	1-13,Black , Clubs	1-13,Red , Diamonds, D1
	1-13,Red, Diamonds	1-13,Red , Diamonds, D2
		1-13,Black , Spade, D1
		1-13, Black , Spade, D2
		1-13, Black , Clubs, D1
		1-13, Black , Clubs, D2

Table 3. Clustering results for decks of cards problem (D1=deck1,D2=deck2)

SN	Clustering Algorithm	Cluster=2 Correctly clustered		Cluster=4 Correctly Clustered		Cluster=8 Correctly clustered
		D1	D2	D1	D2	D2
1	EM	100%	100%	100%	100%	48.07%
2	KM	100%	98%	63.5%	62.5%	56.7%
3	XM	100%	98%	73.1%	62.5%	56.7%
4	Direct	25%	62.5%	38.5%	36.5%	31.7%
5	Repeated Bisection	25%	65.5%	48%	44.2%	31.8%
6	Agglomerative	48%	65.5%	33%	48%	25%
7	Clustering Functions #4, #5, #6 above with FIBCLUS	100%	100%	100%	100%	100%

These were implemented in Weka [6] with both Euclidian and Manhattan distances. Clustering using direct, repeated bisection and agglomerative were used with both the cosine and correlation coefficient similarity measures implemented in gcluto [1]. Only the best results observed are reported for all the methods.

Results clearly show that the mentioned clustering algorithms based on respective similarity measures perform satisfactory with a single deck of cards, but as the complexity increases the clustering performance starts decreasing (Table 3). This problem occurs due to the similarity methods adopted by such algorithms. Such methods are unable to handle the mix of attributes and their inherent relationships. As the number of deck increases from one to two, the distance measures or similarity methods employed by such methods start to overlap distances.

Fig. 1. (a) Agglomerative (4 Clusters)   **Fig. 1. (b)** FIBCLUS with Agglomerative (4 Clusters)

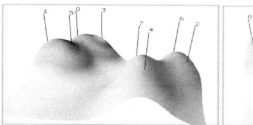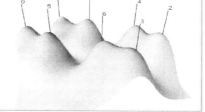

Fig. 1. (c) Agglomerative (8 Clusters)   **Fig. 1. (d)** FIBCLUS with Agglomerative (8 Clusters)

The figures 1(a)-1(d) further visualize the cluster assignments for the 2 deck of cards. For agglomerative algorithm, the best of the cosine and correlation coefficient similarity measures was taken. With FIBCLUS and agglomerative clustering both measures gave same results. From figures 1(a), 1(c) it can be clearly deduced that clusters have overlapping distances, which consequently results in a weak clustering solution. The assignment of same peaks to a set of clusters shows the overlapping and, consequently a weak intra cluster similarity value. However in figures 1(b) and 1(d), with FIBCLUS, the clusters were clearly identifiable. The high peaks in figures 1(b) and 1(d) binding similar instances together confirm that the intra cluster similarity was maximized using the method, hence resulting in the desired and optimal clustering for the underlying problem. Further separate peaks for each of the 8 clusters reflects high inter cluster similarity.

3   Related Work

K means clustering is one of the best known and commonly used algorithm. K means [7] were inherently developed to deal with numerical data, where distances between instances are a factor for clustering them together. The widely used distance measure functions adopted by K means are Euclidean, Manhattan and cosine. Several K means extensions have been developed to cluster categorical data [3],[7]. Authors in [7] developed an efficient algorithm which clusters categorical data using the K means concept. A dissimilarity function based on simple matching, which evaluates the dissimilarity between a categorical instance and the cluster representative is used. The frequencies of all attributes of the instance matching the cluster are used for calculating the dissimilarity. Another approach based on K means to cluster categorical datasets [3] uses simple matching scheme, replaces means of clusters by modes and uses frequency to solve for the best clustering outputs.

A further classification of similarity evaluation for categorical data based on neighbourhood [8],[9],[10] and learning algorithms [11],[12] is discussed in [5]. Mostly neighbourhood based evaluation methods use similarity methods as adopted by the overlap measures [5]. Some of them are Eskin, Goodall, IOF, OF, Lin, Burnaby [2]. Unlike the overlap measure, these measures consider both similarity and dissimilarity between instances, assigning 1 for a perfect match and arbitrary small values for a mismatch. Rock [10] and Cactus [11] are some of the popular agglomerative hierarchical algorithms which are used for categorical data clustering. Rock clusters instances in an agglomerative way maximizing the number of links in a cluster whereas Cactus utilises co-occurrence of pairs of attributes values to summarise the data and to achieve linear scaling. Birch [13] and Coolcat [14] are other popular clustering methods used for clustering categorical data. Birch uses a balanced tree structure (CF tree) which preserves the attribute relationships within different instances as leaf nodes and then clustering is done on these leaf nodes. Coolcat is an incremental algorithm which achieves clustering by trying to minimize the entropy values between clusters. An approach [15] to cluster categorical and mix data uses a distance based similarity. A weighting scheme is adopted by the authors which utilizes the relative importance of each instance. Once distance between instances is evaluated a modified version of similarity metrics defined by [16] as $S(X_i, X_j) = 1 - d_p(X_i, X_j)$ is used to find instances similarities.

Simple similarity measures such as overlap suffer from the problem of clustering dissimilar instances together when the number of attributes matched is same, but attributes that are matched are different. Moreover, these similarity measures may perform well with categorical data, but in the case of mixed data which contains both numerical and categorical data the performance declines as the complexity within clusters increases.

4   The Fibonacci Series and Golden Ratio

The proposed FIBCLUS (Fibonacci based Clustering) uses the Fibonacci series to determine a global score for each instance and then utilizes the aggregate distance as a similarity function. Fibonacci series is a sequence of numbers $\{F_n\}_{n=1}^{\infty}$ defined by the linear recurrence equation $F_n = F_{n-1} + F_{n-2}$. The first two Fibonacci numbers are 0 and 1, and each subsequent number is the sum of the previous two. The Fibonacci series has been applied in many scientific and real life fields [17] from analysis of financial markets, to development of computer algorithms such as the Fibonacci search technique and the Fibonacci heap data structure [18]. One of the prominent properties of Fibonacci series is that the ratio of two successive numbers F_n / F_{n-1}, where $n \geq 7$ tends towards 1.6 or φ, as n approaches infinity [17]. This value of φ is also called as the golden ratio.

The primary purpose of using Fibonacci series is, since each similar attribute of all instances are multiplied by a distinct successive Fibonacci number, only similar attributes in different instances will have same values and will be clustered appropriately. If there are m categorical attributes in an instance which have been converted into equivalent numerical attributes then as we do Fibonacci transformation of the attribute from $1...m$ the ratio between $\frac{x_{i,2}}{x_{i,1}}, \frac{x_{i,3}}{x_{i,1}},.. \frac{x_{i,m}}{x_{i,1}}$ will increase significantly, however for two successive attributes, it will always have a minimum values as φ. Due to this transformation property the ordering of attributes will have no major effect on the clustering solution, as the global scores per instance will be compared with each other when performing the clustering solution.

5   The Proposed FIBCLUS Method

The aim of using FIBCLUS with numeric data is to generate a search space in which the input instances are clearly distinguishable. FIBCLUS represents each instance as an aggregate global value compromising of various attributes. In other words, if there are n numeric instances and m number of attributes then the FIBCLUS reduces the search space for each $X = \{X_1, X_2, ..., X_n\}$ from m to 1:

$$\mathbb{R}^n = \{(x_{n,1}, x_{n,2}..x_{n,m})\} \rightarrow \{(x_{n,1})\} \tag{2}$$

For categorical and mix data the aim of FIBCLUS is to identify the best possible similarity that exists between a pair of instances by considering all the attributes. The score of all attributes in this case is also represented as an aggregate global score. Given the set of instances $X = \{X_1, X_2, .., X_n\}$ with m number of attributes $(x_{i,1}, x_{i,2}, ...x_{i,m})$, a Fibonacci number is initialized for each attribute maintaining the golden ratio φ. Let $F = \{F_1, F_2...F_m\}$ be the set of Fibonacci numbers chosen corresponding to m number of attributes where each successive Fibonacci number F_{j+1} maintains the golden ratio φ with the preceding number F_j. In the experiments F1 is initialized as $F_1 = 5$ because the series starts to get closer and closer to φ after this number. Consider an example for the dataset of four attributes $x_{i,1}, x_{i,2}, x_{i,3}, x_{i,4}$, where $F = \{5,8,13,21\}$ is the set of Fibonacci numbers. In this case, $F_1 = 5$ is used to transform $x_{i,1}$ and $F_2 = 8$ is used to transform $x_{i,2}$ and so on. A value in F maintains the golden ratio as $F_2 / F_1, F_3 / F_2, F_4 / F_3 \cong 1.6$.

There are three cases, which have to be considered while clustering with FIBCLUS. **Case 1:** Clustering pure numeric attributes. In this case the maximum value of each attribute $\max(x_{i,1}), \max(x_{i,2}), ... \max(x_{i,m})$ is used for normalizing the attribute values. Normalization is done to scale the values in a constant range so that the Fibonacci number chosen for that attribute does not drastically change the golden ratio ϕ, which separates the values of one attribute from another. **Case 2:** For clustering pure categorical attributes each categorical attribute values are mapped into numeric values. Each instance X_i with attributes as $(x_{i,1}, x_{i,2}, ...x_{i,m})$, and Fibonacci mapped value $F_j x_{i,j}$ is assigned a score. Each instance is compared for similarity with other instances. **Case 3:** In this case for clustering mix of both numeric and categorical attributes, let k be the number of categorical attributes, and l be the number of numeric attributes, where $k + l = m$. The score of each instance is determined separately based on the values of both numeric and categorical attributes (case 1 and case 2) as shown in step 3 of algorithm (figure 2).

$$Score(X_i) = \sum_{1}^{k}(x_{i,k} \times F_k) + \sum_{1}^{m}\frac{(x_{i,l})}{\max(x_{i,l})} \times F_l. \qquad (3)$$

Input:

$X = \{X_1, X_2, ..., X_n\};$ // Datasets instances with m attributes as $x_{i,1}, x_{i,2}, ..., x_{i,m}$.

$F = \{F_1, F_2 .. F_m\}.$ //Successive Fibonacci numbers F corresponding to each $1..m$ attributes.

$F_j x_{i,j} = $ Categorical attribute values, mapped into numeric value.

Output:

    $\mathbb{R}^n = \{(x_{n,1})\}$ // Numeric instances: Global score

    $A = [n \times n]$ // Categorical or Mix: Similarity Matrix.

Begin:

    Step 1. $F_1 = 5.$ // Initialize Fibonacci series.

    Step 2.   // For numeric attribute $\max(x_m)$ finds maximum attribute value from

                  instances.

        For each $j=1$ to m;

          $\max(x_j)$

    Step 3.   // Evaluate scores for each instance.

        For each $i=1$ to n;

        $Score(X_i) = 0.0;$

          For each $j=1$ to m;

          If domain $(x_{ij}) = $ Numeric

$$Score(X_i) = Score(X_i) + \frac{x_{i,j}}{\max(value_j)} \times F_j.$$

          Else domain $(x_{ij}) = $ Categorical

            $Score(X_i) = Score(X_i) + F_j x_{i,j}.$

    Step 4. // Calculate similarity between instances.

        For each $i=1..n$;

        For each $j=1..n$;

         If $((Score(X_i) <= Score(X_j))$

$$Similarity(X_i, X_j) = \frac{X_i \cap X_j}{m} + \frac{Score(X_i)}{Score(X_j)}$$

      Return $\mathbb{R}^n = \{(x_{n,1})\}$ or $A = [n \times n];$

End.

Fig. 2. Complete FIBCLUS Algorithm

Finally, the instance similarity between two instances X_i, X_j is evaluated based on equation (3) as shown in equation (4) and figure 2 (Step 4), where $X_i \cap X_j$ is the number of similar categorical instances between the two instances and $Score(X_i) <= Score(X_j)$. This condition makes sure that the similarity calculation is done only once between pair of instances.

$$Similarity(X_i, X_j) = \frac{X_i \cap X_j}{m} + \frac{Score(X_i)}{Score(X_j)} \tag{4}$$

The pair wise similarity matrix between all instances denoted as $\mathbf{A} = [n \times n]$ becomes input to a clustering algorithm.

6 Empirical Analysis

The objective of experiments was to evaluate the quality of clustering results obtained using the proposed FIBCLUS similarity scores, adopted in the different clustering algorithms. Standard evaluation criteria such as Entropy, Purity and F-Score were used to assess the quality. For numeric datasets FIBCLUS was used with Expectation Minimization (EM), K means (KM) and Extended K means (XM) [6] shown as #1, #2, #3 respectively. For categorical and mix data we used direct, repeated bisection and agglomerative clustering methods implemented in gcluto [1] and shown as #1, #2 ,#3 in all results table(5,6,7). Correlation coefficient and cosine similarity were taken as similarity evaluation methods and the best results were taken. The test datasets were obtained from the UCI repository except Medical[1] as detailed in Table 4. A total of 9 datasets, three of each category were used in experiments. These datasets were taken due to clear class definitions of each instance, which could be compared accurately against results of various clustering methods.

Table 4. Clustering test datasets details

SN	Dataset	Attribute Type	No. of Attribute	No. of class	No. of instance
1	Liver	Numeric	6	2	345
2	Wine	Numeric	13	3	178
3	IRIS	Numeric	4	3	150
4	Soybean	Categorical	35	4	47
5	Balance	Categorical	4	3	625
6	SpectHeart	Categorical	22	2	267
7	Teaching	Mix	5	3	151
8	Medical	Mix	8	3	90
9	Hepatitis	Mix	19	2	155

---

[1] Creators: Sharon Summers, School of Nursing, University of Kansas Medical Center, Kansas City, KS 66160,Linda Woolery, School of Nursing, University of Missouri, Columbia, MO 65211, Donor: Jerzy W. Grzymala-Busse (jerzy@cs.ukans.edu).

Table 5. Results of Purity of Clustering of all datasets

| | Purity of clustering results | | | | | |
| | Without FIB Values | | | FIB Values With | | |
Datasets	EM	KM	XM	#1	#2	#3
Liver	0.507	0.542	0.557	0.513	0.536	0.536
Wine	0.376	0.433	0.433	0.719	0.719	0.719
IRIS	0.907	0.887	0.880	0.960	0.960	0.960
Soybean	1.000	0.979	0.979	0.979	0.979	0.979
Balance	0.526	0.494	0.538	0.549	0.549	0.549
SpectHeart	0.528	0.614	0.614	0.772	0.772	0.772
Teaching	0.417	0.437	0.437	0.424	0.404	0.430
Medical	0.6	0.422	0.422	0.478	0.478	0.478
Hepatitis	0.516	0.542	0.542	0.775	0.763	0.755
Average	**0.597**	**0.594**	**0.6**	**0.685**	**0.684**	**0.686**

Table 6. Results of Entropy of Clustering of all datasets

| | Entropy of clustering | | | | | |
| | Without FIB Values | | | FIB Values With | | |
Datasets	EM	KM	XM	#1	#2	#3
Liver	0.233	0.21	0.184	0.255	0.231	0.231
Wine	0.372	0.377	0.377	0.209	0.209	0.209
IRIS	0.103	0.128	0.141	0.05	0.05	0.05
Soybean	0	0.025	0.025	0.041	0.041	0.041
Balance	0.446	0.458	0.437	0.41	0.41	0.41
SpectHeart	0.195	0.188	0.188	0.012	0.012	0.012
Teaching	0.472	0.449	0.449	0.469	0.475	0.471
Medical	0.231	0.454	0.454	0.306	0.306	0.306
Hepatitis	0.300	0.297	0.297	0.01	0.017	0.021
Average	**0.261**	**0.287**	**0.284**	**0.196**	**0.195**	**0.195**

Table 7. Results of F-Score of Clustering of all datasets

| | F-Score of clustering | | | | | |
| | Without FIB Values | | | FIB Values With | | |
Dataset	EM	KM	XM	#1	#2	#3
Liver	0.438	0.459	0.457	0.461	0.467	0.469
Wine	0.336	0.376	0.376	0.713	0.713	0.713
IRIS	0.907	0.887	0.88	0.96	0.96	0.96
Soybean	1	0.985	0.985	0.975	0.975	0.975
Balance	0.456	0.437	0.484	0.448	0.448	0.448
SpecHeart	0.517	0.422	0.422	0.436	0.436	0.436
Teaching	0.420	0.433	0.433	0.425	0.408	0.433
Medical	0.333	0.301	0.301	0.332	0.332	0.332
Hepatitis	0.437	0.467	0.467	0.436	0.436	0.436
Average	**0.538**	**0.53**	**0.534**	**0.576**	**0.575**	**0.578**

Overall as can be seen, the performance of all clustering algorithms improves when FIBCLUS based global scores and similarity scores are used. This happens due to the separation ratio that is actively bringing similar instances together (in hence making the intra-cluster similarity larger) and separating dissimilar instances more further from each other (in hence making the inter-cluster similarity lower). Independent of the type of attributes and the clustering process used, FIBCLUS is able to produce clustering solutions of high accuracy.

When each cluster is visualized for its purity in figures 3(a)-3(f), the standard EM, KM and XM methods without any space mapping derives clusters with varied purity. For datasets like Iris and Soybean EM performed exceptionally well when compared to distance based algorithm like KM and XM. However when such datasets were used with FIBCLUS in general it was found out that the distance based algorithms like KM and XM performed much better than the density based algorithm like EM. This observation indicates that FIBCLUS has the ability to improve inter and intra cluster distances in any type of clustering method. For numeric datasets FIBCLUS works reasonably well. This is because the aggregate global score computed by FIBCLUS for each instance, is able to map various attributes to a greater extent. Since each attribute is well separated by the golden ratio, the overall score of similar instances is more similar. For some datasets like IRIS, unsupervised clustering using FIBCLUS is able to get 96% accuracy which is equal to some supervised learning methods like J48 [19]. This shows that the reduced search map obtained using the global score calculated using Fibonacci numbers is able to decrease the complexity of the grouping process. For the Wine dataset, results are exceptionally well. The performance improvement in clustering using FIBCLUS (#1, #2, #3) is nearly 50%. For the Liver

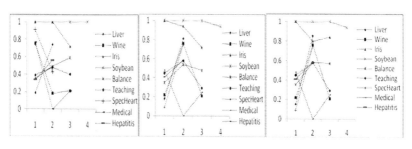

Fig. 3. (a) Purity EM   **Fig. 3. (b)** Purity KM   **Fig. 3. (c)** Purity XM

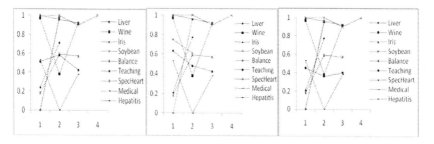

Fig. 3. (d) Purity FIBCLUS(#1) **Fig. 3. (e)** Purity FIBCLUS(#2) **Fig. 3. (f)** Purity FIBCLUS(#3)

dataset, results are nearly comparable, however the clustering achieved using it has better clusters which is evident from the purity and entropy measures.

Overall the average results as percentage of various evaluation metrics are summarized in table 8.

Table 8. Summary of Results on test datasets

Overall Percent(%) improvements in clustering using FIBCLUS		
Similarity Measure	**Best Case-**Best of all Clustering results Versus best FIBCLUS results is taken	**Worst Case-** Worst of all Clustering results Versus worst FIBCLUS result is taken
Purity	14%	15.15%
Entropy	25.29%	31.7%
F Score	7.4%	8.5%

7 Conclusion

This paper proposed an innovative clustering method that reduces the search map for clustering to be performed. An aggregate global score is calculated for each instance using the novel idea of Fibonacci series. Similarity functions are proposed by using the aggregate global score for instances with numerical, categorical or mix attributes. The use of Fibonacci numbers is able to separate the instances effectively and, in hence, enables a higher intra-cluster similarity and a lower inter-cluster similarity. The proposed FIBCLUS method is applied on a wide variety of datasets with categorical, numerical and mix attributes. FIBCLUS is compared with the existing algorithms that are widely used to cluster numeric, categorical and mix data types.

Empirical analysis shows that FIBCLUS is able to produce better clustering solutions in terms of entropy, purity, F-score etc in comparison to existing algorithms such as k-means, x-means, expected maximization and hierarchical algorithms. However the extra overhead in terms of time and space due to the additional step of calculating the similarity scores between instances in case of instances containing mix or categorical, is compensated by the reduced search map during the clustering process. Moreover, clustering usually is an offline process and is more affected by accuracy than such measures.

Acknowledgment

This research has been funded by CRC (Co-operative Research Centre), Australia and Queensland University of Technology, Brisbane Australia under the CRC Smart Services Project 2009-10.

References

1. Rasmussen, M., Karypis, G.: Gcluto: An Interactive Clustering, Visualization, and Analysis System, vol. 21. Citeseer (2008)
2. Liao, H., Ng, M.K.: Categorical data clustering with automatic selection of cluster number. Fuzzy Information and Engineering 1, 5–25 (2009)

3. Huang, Z.: Extensions to the k-means algorithm for clustering large data sets with categorical values. Data Mining and Knowledge Discovery 2, 283–304 (1998)
4. Stanfill, C.: Toward memory-based reasoning. Communications of the ACM 29, 1213–1228 (1986)
5. Boriah, S., Chandola, V., Kumar, V.: Similarity measures for categorical data: A comparative evaluation, vol. 30, p. 3. Citeseer (2007)
6. Ian, H., Witten, E.F.: Data Mining: Practical machine learning tools and techniques, 2nd edn. Morgan Kaufmann, San Francisco (2005)
7. San, O.M., Huynh, V.N., Nakamori, Y.: An alternative extension of the k-means algorithm for clustering categorical data. Internation Journal of Applied Mathematics and Computer Science 14, 241–248 (2004)
8. Ahmad, A., Dey, L.: A method to compute distance between two categorical values of same attribute in unsupervised learning for categorical data set. Pattern Recognition Letters 28, 110–118 (2007)
9. Le, S.Q., Ho, T.B.: An association-based dissimilarity measure for categorical data. Pattern Recognition Letters 26, 2549–2557 (2005)
10. Guha, S., Rastogi, R., Shim, K.: Rock: A robust clustering algorithm for categorical attributes* 1. Information Systems 25, 345–366 (2000)
11. Ganti, V., Gehrke, J., Ramakrishnan, R.: CACTUS—clustering categorical data using summaries. In: Fifth ACM SIGKDD International Conference on Knowledge Discovery and Data Mining, San Diego, California, United States, pp. 73–83 (1999)
12. Gibson, D., Kleinberg, J., Raghavan, P.: Clustering categorical data: An approach based on dynamical systems. The VLDB Journal 8(3), 222–236 (2000)
13. Zhang, T., Ramakrishnan, R., Livny, M.: BIRCH: an efficient data clustering method for very large databases. In: ACM SIGMOD, International Conference on Management of Data, pp. 103–114 (1996)
14. Barbará, D., Li, Y., Couto, J.: COOLCAT: an entropy-based algorithm for categorical clustering. In: 11th International Conference on Information and knowledge Management, pp. 582–589 (2002)
15. Rendón, E., Sánchez, J.: Clustering based on compressed data for categorical and mixed attributes. In: Yeung, D.-Y., Kwok, J.T., Fred, A., Roli, F., de Ridder, D. (eds.) SSPR 2006 and SPR 2006. LNCS, vol. 4109, pp. 817–825. Springer, Heidelberg (2006)
16. Ichino, M., Yaguchi, H.: Generalized Minkoeski metrics for mixed feature-type data analysis. IEEE Transaction on Systems,Man and Cybernitics 24, 694–708 (1994)
17. Chandra, P., Weisstein, E.W.: Fibonacci Number. In: MathWorld–A Wolfram Web Resource, http://mathworld.wolfram.com/FibonacciNumber.html
18. Fredman, M.L., Tarjan, R.E.: Fibonacci heaps and their uses in improved network optimization algorithms, vol. 34, pp. 596–615. ACM, New York (1987)
19. Lacueva-Pérez, F.J.: Supervised Classification Fuzzy Growing Hierarchical SOM. In: Corchado, E., Abraham, A., Pedrycz, W. (eds.) HAIS 2008. LNCS (LNAI), vol. 5271, pp. 220–228. Springer, Heidelberg (2008)

Exploiting Explicit Semantics-Based Grouping for Author Interest Finding

Ali Daud

Department of Computer Science, International Islamic University, Islamabad, Pakistan 44000
ali.daud@iiu.edu.pk

Abstract. This paper investigates the problem of finding author interest in co-author network through topic modeling with providing several performance evaluation measures. Intuitively, there are two types of explicit grouping exists in research papers (1) authors who have co-authored with author A in one document (subgroup) and (2) authors who have co-authored with author A in all documents (group). Traditional methods use graph-link structure by using keywords based matching and ignored semantics-based information, while topic modeling considered semantics-based information but ignored both types of explicit grouping e.g. State-of-the-art Author-Topic model used only one kind of explicit grouping single document (subgroup) for finding author interest. In this paper, we introduce Group-Author-Topic (GAT) modeling which exploits both types of grouping simultaneously. We compare four different topic modeling methods for same task on large DBLP dataset. We provide three performance measures for method evaluation from different domains which are; perplexity, entropy, and prediction ranking accuracy. We show the trade of between these performance evaluation measures. Experimental results demonstrate that our proposed method significantly outperformed the baselines in finding author interest. The trade of between used evaluation measures shows that they are equally useful for evaluating topic modeling methods.

Keywords: Author Interest, Subgroup and Group, Co-author Network, Performance Measures, Topic Modeling.

1 Introduction

Social network analysis has been an active area of research with the proliferation of social applications in different social networks, e.g. Academic social networks such as DBLP and CiteSeer, tagging networks such as Bibsonomy and Delicious, video sharing networks such as Flicker and YouTube, blogging networks such as Blogger and WordPress. Academic social network or Co-author network have several knowledge discovery problems which are useful for fulfilling different suggestion or recommendation tasks. Author interest finding is one of the interesting problems useful for suggesting reviewers for papers, finding collaborators for projects, finding supervisors, finding program committee members for conferences etc.

Co-author network provide the basis for exploiting author interest. Intuitively, there is two type of natural grouping exists in co-author networks (1) Authors who

X. Du et al. (Eds.): APWeb 2011, LNCS 6612, pp. 41–52, 2011.
© Springer-Verlag Berlin Heidelberg 2011

have co-authored with author A in a document (subgroup) e.g. Fig. 1 shows a single document or subgroup which consists of paper title words and co-authors of that paper and (2) authors who have co-authored with author A in all documents (group) which consists of all papers title words and all the co-authors who have written papers with Author A in those papers. For example Fig. 1 show a group of 4 papers title words and co-authors for author A.

Subgroup

- Paper Title of Author A Paper P1
  Coauthors of A in P1

Group

- Paper Title of Author A Paper P1
  Coauthors of A in P1
- Paper Title of Author A Paper P2
  Coauthors of A in P2
- Paper Title of Author A Paper P3
  Coauthors of A in P3
- Paper Title of Author A Paper P4
  Coauthors of A in P4

Fig. 1. An illustration of Group and Subgroup

Previously, three major frameworks used to identify the author interest are (1) stylistic features (such as sentence length), author attribution and forensic linguistics based methods to identify what author wrote a given piece of text [7,9] (2) graph-link structure based methods by using keywords as a basis for representation and analysis for relationships among authors [12,16] and (3) topic modeling based methods for capturing semantics-based intrinsic structure of words presented between subgroups [11,14,15]. Above mentioned frameworks based on writing styles and network connectivity ignored the semantics-based intrinsic structure of words, while semantics-based topic modeling methods exploited grouping at only subgroup and ignored grouping at group level.

In this paper, we investigate the problem of author interest finding by proposing GAT which models the author interest and relationships by considering both type of explicit grouping. Experimental results and discussions elaborate the significance of problem and usefulness of our method. We should mention that exploitation of author interest (writing habits without considering his research level) and expert finding (writing habits with considering his research level) are notably two different knowledge discovery problems [4].

The major contributions of our work described in this paper are the followings:

(1) formulization of author interest finding problem from subgroup to group level
(2) demonstrate that perplexity and entropy (for train and test data) is equally useful for evaluating topic models performance with the fact that entropy provides more lucid results

To the best of our knowledge, we are the first to deal with modeling author interest finding problem by proposing group level method and experimentally showing the relationship between perplexity and entropy.

The rest of the paper is organized as follows. Section 2 illustrates our proposed method and related methods for finding author interest. Section 3 discusses corpus, parameter settings, performance measures, baseline methods, and results and discussions and section 4 concludes this paper.

2 Author Interest Topic Modeling

In this section, before describing our group author interest topic modeling, we first briefly introduce topic modeling idea followed by author-topic model, inverse-author-topic model and conditionally-independent-author-topic model.

2.1 Topic Modeling

Topic modeling brought new notion to the unsupervised learning methods by providing soft clusters of data. Instead of using just a keyword as a measure of relationship for collection of documents in traditional language models fundamental topic modeling assumes that there is a hidden topic layer $Z = \{z_1, z_2, z_3, ..., z_t\}$ between the word tokens and the documents, where z_i denotes a latent topic and each document d is a vector of N_d words w_d. A collection of D documents is defined by $D = \{w_1, w_2, w_3, ..., w_d\}$ and each word w_{id} is chosen from a vocabulary of size W. For each document, a topic mixture distribution is sampled and a latent topic Z is chosen with the probability of topic given document for each word with word having generated probability of word given topic [2,10]. The generating probability of word w for a document d for the state-of-the-art topic model Latent Dirichlet Allocation is given in Eq. 1.

$$P(w|d, \emptyset, \theta) = \sum_{z=1}^{T} P(w|z, \emptyset_z)P(z|d, \theta_d) \tag{1}$$

2.2 Author-Topic Model (AT)

AT [15] is a two way stochastic process which is based on the idea that author thinks about a topic and starts writing a paper on that topic with the help of co-authors. In AT a randomly chosen author from a subgroup is responsible for generating words of a document. In AT, each author (from set of A authors) of a document d is associated with a multinomial distribution θ_a over topics which is sampled from Dirichlet α and each topic is associated with a multinomial distribution Φ_z which is sampled from Dirichlet β over words of a document for that topic. The generating probability of word w for author r of a document d is given in Eq. 2. AT has successfully discovered topically related authors but did not consider explicit group information.

$$P(w|r, d, \emptyset, \theta) = \sum_{z=1}^{T} P(w|z, \emptyset_z)P(z|r, \theta_r) \tag{2}$$

2.3 Inverse-Author-Topic Model (IAT)

IAT is a two way stochastic process which is based on the idea that a randomly chosen word from a subgroup is responsible for generating authors of a document. This idea is opposite to the basic idea of AT. In IAT, each word (from set of W words) of a document d is associated with a multinomial distribution θ_w over topics which is sampled from Dirichlet α and each topic is associated with a multinomial distribution Φ_z which is sampled from Dirichlet β over authors of a document for that topic. The generating probability of author r for word w of a document d is given in Eq. 3. IAT did not consider explicit group information.

$$P(r|w, d, \emptyset, \theta) = \sum_{z=1}^{T} P(r|z, \emptyset_z) P(z|w, \theta_w) \qquad (3)$$

2.4 Conditionally-Independent-Author-Topic Model (CIAT)

CIAT is based on the idea that words and authors are independently generated from a subgroup which is a variation of GM-LDA used for image annotation [3]. AT and IAT assumes that randomly chosen author or word generates a topic, respectively. On the contrary CIAT assumes that authors and words are independently generated by the topic. In CIAT, topics are sampled from multinomial distribution θ with Dirichlet α over words and authors and each word and author is associated with a multinomial distribution Φ_z which is sampled from Dirichlet β over words and a multinomial distribution ψ_z which is sampled from Dirichlet μ over authors of a document, respectively. The generating probability of word w and author r of a document d is given in Eq. 4. CIAT did not consider explicit group information.

$$P(w, r|d, \emptyset, \theta, \psi) = \sum_{z=1}^{T} P(w|z, \emptyset_z) P(r|z, \psi_z) \qquad (4)$$

2.5 Group-Author-Topic Modeling (GAT)

GAT is a two way stochastic process which is based on the idea that author thinks about a topic and his thinking is influenced by all co-authors of his papers. In GAT a randomly chosen author from a group is responsible for generating words of a group. In the proposed approach, we viewed a group as a composition of authors all documents (subgroups). Symbolically, for a group G we can write it as: $G = \{(\mathbf{w}_1, \mathbf{a}_{d1}) + (\mathbf{w}_2, \mathbf{a}_{d2}) + (\mathbf{w}_3, \mathbf{a}_{d3}) + \ldots + (\mathbf{w}_i, \mathbf{a}_{di})\}$, where d_i is a subgroup of a group and a_{di} are the author (s) of subgroup d_i.

Subgroup based methods considers that an author is responsible for generating latent topics of the document, while, group based method considers that an author is responsible for generating latent topics of the group (please see Fig. 1 and 3). In GAT, each author (from set of A authors) of a group g is associated with a multinomial distribution θ_u over topics which is sampled from Dirichlet α and each topic is associated with a multinomial distribution Φ_z which is sampled from Dirichlet β over words of a group for that topic. The generating probability of word w for author r of a group g is given in Eq. 5.

$$P(w|r, g, \emptyset, \theta) = \sum_{z=1}^{T} P(w|z, \emptyset_z) P(z|r, \theta_r) \qquad (5)$$

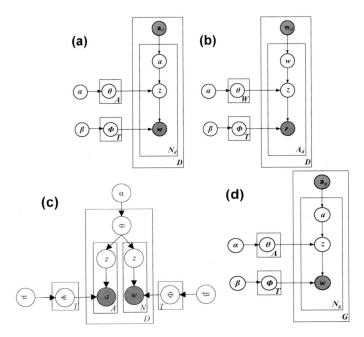

Fig. 2. GAT (d) is shown with three related Models (a) Author-Topic (AT) Model, (b) Inverse-Author-Topic (IAT) Model, and (c) Conditionally-Independent-Author-Topic (CIAT) Model

The generative process of GAT is as follows:

For each author $r = 1,\ldots, K$ of group g
    Choose θ_r from Dirichlet (α)
For each topic $z = 1,\ldots, T$
    Choose Φ_z from Dirichlet (β)
For each word $w = 1,\ldots, N_g$ of group g
    Choose an author r uniformly from all authors $\mathbf{a}_g$
    Choose a topic z from multinomial (θ_r) conditioned on r
    Choose a word w from multinomial (Φ_z) conditioned on z
Gibbs sampling is utilized [1] to solve all related methods and our proposed method which has two latent variables z and r; the conditional posterior distribution for z and r is given by:

$$P(z_i = j, r_i = k \mid w_i = m, \mathbf{z}_{-i}, \mathbf{r}_{-i}, \mathbf{a}_g) \propto \frac{n_{-i,j}^{(wi)} + \beta}{n_{-i,j}^{(\cdot)} + W\beta} \frac{n_{-i,j}^{(ri)} + \alpha}{n_{-i,\cdot}^{(ri)} + A\alpha} \qquad (6)$$

where $z_i = j$ and $r_i = k$ represent the assignments of the ith word in a group to a topic j and author k respectively, $w_i = m$ represents the observation that i^{th} word is the mth word in the lexicon, and z_{-i} and r_{-i} represents all topic and author assignments not including the ith word. Furthermore, $n_{-i,j}^{(wi)}$ is the total number of words associated with topic j, excluding the current instance, and $n_{-i,j}^{(ri)}$ is the number of times author k is assigned to topic j, excluding the current instance, W is the size of the lexicon and A is the number of authors. "." Indicates summing over the column where it occurs and $n_{-i,j}^{(.)}$ stands for number of all words that are assigned to topic z excluding the current instance.

For parameter estimation model needs to keep track of W x Z (word by topic) and Z x A (topic by author) count matrices for group. From these count matrices, topic-word distribution Φ and author-topic distribution θ can be calculated as given in Eq. 7 and Eq. 8, where, $\emptyset_{zw}$ is the probability of word w in topic z and θ_{rz} is the probability of topic z for author r. These values correspond to the predictive distributions over new words w and new topics z conditioned on w and z.

$$\emptyset_{zw} = \frac{n_{-i,j}^{(wi)}+\beta}{n_{-i,j}^{(.)}+W\beta} \tag{7}$$

$$\theta_{rz} = \frac{n_{-i,j}^{(ri)}+\alpha}{n_{-i,.}^{(ri)}+A\alpha} \tag{8}$$

3 Experiments

3.1 Corpus

We downloaded five years paper corpus of conferences from DBLP database [6], by only considering conferences for which data was available for the years 2003-2007. In total, we extracted 112,317 authors and 90,124 papers. We then processed corpus by (a) removing stop-words, punctuations and numbers (b) down-casing the obtained words of papers, and (c) removing words and authors that appear less than three times in the corpus. This led to a vocabulary size of V=10,872, a total of 572,592 words and 26,078 authors in the corpus.

There is certainly some noise in data of this form especially author names which were extracted automatically by DBLP from PDF, postscript or other document formats. For example, for some very common names there can be multiple authors (e.g. L Ding or J Smith). This is a known as limitation of working with this type of data (please see [13] for details). There are algorithmic techniques for name disambiguation that could be used to automatically solve these kinds of problems; however, in this work we do not focus on name disambiguation problems.

3.2 Parameter Settings

Estimation of hyper-parameters α and β is done by using Gibbs sampling algorithm [8]. For some applications topic models are sensitive to the hyper parameters and need to be optimized. For application in this paper, we found that our topic model based methods are not sensitive to the hyper parameters. In our experiments, for different number of topics Z the hyper-parameters α, β and μ were set at 50/Z, 0.01 and 0.01.

3.3 Performance Measures

We used three performance measures for evaluating the performance of methods from different domains (1) perplexity; a standard performance measure for evaluating topic modeling (soft clustering) by examining the generative power of trained model on unseen dataset. Lower values of perplexity indicate better generalization power of model on the words of test documents by the trained topics. For a test set of D documents the perplexity is given in Eq. 9, (2) entropy (under root of perplexity) for training set and testing set to measure the quality of discovered topics, which reveals the purity of topics is given in Eq. 10, less intra-topic entropy is better; a performance measure for evaluating traditional clustering (hard clustering), and (3) Prediction ranking accuracy; a performance measure for evaluating recommendation is given in Eq. 11. We employ the top-k recommendations, that is, each ranking algorithm needs to recommend the top k objects (words and authors) for documents by ranking randomly withhold objects from the original set mixed with objects not from the original set.

$$perplexity\ (D_{test}) = \exp\ \{-\frac{\sum_{d=1}^{D} \log p(\mathbf{w}_d)}{\sum_{d=1}^{D} N_d}\} \tag{9}$$

$$Entropy\ of\ (Topic) = -\sum_z P(z) log_2[P(z)] \tag{10}$$

$$Prediction\ Ranking\ Accuracy = \frac{k\ value + 1 - rank\ of\ object}{k\ value} \tag{11}$$

3.4 Baseline Methods

We compare our proposed method GAT with AT [15], which considers that words and authors of a document are dependent on each other and authors are responsible for generating words of a document, IAT, which considers that words and authors of a document are dependent on each other and words are responsible for generating authors of a document, and CIAT, which considers that words and authors of a document are independent of each other and topics are responsible for generating words and authors of a document.

3.5 Results and Discussions

We extracted authors related to a specific area of research on the basis of semantics-based similarity of topic so called latent topics. Table 1 shows authors' interests for different topics by using GAT. It illustrates 3 topics out of 150, discovered from the 1000th iteration of the particular Gibbs sampler run. The words and authors associated with each topic are quite precise and depict a real picture of specific area of research. For example, topic # 19 "Semantic Web" shows quite specific and meaningful vocabulary (semantic, web, ontology, owl, rdf, annotation, semantics, and knowledge) when a user is searching for semantic web related documents or authors. Other topics, such as "Pattern Mining" and "Information Retrieval" are quite descriptive that shows the ability of GAT to discover precise topics. We have analyzed and found that authors related to different topics are typically writing for that area of research. For example, in case of topic 74 "Semantic Web" top ranked authors web pages shows their interest in semantic web research topic and they are mostly publishing on this topic.

Table 1. Illustration of 3 topics with related authors. Tiles are assigned to clustered words manually.

Topic 18 "Pattern Mining"		Topic 8 "Information Retrieval"		Topic 74 "Semantic Web"	
Word	**Prob.**	**Word**	**Prob.**	**Word**	**Prob.**
mining	0.242013	retrieval	0.160582	semantic	0.260961
patterns	0.101704	text	0.116067	web	0.138429
pattern	0.067878	document	0.074017	ontology	0.124851
frequent	0.047791	extraction	0.050939	ontologies	0.060605
privacy	0.035545	documents	0.045187	owl	0.033670
preserving	0.034873	information	0.031594	rdf	0.026936
discovery	0.034201	relevance	0.029353	annotation	0.018105
discovering	0.025315	categorization	0.026665	semantics	0.016670
databases	0.022626	topic	0.025694	approach	0.014352
sequential	0.020984	feedback	0.025619	knowledge	0.012365
Author	**Prob.**	**Author**	**Prob.**	**Author**	**Prob.**
Jian Pei	0.011381	Wei-Ying Ma	0.008950	Carole A Goble	0.018056
Jiawei Han	0.010317	ChengXiang Zhai	0.005378	Robert Stevens	0.014153
Wei Wang	0.007429	W. Bruce Croft	0.004808	Peter Haase	0.013177
Philip S. Yu	0.006061	Charles L. A. Clarke	0.003439	Amit P Sheth	0.012201
Hui Xiong	0.005263	Mounia Lalmas	0.003249	Steffen Staab	0.011714
Ada Wai-Chee Fu	0.005111	Weiguo Fan	0.003135	Phillip W Lord	0.011714
Srinivasan Parth.	0.004237	Tat-Seng Chua	0.002679	Luc Moreau	0.010738
Ke Wang	0.004161	David A. Grossman	0.002527	Anupam Joshi	0.010250
Kotagiri Rama.	0.003781	Xuanhui Wang	0.002527	Ian Horrocks	0.009762
Jianyong Wang	0.003667	James Allan	0.002451	David DeRoure	0.009762

GAT also discovered several other topics such as image retrieval, neural networks, business process modeling, semi-supervised learning and XML databases. In addition, by doing analysis of authors' home pages and DBLP [10], we have found that all authors assigned with higher probabilities have published many papers on their relevant topics. In the following we provide the links to the home pages of top five authors related to semantic web topic for authentication.

http://www.cs.manchester.ac.uk/~carole/
http://www.cs.manchester.ac.uk/~stevensr/
http://semanticweb.org/wiki/Peter_Haase
http://knoesis.wright.edu/amit/
http://www.uni-koblenz.de/~staab/

3.5.1 Perplexity Based Comparison

Perplexity is a standard measure for estimating the performance of probabilistic topic models. It shows generalization power of a topic model for the test dataset; with lower perplexity corroborate better performance.

Fig. 3 on the left side presents the perplexity for each method for different values of Z. GAT performs better than the subgroup based AT, CIAT and IAT. As the performance difference between the GAT, AT and ACIT is not very clear, we take the under root of perplexity which is shown in right side of Fig.4. It shows the performance difference between methods more clearly and proves the dominance of GAT over baseline methods. Fig. 3 may suggest that exploitation of both types of grouping for author interest finding task results in the better generalization power of model on the unseen dataset.

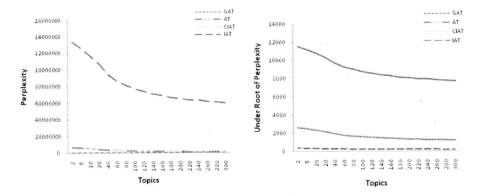

Fig. 3. Perplexity for different number of topics

3.5.2 Entropy Based Comparison

Fig. 4 provides a quantitative comparison between proposed GAT, AT, CIAT, and IAT. Fig. 4 (left) shows the average entropy of topic-word distribution of training data for all topics calculated by using Eq. 10. Lower entropy for different number of topics $T = 20,40, …300$ proves the effectiveness of GAT for obtaining dense topics when compared to baselines. We see when number of topics are less than 40 the performance of GAT, AT and CIAT is same but when the number of topics increases one can see a clear performance difference of GAT with baselines. GAT exploits subgroup and group structures both so able to produce dense topics which results in better performance of method [5].

Fig. 4 (right) shows the average entropy of topic-word distribution of test data for all topics calculated by using Eq. 10. Similar results are obtained for test data except IAT entropy is lower than other methods for number of topics less than 40. We again see that when number of topics are less than 40 the performance of GAT, AT and CIAT is same but when the number of topics increases a clear performance difference of GAT with baselines is observed.

3.5.3 Prediction Accuracy Based Comparison

We show quantitatively the effectiveness of our proposed method GAT for predicting words and authors of documents in Table 2. GAT performed better when compared

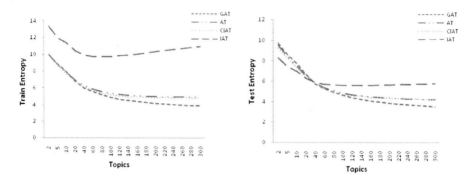

Fig. 4. Average Entropy curve as a function of different number of topics for training and test dataset

with AT, CIAT and IAT for words ranking prediction with values of k= 2,5,10 and for number of topics varied from 2, 5, 10, 20, 40,…,300 shown in Table 1 are 0.56 for GAT, 0.49 for AT, 0.49 for CIAT and 0.43 for IAT. It shows that GAT performed 7% better than AT and CIAT and 13% better than IAT in terms of ranking accuracy which show the better performance of our proposed method. The average ranking accuracy results for author prediction is 0.54 for GAT, 0.44 for AT, 0.45 for CIAT and 0.50 for IAT which show that GAT performed 10% better than AT, 9% better than CIAT, and 4% better than IAT which is significant. Collectively one can say that exploiting subgroup and group level structure together not only increases the generative power of topic model but also helps to have increased ranking accuracy for predicting words and authors.

Table 2. Ranking accuracy for words and authors prediction

Words Prediction			
Words	K=2	K=5	K=10	Average
GAT	0.504506	0.507626	0.689609	0.567247
AT	0.484855	0.426692	0.563054	0.491534
CIAT	0.501782	0.425927	0.556542	0.49475
IAT	0.474909	0.30546	0.511416	0.430595
Authors Prediction			
Authors	K=2	K=5	K=10	Average
GAT	0.5	0.433	0.7	0.544333
AT	0.481	0.321	0.526	0.442667
CIAT	0.514841	0.328181	0.527581	0.456867
IAT	0.49961	0.405338	0.619148	0.508032

4   Conclusions

This study deals with the problem of finding author interest. Two types of natural grouping existing in co-author networks is considered in this paper and found to be effective. GAT uses both type of explicit networks and performed better than baselines for several performance measures from different domains. We can say that both explicit grouping structures are important and should be considered simultaneously. We conclude that perplexity and entropy are equally useful for evaluating generative power of topics models with the fact that entropy based results are more understandable with the increasing number of topics. Exploitation of both types of explicit grouping structures also results in increased prediction ranking accuracy for words and authors.

Future work includes the formulization of grouping structure exits in other social networks and structures exploitation for finding their usefulness of social application on the Web by using novel methods.

Acknowledgements

The work is supported by the Higher Education Commission (HEC), Islamabad, Pakistan.

References

[1] Andrieu, C., de Freitas, N., Doucet, A., Jordan, M.I.: An introduction to MCMC for Machine Learning. Machine Learning 50, 5–43 (2003)

[2] Blei, D.M., Ng, A.Y., Jordan, M.I.: Latent Dirichlet Allocation. Journal of Machine Learning Research 3, 993–1022 (2003)

[3] Blei, D.M., Jordan, M.I.: Modeling annotated data. In: Proceedings of the Annual Conference on Research and Development in Information Retrieval, SIGIR (2003)

[4] Daud, A., Li, J., Zhu, L., Muhammad, F.: Temporal Expert Finding through Generalized Time Topic Modeling. Knowledge-Based Systems (KBS) 23(6), 615–625 (2010)

[5] Daud, A., Li, J., Zhou, L., Muhammad, F.: Conference Mining via Generalized Topic Modeling. In: Buntine, W., Grobelnik, M., Mladenić, D., Shawe-Taylor, J. (eds.) ECML PKDD 2009. LNCS (LNAI), vol. 5781, pp. 244–259. Springer, Heidelberg (2009)

[6] DBLP Bibliography Database,
     http://www.informatik.uni-trier.de/~ley/db/

[7] Diederich, J., Kindermann, J., Leopold, E., Paass, G.: Authorship Attribution with Support Vector Machines. Applied Intelligence 19(1) (2003)

[8] Griffiths, T.L., Steyvers, M.: Finding scientific topics. Proceedings of the National Academy of Sciences, 5228–5235 (2004)

[9] Gray, A., Sallis, P., MacDonell, S.: Softwareforensics: Extending Authorship Analysis Techniques to Computer Programs. In: Proceedings of the 3rd IAFL, Durham NC (1997)

[10] Hofmann, T.: Probabilistic Latent Semantic Analysis. In: Proceedings of the 15th Annual Conference on Uncertainty in Artificial Intelligence (UAI), Stockholm, Sweden, July 30-August 1 (1999)

[11] Mimno, D., McCallum, A.: Expertise modeling for matching papers with reviewers. In: Proceedings of KDD, pp. 500–509 (2007)

[12] Mutschke, P.: Mining Networks and Central Entities in Digital Libraries: A Graph Theoretic Approach Applied to Co-author Networks. Intelligent Data Analysis, 155–166 (2003)

[13] Newman, M.E.J.: Scientific collaboration networks: I. Network construction and funda-mental results. Physical Review E 64, 016131 (2001)
[14] Kawamae, N.: Author Interest Topic Model. In: Proceedings of SIGIR, July 19–23, pp. 887–888 (2010)
[15] Rosen-Zvi, M., Chemudugunta, C., Griffiths, T., Smyth, P., Steyvers, M.: Learning Author-Topic Models from Text Corpora. ACM Transactions on Information Systems, 1–38 (March 2009)
[16] White, S., Smyth, P.: Algorithms for Estimating Relative Importance in Networks. In: Proceedings of the 9th ACM SIGKDD International Conference on Knowledge Discovery and Data Mining, pp. 266–275 (2003)

Top-K Probabilistic Closest Pairs Query in Uncertain Spatial Databases

Mo Chen, Zixi Jia, Yu Gu, and Ge Yu

Northeastern University, China
{chenmo,jiazixi,guyu,yuge}@ise.neu.edu.cn

Abstract. An important topic in the field of spatial data management is processing the queries involving uncertain locations. This paper focuses on the problem of finding probabilistic K closest pairs between two uncertain spatial datasets, namely, *Top-K probabilistic closest pairs* (Top$K$-PCP) query, which has popular usages in real applications. Specifically, given two uncertain datasets in which each spatial object is modeled by a set of sample points, a Top$K$-PCP query retrieves the pairs with top K maximal probabilities of being the closest pair. Due to the inherent uncertainty of data objects, previous techniques to answer K-*closest pairs* (K-CP) queries cannot be directly applied to our Top$K$-PCP problem. Motivated by this, we propose a novel method to evaluate Top$K$-PCP query effectively. Extensive experiments are performed to demonstrate the effectiveness of our method.

1 Introduction

Uncertainty is an inherent property in many modern applications. For example, the location values acquired from mobile devices usually have measurement errors. Furthermore, the uncertainty of location values is in natural due to the delay on data updates in the area of Location-Based Services (LBS). Specifically, the server always knows that the location values of each object don't exceed a certain threshold value from the last sent position, which are usually handled as uncertain data. Uncertain data can also be derived from sensor databases, privacy preservation and so on.

The formal definition of K *closest pairs* (K-CP) query was first described in [1], which is considered as a combination of join and nearest neighbor query. By definition, given two data sets A and B, their K-CP join is a one to one assignment of objects from the two sets, such that K result pairs in $A \times B$ has the minimum distance. K-CP queries are widely used in the applications involving spatial data for decision making, especially in the applications involving uncertain data [2]. For example, consider a case where K-closest pairs over moving objects are monitored, which aims at maintaining closest pair results while the underlying objects change their positions. Queries in such a case are often issued as ($K = 1$) *"return a pair of taxi stand and taxi that have the smallest distances"* or *"return a pair of sonar tracking station and ship that are closest to each other"*, where the locations of these moving objects are uncertainty.

X. Du et al. (Eds.): APWeb 2011, LNCS 6612, pp. 53–64, 2011.

Therefore, K-CP query on uncertain data is very important in many real world applications. In this paper, we focus on this important type of query in uncertain spatial databases, which, to the best of our knowledge, no other work has studied before. The approach presented in this paper deals with spatial uncertain data which are modeled by discrete values based on the concept of monte-carlo sampling. This is because positional uncertainties of objects are often given in the form of discrete values. Even if the uncertain object is described by a continuous probability density function (pdf), we can easily derive a set of samples according to the function.

The rest of the paper is organized as follows. Section 2 briefly overviews previous methods to retrieve the closest pairs over precise spatial data objects and related works on query processing in uncertain databases. Section 3 defines the problem of finding top K closest pairs in uncertain databases. The details of the query processing are presented in Section 4. Section 5 describes the experimental results of the proposed method. The paper is concluded with a discussion about future work in Section 6.

2   Related Work

Section 2.1 reviews previous methods to answer closest pair queries on precise spatial data. Section 2.2 illustrates query processing in uncertain databases.

2.1   Closed Pair Queries in Spatial Databases

The computation of closest pairs queries in spatial databases have been studied for several years. Hjaltason and Samet [3] proposed distance-join algorithms for closest pair queries in spatial databases, which are based on a priority queue and take up a lot of main memory. This is because the pair items store both node pairs and object pairs in priority queue. The algorithms in [3] is incremental, i.e., report the result pair in ascending order of distance in one-by-one fashion, which leads to a fact that these algorithms are incompetent when a large quantity of elements of the result is needed. To solve the above problems, Corral et al. [1] proposed several non-incremental algorithms to process closest pair queries, which reports all the elements of the result all together at the end of the algorithm. The heap algorithm proposed in [1] stores pair items that are only node pairs since it can significantly reduce the size of the queue. Afterwards, Corral et al. extended and enhanced the work with respect to the design of branch-and-bound algorithms in a non-incremental way [4] .

However, the methods in [1],[3],[4] cannot work well in cases that the two data sets 'overlap'. Thus, Yang and Lin [5] proposed a new index structure by pre-computing and storing the nearest neighbor information (with respect to the other data set) in the node entry of R-tree. They showed that the index structure is also very efficient on other types of queries, such as spatial join query and closest nearest neighbor pair query. Leong et al. [6] identified the ECP (*exclusive closest pairs*) problem, which is a spatial assignment problem.

In [6], main-memory algorithms were proposed for solving the static version of the problem. The problem of monitoring ECP pairs in a dynamic environment was also tackled in that paper.

2.2   Query Processing on Uncertain Data

Recently, a lot of work has been done in the field of query processing with the focus on management and processing on uncertain data. A survey of the research area concerning uncertainty and incomplete information in databases is given in [7]. Tuple uncertainty and attribute uncertainty are usually considered as two categories of data uncertainty [8]. In particular, tuple uncertainty is used to model the probabilistic relational data in probabilistic databases [9],[10],[11]. Compared with tuple uncertainty, attribute uncertainty is introduced in the context where each uncertain object is modeled by a distribution within an uncertain region. The distribution can be represented by either *pdf* [12],[13],[14] or discrete samples [15],[16],[17]. In this paper, we use attribute uncertainty with sample representation.

Usually, queries on uncertain data need to consider the confidence guarantee of query answers due to the uncertainty. Thus, many types of queries have to be re-defined in the uncertain scenario. Specifically, previous works studied queries in uncertain databases include *probabilistic nearest neighbor* (PNN) query [16],[18], *probabilistic range* (PR) query [13], *probabilistic reverse nearest* (PRNN) query [19], *probabilistic top-K* (PTK) query [20],[21], *probabilistic group nearest neighbor* (PGNN) query [22] and so on. However, to our best knowledge, no work has addressed the problem of processing closed pair queries on uncertain spatial data.

3   Problem Definition

We next describe the uncertainty model adopted in this paper followed by our formal query definition.

3.1   Uncertainty Model

Positional uncertainties of spatial objects are often obtained in form of discrete values, especially if the locations are derived from different observations. Furthermore, we can easily sample according to the *pdf*, even when the uncertain object is specified by a smooth *pdf*. Let $P = \{P_1, P_2, \ldots, P_{NP}\}$ and $Q = \{Q_1, Q_2, \ldots, Q_{NQ}\}$ be two uncertain object sets. We assume each uncertain object in the object sets is represented by a set of *sr* sample points, i.e., $P = \{\{p_{1,1}, p_{1,2}, \ldots, p_{1,sr}\}, \{p_{2,1}, p_{2,2}, \ldots, p_{2,sr}\}, \cdots, \{p_{NP,1}, p_{NP,2}, \ldots, p_{NP,sr}\}\}$, $Q = \{\{q_{1,1}, q_{1,2}, \ldots, q_{1,sr}\}, \{q_{2,1}, q_{2,2}, \ldots, q_{2,sr}\}, \cdots, \{q_{NQ,1}, q_{NQ,2}, \ldots, q_{NQ,sr}\}\}$. In order to reduce the computation cost of query evaluation, we process the query on the groups of samples. We apply k-means clustering algorithm individually to each sample set $\{p_{g,1}, p_{g,2}, \ldots, p_{g,sr}\}$ $(g = 1, 2, \ldots, NP)$ and $\{q_{h,1}, q_{h,2}, \ldots, q_{h,sr}\}$ $(h = 1, 2, \ldots, NQ)$. Therefore, an uncertain object in this paper is approximated by k clusters that contain all of the samples. Similar to [23], these clustered objects are stored in R*-tree [24], where each cluster is bounded by a *minimum bounding rectangle* (MBR).

3.2   Query Definition

In "certain" databases containing precise data objects, K-CP query [1] discovers K pairs of spatial objects formed from the two data sets that have the K smallest distances between them, where $K \geq 1$. In uncertain databases, the attribute data of each object are "probabilistic". Therefore, we define a novel closest pair query in uncertain databases as follows.

Definition 1. Top-K Probabilistic Closed Pairs (Top$K$-PCP) Query.
Given two uncertain object sets $P = \{P_1, P_2, \ldots, P_{NP}\}$ and $Q = \{Q_1, Q_2, \ldots, Q_{NQ}\}$, a Top$K$-PCP query returns K ordered pairs

$$(P_{(1)}, Q_{(1)}), (P_{(2)}, Q_{(2)}), \cdots, (P_{(K)}, Q_{(K)})$$

$$P_{(1)}, P_{(2)}, \ldots, P_{(K)} \in P \wedge Q_{(1)}, Q_{(2)}, \ldots, Q_{(K)} \in Q$$

such that

$$\Pr(P_{(1)}, Q_{(1)}) \geq \Pr(P_{(2)}, Q_{(2)}) \geq \cdots \geq \Pr(P_{(K)}, Q_{(K)}) \geq \Pr(P_w, Q_v)$$

$$\forall (P_w, Q_v) \in (P \times Q - \{(P_{(1)}, Q_{(1)}), (P_{(2)}, Q_{(2)}), \ldots, (P_{(K)}, Q_{(K)})\}), K \leq |P| \cdot |Q|$$

where we denote with $\Pr(P_g, Q_h)(g = 1 \ldots NP, h = 1 \ldots NQ)$ the probability of pair (P_g, Q_h) to be the closest pair. The value of $\Pr(P_g, Q_h)$ is defined as follows:

$$\Pr(P_g, Q_h) = \frac{\sum_{l=1}^{sr} \sum_{f=1}^{sr} F_s(p_{g,l}, q_{h,f})}{sr^2} \tag{1}$$

$F_s(p_{g,l}, q_{h,f})$ is the probability of $(p_{g,l}, q_{h,f})$ being the closest sample pair among sample pairs of all the other objects. The average value of these independent sr^2 probabilities is equal to $\Pr(P_g, Q_h)$. $F_s(p_{g,l}, q_{h,f})$ can be computed by:

$$F_s(p_{g,l}, q_{h,f}) =$$

$$\prod_{\substack{\forall (P_{g'}, Q_{h'}) \in P \times Q \\ \wedge (P_g, Q_h) \neq (P_{g'}, Q_{h'})}} 1 - \frac{\left| \{(p_{g',l}, q_{h',f}) | D(p_{g',l}, q_{h',f}) < D(p_{g,l}, q_{h,f})\} \right|}{sr^2} \tag{2}$$

where $\frac{\left| \{(p_{g',l}, q_{h',f}) | D(p_{g',l}, q_{h',f}) < D(p_{g,l}, q_{h,f})\} \right|}{sr^2}$ $(l, f = 1 \ldots sr)$ is the probability that the sample pairs of object pair $(P_{g'}, Q_{h'})$ is closer than pair $(p_{g,l}, q_{h,f})$, $D(p_{g,l}, q_{h,f})$ is the Euclidean distance between two sample points $p_{g,l}$ and $q_{h,f}$.

4   Top$K$-PCP Query Processing

In this section, we firstly present some useful metrics and lemmas, which is followed by the query processing algorithm based on these metrics and lemmas.

4.1   Useful Metrics and Lemmas

Given two indexing R*-trees, R_P and R_Q, we assume that MBR(I_P) and MBR(I_Q) respectively denote the MBRs of the nodes in R_P and R_Q. We denote with $Min\_MinD(I_P, I_Q)$ and $Max\_MaxD(I_P, I_Q)$ the minimum distance and the maximum distance between MBR(I_P) and MBR(I_Q). Furthermore, by $Min\_MaxD(I_P, I_Q)$ we denote the minimum maximum distance. The above metrics can be calculated by the similar formulae presented in [1], which are depicted as follows:

$$Min\_MinD(\text{MBR}(I_P), \text{MBR}(I_Q)) = \min_{n,m=1,2,3,4}\{MinD(a_n, b_m)\} \qquad (3)$$

$$Min\_MaxD(\text{MBR}(I_P), \text{MBR}(I_Q)) = \min_{n,m=1,2,3,4}\{MaxD(a_n, b_m)\} \qquad (4)$$

$$Max\_MaxD(\text{MBR}(I_P), \text{MBR}(I_Q)) = \max_{n,m=1,2,3,4}\{MaxD(a_n, b_m)\} \qquad (5)$$

where a_1, a_2, a_3 and a_4 are four edges of MBR(I_P), b_1, b_2, b_3 and b_4 are the edges of MBR(I_Q). $MinD(a_n, b_m)$ and $MaxD(a_n, b_m)$ respectively denote the minimum and maximum Euclidean distance between two points falling on a_n and b_m.

Lemma 1. *Let $I_{P,1}, I_{P,2}, \ldots$ and $I_{Q,1}, I_{Q,2}, \ldots$ be the nodes in the same level of R_P and R_Q. If*

$$Max\_MaxD(\text{MBR}(I_{P,i}), \text{MBR}(I_{Q,j})) \leq$$
$$\min_{\substack{\forall(I_{P,i'}, I_{Q,j'})\in R_P\times R_Q \\ \wedge(I_{P,i}, I_{Q,j})\neq(I_{P,i'}, I_{Q,j'})}} Min\_MinD(\text{MBR}(I_{P,i'}), \text{MBR}(I_{Q,j'})) \qquad (6)$$

the probability of pair $(I_{P,i}, I_{Q,j})$ being the closest pair among all the other node pairs equals to 1, which can be represented as $F_s'(I_{P,i}, I_{Q,j})$.

Proof. Similar to Equation 2, we have

$$F_s'(I_{P,i}, I_{Q,j}) =$$
$$1 - \frac{\left|\{(I_{P,i'}, I_{Q,j'})|D(I_{P,i'}, I_{Q,j'}) < D(I_{P,i}, I_{Q,j}) \wedge (I_{P,i}, I_{Q,j}) \neq (I_{P,i'}, I_{Q,j'})\}\right|}{\left|\{(I_{P,i'}, I_{Q,j'})|(I_{P,i}, I_{Q,j}) \neq (I_{P,i'}, I_{Q,j'})\}\right|}$$

By Inequality 6, there exists no pair satisfying $D(I_{P,i'}, I_{Q,j'}) < D(I_{P,i}, I_{Q,j})$. Thus, the cardinality of $\{(I_{P,i'}, I_{Q,j'})|D(I_{P,i'}, I_{Q,j'}) < D(I_{P,i}, I_{Q,j}) \wedge (I_{P,i}, I_{Q,j}) \neq (I_{P,i'}, I_{Q,j'})\}$ is 0, which leads to $F_s'(I_{P,i}, I_{Q,j}) = 1$.

Lemma 1 indicates that the node pair $(I_{P,i'}, I_{Q,j'})$ can be pruned if it satisfies Inequality 6. Considering the types of the elements indexed by R*-trees, we extract four types of pairs: $(I_P.obj, I_Q.obj)$, $(I_P.clu, I_Q.clu)$, $(I_P.ord, I_Q.ord)$ and $(I_P.sam, I_Q.sam)$, which are respectively constituted by object nodes (i.e., internal nodes stand for objects), cluster nodes (i.e., internal nodes stand for clusters), ordinary nodes (i.e., the other internal nodes) and sample nodes (i.e., leaves). Based on the above classifications, we derive the following lemma.

Lemma 2. *Let $I_{P,i.chi}$, $I_{Q,j.chi}$ be the child entries of $I_{P,i}$, $I_{Q,j}$, and let $I_{P,i.chi.chi}$, $I_{Q,j.chi.chi}$ be the child entries of $I_{P,i.chi}$, $I_{Q,j.chi}$. If the type of $(I_{P,i}, I_{Q,j})$, $(I_{P,i.chi}, I_{Q,j.chi})$ and $(I_{P,i.chi.chi}, I_{Q,j.chi.chi})$ are $(I_P.obj, I_Q.obj)$, $(I_P.clu, I_Q.clu)$ and $(I_P.sam, I_Q.sam)$, respectively, the following inequality holds for any object node pair $(I_{P,i}, I_{Q,j})$:*

$$Min\_MinD(\mathrm{MBR}(I_{P,i.chi.chi}), \mathrm{MBR}(I_{Q,j.chi.chi})) \geq$$
$$Min\_MinD(\mathrm{MBR}(I_{P,i.chi}), \mathrm{MBR}(I_{Q,j.chi})) \geq \qquad (7)$$
$$Min\_MinD(\mathrm{MBR}(I_{P,i}), \mathrm{MBR}(I_{Q,j}))$$

Proof. Due to the fact that $\mathrm{MBR}(I_{P,i.chi})$ and $\mathrm{MBR}(I_{Q,j.chi})$ are respectively bounded by $\mathrm{MBR}(I_{P,i})$ and $\mathrm{MBR}(I_{Q,j})$, the minimum distance between the closest edges of $\mathrm{MBR}(I_{P,i.chi})$ and $\mathrm{MBR}(I_{Q,j.chi})$ is smaller or equal to that of $\mathrm{MBR}(I_{P,i})$ and $\mathrm{MBR}(I_{Q,j})$. So based on Equation 3, we have $Min\_MinD(\mathrm{MBR}(I_{P,i.chi}), \mathrm{MBR}(I_{Q,j.chi})) \geq Min\_MinD(\mathrm{MBR}(I_{P,i}), \mathrm{MBR}(I_{Q,j}))$. Moreover, we can prove $Min\_MinD(\mathrm{MBR}(I_{P,i.chi.chi}), \mathrm{MBR}(I_{Q,j.chi.chi})) \geq Min\_MinD(\mathrm{MBR}(I_{P,i.chi}), \mathrm{MBR}(I_{Q,j.chi}))$ in the similar way. Proof details are omitted due to space limitation.

4.2   Top$K$-PCP Query Algorithm

Algorithm 1 depicts our query algorithm. We assume R_P and R_Q have the same heights for simplicity. The algorithm can be easily modified to deal with the trees of different heights by the approaches in [1], which are omitted here due to space limitation. The algorithm maintains a priority queue H_queue which contains the entries of form $((I_{P,i}, I_{Q,j}), Min\_MinD(\mathrm{MBR}(I_{P,i}), \mathrm{MBR}(I_{Q,j})))$, The entry with the minimum $Min\_MinD(\mathrm{MBR}(I_{P,i}), \mathrm{MBR}(I_{Q,j}))$ has the highest priority. Firstly, H_queue is initialized (line 1). We use a set U_set with entries (P_g, Q_h) to store result pairs. The set is initialized as an empty set (line 2). Moreover, a list L_list is employed to record the information of the probabilistic closest pairs found up-to-now, which is in the form of $((P_g, Q_h), \Pr(P_g, Q_h), num_{gh})$, num_{gh} is the total number of the sample pairs of (P_g, Q_h) that we have already processed (line 3). Starting from the roots of R_P and R_Q, we insert $((R_P.root, R_Q.root),$ $Min\_MinD(\mathrm{MBR}(R_P.root), \mathrm{MBR}(R_P.root))$ into H_queue. Threshold Z is set to ∞, which denotes the minimum distance of the pairs found so far (lines 4 and 5). Each time we pop out the first entry Fir from H_queue (line 7), and process it according to the type of $Fir.(I_{P,i}, I_{Q,j})$ as follows.

If the type is ordinary node pair, we derive the minimum value of Min_MaxD for all possible MBR pairs of the child nodes of $I_{P,i}$ and $I_{Q,j}$. If this minimum Min_MaxD is smaller than Z, a smaller Z is obtained by being updated as $\min_{\substack{\forall I_{P,i.chi} \\ \forall I_{Q,j.chi}}} (Min\_MaxD(\mathrm{MBR}(I_{P,i.chi}), \mathrm{MBR}(I_{Q,j.chi})))$ (lines 8-10). For each child pair, insert it into H_queue if $Min\_MinD(\mathrm{MBR}(I_{P,i.chi}), \mathrm{MBR}(I_{Q,j.chi})) \leq Z$ (lines 11-13).

If the type is object node pair, we check whether the second entry in H_queue has $Min\_MinD(\mathrm{MBR}(I_{P,i'}), \mathrm{MBR}(I_{Q,j'}))$ smaller than $Min\_MinD(\mathrm{MBR}(I_{P,i}),$

Algorithm 1. Top$K$-PCP Query Processing

Input: two uncertain data sets P and Q
Output: all the object pairs in U_set
1: initialize H_queue accepting entries $((I_{P,i}, I_{Q,j}), Min\_MinD(\mathrm{MBR}(I_{P,i}), \mathrm{MBR}(I_{Q,j})))$

2: initialize U_set accepting entries in the form (P_g, Q_h)
3: initialize L_list accepting entries in the form $((P_g, Q_h), \Pr(P_g, Q_h), num_{gh})$
4: start from the roots of R_P and R_Q, set Z to ∞
5: insert $((R_P.root, R_Q.root), Min\_MinD(\mathrm{MBR}(R_P.root), \mathrm{MBR}(R_Q.root))$ into H_queue
6: **while** $|U\_set| \neq K$ **do**
7:    pop the first entry Fir of H_queue
8:    **if** the type of $Fir.(I_{P,i}, I_{Q,j})$ is $(I_P.ord, I_Q.ord)$ **then**
9:       **if** $\min_{\forall I_{P,i.chi}} (Min\_MaxD(\mathrm{MBR}(I_{P,i.chi}), \mathrm{MBR}(I_{Q,j.chi}))) < Z$ **then**
         $\quad\forall I_{Q,j.chi}$
10:         update Z
11:         **for** each child pair $(I_{P,i.chi}, I_{Q,j.chi})$ **do**
12:            **if** $Min\_MinD(\mathrm{MBR}(I_{P,i.chi}), \mathrm{MBR}(I_{Q,j.chi})) \leq Z$ **then**
13:               insert $((I_{P,i.chi}, I_{Q,j.chi}), Min\_MinD(I_{P,i.chi}, I_{Q,j.chi}))$ into H_queue
14:    **if** the type of $Fir.(I_{P,i}, I_{Q,j})$ is $(I_P.obj, I_Q.obj)$ **then**
15:       Pop the second entry $((I_{P,i'}, I_{Q,j'}), Min\_MinD(\mathrm{MBR}(I_{P,i'}), \mathrm{MBR}(I_{Q,j'})))$
16:       **if** $Min\_MinD(I_{P,i'}, I_{Q,j'}) \geq Max\_MaxD(I_{P,i}, I_{Q,j})$ **then**
17:          calculate $\Pr\_o(I_{P,i}, I_{Q,j})$
18:          update L_list
19:          add (P_g, Q_h) to U_set
20:       **else**
21:          repeat 9-13
22:    **if** the type of $Fir.(I_{P,i}, I_{Q,j})$ is $(I_P.clu, I_Q.clu)$ **then**
23:       Pop the second entry $((I_{P,i'}, I_{Q,j'}), Min\_MinD(\mathrm{MBR}(I_{P,i'}), \mathrm{MBR}(I_{Q,j'})))$
24:       **if** $Min\_MinD(I_{P,i'}, I_{Q,j'}) \geq Max\_MaxD(I_{P,i}, I_{Q,j})$ **then**
25:          calculate $\Pr\_c(I_{P,i}, I_{Q,j})$
26:          update L_list
27:          **if** $num_{gh} = sr^2$ **then**
28:             add (P_g, Q_h) to U_set
29:       **else**
30:          repeat 11-13
31:    **if** the type of $Fir.(I_{P,i}, I_{Q,j})$ is $(I_P.sam, I_Q.sam)$ **then**
32:       calculate $\Pr\_s(I_{P,i}, I_{Q,j})$
33:       update L_list
34:       **if** $num_{gh} = sr^2$ **then**
35:          add (P_g, Q_h) to U_set
36: return U_set

$\mathrm{MBR}(I_{Q,j}))$ (lines 14-16). If the answer is negative, each of the other pairs has no chance of being the closest pair. Thus, we calculate the likelihood that $Fir.(I_{P,i}, I_{Q,j})$ is the closest pair, which is denoted by $\Pr\_o(I_{P,i}, I_{Q,j})$. In particular,

$$\Pr{\_o}(I_{P,i}, I_{Q,j}) = \prod_{\forall L\_list.(P_x, Q_y)} (1 - \frac{num_{xy}}{sr^2}) \tag{8}$$

where (P_x, Q_y) denotes the pair already exists in L_list, $(I_{P,i}, I_{Q,j}) = (P_g, Q_h)$ (line 17). We update L_list by adding the entry $((P_g, Q_h), \Pr(P_g, Q_h), num_{gh})$, of which $\Pr(P_g, Q_h) = \Pr{\_o}(I_{P,g}, I_{Q,h})$ and $num_{gh} = sr^2$ (line 18). Note that $num_{gh} = sr^2$, we add (P_g, Q_h) to U_set (line 19). Otherwise, we refine the pair $Fir.(I_{P,i}, I_{Q,j})$ by repeating the steps in lines 9-13 (line 21).

If the type is cluster node pair, we check whether the pair $Fir.(I_{P,i}, I_{Q,j})$ needs to be refined in a similar way as above (lines 22-24). The likelihood indicates that $Fir.(I_{P,i}, I_{Q,j})$ being the closest pair is computed by

$$\Pr{\_c}(I_{P,i}, I_{Q,j}) = \frac{Num(I_{P,i}) \cdot Num(I_{Q,j})}{sr^2} \cdot \prod_{\forall L\_list.(P_x, Q_y) \wedge (P_x, Q_y) \neq (P_g, Q_h)} (1 - \frac{num_{xy}}{sr^2}) \tag{9}$$

where $Num(I_{P,i})$ and $Num(I_{Q,j})$ denote the number of the samples in cluster $I_{P,i}$ and $I_{Q,j}$, respectively, and $I_{P,i} \in P_g, I_{Q,j} \in Q_h$ (line 25). Then we update L_list according to $\Pr(P_g, Q_h) = \Pr{\_o}(I_{P,i}, I_{Q,j}) + \Pr(P_g, Q_h)$ and $num_{gh} = num_{gh} + Num(I_{P,i}) \cdot Num(I_{Q,j})$ (line 26). The default value of $\Pr(P_g, Q_h)$ is 0. If num_{gh} is equal to sr^2, (P_g, Q_h) is added to U_set (lines 27-28). Otherwise, the repeating procedure is implemented (line 30).

If the type is sample pair, $\Pr{\_s}(I_{P,i}, I_{Q,j})$ is calculated by

$$\Pr{\_s}(I_{P,i}, I_{Q,j}) = \frac{1}{sr^2} \cdot \prod_{\forall L\_list.(P_x, Q_y) \wedge (P_x, Q_y) \neq (P_g, Q_h)} (1 - \frac{num_{xy}}{sr^2}) \tag{10}$$

which is the likelihood that $Fir.(I_{P,i}, I_{Q,j})$ is the closest pair (lines 31 and 32). Here, we have $I_{P,i} \in P_g, I_{Q,j} \in Q_h$. The updating is done by setting $\Pr(P_g, Q_h) = \Pr{\_o}(I_{P,i}, I_{Q,j}) + \Pr(P_g, Q_h)$ and $num_{gh} = num_{gh} + 1$ (line 33). After that, we need to make a decision on adding (P_g, Q_h) to U_set or not according to the value of num_{gh} (lines 34 and 35).

The algorithm terminates if the cardinality of U_set is K. The pairs in U_set, which are returned finally (line 36), are the K pairs with the highest probabilities of being the closest pair.

5   Experimental Evaluation

All our experiments are conducted on a 1.86 GHz Intel Core 2 6300 CPU and 2 GB RAM. An open source R*-tree implementation[1] is used for indexing. Two real spatial datasets of geographical objects in Germany[2], namely RU (utility),

---

[1] http://www.research.att.com/~marioh/spatialindex/
[2] http://www.rtreeportal.org/

RR (roads), with respective sizes 17K, 30K are used in the experiments. The MBRs in the datasets are employed to represent rectangular uncertain regions. For each uncertain region, sr points are sampled and the spatial distribution follows *uniform* distribution. A group of synthetic datasets of cardinality 100K, 200K and 400K points (respectively denoted as SD1, SD2 and SD3) are generated, in which the points are considered as the centers of uncertain regions. The data in these uncertain regions have the same uncertainty *pdf* information as that of the real data. Both of the two dimensions in these datasets are normalized to domain [0, 1]. We split each of the above datasets into two sets which respectively contain 50% of the objects. The splitting processing is implemented randomly for each time, thus each point in the following graph is an average of the results for 100 Top$K$-PCP queries.

Effect of sample rate. In the first group of experiments, we examine the performance of our method by varying the number of the samples per object, which is considered as the sample rate sr. We proceed with our algorithm as a function of sample rate varies from 5 to 20. Fig. 1a illustrates the number of disk accesses for Top$K$-PCP query over two real datasets. Fig. 1c shows the same metric for the synthetic datasets. In particular, when the sample rate becomes larger, the disk accesses increases. The reason for the above observation is that a larger sample rate leads to more computations and a larger search space, which requires more I/O activities. Furthermore, comparing the datasets with larger size to those with smaller size, we can observe that a larger dataset leads to a worse I/O performance.

Fig. 1b and 1d show the accuracy of the algorithm for a varying sample rate over real data and synthetic data, respectively. Note that the results do not change considerably when $sr \geq 80$, we denote with R_{exa} the exact result based

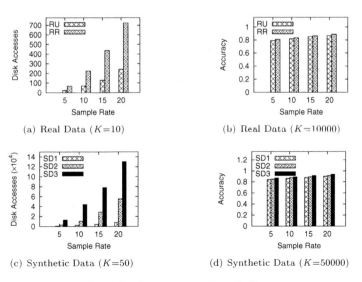

(a) Real Data (K=10)   (b) Real Data (K=10000)

(c) Synthetic Data (K=50)   (d) Synthetic Data (K=50000)

Fig. 1. Performance vs. Sample Rate

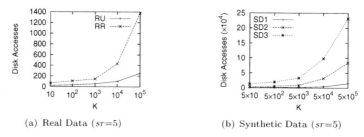

(a) Real Data (sr=5)   (b) Synthetic Data (sr=5)

Fig. 2. Performance vs. K

on 80 samples. R_{exa} is used as the reference result for measuring the accuracy of the query result derived in the case of $sr < 80$, which is denoted by R_{app}. Therefore, Accuracy is defined by the formula: $\frac{|R_{exa} \cap R_{app}|}{|R_{exa}|}$. As expected, the experiment results shows that accuracy increases as the sample size gets larger. Thus, a trade-off between the cost and accuracy can be achieved when the query is processing.

Effect of K. The scalability of our algorithm is measured by taking into account the behavior with increasing K values. Fig. 2a (real data) and Fig. 2b (synthetic data) show the number of disk accesses with different K values. Fig. 2a shows that the number of disk accesses increases rapidly with the increase of K for real datasets, e.g., from 431 for K=10000 to 1379 for K=100000 (RR). The trend for synthetic datasets is very similar. Obviously, the rapid increase of disk accesses is due to the fact that a larger K needs more computations of probabilistic closest pairs.

Effect of buffer size. In Fig. 3, we present how the LRU-buffer size affects the performance of our algorithm. We use the buffer sizes of B=0, 16, 64, 256, 1024 pages (dedicated to each R*-tree as two equal portions of B/2 pages) in this set of experiments. It is observed that the performance is getting improved as long as the buffer size grows. However, we can see that the improvement is not very drastic. For instance, as the buffer size increases from 64 to 256, only a 13.5% improvement is obtained (RR). As a conclusion, our algorithm is not so sensitive to buffer size. This is because our algorithm processes the query by following a best-first traversal pattern using a priority queue.

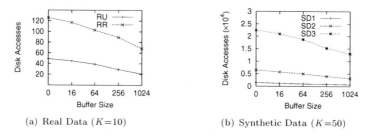

(a) Real Data (K=10)   (b) Synthetic Data (K=50)

Fig. 3. Performance vs. Buffer Size

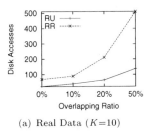
(a) Real Data (K=10)

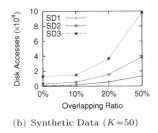
(b) Synthetic Data (K=50)

Fig. 4. Performance vs. Overlapping Ratio

Effect of overlapping ratio. In the following, we investigate I/O activities of the algorithm with the portion of overlapping ranging from 0% to 50%. From Fig. 4, we can see that for small overlap (e.g., 10% in SD3), a significant improvement is about 6 times better than the case of 50% overlapping. The big gap is due to the inability of nodes pruning, and both of the two R*-trees are likely to be completely traversed more times. This shows that the overlapping between the data sets is crucial for Top$K$-PCP query. Thus, the parameter of overlapping must be taken into account very seriously when performing the algorithm.

6   Conclusion

Query processing on uncertain data has recently attracted a lot of research interest. In this paper, we studied the evaluation of a Top$K$-PCP query, which, to the best of our knowledge, no other work has studied before. Specifically, each uncertain object is modeled by a set of sample points, based on which the query is effectively evaluated. We conducted extensive experiments to confirm the efficiency and effectiveness of our proposed approaches. In the future, we will study how the probabilistic algorithms can be extended to support other types of closest pairs queries, e.g., exclusive closest pairs query.

Acknowledgment. This research was supported by NSFC No.60933001 and No.61003058, National High-tech R&D Program of China No.2009AA01Z131 (863 program), Research and Innovation Fund for Young Teachers N090304003.

References

1. Corral, A., Manolopoulos, Y., Theodoridis, Y., Vassilakopoulos, M.: Closest pair queries in spatial databases. In: SIGMOD, pp. 189–200 (2000)
2. Zhu, M., Lee, D.L., Zhang, J.: k-closest pair query monitoring over moving objects. In: MDM, vol. 14 (2006)
3. Hjaltason, G.R., Samet, H.: Incremental distance join algorithms for spatial databases. In: SIGMOD, pp. 237–248 (1998)
4. Corral, A., Manolopoulos, Y., Theodoridis, Y., Vassilakopoulos, M.: Algorithms for processing k-closest-pair queries in spatial databases. Data and Knowledge Engineering 49, 67–104 (2004)

5. Yang, C., Lin, K.I.: An index structure for improving nearest closest pairs and related join queries in spatial databases. In: IDEAS, pp. 140–149 (2002)
6. Leong Hou, U., Mamoulis, N., Yiu, M.L.: Computation and monitoring of exclusive closest pairs. IEEE Trans. Knowl. Data Eng (TKDE) 20(12), 1641–1654 (2008)
7. Aggarwal, C.C., Yu, P.S.: A survey of uncertain data algorithms and applications. IEEE Trans. Knowl. Data Eng (TKDE) 21(5), 609–623 (2009)
8. Singh, S., Mayfield, C., Shah, R., Prabhakar, S., Hambrusch, S.E., Neville, J., Cheng, R.: Database support for probabilistic attributes and tuples. In: ICDE, pp. 1053–1061 (2008)
9. Bosc, P., Pivert, O.: Modeling and querying uncertain relational databases: a survey of approaches based on the possible worlds semantics. International Journal of Uncertainty, Fuzziness and Knowledge-Based Systems (IJUFKS) 18(5), 565–603 (2010)
10. Dalvi, N.N., Suciu, D.: Efficient query evaluation on probabilistic databases. VLDB J. 16(4), 523–544 (2007)
11. Chang, L., Yu, J.X., Qin, L.: Query ranking in probabilistic xml data. In: EDBT, pp. 156–167 (2009)
12. Xu, J., Zhang, Z., Tung, A.K.H., Yu, G.: Efficient and effective similarity search over probabilistic data based on earth mover's distance. PVLDB 3(1), 758–769 (2010)
13. Cheng, R., Kalashnikov, D.V., Prabhakar, S.: Evaluating probabilistic queries over imprecise data. In: SIGMOD, pp. 551–562 (2003)
14. Xu, C., Wang, Y., Lin, S., Gu, Y., Qiao, J.: Efficient fuzzy top-k query processing over uncertain objects. In: Bringas, P.G., Hameurlain, A., Quirchmayr, G. (eds.) DEXA 2010. LNCS, vol. 6261, pp. 167–182. Springer, Heidelberg (2010)
15. Pei, J., Jiang, B., Lin, X., Yuan, Y.: Probabilistic skylines on uncertain data. In: VLDB, pp. 15–26 (2007)
16. Kriegel, H.-P., Kunath, P., Renz, M.: Probabilistic nearest-neighbor query on uncertain objects. In: Kotagiri, R., Radha Krishna, P., Mohania, M., Nantajeewarawat, E. (eds.) DASFAA 2007. LNCS, vol. 4443, pp. 337–348. Springer, Heidelberg (2007)
17. Yang, S., Zhang, W., Zhang, Y., Lin, X.: Probabilistic threshold range aggregate query processing over uncertain data. In: APWeb/WAIM, pp. 51–62 (2009)
18. Beskales, G., Soliman, M.A., Ilyas, I.F.: Efficient search for the top-k probable nearest neighbors in uncertain databases. PVLDB 1(1), 326–339 (2008)
19. Lian, X., Chen, L.: Efficient processing of probabilistic reverse nearest neighbor queries over uncertain data. VLDB J. 18(3), 787–808 (2009)
20. Soliman, M.A., Ilyas, I.F., Chang, K.C.-C.: Top-k query processing in uncertain databases. In: ICDE, pp. 896–905 (2007)
21. Jin, C., Yi, K., Chen, L., Yu, J.X., Lin, X.: Sliding-window top-k queries on uncertain streams. PVLDB 1(1), 301–312 (2008)
22. Lian, X., Chen, L.: Probabilistic group nearest neighbor queries in uncertain databases. IEEE Trans. Knowl. Data Eng (TKDE) 20(6), 809–824 (2008)
23. Kriegel, H.-P., Kunath, P., Pfeifle, M., Renz, M.: Probabilistic similarity join on uncertain data. In: Li Lee, M., Tan, K.-L., Wuwongse, V. (eds.) DASFAA 2006. LNCS, vol. 3882, pp. 295–309. Springer, Heidelberg (2006)
24. Beckmann, N., Kriegel, H.-P., Schneider, R., Seeger, B.: The r*-tree: An efficient and robust access method for points and rectangles. In: SIGMOD, pp. 322–331 (1990)

CkNN Query Processing over Moving Objects with Uncertain Speeds in Road Networks

Guohui Li[1], Yanhong Li[1], LihChyun Shu[2], and Ping Fan[1]

[1] School of Computer Science and Technology,
Huazhong University of Science and Technology, Wuhan, P.R. China
[2] College of Management, National Cheng Kung University
Tainan City, Taiwan 701, R.O.C.
anddylee@163.com

Abstract. This paper focuses on processing continuous k nearest neighbor queries over objects moving at uncertain speeds ($CUkNN$) in road networks. We present a novel model to estimate the distances between objects and a query, both of which move at variable speeds in the road network. Based on the proposed distance model, we present a $CUkNN$ query monitoring method to continuously find the objects that could potentially be the k-nearest neighbors (kNN) of the query. We propose an efficient method to calculate the probability of each object being a kNN of a query. The key thing about the method is that the probability of an object being a kNN of query q is shown to be equivalent to the probability of a special line segment being one of the k-nearest lines from q, which greatly simplifies the probability calculation.

Keywords: $CkNN$ queries, uncertain speed, $CUkNN$, road network.

1 Introduction

Continuous k-nearest neighbor ($CkNN$) query processing aims to efficiently retrieve k-nearest neighbors (kNN) of a moving query from among a number of moving objects over continuous time intervals. Recently, several methods have been proposed to process $CkNN$ queries [1-8]. To date, most of the existing $CkNN$ query processing methods suffer from at least one of the following limitations: (1) they are limited to Euclidean space [8]; (2) objects are assumed to move at constant speed [3] [4]; (3) they assume that object location updates are discrete and thus they would return invalid results between two successive update time instants [2]. These limitations imply that most of the existing solutions have made simplified but less realistic assumptions in one way or another.

In this paper, our aim is to overcome the limitations of the previous studies mentioned above and focus on processing $CkNN$ queries over objects that move at uncertain speeds in road networks, which are called $CUkNN$ queries. Specifically, we address the $CUkNN$ problem with the following three requirements: (1) all objects and queries move continuously in the road network, and the distance between an object and a query is network distance; (2) we return valid

X. Du et al. (Eds.): APWeb 2011, LNCS 6612, pp. 65–76, 2011.
© Springer-Verlag Berlin Heidelberg 2011

result at any monitored time instant; (3) the speeds of both the moving objects and queries are allowed to vary within pre-determined ranges. Hereinafter, we call kNN queries over objects with uncertain speeds $UkNN$ queries. Similarly, we call the objects which belong to the result of a $UkNN$ query $UkNN$ objects.

To achieve our goals, the first problem we address is how to efficiently calculate the distances between moving objects and queries. In a road network, moving objects and queries are restricted by the connectivity of the road network, and it is difficult to estimate the distances between objects and queries when they can continuously move on the roads with uncertain speeds. We propose the maximal and minimal distance functions, called $MaxD_{q,o}(t)$ and $MinD_{q,o}(t)$, to bound the distance between a query q and an object o at time t. By categorizing different location relationships between q and o and distinguishing the relationship of the two roads where q and o roam into two different states, we obtain the $MaxD_{q,o}(t)$ and $MinD_{q,o}(t)$ functions in an efficient manner. Both $MaxD_{q,o}(t)$ and $MinD_{q,o}(t)$ are polylines consisting of one to three line segments, which will greatly facilitate subsequent calculation of probabilities of candidate objects.

Based on these two distance functions, we present a $CUkNN$ query method to continuously obtain the $UkNN$ result of query q. In particular, the $UkNN$ result consists of one or more $UkNN$ objects, each with its associated probability being one of the kNNs of query q. Thus, we are able to determine which objects are more likely to be the kNNs of query q in order to provide useful kNNs information to users. Our $CUkNN$ method consists of three stages, pruning phase, refining phase, and probability evaluation phase. In the pruning phase, by calculating the pruning distance we efficiently prune the objects that are impossible to be the kNN result of query q and form the candidate set. Then in the refining phase, by constructing the kNN-MaxD-PL of query q, we decide the possible kNN objects of query q at time period $[t_s, t_e]$, depending on the candidate set determined at the pruning phase. Then in the probability evaluation phase, by introducing the possible-line-segment(PLS) to represent the possible distance area of each $UkNN$ object, the probability calculation is greatly simplified.

2   Data Structures and Network Distance Calculation

Data structures used. We use an undirected weighted graph to represent a road network which includes an edge set and a node set. And four in-memory tables are constructed to store information about edges, nodes of a network, moving objects and queries. We propose the following definition to simplify network distance calculation.

Definition 1. *Given two different edges e_i and e_j, if the shortest distance between every pair of two points, one from e_i and another from e_j, is always determined by the same path, then we say e_i and e_j are distance determinate; otherwise, they are distance indeterminate.*

By categorizing edge pairs into these two states, we can construct a matrix DD in which DD_{ij} represents that e_i and e_j are *distance determinate* or not.

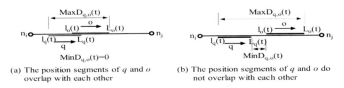

(a) The position segments of q and o overlap with each other

(b) The position segments of q and o do not overlap with each other

Fig. 1. q and o move on the same edge

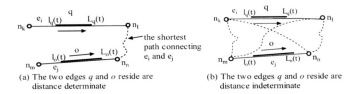

(a) The two edges q and o reside are distance determinate

(b) The two edges q and o reside are distance indeterminate

Fig. 2. q and o move no two different edges

Besides, another matrix, DN, is adopted to store the shortest distance between any two nodes in which DN_{ij} is the shortest distance between nodes n_i and n_j.

Network distance calculation. We assume an object can not change its moving direction arbitrarily except at the road junctions. Thus an object o moves toward a fixed direction before it reaches an endpoint of the road on which it travels and its speed lies between $o.v_{min}$ (the minimal speed of o) and $o.v_{max}$ (the maximal speed of o). Thus we obtain two locations $l_o(t)$ (the nearest possible location) and $L_o(t)$ (the farthest possible location) of o at time t, respectively. Similarly, we can obtain two locations $l_q(t)$ and $L_q(t)$ of query q at time t. At any time t, object o should be on the line segment $l_o(t)L_o(t)$. Henceforth, we call this line segment $l_o(t)L_o(t)$ the *position segment* of o.

Then, we use the minimal and the maximal distance to bound the distance between o and q. In particular, we use two functions, $MinD_{q,o}(t)$ and $MaxD_{q,o}(t)$, to represent the minimal and the maximal distance between o and q at time t, respectively.

Case 1. q and o move on the same edge e.

Assume that both q and o are moving on edge e which starts from n_i and ends at n_j. As shown in Figure 1, there are two possible relations between the locations of the position segments of q and o: *overlap* and *not overlap*. When these two segments overlap with each other, $MinD_{q,o}(t)$ equals zero and $MaxD_{q,o}(t)$ is a linear function of time t; otherwise, both $MinD_{q,o}(t)$ and $MaxD_{q,o}(t)$ are polylines consisting of three line segments.

Case 2: q and o move on two different edges.

Assume that q is moving on edge e_i which starts from n_k and ends at n_l, and o is on edge e_j which starts from n_m and ends at n_n. There are two subcases:

(1) e_i and e_j are distance determinate. There is a shortest path connecting e_i and e_j. Without loss of generality, assume that the two nodes on this shortest path are n_l and n_n respectively. As shown in Figure 2(a), $MinD_{q,o}(t)$ is

equal to the sum of $min(d(l_q(t), n_l), d(L_q(t), n_l))$, $DN_{l,n}$, and $min(d(l_o(t), n_n)$, $d(L_o(t), n_n))$, where $d(l_q(t), n_l)$ is the distance from $l_q(t)$ to n_l, $DN_{l,n}$ is the distance between n_l and n_n. Moreover, $MaxD_{q,o}(t)$ is equal to the sum of $max(d(l_q(t), n_l), d(L_q(t), n_l))$, $DN_{l,n}$, $max(d(l_o(t), n_n), d(L_o(t), n_n))$. In this case, both $MinD_{q,o}(t)$ and $MaxD_{q,o}(t)$ are liner functions of time t.

(2) e_i and e_j are *distance indeterminate*. Note that the possible locations of q and o are within two segments, $l_q(t)L_q(t)$ and $l_o(t)L_o(t)$, respectively. Thus as shown in Figure 2(b), the shortest path from any point in these two segments to the outside network should pass through their terminals. Therefore, $MinD_{q,o}(t)$ must be the distance between one terminal of $l_q(t)L_q(t)$ and one terminal of $l_o(t)L_o(t)$. As both $l_q(t)L_q(t)$ and $l_o(t)L_o(t)$ have two terminals, we need to take into account four network distances, i.e., $d(l_q(t), l_o(t))$, $d(l_q(t), L_o(t))$, $d(L_q(t), l_o(t))$, and $d(L_q(t), L_o(t))$. $MinD_{q,o}(t)$ is simply the minimum of these four distances. Since it is difficult to calculate the exact value of $MaxD_{q,o}(t)$, we use an approximate value, $app\_MaxD_{q,o}(t)$, to serve as $MaxD_{q,o}(t)$. First, $app\_MaxD_{q,o}(t)$ must be no smaller than the exact value of $MaxD_{q,o}(t)$ at any time instant. Second, $app\_MaxD_{q,o}(t)$ should be as close as possible to the exact value of $MaxD_{q,o}(t)$. Hence we set $app\_MaxD_{q,o}(t)$ to be the sum of $MinD_{q,o}(t)$, the length of $l_q(t)L_q(t)$, and the length of $l_o(t)L_o(t)$. In this case, $MinD_{q,o}(t)$ and $MaxD_{q,o}(t)$ are polylines consisting of two line segments.

3 CUkNN Algorithm

In this section, we discuss how to process $CUkNN$ queries in road networks. To illustrate this problem clearly, we consider an example in Figure 3, where a set of objects o_1 to o_6 and a query q move with uncertain speeds in a road network. Assume that $2NNs$ of q are required. Moving objects are denoted by black spots and a query is denoted by a black triangle, and arrows indicate their moving directions. Here an object or a query is bounded by a minimal and a maximal speed enclosed in parentheses and the speed follows a uniform distribution. Assume the positions of objects and the query at time $t=0$ are shown in Figure 3(a). Then the possible location of an object or query at time $t=0.5$ is within a line segment which is denoted by a dark line in Figure 3(b).

Firstly, we use an efficient static kNN method [9] [10] [11] [12] to calculate the kNN result at the start time t_s. The main issue is how to continuously monitor the kNN result after t_s. As the objects and the query may choose different edges in the road network to travel or change their directions at road intersections, we only consider the time period from t_s to the earliest time instant t_e, when one of the monitored objects or the query reaches an intersection. As for the time after t_e we consider it a new time period. Our $CUkNN$ method consists of the pruning phase, the refining phase and the probability evaluation phase.

3.1 Pruning Phase

The main goal of the pruning phase is to efficiently prune the objects that are impossible to be the $kNNs$ of query q. The pruning phase of our $CUkNN$

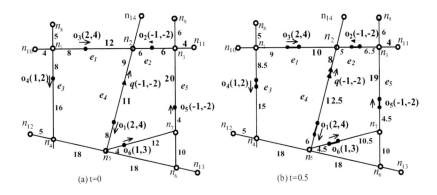

Fig. 3. An example of CUkNN query in road network

Algorithm 1. Pruning Phase

---

   input : the $kNNs$ at t_s, a number k, and a time peroid $[t_s, t_e]$
   output: the $Cand\_Set$
1 **begin**
2 | **for** *each object o being one of the kNNs at t_s* **do**
3 | | compute $MaxD_{q,o}([t_s, t_e])$ using $MaxD_{q,o}(t)$;
4 | let $D_{pruning}$ to be the Kth smallest of $MaxD_{q,o}(t_s, t_e)$;
5 | $D_{pruning} = D_{pruning} + max.vlimit \times (t_e - t_s)$ // $max.vlimit$ is the largest
   | $vlimit$ of the edges within $D_{pruning}$ of q ;
6 | any objects which are within $D_{pruning}$ of q at time instant t_s are regarded
   | as candidate object and put into $Cand\_Set$;
7 | return $Cand\_Set$;
8 **end**

---

algorithm is shown in Algorithm 1. We first calculate the maximal distance $MaxD_{q,o}$ of each object o within the time period $[t_s, t_e]$, where o is one of the $kNNs$ of query q at time t_s (line 4 of Algorithm 1). Then we sort these objects in an ascending order of their maximal distances $MaxD_{q,o}$, and select the kth object's $MaxD_{q,o}$, denoted MD, as the pruning distance (line 5). Note that objects outside the pruning distance of q may move into the distance range MD within the entire period $[t_s, t_e]$, thus we enlarge the pruning distance by $max.vlimit \times (t_e - t_s)$. Here, $max.vlimit$ is the largest velocity limit of the edges where the previous kNN objects reside at time t_s. Now, the pruning distance, $D_{pruning}$, is set to be $MD + max.vlimit \times (t_e - t_s)$ (line 5). Thus all the objects within $D_{pruning}$ of query q at time t_s are regarded as the candidate objects and put into $Cand\_Set$, and all other objects can be pruned safely (line 6).

   We use query q in Figure 3 as an example and calculate the value of $D_{pruning}$ which equals 31. Thus, there are only four objects o_1, o_2, o_3 and o_6 within the pruning range and the $Cand\_Set$ is $\{o_1, o_2, o_3, o_6\}$.

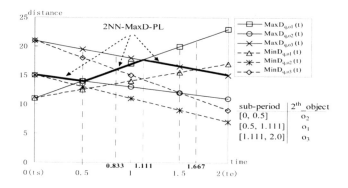

Fig. 4. 2NN-MaxD-PL of query q in Figure 3

3.2   Refining Phase

The refining phase is to decide possible kNN objects of $UkNN$ query q within the time period $[t_s, t_e]$. As we know, for query q and time instant t_s, there exists an object whose maximal distance to query q ranks the kth smallest among all the objects' maximal distances to q; we call this object k^{th}_obj. The k^{th}_obj at t_s may be replaced by another object o' at some time instant t'. Similarly, o' may be replaced by another object o'' at yet another time instant t''. All these time instants, including t', t'', divide the time period $[t_s, t_e]$ into several sub-periods, each with its own k^{th}_obj. For each sub-period, we draw the maximal distance line of its k^{th}_obj. By combining all these maximal distance lines together in an ascending order of time, we can get a polyline called the kNN-MaxD-PL of q.

The refining phase of our algorithm is shown in Algorithm 2. It includes two main steps: (1) constructing the kNN-MaxD-PL of query q (lines 2-8); (2) finding $UkNN\_Set$ for each sub-period $[t_i, t_j]$ within $[t_s, t_e]$ (lines 9-20).

We first continue our example to illustrate how to construct kNN-MaxD-PL. As shown in Figure 4, the k^{th}_obj at time t_s is o_2. Then we should make sure that at which time instant o_2 will not be the k^{th}_obj any more. We consider each other object o in $Cand\_Set$ by solving the equation $MaxD_{q,o_2}(t) = MaxD_{q,o}(t)$ (lines 3-6). In this way we determine that o_1 replaces o_2 as the k^{th}_obj at time $t=0.5$, and we obtain the first tuple of kNN-MaxD-PL, i.e., $([0, 0.5], MaxD_{q,o_2}(t))$ (line 7). Then we use o_1 as the k^{th}_obj at time $t=0.5$ and continue the construction of kNN-MaxD-PL of q until time period $[t_s, t_e]$ is all processed. Finally, kNN-MaxD-PL consists of three tuples: $< [0, 0.5], MaxD_{q,o_2}(t) >$, $< [0.5, 1.111], MaxD_{q,o_1}(t) >$ and $< [1.111, 2.0], MaxD_{q,o_3}(t) >$.

Note for an object o, if $MinD_{q,o}$ is larger than $MaxD_{q,k^{th}\_obj}$ at a time instant t, o can not be a kNN of q at time t; otherwise, o can possibly be a kNN of q at time t. Moreover, kNN-MaxD-PL characterizes $MaxD_{q,k^{th}\_obj}$ within the time period $[t_s, t_e]$. Hence, for an object o whose minimal distance line intersects kNN-MaxD-PL at time t, if o is not in $UkNN\_Set$, o will be included into $UkNN\_Set$; otherwise, o will be deleted from $UkNN\_Set$. That is, whenever the minimal distance line of an object intersects kNN-MaxD-PL, $UkNN\_Set$ will change. We call

---

Algorithm 2. Refining Phase

---

    input   : the kNNs at t_s, the $Cand\_Set$, the time period $[t_s, t_e]$
    output: the $UkNN\_interim\_Set$

1 **begin**
2   let o_k to be the kth-NN at t_s; kNN-MaxD-PL=$\varnothing$; $t_a = t_s$; $t_b = t_s$; $t_{temp} = t_e$
      while $t_b < t_e$ **do**
3   **for** *each object* $o \in Cand\_Set$ *and* $o <> o_k$ **do**
4   **if** $MaxD_{q,o}(t)$ *equals* $MaxD_{q,o_k}(t)$ *at* t & $t \in [t_b, t_{temp}]$ **then**
5             $t_{temp} = t$;
6          $t_a = t_b$; $t_b = t_{temp}$; $t_{temp} = t_e$;
7   insert $< [t_a, t_b], MaxD_{q,o_k}(t) >$ into kNN-MaxD-PL; $o_k = o$;
8   insert $< [t_a, t_e], MaxD_{q,o_k}(t) >$ into kNN-MaxD-PL ;
9   let $UkNN\_Set$ to be objects whose $MinD_{q,o}(t)$ are below kNN-MaxD-PL at t_s; $t_a = t_s$; $t_b = t_s$; $Q = \varnothing$; $UkNN\_Result\_Set = \varnothing$;
10   **for** *each object* $o \in Cand\_Set$ **do**
11   **if** $MinD_{q,o}(t)$ *intersect* kNN-MaxD-PL *at* t & $t \in [t_a, t_b]$ **then**
12   insert the tuple $< t, o >$ into queue Q in an ascending order of time;

13   **while** Q *is not empty* **do**
14   de-queue$< t, o >$; $t_a = t_b$; $t_b = t$;
15   insert the tuple $< [t_a, t_b], \{UkNN\_Set\} >$ into $UkNN\_interim\_Set$;
16   **if** o *is not in* $UkNN\_Set$ **then**
17   insert o into $UkNN\_Set$;

18   **else**
19   delete o from $UkNN\_Set$;

20   insert the tuple $< [t_b, t_e], \{UkNN\_Set\} >$ into $UkNN\_interim\_Set$;
21   return $UkNN\_interim\_Set$;
22 **end**

---

this time instant Result Change Time (RCT) and all RCTs within $[t_s, t_e]$ should be found and inserted into a queue(lines 10-12). As shown in Figure 4, $MinD_{q,o_3}$ intersects kNN-MaxD-PL at time $t=0.833$, and o_3 is not in $UkNN\_Set$, thus o_3 is inserted into $UkNN\_Set$(line 17). However, $MinD_{q,o_1}$ intersects kNN-MaxD-PL at time $t=1.667$, and o_1 is in $UkNN\_Set$, thus o_1 is deleted from $UkNN\_Set$ (line 19). Here time instants $t=0.833$ and 1.667 are RCTs and all these RCTs within time period $[0,2]$ divide this time period into several sub-periods. All sub-periods together with their $UkNN\_Sets$ form the $UkNN\_interim\_Set$ of q (line 15), which includes three tuples: $< [0, 0.833], \{o_1, o_2\} >$, $< [0.833, 1.667], \{o_1, o_2, o_3\} >$, and $< [1.667, 2.0], \{o_2, o_3\} >$.

3.3 Probability Evaluation Phase

We denote the probability of o being one of the kNNs of q within $[t_i, t_j]$ as $P_o([t_i, t_j])$. Algorithm 3 gives the pseudo code of this phase. Continuing our example, there are two objects, o_1 and o_2, in $UkNN\_Set$ within the time period $[0, 0.833]$, and thus $P_{o_1}([0, 0.833])$ and $P_{o_2}([0, 0.833])$ are certainly 1 (line 6).

There are, however, three objects, o_1, o_2 and o_3, in $UkNN\_Set$ within [0.833,1.667], and $P_{o_i}([0.833, 1.667])$, $1 \leq i \leq 3$, must be determined properly.

For any time period $[t_i, t_j]$ whose $UkNN\_Set$ includes more than k objects, if one of the following two conditions holds, then $P_o([t_i, t_j])$ may change for some object o in $UkNN\_Set$, and in this case we divide $[t_i, t_j]$ into two sub-intervals $[t_i, t_u]$ and $[t_u, t_j]$ (line 3 of Algorithm 3): (1) The $MaxD_{q,o}(t)$ or $MinD_{q,o}(t)$ function of some object o in $UkNN\_Set$ has a turning point t_u in $[t_i, t_j]$; (2) $MaxD_{q,o_i}(t)$ (or $MinD_{q,o_i}(t)$) intersects with $MaxD_{q,o_j}(t)$ (or $MinD_{q,o_j}(t)$) at time t_u in $[t_i, t_j]$ for two different objects o_i and o_j. Consider Figure 4 again, $MinD_{q,o_1}(t)$ intersects $MinD_{q,o_3}(t)$ at time t=1.111, and $MaxD_{q,o_2}(t)$ intersects $MinD_{q,o_3}(t)$ at time t=1.5, thus the time period [0.833, 1.667] is divided into 3 sub-periods: [0.833, 1.111], [1.111, 1.5], and [1.5, 1.667].

For an object o and a divided time period $[t_i, t_j]$, the area bounded by $MaxD_{q,o}(t)$, $MinD_{q,o}(t)$, and the two vertical lines $t = t_i$ and $t = t_j$ gives the *possible distance area* in which each point represents a possible distance between o and q. Note that $MaxD_{q,o}(t)$ does not intersect with $MinD_{q,o}(t)$, and each only includes one line segment within $[t_i, t_j]$, thus the possible distance area of o forms a trapezoid. Figure 5(a) shows the possible distance area of o_2 in Figure 3, which is bounded by the following four lines: $MaxD_{q,o_2}(t)$, $MinD_{q,o_2}(t)$, and the two vertical lines $t = 1.111$ and $t = 1.5$. Following we give a definition and a lemma for the purpose of simplifying the probability calculation.

Definition 2. *Given the possible distance area of object o in $UkNN\_Set$ within $[t_i, t_j]$, we call the middle line segment $[MinD_{q,o}((t_i + t_j)/2), MaxD_{q,o}((t_i + t_j)/2)]$, which traverses the possible distance area of o at its half-height (i.e., at time $t=(t_i + t_j)/2$), the possible-line-segment (PLS for short) of o.*

Lemma 1. *Given the possible distance area of object o in $UkNN\_Set$ within a divided time interval $[t_i, t_j]$, $P_o([t_i, t_j])$ equals the probability of the PLS of o being one of the k-nearest lines from q.*

The proof of Lemma 1 is omitted during to space constraints.

The calculation of the PLS of o_2 is shown in Figure 5(b). Similarly, we can get the PLSs of o_1 and o_3. As shown in Figure 5(c), these three PLSs span the distance interval between 9.706 and 18.836.

Now, we divide the distance interval covered by the PLSs of all the objects in $UkNN\_Set$ into one or more sub-intervals, e.g., $D_1, D_2, ..., D_n$, by the end points of each PLS (line 10). Hence, the distance interval [9.706,18.836] in Figure 5(c) is divided into five sub-intervals: [9.706,12.388], [12.388,13.164], [13.164,14.918], [14.918,17.082], and [17.082,18.836]. To calculate $P_o[t_i, t_j]$, we first consider those distance sub-intervals that overlap with the PLS of o, say $D_u, ..., D_v$. Then for each distance sub-interval D_i, $u \leq i \leq v$, we calculate the probability that the distance between o and q falls in D_i, which is denoted as $P_I(o, D_i)$, and the probability that o is a kNN if its distance to q falls in D_i, which is denoted as $P_{knn}(o, D_i)$. With these two probabilities available, we could calculate the probability of o being a kNN of q within $[t_i, t_j]$ as follows:

$$P_o([t_i, t_j]) = \sum_{i=u}^{v} P_I(o, D_i) \times P_{knn}(o, D_i) \quad (line\ 20).$$

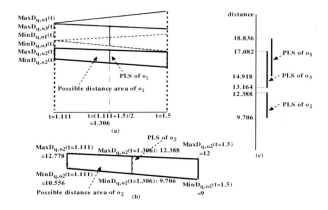

Fig. 5. An illustration of the probability evaluation phase

For ease of computing $P_o([t_i, t_j])$, we consider each distance interval in an ascending order of distance. First, for each sub-interval, we use two variables cnt_b and cnt_c, to represent the number of objects whose PLSs are completely below this sub-interval and the number of objects whose PLSs overlap with this sub-interval, respectively (lines 12-13). Then, as the probability of the distance between o and q falling in any point of the PLS of o is equal, we thus have

$P_I(o, D_i) = \frac{\text{the length of } D_i}{\text{the length of PLS(o)}}$ (*line* 19).

The calculation of $P_{knn}(o, D_i)$ is divided into three cases: (1) if $cnt_b + cnt_c \leq k$, $P_{knn}(o, D_i)$ is 1 (line 16); (2) if $cnt_b + cnt_c > k$ *and* $cnt_b < k$, the calculation is a little more complicated. Observe that for D_i, the inequality $cnt_c > k - cnt_b$ holds, which means the number of objects whose PLSs overlap with D_i is larger than the number of kNNs that need to be located. As each object whose PLS overlaps with D_i has equal probability of being a kNN, thus $P_{knn}(o, D_i)$ equals $(k - cnt_b)/cnt_c$ (line 18); (3) if $count_b \geq k$, $P_{knn}(o, D_i)$ equals 0. Thus we have

$$P_{knn}(o, D_i) = \begin{cases} 1 & cnt_b + cnt_c \leq k \\ \frac{k - cnt_b}{cnt_c} & cnt_b + cnt_c > k \text{ and } cnt_b < k \\ 0 & cnt_b \geq k \end{cases}$$

4   Performance Evaluation

We evaluate the performance of the proposed $CUkNN$ algorithm through simulation experiments, and compare our $CUkNN$ algorithm with the IMA algorithm [2] to evaluate the efficiency and precision of these two algorithms.

We test our method on a real road network. To model the real-world road network, we use the real data of the traffic network of San Francisco Bay Area in America[13] and construct a sub-network with 20K edges. The objects and queries move in the road network with uncertain speeds which have the maximal and the minimal speed limits. IMA reevaluates a kNN query when objects' location updates occur, and the update interval (UI) is set to be 10 time units.

Algorithm 3. Probability evaluation phase

---

  input : $UkNN\_interim\_Set$
  output: $UkNN\_Final\_Set$

1   **begin**
2      $UkNN\_Final\_Set = \varnothing$;
3   Further divide the time period to make sure that, $MaxD_{q,o}(t)$ and $MinD_{q,o}(t)$ of each object o in $UkNN\_Set$ not have a turning point and they do not intersect with each other, within each divided sub-period;
4   **for** *each tuple* $< [t_i, t_j], UkNN\_Set >$ *of* $UkNN\_interim\_Set$ **do**
5   **if** *the size of* $UkNN\_Set \leq k$ **then**
6   set the P_o of each object o in $UkNN\_Set$ to be 1;
7   **else**
8   **for** *each object* o *in* $UkNN\_Set$ **do**
9             $P_o = 0$;PLS(o)=$[MinD_{q,o}((t_i + t_j)/2), MaxD_{q,o}((t_i + t_j)/2)]$;
10   divide the distance range, where all PLSs covers, into sub-intervals: $D_1, D_2, ..., D_n$; cnt_b=0; cnt_c=0 ;
11   **for** $i=1$; $i \leq n$ and $cnt_b < k$; i++ **do**
12   cnt_c= the number of objects whose PLSs overlap D_i;
13   cnt_b= the number of objects whose PLSs are entirely below D_i;
14   **for** *each object* $o \in UkNN\_Set$ *whose PLS overlaps* D_i **do**
15   **if** $cnt_b + cnt_c \leq k$ **then**
16                  $P_{knn}(o, D_i)$=1;
17   **else**
18                  $P_{knn}(o, D_i) = (k - cnt_b)/cnt_c$;
19                $P_I(o, D_i)$= (the length of D_i)/ (the length of PLS(o));
20   P_o+= $P_I(o, D_i) \times P_{knn}(o, D_i)$;
21   insert $< [t_i, t_j], \{(o_1, P_{o_1}), (o_2, P_{o_2}), ..., (o_n, P_{o_n})\} >$ into $UkNN\_Final\_Set$;
22   output $UkNN\_Final\_Set$;
23   **end**

We first evaluates the effect of query time interval on the performance of $CUkNN$ and IMA. As shown in Figure 6(a), the CPU time of these two algorithms increases with the increasing query interval length. Clearly, $CUkNN$ has a better performance at all time intervals, compared to IMA. Figure 6(b) shows that the precision of $CUkNN$ at each time interval is above 85%. While the precision of IMA is low, and it is at best 45%.

We compare the performance of $CUkNN$ and IMA by varying the value of k from 1 to 20. Figure 7(a) shows that the CPU overhead for these two algorithms grows when k increases and $CUkNN$ outperforms its competitor in all cases. Figure 7(b) shows that the precision of $CUkNN$ increases slightly as k increases and it is above 80 % in all cases. While the precision of IMA is at best 42%.

We then study the effect of the query cardinality on the performance of the algorithms. As shown in Figure 8(a), the running time of these two methods

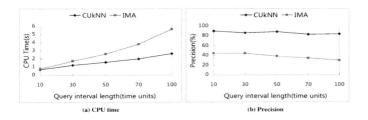

(a) CPU time

(b) Precision

Fig. 6. Effect of query interval length

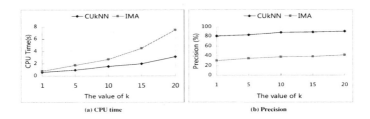

(a) CPU time

(b) Precision

Fig. 7. Effect of k

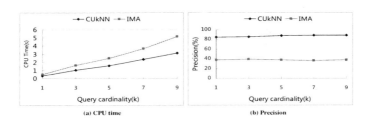

(a) CPU time

(b) Precision

Fig. 8. Effect of query cardinality

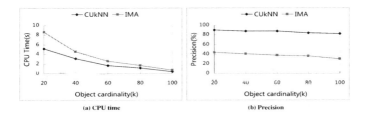

(a) CPU time

(b) Precision

Fig. 9. Effect of object cardinality

increases almost linearly as the number of query points increases. Figure 8(b)
shows that the query cardinality has little effect on the precision of the methods.

Finally, we examine the effect of changing the number of moving objects
on the performance of the algorithms. Figure 9(a) shows that the running time

of these two methods decreases as the number of moving objects increases. As shown in Figure 9(b), the precision of these two methods decreases slightly as the number of moving objects increases.

5   Conclusion

This paper presents an efficient method for processing $CkNN$ queries over moving objects with uncertain speeds ($CUkNN$) in road networks. We present an uncertain network distance model to calculate the distances between moving objects and a submitted query, both of which move with variable speeds in the road network. Based on the distance model, we propose a continuous $UkNN$ query monitoring method to get the $UkNN$ result set of the query during the time period being monitored. Experimental results show that our approach is efficient and precise.

References

1. Cho, H.J., Chung, C.W.: An efficient and scalable approach to CNN queries in a road network. In: VLDB, pp. 865–876 (2005)
2. Mouratidis, K., Yiu, M.L., Papadias, D., Mamoulis, N.: Continuous Nearest Neighbor Monitoring in Road Networks. In: VLDB, pp. 43–54 (2006)
3. Huang, Y.-K., Chen, Z.-W., Lee, C.: Continuous K-Nearest Neighbor Query over Moving Objects in Road Networks. In: Li, Q., Feng, L., Pei, J., Wang, S.X., Zhou, X., Zhu, Q.-M. (eds.) APWeb/WAIM 2009. LNCS, vol. 5446, pp. 27–38. Springer, Heidelberg (2009)
4. Liao, W., Wu, X., Yan, C., Zhong, Z.: Processing of continuous k nearest neighbor queries in road networks. In: Lee, R., Ishii, N. (eds.) Software Engineering, Artificial Intelligence, Networking and Parallel/Distributed Computing. SCI, vol. 209, pp. 31–42. Springer, Heidelberg (2009)
5. Safar, M.: Enhanced Continuous KNN Queries Using PINE on Road Networks. In: ICDIM, pp. 248–256 (2006)
6. Trajcevski, G., Tamassia, R., Ding, H., Scheuermann, P., Cruz Isabel, F.: Continuous probabilistic nearest-neighbor queries for uncertain trajectories. In: EDBT, pp. 874–885 (2009)
7. Chen, J.C., Cheng, R., Mokbel, M., Chow, C.Y.: Scalable processing of snapshot and continuous nearest-neighbor queries over one-dimensional uncertain data. The VLDB Journal 18, 1219–1240 (2009)
8. Huang, Y.K., Liao, S.J., Lee, C.: Evaluating continuous K-nearest neighbor query on moving objects with uncertainty. Information Systems 34(4-5), 415–437 (2009)
9. Papadias, D., Zhang, J., Mamoulis, N., Tao, Y.: Query Processing in Spatial Network Databases. In: VLDB, pp. 802–813 (2003)
10. Kolahdouzan, M.R., Shahabi, C.: Voronoi-Based K Nearest Neighbor Search for Spatial Network Databases. In: VLDB, pp. 840–851 (2004)
11. Huang, X., Jensen, C.S., Šaltenis, S.: The islands approach to nearest neighbor querying in spatial networks. In: Anshelevich, E., Egenhofer, M.J., Hwang, J. (eds.) SSTD 2005. LNCS, vol. 3633, pp. 73–90. Springer, Heidelberg (2005)
12. Wang, H., Zimmermann, R.: Location-based Query Processing on Moving Objects in Road Networks. In: VLDB, pp. 321–332 (2007)
13. http://www.fh-oow.de/institute/iapg/personen/brinkhoff/generator/

Discrete Trajectory Prediction on Mobile Data

Nan Zhao, Wenhao Huang, Guojie Song*, and Kunqing Xie

Key Laboratory of Machine Perception and Intelligence,
Peking University, Beijing, China, 100871
{znzn007,rubio8741}@gmail.com, {gjsong,kunqing}@cis.pku.edu.cn

Abstract. Existing prediction methods in moving objects databases cannot work well on fragmental trajectory such as those generated by mobile data. Besides, most techniques only consider objects' individual history or crowd movement alone. In practice, either individual history or crowd movement is not enough to predict trajectory with high accuracy. In this paper, we focus on how to predict fragmental trajectory. Based on the discrete trajectory obtained from mobile billing data with location information, we proposed two prediction methods: Crowd Trajectory based Predictor which makes use of crowd movement and Individual Trajectory based Predictor uses self-habit to meet the challenge. A hybrid prediction model is presented which estimates the regularity of user's movements and find the suitable predictor to gain result. Our extensive experiments demonstrate that proposed techniques are more accurate than existing forecasting schemes and suggest the proper time interval when processing mobile data.

Keywords: Trajectory Prediction, Spatial-Temporal Behavior, Location.

1 Introduction

Whether in research or application, it is essential to know the exact region of a user. With a user's history movements, trajectory prediction is dedicated to find the probable location of the user at some future time. The researchers have done a lot of trajectory prediction works on GPS data. The position of a user is collected every 3~5 minutes so that the trajectory from a moving object approximate the object's movements, which is a continuous function from time to space. The trajectory prediction work is of great value to location based services, city planning, resource management and many other related areas[1].

However, in reality, not all the datasets could provide satisfying continuous location data as the real trajectory. On one hand, GPS information may be incomplete during some time periods which will make the trajectory discrete in some sense[2]. On the other hand, many location information datasets cannot provide continuous trajectory such as the mobile telecommunication system. The location of a user is recorded when a call is made in the mobile billing data by mobile operators. The trajectory obtained from the mobile billing data is

---

* Corresponding author.

X. Du et al. (Eds.): APWeb 2011, LNCS 6612, pp. 77–88, 2011.

fragmental because the call frequency is much lower than the GPS refreshing frequency.

The swift progress of modern wireless communication made mobile phone more and more popular for everyone. The coverage of mobile phones could reach 90% in developed countries and more than 70% globally[3]. The result of the trajectory prediction of the large crowds on "discrete"mobile billing data is meaningful because of the high coverage of the mobile users.

In this paper, we proposed three different trajectory predicting methods. The main contributions of this paper could be summarized as follows:

1. We proposed a universal framework of trajectory prediction on discrete trajectory.
2. We presented the definition of discrete entropy which is used to measure people's predictability on discrete trajectory. People's movements are regular if their discrete entropies are low and vice versa.
3. According to the different discrete entropy, we proposed the Individual Trajectory Predictor (ITP) to predict people whose movements are with high regularity and the Crowd Trajectory Predictor (CTP) to find people's future location follow the trends of the large crowds.
4. A practical hybrid predicator with high accuracy and predictability is proposed.

The remaining of the paper is organized as follow: In section 2, some related work and background knowledge has been prepared. In section 3, we make some definitions and statements of the problem. Section 4 details the general prediction framework and introduced two predictors. Further, the hybrid predicator is proposed to improve the accuracy. Section 5 reports on the dataset introduction and experimental study. We conclude our work and discussed the prospects of the study in Section 6.

2   Related Work

There are three popular positioning methods, GPS, Wi-Fi, and mobile data. The trajectory prediction work is mainly based on these three distinct datasets.

GPS possesses the highest position precision which can be near 1~10m. In [1], Monreale introduced a new method of predicting in the city by GPS data. By using crowd data, they can predict 60% trajectories averagely. However, due to the cost of GPS devices, the sample rate is limited. Generally, the sample rate would not be over 1000 individuals what makes the GPS trajectory prediction work not convincing enough.

With Wi-Fi mac address, a user can be located in the registered Wi-Fi base stations. Therefore, the precision of the location is relatively high. [4][5] has made some meaningful attempts. However, the Wi-Fi devices are not so popular all over the world. The trajectory prediction work could only be applied in some special areas such as a campus or a community.

Mobile phones are more and more widespread all over the world. It makes us possible to do some trajectory prediction works on large mobile datasets. Though the precision of the location areas obtained from the mobile data is not very high, usually 20~500m in the city. The large amounts of mobile users could make the result not only limited in the single individuals but also the large crowd in the city. In [2], Lei did meaningful attempts on 10,000,000 users' phone records. They demonstrate that the upper limit of the prediction accuracy on mobile datasets is 93%. But they do not propose any specific prediction method.

3   Problem Statement

We proceed to cover the data model used in this paper and the problem addressed.

3.1   Data Model

Using location data, the trace of a moving object can be described as a trajectory with time-stamped locations. The location is abstracted by using ordinary Cartesian coordinates, as formally stated by the following form:

Definition 1. *A Positioning data is a sequence of triples* $P = < x_1, y_1, t_1 >$ *, ...,* $< x_n, y_n, t_n >$*, where* $t_i(i = 1...n)$ *denotes a timestamp such that* $\forall_{0<i<n}$ $t_i < t_i + 1$ *and* (x_i, y_i) *are point in* $\mathbb{R}^2$*.*

Usually, it is hard to proceed with Cartesian coordinates. So the range is divided into regions and there is no need to know the exact coordinate of the object, which indicates that knowing the region that the object belonging to is enough. In order to study activity regularity, time is segmented into equal-length intervals. The records is changed into time series with region tabs according to time sequence. We call it trajectory, defined as following:

Definition 2. *A Trajectory is a time series* $T = < R_0, ..., R_n >$ *where time is divided into n equal-length intervals, and* $\forall_{0<i<n} R_i \in S$*.* $S = \{R_0, ..., R_n\}$ *is a set of regions, each position* (x_i, y_i) *corresponds to a region* R_i*.*

In the "blind" area where we cannot find out the region the user stays, we simply mark the region with a "?". The trajectory is fragmental or "discrete", described as $T = < R_0, ..., R_n >$, where $\forall_{0<i<n} R_i \in S \cup ?$. This kind of trajectory is no longer "continious".

In this paper, we consider discovering a user's periodic patterns from his historical trajectory. Given a time period C, which is the number of timestamps that a pattern may re-appear, a user's trajectory is divided into $\lfloor \frac{n}{C} \rfloor$ sub-patterns G. G_i represents the sub-pattern in the ith of T. T can be represented as $T = \{G_0, G_1, ..., G_{\lfloor \frac{n}{C} \rfloor}\}$ and $G_i = \{R_i, ..., R_{i+C}\}$. In our work, the time period C is set as one day.

3.2   Problem Characterization

We aim to predict a user's future trajectory based on history trajectory and last-few-hour trajectory.

There are two kind of prediction cases:

1. Given last few hours trajectory of the day, predict next few hours location which is the most common case. By positioning devices, a user's recent trajectory is known, and we want to know where he is in next few hours.
2. Predict next few days location without last few hours trajectory. Usually, there is no need to predict this kind of case. But in mobile data, it is possible for a user not calling for several days. So we have to take this case into account.

Both two cases are required to be successfully predicted. Here we try to define a predictor:

Problem Statement. *A predictor is a function or an algorithm, with the history or common sense, predicts the region of the object staying at a given time. Formally written as: $r = Predictor(sub - trajectory, time)$ where sub-trajectory is the recent trajectory of the day, or is ϕ, and time is the predicting time.*

The standard to judge a predictor is prediction accuracy, but a predictor will not always be the best with the variation of parameters.

4   Proposed Method

In this section, we first demonstrate the framework of prediction.The strategy of evaluating prediction accuracy is introduced. A parameter called Discrete Entropy is introduced to measure the predictability of a user. Given a threshold, users are divided into two groups. In practice, Crowd Trajectories based Predictor(CTP) works well with users of high Discrete Entropy. And Individual Trajectory based Predictor(ITP) is better for users of low Discrete Entropy.

4.1   General Prediction Framework

Because of the nature of mobile data, the trajectory we get is fragmental. When verifying result, we only calculate time i in which the corresponding region R_i is not equal to "?". In other words, we know the exact region of time i. We compared the result of prediction with R_i, and gather all the instances to calculate prediction accuracy.

Entropy is probably the most fundamental quantity measuring the regularity of user's movements. According to incomplete history, we can judge whether an object is hard to predict or not. We call the parameter Discrete Entropy, defined as follows:

Definition 3. *A Discrete Entropy is a parameter to measure the degree of pre-dictability, writtern as*

$$E_d = -\sum_{i=1}^{n} p(R_i) log_2 p(R_i) \tag{1}$$

where $p(R_i)$ is the probility of R_i exists in T' (T' is the sub-trajectory of T whose element of "?"is remove), and i is the timestamp.

Fig. 1 shows the distribution of discrete entropy in the crowd. From the figure, we can see that it is close to normal distribution. The users whose entropy is between 0 and 0.1 hold a high proportion which is because in our datasets a group of people call so few merely 1 or 2 calls during two weeks.

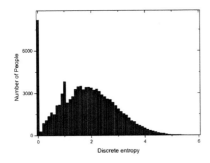

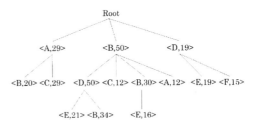

Fig. 1. Discrete Entropy   **Fig. 2.** An ultra-pattern tree

We use E_d, a simple way to measure the degree of the predictability. If the E_d is low, it is easy to predict by the existing history T'. If the E_d is high, it is hard to predict.

4.2   Individual Trajectory Based Predictor

Ultra-pattern Tree Building. In this section, we first tried to combine similar sub-trajectories of a user which aimed at reducing user's history. Then we use aproiriate algorithm to calculate the frequent patterns of the user. At last, we build so called ultra-pattern tree to get result.

Due to the lack of continues trajectory, the trajectory by mobile data cannot reflects the true movement of a user. So it is impossible to rebuild the whole trajectory. Normal data mining algorithm based on frequent trajectory cannot work.

To combine similar sub-trajectories, a parameter which characterize the simi-larity of two trajectories is required. The average distance of two sub-trajectories can measure the similarity of movement which is able to measure the similarity of sub-trajectories.

Definition 4. *The similarity of two trajectories is*

$$S_{(G_i,G_j)} = \frac{1}{\sum_{k=1}^{C} distance(G_{i,k}, G_{j,k})} \tag{2}$$

where G_i and G_j are two sub-trajectories of a user. And $G_{i,k}$ is the region in time offset k of sub-trajectory G_i.

$$distance(G_{i,k}, G_{j,k}) = \sqrt{(x_{G_{i,k}} - x_{G_{j,k}})^2 + (y_{G_{i,k}} - y_{G_{j,k}})^2} \tag{3}$$

If $G_{p,q}$ is "?", we use the most frequent region of the sub-trajectory instead.

If the distance is small enough, we believe that the two-day movements are similar. Take the office employee as an example. In the working days, he may go to work by the same way, and do similar things. So the trajectories of working days are similar, and we could combine these sub-trajectories represented as $G_{(i,j)}^2$.

Input: A set of t-patterns $\mathcal{P}_{set}$
Output: A ultra-pattern tree $\mathcal{T}_u$
1 $\mathcal{U}=$ new ultra-pattern tree();
2 **foreach** p *in* $\mathcal{P}_{set}$ **do**
3   node= Root($\mathcal{T}_u$);
4   **foreach** (i, r) *in* p **do**
5   n= findChild(node, r);
6   **if** $\nexists n$ **then**
7   v= new Node(r);
8   v.support = p.supp;
9   node.appendChild(v,i);
10   node=v;
11   **end**
12   **else**
13   UpdateSupport(n,p.supp);
14   node=childNode;
15   **end**
16   **end**
17 **end**
18 **return** $\mathcal{T}_u$;

Algorithm 1. Ultra-pattern tree building

Definition 5. *The strategy of combining two sub-trajectories is: according to time order we combine G_i and G_j, if the two regions of two sub-trajectories are the same, just reserve the region as the result. If one region is known and the other is '?', the result is the known region. If two region are known, choose one randomly.*

Example 1. Two sub-trajectories is : $G_1=\{A, B, ?, C, A, A\}$ $G_2=\{A, ?, ?, A, C, ?\}$ One case of combination result is : $G_{1,2}^2=\{A, B, ?, C, C, A\}$.

The method of combination is, we first calculate the $S_{(G_i,G_j)}$ of two day's sub-trajectories in T, then choose the smallest $\rho\%$ as candidate and represented as G^2. We calculate the $S_{(G^2_{(i,j)},G^2_{(p,q)})}$ of two sub-trajectories in G^2, then choose the smallest $\rho\%$ as candidate and represented as G^3. The work is finished until $G^{\lfloor\rho\times C\rfloor}$. At last, we collect all the sub-trajectories and combinatioing these trajectoryies as U^G. Then we delete all '?' elements of U^G which is represented as $U^{G'}$.

Here we have restored all the trajectories of a user and the combinated trajectories, and then we use $U^{G'}$ to build t-patterns which is describe in [6].

Because our data is not the real trajectory of a user, we call our tree as ultra-pattern tree. Fig. 2 is an example of an ultra-Pattern Tree. The *ultra-pattern tree building* algorithm (Algorithm. 1) describes how to build an ultra-pattern tree given a set of t-patterns [1].

Each t-pattern belonging to $\mathcal{P}_{set}$ is inserted into the ultra-pattern tree $\mathcal{U}$. We try to search for the path that corresponds to the longest prefix of p. Next, we append a branch to represent the rest of the elements of p in this path. If this tree is a prefix of p, a p is appended to the path. The $findChild(node, r)$ function returns the child of *node* that has the region equal to r and a connection between them. The $UpdateSupport(n, p.supp)$ procedures updates the support of *node*.

Prediction Strategy. The main idea behind our prediction is to find the best path on the tree, namely the best ultra-pattern tree that matches the given trajectory. Hence, for a given trajectory we compute the best matching score of all the admissible paths of the t-pattern tree. The children of the best node that produces a prediction are selected as next possible regions.Here we use similarity of two trajectories S to score the degree of matching.

Given a sub-trajectory *sub* and an ultra-pattern tree P, we computes the similarity score ,which is the the similarity of sub-trajectory (Definition 5) for each path of the P relative to p. When the tree has been completely visited, the best match is computed by selecting those candidates with the smallest score, and the region associated with the candidates is returned as the prediction. There may be several predictions with the same score corresponding to all the children in the last node of the path with the best score.

4.3   Crowd Trajectory Based Predictor

In the previous part, we discuss the prediction on individual history. However, some users' movements are impossible to predict with ultra-pattern tree. There's no pattern that corresponds the given sub-trajectory.

In the city, people travel by public traffic and the city is often divided into different function districts which means when moving in the city, people pass the same streets, the same buildings and the same cellular towers. So the patterns of moving are similar. To high-E_d objects, we could do the prediction work based on the crowd trajectory.

Modeling of Dynamic Bayesian Network. The dynamic Bayesian network (DBN) is defined as a pair $(B_1, B_\rightarrow)$,where B_1 is a Bayesian network which defines the prior $P(Z_1)$, and $B_\rightarrow$ is a two-slice temporal Bayes net which defines $P(Z_t|Z_{t-1})$ by means of a DAG (directed acyclic graph) as follows:

$$P(Z_t|Z_{t-1}) = \prod_{i=1}^{N} P(Z_t^i|Pa(Z_t^i)) \tag{4}$$

where Z_t^i is the i'th node at time t, which could be a component of X_t, and $P_a(Z_t^i)$ are the parents of Z_t^i in the graph.

In our model, $Pa(Z_t^i) = Z_{t-1}$. That means we only consider the previous state. Experiment in Section 5 shows that it is enough to predict well. $Z_t = (U_t, X_t, Y_t)$ represents the input, hidden and output variables of a state-space model. In our model, $X_t = \emptyset, Y_t = \emptyset$. U_t represents the region exist in trajectory, $U_t \subset S$. The transfer frequency between regions is 0 in the time t. That means we do not consider object moving between regions in a time interval.

We compute $P(Z_t|Z_{t-1})$, which suggest the travel trend between time $t - 1$ and time t. There are C slices in DBN model. Because we construct the model with trajectory, $P(Z_C|Z_0)$ is allowed.

Input: The set of crowd trajectories $S = \{T_1, T_2, ..., T_n\}$
Output: A corresponding DBN DBN
1 DBN=new dynamic Bayesian network;
2 **for** *time from 1 to C* **do**
3   **for** *i from 1 to n* **do**
4         $DBN.P(Z_{(i+1)\%C}|Z_i)$=computeTranfer(G_i,$G_{(i+1)\%C}$);
5   **end**
6 **end**
7 **return** DBN

Algorithm 2. Build a DBN

Algorithm. 2 shows the steps to construct a DBN. In the model, the trend of object flowing is represented by $P(Z_t|Z_{t-1})$. Fig. 3 is a visual example. The function computeTranfer(G_i,$G_{(i+1)\%C}$) collect all instances like G_i,$G_{(i+1)\%C}$, and compute the transfer frequency matrix of S.

Prediction Algorithm. The model is based on previous-timeslice status. But when you want to predict next two or more time slices' location, the model cannot directly give the result.

So the strategy here is : find the last region which is not "?", then put it into DBN, follow the crowd choice, simulate last n steps.

5   Experimental Evaluation

5.1   Dataset

The work of this paper is based on a China Mobile billing dataset of a medium-sized city in China. It included over 100,000 anonymous mobile phone subscribers' phone call records during 14 days. We transform the phone call records to personal trajectory according to the call time and call cellular tower location. Then we segment the time to one hour intervals. If several calls were made in one hour, we choose the most frequent cell tower existing in the call list. If there is no call records in one hour, we mark it with a "?". By this way, we build a pattern for each user, whose length is L=24 x14 = 336 with cell tower ID and "?". In the experiment, we use the history of former 13 days to build our predictors. The real call records of last day to measure the accuracy.

In the article [2], the author introduced a function R(t), measuring the fraction of instances when the user is found in his or her most visited location during the corresponding hour-long period. R represents a lower bound for predictability. We redo the experiment and find that the result is restricted by history as Fig. 4. With 13-days long history, the accuracy is 42%, which can be seen as the lower bound in our article.

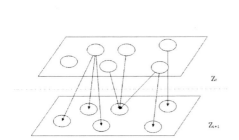

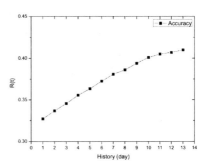

Fig. 3. A visual example of DBN   **Fig. 4.** History vs R(t)

5.2   Features of Predictors

We evaluated the accuracy and the step size (the accuracy when predict next several time slices) of the individual trajectory predictor and crowd trajectory predictor respectively. We also proposed a hybrid predictor which combined the ITP and CTP together.

Accuracy. Before analyzing the features of predictors, we first define some parameter of algorithm. As Fig. 5 shows, when the combination proportion ρ is 60%, the accuracy of prediction is at the highest point, which proves the view that 60% of human trajectories are self similar in [7]. We define $\rho = 0.6$ as

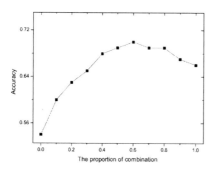
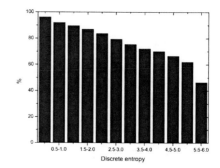

Fig. 5. The effect of combination   **Fig. 6.** The percentage of predictability

the combination proportion in the ITP. For the convenience of comparing with existing methods, the time interval is set at one hour in the article.

In Fig. 6, we see that most of the people are predictable especially whose discrete entropy is small. Limited by ultra-pattern tree, as the discrete entropy increases, people become hard to predict by the individual trajectory predictor. For whose discrete entropy are bigger than 6.0 which means those people's movements are full of randomness, only half of them are predictable using individual trajectory predictor.

For crowd trajectory predictor, all the users are predictable. As introduced, we utilize the real phone call records one day to calculate the accuracy of the people who are predictable. The result is shown in Fig. 7. The figure reflects the predicting accuracy in different time slice. As the figure shows, the accuracy in different hour varies slightly. The abnormal peak during 3 a.m. to 5 a.m. was because only few samples have phone calls during that time periods.

Step Size. From Fig. 8, the accuracy does not drop dramatically with the increase of step size which means the predictor could work well when predicting a user's location whose last call was made many hours before. The vibrating of

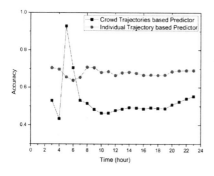
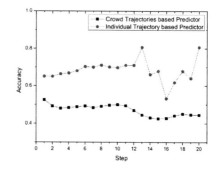

Fig. 7. Accuracy over a day long period   **Fig. 8.** Step Size Accuracy

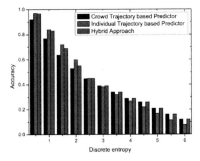

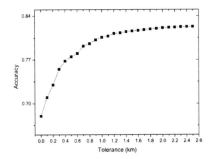

Fig. 9. Accuracy of different predictors   **Fig. 10.** Tolerant Distance

the accuracy during 14∼22 hours later was also because of the small sample rate. It is obvious that individual trajectory predictor performs better than the crowd trajectory predictor since the ITP utilizes the user's history in a long period of time while the CTP only utilize the history of transfer matrix.

Hybrid Predictor. From Fig. 7 and Fig. 8, we can see the accuracy of the individual trajectory predictor whose average accuracy is 68% is higher than the 54% of crowd trajectory predictor. But the ITP suffers the problem that it is suitable to only parts of the people. If those people who are not predictable by using ITP are including when calculating the accuracy, the accuracy will decrease in some sense as Fig. 9 shows.

It is a natural idea to combine the two predictors together. As Fig. 9 shows, discrete entropy of 2.5 is the turning point of the performance for two predictors. That is to say, ITP is good at predicting the users whose activity is regular while the CTP is suit for people moves randomly. Therefore, we set 2.5 as the threshold. If the discrete entropy is smaller than 2.5, ITP is chosen in the hybrid predictor and when it is bigger than 2.5 the CTP is chosen.

The accuracy of the hybrid predictor is also shown in Fig. 9. The average accuracy is 62% and it is suitable to all the mobile users. It is a big increase from the low bound of the predicting accuracy of 40% to 62%.

Tolerant Distance. The tolerant distance should be considered in our experiment. If the location obtained from the predictor is near with the real location, it may be the oscillation effect caused by the cell towers while the locations are the same and if not the same location the near location is also meaningful in the predicting work. We calculate the accuracy in different tolerant distance as Fig. 10 shows. The distance between two regions is calculated as the distance between the two cell towers which are the center of the region.

6 Conclusions and Future Works

In this paper, we presented an approach which predicting a user's trajectory in a hybrid manner utilizing not only individual history but also crowd movement.

Specifically, patterns were defined and fragmental trajectory can be modeled. We then introduced ultra-pattern tree to query on individual trajectory and Dynamic Bayesian Network is introduced to predict on crowd trajectories. We also proposed a hybrid prediction algorithm which could provide result of high accuracy.

With the development of the wireless communication as well as the world wide spread of the mobile phones, our approach of trajectory predicting based on the mobile data will cause more attention. The approximate locations of the large mobile users will be significant in future LBS related applications and other related areas.

Acknowledgement. Research was supported by the National Natural Science Foundation of China under Grant No. 60703066 and No.60874082; Supported by Beijing Municipal Natural Science Foundation No. 4102026.

References

1. Monreale, A., Pinelli, F., Trasarti, R., Gianotti, F.: WhereNext: a location predictor on trajectory pattern mining. In: SIGKDD, pp. 637–645 (2009)
2. Song, C., Qu, Z., Blumn, N., Barabasi, A.L.: Limits of predictability in human mobility. Science 327, 1018–1021 (2010)
3. Ministry of Industry and Information Technology of the People's Republic of China, http://www.miit.gov.cn/
4. Tuduce, C., Gross, T.: A Mobility Model Based on WLAN Traces and its Validation. In: IEEE INFOCOM, vol. 1, pp. 664–674 (2005)
5. Petzold, J., Pietzowski, A., Bagci, F., Trumler, W., Ungerer, T.: Prediction of indoor movements using bayesian networks. In: Strang, T., Linnhoff-Popien, C. (eds.) LoCA 2005. LNCS, vol. 3479, pp. 211–222. Springer, Heidelberg (2005)
6. Gianotti, F., Nanni, M., Pinelli, F., Pedreschi, D.: Trajectory pattern mining. In: SIGKDD, pp. 330–339 (2007)
7. Song, C., Koren, T., Wang, P., Barabasi, A.L.: Modelling the scaling properties of human mobility. Nature Physics 6, 818–823 (2010)

Cleaning Uncertain Streams for Query Improvement

Qian Zhang[1,2], Shan Wang[1,2], Biao Qin[1,2], and Xiao Zhang[1,2]

[1] Key Laboratory of the Ministry of Education for Data Engineering and
Knowledge Engineering, Renmin University of China, Beijing 100872, China
[2] School of Information, Renmin University of China, Beijing 100872, China
{zhangqian15,swang,qinbiao,zhangxiao}@ruc.edu.cn

Abstract. Real-world applications confront uncertain streams derived from unreliable data acquisition equipments and/or defective processing algorithms.
However, application context covers specific cleaning rules to bring data close
to the reality (i.e. data quality), and query features can filter data for the
efficiency (i.e. data volume). In this paper, we propose a framework for cleaning uncertain data for query effectiveness and efficiency, which processes
high-volume streams in parallel, and append new cleaning rules & queries
seamlessly. We implement a prototype for video surveillance application over
the architecture.

Keywords: uncertain data cleaning, parallelized probabilistic graphic model,
uncertain streams.

1 Introduction

Real-world applications such as sensor and RFID networks [1] generate uncertain
streams inevitably. In the applications, data uncertainty is mostly due to some external
(as to application context itself) factors like unreliable data acquisition equipments
and/or defective data processing algorithms. Considering video surveillance application, face identification cannot always provide correct results coinciding with the
reality, and object tracking in videos is still an open problem of computer vision.
Uncertainty exists widely and cannot be eliminated in the near future.

Query processing in the probabilistic database is on the basic of Possible World
semantics, and possible worlds (i.e. the processing cost) increases exponentially with
the tuples involved. Research work is proposed for evaluating probabilistic queries in
a polynomial time (*safe query*) [6], when possible worlds need not to be enumerated
in the processing. While queries may be not *safe*, and data transmission is a potential
bottleneck for applications, cleaning uncertain data is an impressive way to improve
the efficiency in streaming environment.

In this paper, we propose a systemic solution to clean uncertain streams for data
quality and volume by Probabilistic Graphical Models and probabilistic queries
respectively. Our contributions include:

- We propose a parallelized framework for high-volume stream environment, as well
  as a generalized one for users to append queries and Probabilistic Graphical models
  implied in the application context.

X. Du et al. (Eds.): APWeb 2011, LNCS 6612, pp. 89–94, 2011.
© Springer-Verlag Berlin Heidelberg 2011

- We propose a set of fundamental theorems to guarantee the quality of cleaning uncertain data by probabilistic queries.
- We implement a prototype of video surveillance application with two cleaning rules over the framework.

2 Related Work

Data Cleaning. In the traditional data management, data cleaning is responsible for improving data quality. It makes the conversion from imprecise/inaccurate data to accurate one, and processing technologies for certain data are able to be applied subsequently. There are three popular cleaning methods, which make use of data dependency [1], statistics and entropy [4] respectively. Probabilistic Graphical model can be viewed as a kind of data dependency. Two shortcomings of traditional data cleaning technologies are that, data conversion loses characteristics of original data, and cleaning technology is often specified for particular dataset, which makes it difficult to be ported to other applications.

Probabilistic Data Management. Recent research work of probabilistic databases is mostly concerned about data structure and query processing. Two basic types of probabilistic data structure are *tuple existence uncertainty* and *attribute value uncertainty* [5]. Query processing in probabilistic database is based on Possible World semantic, and research work is proposed for polynomial-time evaluation [6].

We adopt *x-relation* data model in our prototype [5]. An *x-relation* is an uncertain data set consisting of *x-tuples* which are assumed independent from each other. An *x-tuple* τ contains a finite set of tuples as *alternatives* with the constraint that $\sum_{t \in \tau} p(t) \leq 1$, where $p(t)$ represent the probability of tuple t. X-tuple τ takes the value t with the probability $p(t)$, or does not exist at all with probability $1 - \sum_{t \in \tau} p(t)$. Fig. 1 illustrates two kinds of *x-tuples* in video surveillance application, with the same *time t* and *position (posx, posy)*, or the same *t* and *person id (pid)*.

Probabilistic Graphical Model (PGM) is used in modeling and reasoning about probabilistic data [1]. Some research adopts the *First-Order (FO)* model [3], an evolution of traditional PGM, in which random variables represent sets of random parameters. PGMs are always implemented as user views for hiding data details from users, and then queries are evaluated over them [2].

t	posx	posy	pid	pro	tid	t	pid	posx	posy	pro	tid
			A	0.158342	s_{01}			30	87	0.158342	t_{01}
0	30	87	B	0.385905	s_{02}	0	A	98	73	0.527037	t_{02}
			C	0.299473	s_{03}			116	87	0.314621	t_{03}
			A	0.136924	s_{11}			30	87	0.385905	t_{11}
1	99	73	B	0.153887	s_{12}	0	B	99	73	0.277734	t_{12}
			C	0.634996	s_{13}			96	3	0.185342	t_{13}
			A	0.240111	s_{21}			99	73	0.136924	t_{21}
2	99	90	B	0.252548	s_{22}	1	A	85	90	0.769932	t_{22}
			C	0.346263	s_{23}			20	36	0.093144	t_{23}
		$\vdots$						$\vdots$			

Fig. 1. X-tuples in the prototype of video surveillance application

Query processing over uncertain streams gains much attention recently [1] [7]. The main effort in the work is to evaluate queries effectively and efficiently, while our work aims at cleaning uncertain streams by application context to improve data quality (query effectiveness), and by query characteristic to reduce the volume (query efficiency).

3   Architecture

Fig. 2 (a) shows that the framework consists of PGM and Query cleaning processes, and (b) shows our prototype of video surveillance application over the architecture.

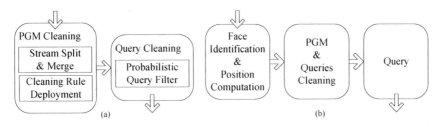

Fig. 2. (a) The architecture of uncertain stream cleaning. (b) The prototype of video surveillance application.

3.1 Module Description

Stream Merge & Split generates specified data streams for parallelizing the cleaning rule deployment. The reason of stream reorganization is that, original uncertain streams in applications are not always ready for parallelization. Considering the video surveillance application, for uncertain streams are generated by face identification over the videos derived from cameras, records for the same person may appear across the streams. Then we assign PersonID as the divided attribute, and produce new streams corresponding to individual persons, which can be processed by different threads in parallel.

Cleaning Rule Deployment parallelizes cleaning rule evaluation on reorganized uncertain streams, and outputs cleaned streams. A cleaning rule is submitted by users as a java file implementing specified APIs, and can share identical algorithms among them. If there are dependent rules to be processed, we also need to define the processing order of them, or system executes all rules simultaneously.

Probabilistic Query Filter generates cleaned uncertain dataset specific for probabilistic queries submitted by users. By foundational theorems proposed in Section 3.3, the module produces a shrunken dataset for each query to reduce the cost of subsequent processing. It takes cleaned streams processed by PGM Cleaning as input, and generates further cleaned streams for each query.

3.2 Cleaning Rules in Video Surveillance Application

In video surveillance application, system gets video streams from cameras, and processes them by face identification and position computation. Specifically, each person-like object in a video frame is represented by several uncertain tuples with the same time $t(t)$, the same position $posx(t)$ and $posy(t)$, as well as different person id $pid(t)$ and confidence $pro(t)$. Meanwhile in the application, the physical function of human beings restricts people's moving speed. If person A appears at position I and his possible position in the next second is II, we suppose that the *probability distribution function* of the distance between I and II obeys *Normal Distribution*: $N(\mu, \delta^2)$. $\mu = 0$ if A is most likely to stay still over time, and δ can be viewed as the moving speed and differs in various environments. Moreover, given A's position II at $t+1$, his position I at t is restricted by II in the same way. In the following statements, Pos_T is A's position at time T, and $Dist_{t+1}$ $Dist'_{t+1}$ represents the distance between Pos_T and Pos_{T+1}. Fig. 3 shows the corresponding PGMs.

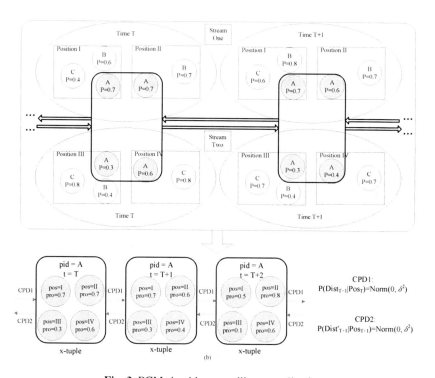

Fig. 3. PGMs in video surveillance application

Rule 1. $\forall t'$, if $\exists T\{ t \in T: t(t) = t(t')-1, pid(t) = pid(t')\} \neq \varnothing$, $dist(t, t') :=$ The distance function of $(pos(t), pos(t'))$, $dist(t, t') \sim N(0, \delta^2)$.

Rule 2. $\forall t'$, if $\exists T\{ t \in T: t(t) = t(t')+1, pid(t) = pid(t')\} \neq \varnothing$, $dist(t, t') :=$ The distance function of $(pos(t), pos(t'))$, $dist(t, t') \sim N(0, \delta^2)$.

3.3 Foundational Theorems for Query Cleaning

We propose three theorems to support our cleaning approach, by which uncertain data can be shrunken without loss for specific queries. If there is a selective query Q (SQL query with a selective WHERE clause) over a uncertain dataset D of x-tuples, we define the result set of Q on D as Q(D). We denote an x-tuple T in D as $T\{t_1, t_2, ..., t_n\}$ and t_i is a tuple with existence uncertainty.

Theorem 1. If $T\{t_1, t_2, ..., t_n\}$ does not satisfy Q, which means all t_i in T does not satisfy Q's WHERE clause, we have that Q(D) = Q(D-T) (D-T denotes the dataset of all x-tuples in D except T).

Proof. If Q(D) ≠ Q(D-T), there must be a result r in Q(D) but not in Q(D-T), or the converse. The first case is that r must result from T (or partly from). While all t_i in T does not satisfy Q, we get the contradiction. In the second case, for Q is a selective query, r satisfies Q's WHERE clause and r is in Q(D). We get the contradiction.   □

Theorem 2. Generally, if $t_1, t_2, ..., t_m$ of $T\{t_1, t_2, ..., t_m, t_{m+1}, t_{m+2}, ..., t_n\}$ satisfy Q, and $t_{m+1}, t_{m+2}, ..., t_n$ does not satisfy Q, we have that Q(D) = Q(D'), while D' denotes the dataset of all tuples in D except for $t_{m+1}, t_{m+2}, ..., t_n$ (T converts to $\{t_1, t_2, ..., t_m\}$).

Proof. If Q(D) ≠ Q(D'), there is a result r in Q(D) but not in Q(D'), or the converse. Set T'=$\{t_1, t_2, ..., t_m\}$. In the first case, r must result from T (or partly from) but not T'. While $t_{m+1}, t_{m+2}, ..., t_n$ does not satisfy Q, all results derived from T can also be derived from T'. We get the contradiction. In the second case, for Q is a selective query, r satisfies Q's WHERE clause and r is also in Q(D). We get the contradiction.□

Theorem 3. Generally, if Q generates result $\{r_1, r_2, ..., r_m\}$ on $T\{t_1, t_2, ..., t_n\}$ (or partly), when t_i and t_j (i ≠ j) produce the same result r_k, set t' = t_i (or t_j) with the existence probability equaling to the sum of t_i's and t_j's, and T' = T-t_i-t_j+t', we have that Q(D) = Q(D-T+T').

Proof. If Q(D) ≠ Q(D-T+T'), we suppose that there is a result r in Q(D) but not in Q(D-T+T'), or the converse. In the first case, r must result from T but not T' (or partly from). For t_i, t_j satisfy Q and Q is a selective query, t' also satisfies Q and r can result from T' as well. We get the contradiction. In the second case, for Q is a selective query, r satisfies Q's WHERE clause. If r results from t', r also results form t_i (or t_j) and r is in Q(D). If r is irrespective of t', r is also in Q(D).We get the contradiction as well. If there is a result r in Q(D) and r' in Q(D-T+T'), r and r' is the identical result except for existence probabilities. By Possible World definition, we have that all possible worlds including t_i (t_j) can convert to possible worlds including t', with the latter's probabilities is the sum of the former's correspondingly. That is, r and r' have the identical existence probabilities. We get the contradiction as well.   □

4 Conclusion

Uncertain streams are generated widely in real-world applications, while most of the probabilities can be revised with application context restrictions. On the other side, query on uncertain data based on Possible World semantic makes great processing cost, for possible worlds increase with related tuples exponentially. In this paper, we propose a novel architecture for cleaning uncertain streams by application context and probabilistic queries, and we implement a prototype of video surveillance application over the framework.

Acknowledgments. This work is partly supported by the Important National Science & Technology Specific Projects of China (HGJ Projects, Grant No.2010ZX01042-001-002), the National Natural Science Foundation of China (Grant No.61070054), the Fundamental Research Funds for the Central Universities (the Research Funds of Renmin University of China, Grant No.10XNI018), Renmin University of China (Grant No.10XNB).

References

1. Kanagal, B., Deshpande, A.: Online Filtering, Smoothing and Probabilistic Modeling of Streaming data. In: ICDE 2008, Cancún, Mexico (2008)
2. Deshpande, A., Madden, S.: MauveDB: Supporting Model-based User Views in Database Systems. In: SIGMOD 2006, Chicago, Illinois, USA (2006)
3. Wang, D.Z., Michelakis, M., Garofalakisy, M., Hellerstein, J.M.: BAYESSTORE: Managing Large, Uncertain Data Repositories with Probabilistic Graphical Models. In: VLDB 2008, Auckland, New Zealand (2008)
4. Cheng, R., Chen, J.C., Xie, Y.K.: Cleaning Uncertain Data with Quality Guarantees. In: PVLDB 2008, Auckland, New Zealand (2008)
5. Benjelloun, O., Sarma, A.D., Halevy, A., Widom, J.: ULDBs: databases with uncertainty and lineage. In: VLDB 2006, Seoul, Korea (2006)
6. Dalvi, N., Suciu, D.: Efficient Query Evaluation on Probabilistic Databases. In: VLDB 2004, Toronto, Canada (2004)
7. Lian, X., Chen, L.: Efficient Join Processing on Uncertain Data Streams. In: CIKM 2009, Hong Kong, China (2009)

Personalized Web Search with User Geographic and Temporal Preferences

Dan Yang[1,2], Tiezheng Nie[1], Derong Shen[1], Ge Yu[1], and Yue Kou[1]

[1] School of information Science&Engineering, Northeastern University
Shenyang 110004, China
[2] School of Software, University of Science and Technology LiaoNing
Anshan 114051 China
asyangdan@163.com,
{nietiezheng,shenderong,yuge,kouyue}@ise.neu.edu.cn

Abstract. Personalized web search according to user's geographic and temporal preferences can improve search results quality and satisfy user's different information needs. We propose a novel approach to capture user's geographic and temporal preferences in the form of *query profile* and *user preference profile* by mining search results and user click-through data leveraging knowledge bases. Our approach classifies queries into five classes based on decision tree algorithm. When personalizing search results, different weights are set to different query classes to balance among content, geographic and temporal information associated with a query. The experiment evaluation results show the effectiveness of our approach and improvement of the search quality.

Keywords: personalized, Web search, geographic and temporal preferences.

1 Introduction

With the large scale of data in the Internet, how to provide user the most relevant and essential search results according to user's preferences has become the important problem faced with the search engine developers and researchers. When a user issues a query, he/she usually has geographic and temporal preferences in mind. For example query "department discount", user prefers to find department discount information in some place(s) recently or some time in the future (e.g. Christmas). But currently most commercial search engines return roughly the same results to all users not paying attention to user's geographic and temporal intents and preferences. It is important for the search engine to capture user's geographic and temporal preferences, which can provide more satisfied results for the user and impove search quality. Rather than solving the general problem of personalized web search results, we focus on queries that have geographic and temporal dependent preferences in this paper. There are two intuitive ways to personalize web search results: one is by query rewriting and one is by document re-ranking. We propose a novel approach based on document re-ranking to personalize web search results according to user's geographic and temporal preferences by mining search results data and user click-through data leveraging knowledge bases.

X. Du et al. (Eds.): APWeb 2011, LNCS 6612, pp. 95–106, 2011.
© Springer-Verlag Berlin Heidelberg 2011

The experiment results show the effectiveness of our approach in improveing search results quality.

The main contributions of this paper are four-fold:

- First we give a problem definition and propose a novel approach to personalize web search according to user's geographic and temporal preferences. It is to the best of our knowledge that no related work to combine both geographic and temporal preferences to personalize web search results.
- We propose to classify the query into five different classes according to their association with geographic and temporal information using decision tree algorithm.
- We propose a linear rank score function to calculate the score of each document related to a query.
- We present an evaluation to demonstrate the feasibility and utility of the proposed approach. It out-performs strategies which use either content or geographic or temporal preference only.

The remainder of the paper is organized as follows. Section 2 introduces related work. In Section 3 and Section 4 we describe our system in detail. Section5 gives the experimental data setup and evaluating results of our approach. Section 6 summarizes the main contributions of the paper and future work.

2   Related Work

The section overviews some newest research efforts related to user query intent discovery or search results personalization in web search and dataspaces environment. We then conclude differences between our work and the existed works.

[3,13-17] are related work about leveraging temporal information in query process or results clustering. [3] proposed clustering and exploring search results using timeline constructions as an extension to existing ranking techniques. [13] assumed that the timeline is generated from a collection of supporting documents. The components of a timeline are timeline, time frame, and label. [14] proposed an alternative document snippet based on temporal information that can be useful for supporting exploratory search. [15] proposed a approach which integrated the temporal dimension into a language model based retrieval framework. [16] proposed a algorithm to mining implicit year qualified queries. [17] proposed three temporal classes of queries, i.e., atemporal, temporally unambiguous, temporally ambiguous. [4,5,18] are related work in intent discovery or personalization according to geographic location. [4] built a geo intent analysis system to learn a model from large amounts of web-search logs for this discovery. It built a city language model, which is probabilistic representation of language surrounding the mention of a city in web queries. [5] proposed a web search personalization approach that captured the user's interests and preferences in the form of concepts by mining search results and their clickthroughs. [18] built a probabilistic model to identify implicit local intent queries and leverage user's physical location to improve web search results for these queries. [6, 8, 10] are related works done in the query classification. [6] proposed using past user-click behavior and anchor-link

distribution as implicit feature for query goal identification automatically. [8] used query log as a source of unlabeled data to aid in automatic classification. It used a rule-based automatic classifier produced using preferences mined from the linguistic analysis of an unlabeled query log containing hundreds of millions of queries. [10] used click graphs(bipartite-graph representation of click-through data) in improving query intent classifiers. Besides there also some related works which leveraging the external knowledge base such as [12, 20]. [12] proposed using Wikipedia to discovery intent. [20] proposed a intent based categorization of search results using web Q&A corpus.

[7] captured the context, the interests, and the preferences aspects of a user's intent with a context network, an interest set, and a preference network. In [2] a query intent suggestion algorithm was proposed by mapping implicit intent queries to a set of potential explicit intent query suggestions. [1] proposed a clusters refinements algorithm based on their likely underlying user intent by combining document click and session co-occurrence information. The algorithm operates by performing multiple random walks on a Markov graph that approximates user search behavior. [19] Proposed an approach that combined click and reformulation information to find likely user intents using Expand, Filter, Cluster three steps.

At same time there are some existing related works done in the field of data integration and dataspace such as [9, 11]. [9] introduced the concept of iTrails which enables a search engine to rewrite a simple keyword query into a structural query that encodes schema information from a given data source. Other integration system such as Hermes [11] allows a limited set of spatial operations (such as close to, travel time) through its push-button listing-based interface or a form-based interface. However, these user interfaces are not expressive enough and restrict users from specifying their intent in a flexible manner.

The differences between our work and existing works are:

- [4, 5, 22] only consider geographic location intent or preference. [3,13-17] only consider temporal information of document in web search. Our approach considers user's both geographic and temporal information needs associated with the query.
- [4] preprocesses the query before it is submitted to search engine. [5] submits the query to the search engine, then set a middleware to re-ranking the results returned by the search engine. Our approach processes the query both beforehand and after at the same time.
- [5] proposed a user profiling strategy to capture user's content and location preferences. But a single user profile sometimes can't reflect the features within the query. So our appraoch proposes to build *query profile* and *user preference profile* to balance the preference of user and context within a query.
- [9] is the technique used in dataspace which introduces iTrail to confine and accurate user's query requirements and intent. But it needs to define iTrail manually beforehand which is too complicated for common users and not suitable in web environment. Our approach is to discover geographic and temporal preference with the least manual effort by mining search results and user click-through data.

3 System Overview

In this paper we propose a novel approach to capture user's implicit geographic and temporal preferences, which combined query intension feature and user specific preferences feature in the form of *query profile* and *user preference profile* by mining both web search results and user click-through data.

3.1 Problem Definition

Given a *Cube(X, Y, Z)*, where X axis is about geographic information; Y axis is about temporal information; Z axis is about content information, the query results space of query Q satisfying user's specific or implicit geographic and temporal preferences is one cell or more cells, i.e. {$cell_1$, ..., $cell_n$} of the *Cube*, where $n \in$ {natural number}.

3.2 The Process of Our Approach

The general process of our approach is depicted in Figure1. Given a query Q input by the user, firstly with the decomposer tool the system decomposes it into content, geographic and temporal information three parts, namely, $Q=(Q_c, Q_g, Q_t)$, where Q_c is content part, Q_g is geographic part, Q_t is temporal part. Then submit the query to the backend search engine (e.g., one of the top commercial search engines). The returned search results are processed through personalized tool of our system. The content terms and geographic terms, temporal terms are mined online from the top k (in our experiment k is set 100) search results and are stored in query profiles. When the user clicks on a search result, the clicked result together with its associated content terms, geographic terms and temporal terms are collected by the user click-through collector and stored in user preference profile. Finally re-rank the search results according to *query profile* and *user preference profile*.

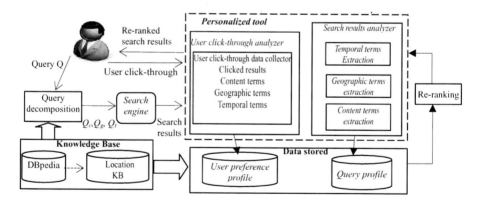

Fig. 1. The general process of user geographic and temporal preference personalization

3.3 Query Decomposition

Query decomposition before the query is submitted to *Search engine* includes three steps, which is showed in Figure 2. **Step1:** *matcher* matches the query into different parts: Q_c, Q_g, Q_t by exploiting two off-line knowledge bases (KBs) i.e., *location KB* and *DBpedia*. The matching results maybe short of geographic part or/and temporal part. **Step2:** *transformer* normalizes the geographic part and temporal part. **Step3:** *classifier* classifies the query into different query class using classifier learned from the query profile.

Fig. 2. The process of query decomposition

Matcher uses mapping functions and match functions to decide which part of the query belongs to each of the content, geographic and temporal part. *Transformer* uses regular expressions to normalize the temporal part due to heterogeneous temporal expressions and formats. Details of *Matcher* and *Transformer* are beyond the scope of this paper. *Classifier* classifies the query into five classes based on decision tree algorithm according to the query content associated with geographic and temporal information, which is described in detailed in section 4.2.

4 Capture Preferences Using Profiles

In our approach there are two profiles: *query profile* and *user preference profile* to capture user's geographic and temporal preferences. The former is got by mining web search results using *Search results analyzer* and the latter is got by mining user click-through data using *User click-through analyzer*.

4.1 Search Results Analyzer

In *Search results analyzer*, *content extraction* tool, *location extraction* tool and *temporal extraction* tool are used to extract content terms (denoted as c_i), geographic terms (denoted as g_i) and temporal terms/expressions (denoted as t_i) from documents of web search results leveraging *Location KB* and *DBpedia*. The content terms, geographic terms and temporal terms extracted from search results indicate a possible concept, geographic and temporal spaces arising from a query which can be maintained along with the user click-through data for further preference adaptation. Table 1 shows an example query "department discount" with the content terms, geographic terms and temporal terms extracted.

DBpedia is used as our off-line knowledge base which currently describes more than 3.4 million things. We use location ontology to maintain the geographical relationships among locations.

Table 1. Concept, geographic and temporal terms associated with query "department discount"

Content Terms	Factory outlet	Business Directory	Retailers	Retail chain store	vouchers
Geographic Terms	New York	Buffalo	Phoenix	San Diego	Long Island
Temporal Terms	2006	2010	2009	Christmas	Columbus Day

Location KB is built and maintained based on ontology in our framework. Previous research has show that a large portion of explicit geo queries contain city level information [21], so we define our user's implicit geographic preference at the city/location level. We define the city level ontology beforehand which we named *geographic location ontology*. We predefined location ontology consisting of country name, province/state and city in *Location KB*.

Content extraction tool extracts all the content terms (c_i) from the top k search result web-snippets arising from each query Q. We adopt the following formula to find terms associated with a query content part.

$$F(c_i) = \frac{f(c_i)}{n} |c_i| \tag{1}$$

Where f is frequency of the term c_i (i.e. the number of documents containing c_i), n is the number of documents returned and $|c_i|$ is the number of term c_i. If the value of F of a term is higher than the threshold t (t=0.04 in our experiments) we treat c_i as a content term for the query Q.

Location extraction tool extracts all the geographic terms (g_i) from the full documents (due to sparse of them) leveraging *location KB*. The types of geographic terms include the followings: place names e.g. New York, Beijing, Shanghai; other locators e.g. postcode, ZIP code; terms descriptive of location e.g. state, country, city; adjectives of place e.g. American, Eastern; geographic features e.g. island, lake. Currently directions and spatial prepositions (e.g. "south", "near", "between" and "north of" etc.) are not considered in our approach.

Temporal extraction tool extracts all the temporal terms/expressions (t_i) such as "May 25, 2010", "Charismas Eve", "October 2010" from the full documents (due to sparse of them) leveraging *DBpedia*. Due to the complexity of temporal feature of document, currently our approach only focuses on the following time granularity, i.e., year, month, and date.

We adopt some existed extraction tools in our experiments section. Details of extraction tools are beyond the scope of this paper.

4.2 Query Profile

Query profile stores the query content intension information which includes content capacity, geographic location capacity and temporal capacity. It is obvious that the query content itself has different geographic and temporal requirements (demanding). For example the correlated geographic information with query "abroad travel" is larger than temporal information. However "newest movies introduction" is a both geographic and temporal-demanding query. But query "java programming" has higher content demanding than geographic and temporal demanding.

We introduce the concepts of *geographic Ratio* (denoted as R_g) and *temporal Ratio* (denoted as R_t) to estimate the ratios of geographic and temporal information related with content information of a query. We adopt *entropy* when calculating the value of R_l and R_t inspired from approach [5]. The information entropy of discrete random variable X is defined as formula (2), where n is possible values $\{x_1, x_2, ..., x_n\}$ of X and $p(x_i)=pr(X=x_i)$. R_g and R_t are defined as formula (3), (4).

$$entropy\ (X) =- \sum_{i=1}^{n} P(x_i) \log p(x_i) \tag{2}$$

$$R_g = entropy\ (g_i)\ /entropy\ (c_i) \tag{3}$$

$$R_t = entropy\ (t_i)\ /entropy\ (c_i) \tag{4}$$

Then we use decision tree to classify the query into five classes (denoted as $C1 \sim C5$) according to values of R_l and R_t, which is showed in Figure 3. Decision trees have the advantage that they can learn conjunctions of features. If the value of R_l is more than a threshold θ_1(defined in our experiment is 0.8) we deem it as high, otherwise it's deemed low. If the value of R_t is more than a threshold θ_2 (defined in our experiment is 0.5) we deem it as high, otherwise low. θ_2 is set smaller than θ_1 because we found in experiment that the number of temporal terms is usually less than geographic terms in the documents. Queries of $C1$ have association with location and temporal such as query "department discount". Queries of $C2$ have high association with geo location information but low temporal information such as query "study abroad". Queries of $C3$ have low association with location information but high temporal information such as query "stock price". Queries of $C4$ have relative low association with location information and temporal information. Queries of $C5$ have no obvious association with location and temporal information, which are not sensitive to temporal or geographic information such as query "health".

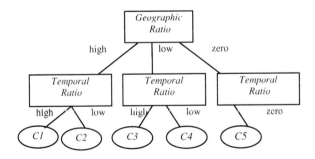

Fig. 3. Decision tree of query content part classification

4.3 User Click-through Analyzer

Click-through data is important in analyzing user preference and behavior on a search engine. It is obvious that different user has different content, geographic and temporal preferences associated with a query. The user's click actions reflect his/her diversity of interest in response to search results. For example , when user submits the query "lotus", the concept terms of it composes of "software", "car", "plant", "vodaka" and so on. If the user's preference is lotus car, then he/she will click the accordingly search results containing content terms "car".

We adopt *clicked entropies* (similar to formula (4)) to indicate user preferences which is defined in section 4.4. When the user clicks on a search reslut, the clicked result together with its assocaited content terms, geographic terms and temporal terms (temporal expressions) are collected by *user click-throug collector* of *Search results analyzer*. The content terms, geographic terms and temporal terms extracted from the user's clicked results indicate a user's possible concept, geographic and temporal preference spaces.

4.4 User Preference Profile

User preference profile stores the user's content preference (denoted as PRE_c), geographic location preference (denoted as PRE_l) and temporal preference (denoted as PRE_t) mined from user's click-through data which are defined as follows (formula(5)~(7)). When user clicked some search result, the content terms, geograhic terms and temporal terms in it are incremented in the content feature vector, geograhpic feature vector and temporal feature vector respectively.

$$PRE_c = - \sum_{i=1}^{n} p(C_i) log\ p(C_i) \tag{5}$$

$$PRE_l = - \sum_{i=1}^{m} p(l_i) log\ p(l_i) \tag{6}$$

$$PRE_t = - \sum_{i=1}^{k} p(t_i) log\ p(t_i) \tag{7}$$

where n is the number of content terms clicked by the user u, $C = \{c_1, c_2, ..., c_n\}$, $|c_i|$ is the number of times that the content term c_i has been clicked by user u, $|C|$ $=|c_1|+|c_2|+...+|c_n|$, $p(c_i) = \frac{|c_i|}{|c|}$, m is the number of location terms $L=\{l_1, l_2, ... l_m\}$ clicked by u, $|l_i|$ is the number of times that the location term l_i is being clicked by the user u, $|L| =|l_1| + |l_2| + ... + |l_k|$, and $p(l_i) = \frac{|l_i|}{|l|}$. k is the number of temporal terms $T=\{t_1, t_2, ... t_k\}$ clicked by u, $|t_i|$ is the number of times that the location term t_i is being clicked by the user u, $|T| =|t_1| + |t_2| + ... + |t_k|$, and $p(ti) = \frac{|ti|}{|t|}$.

If the vaule of *PRE* is high, the user is more likely to be interested in many content (geographic,temporal) terms; otherwise, the user is biased towards certain content (geographic,temporal) terms.

4.5 Re-ranking According to User Geographic and Temporal References

Documents are re-ranked according to rank score $S\ (Q,d)$ which is showed below. We use simple linear model to re-calculate rank score $S\ (Q,d)$ for each document in search results.

$$S(Q,d)= \alpha\ PRE_c\ S(Q_c,d) + \beta\ PRE_l\ S(Q_l,d)+ \lambda\ PRE_t\ S(Q_t,d) \tag{8}$$

$$S(Q_c,d) = \sum \omega_c \phi_c \qquad (9)$$

$$S(Q_l,d) = \sum \omega_l \phi_l \qquad (10)$$

$$S(Q_t,d) = \sum \omega_t \phi_t \qquad (11)$$

Where $S(Q,d)$ is ranking score of document d in response to query Q. $S(Q_c,d)$ is content part score, α PRE_c is the weight associated with $S(Q_c,d)$, $S(Q_l,d)$ is query location part score, β PRE_l is the weight associated with $S(Q_l,d)$, $S(Q_t,d)$ is temporal part score, λ PRE_t is the weight associated with $S(Q_t,d)$. The values of α, β and λ are set according to different query class in our experiment beforehand (see section 5.3). ϕ_c, ϕ_l and ϕ_t represent content feature vector, geographic feature vector and temporal feature vector respectively, and ω_c, ω_l and ω_t are their corresponding weights. Supervised learning is used to estimate the parameter ω_c, ω_l and ω_t to maximize the relevance after personalization.

5 Experiments

To evaluate our approach of personalization web query according to user's geographic and temporal preferences we conduct the experiments. The primary objective of our evaluation here is to show the advantage of our approach and also aim to show that the results personalized by our approach are indeed meaningful.

5.1 Experimental Setup

Because of person privacy problems of using some commercial search engine data, we developed a middleware which use one of the top commercial search engines as the backend search engine. The middleware includes user click-through analyzer and web results analyzer. We use search results data and click-through data of small size from our middleware for our evaluation. Our evaluation involves a set of 500 test queries (see Appendix) submitted by 25 users (each user submits 20 queries) from our lab. Table 2 shows the topical categories of the test queries we have chosen. The statistics of collected click-through data is showed in Table 3.

Table 2. Topical categories of the test queries

Category	Description	Category	Description
1	Movie	6	Computer science&Software
2	Famous people	7	Places
3	History event	8	Sports
4	Traveling	9	Shopping
5	Academic conference	10	Food&plant

Table 3. Statistics of our click-through data set

Number of users	Number of test queries	Number of clicks	Avg. clicks per query
25	500	1875	3.75

5.2 Experimental Data Preprocess

Similar queries Due to sparse of content terms of query we first merge them based on similarity. We considered query having the following similarities of query content parts as *similar query*.

A. The content parts of the two queries are parent-child relationship in the concept of *DBpedia*, For example "fruit" and "peach".
B. Two query content terms coexist a lot on the search results might represent the same topical interest, so if coexist $(c_i, c_j) > \delta$ (δ is a threshold), then c_i and c_j are considered as similar query.

5.3 Experimental Parameters Setting and Evaluation

In our experiment we set α, β and λ of the five query classes respectively, which is showed in Table 4. In each class $(C1 \sim C5)$ there has $\alpha + \beta + \lambda = 1$.

Table 4. The weights assigned to the three parts of the five query classes

Query class	C1	C2	C3	C4	C5
α	0.5	0.5	0.5	0.8	1
β	0.25	0.35	0.15	0.1	0
λ	0.25	0.15	0.35	0.1	0

In the evaluation of ranking quality of the personalization method, we use search results returned by the backend search engine as the baseline. Figure 4 shows comparisons between baseline and our approach on the top 10, 20, 30 and 50 precisions. The experiments results show that the precisions are improved compared to the baseline. Then we do experiments to consider the effectiveness of *query profile* to the personalization method. Namely we treat all the queries with the same content, geographic and temporal weights (α, β and λ are same). Figure 5 shows top 10, 20, 30 and 50 precisions for not considering *query profile* and our approach. We observe that the precisions are not good as our approach, namely considering both *query profile* and *user preference profile*.

Intuitively, a good ranking function or good query results should give high ranking to links that the users want. Thus, the smaller the average rank of the users' clicks, the better the ranking quality. So we measure ranking quality based on the average rank of

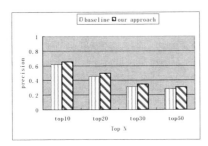
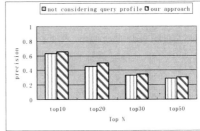

Fig. 4. Top 10, 20, 30 and 50 precisions for baseline and our approach

Fig. 5. Top 10, 20, 30 and 50 precisions for not considering query profile and our approach

Table 5. Comparison of avg. rank clicked on before and after personalized

Avg.c(Q)(before personalized)	Avg.c(Q)(after personalized)	Improvement
3.5	2.8	20%

users' clicks, denoted by $Avg.c\ (Q)$. Thus the smaller of $Avg.c\ (Q)$ indicates the better ranking quality. We evaluate the effectiveness of our approach by comparing $Avg.c\ (Q)$ before and after personalized process. The experiments show the average relevant rank improvement, which is showed is Table 5.

6   Conclusions

In this paper, we propose a novel approach for automatically extracting and learning user's geographic and temporal preferences associated with web query based on search results and user's click-through data. Experimental results confirmed that our approach can provide more personalized results associated with geographic location and temporal preferences comparing to the original results and improve the importance of temporal and geographic features related of documents in the ranking score. Currently our approach set the weights of query class manually and can not adjust them by the system automatically. In the future work we will find an effective learning algorithm to give the proper weights to each query class.

Acknowledgements. This paper is supported by the National Natural Science Foundation of China (No. 60973021, 61003059), the National High Technology Development 863 Program of China (No. 2009AA01Z131).

References

1. Eldar, S., Jayant, M., Lu, W., Alon, H.: Clustering query refinements by user intent. In: WWW (2010)
2. Markus, S., Mark, K., Christian, K.: Intentional query suggestion: making user goals more explicit during search. In: WSCD (2009)

3. Omar, A., Michael, G., Ricardo, B.-Y.: Clustering and exploring search results using time-line constructions. In: CIKM (2009)
4. Yi, X., Hema, R., Chris, L.: Discovering users' specific geo intention in web search. In: WWW (2009)
5. Leung, K.W.-T., Dik, L.L., Wang, C.L.: Personalized web search with location preferences. In: ICDE (2010)
6. Uichin, L., Zhenyu, L., Junghoo, C.: Automatic identification of user goals in web search. In: WWW (2005)
7. Hien, N.: Capturing user intent for information retrieval. In: The Proceedings of the 48th Annual meeting for the Human Factors and Ergonomics Society HFES (2004)
8. Steven, M.B., David, D.L.: Improving automatic query classification via semi-supervised learning. In: ICDM (2005)
9. Marcos, A.V., Jens, D., Lukas, B.: Adding structure to web search with iTrails. In: ICDEW (2008)
10. Li, X., Wang, Y.-Y., Alex, A.: Learning query intent from regularized click graphs. In: SIGIR (2008)
11. Thanh, T., Haofen, W., Peter, H.: Hermes: Data Web search on a pay-as-you-go integration infrastructure. J. Web Sem., 189–203 (2009)
12. Jian, H., Gang, W., Fred, L., Jian-Tao, S., Zheng, C.: Understanding user's query intent with wikipedia. In: WWW (2009)
13. Catizone, R., Dalli, A., Wilks, Y.: Evaluating automatically generated timelines from the web. In: 5th International Conference on Language Resources and Evaluation (2006)
14. Alonso, O., Yates, R.B., Gertz, M.: Effectiveness of temporal snippets. In: WWW (2009)
15. Irem, A., Srikanta, B., Klaus, B.: Time will tell: leveraging temporal expressions in IR. In: WSDM (2009)
16. Donald, M., Rosie, J., Fuchun, P., Ruiqiang, Z.: Improving search relevance for implicitly temporal queries. In: SIGIR (2009)
17. Rosie, J., Fernando, D.: Temporal profiles of queries. ACM Transactions on Information System 25(3), Article 14 (2007)
18. Yumao, L., Fuchun, P., Xing, W., Benoit, D.: Personalize web search results with user's location. In: SIGIR (2010)
19. Filip, R., Martin, S., Nick, C.: Inferring query intent from reformulations and clicks. In: WWW (2010)
20. Yoon, S., Jatowt, A., Tanaka, K.: Intent-Based Categorization of Search Results Using Questions from Web Q&A Corpus. In: Vossen, G., Long, D.D.E., Yu, J.X. (eds.) WISE 2009. LNCS, vol. 5802, pp. 145–158. Springer, Heidelberg (2009)
21. Rosie, J., Wei Vivian, Z., Benjamin, R., Pradhuman, J.: Geographic intention and modification in web search. International Journal of Geographical Information Science (2008)

Web User Profiling on Proxy Logs and Its Evaluation in Personalization

Hiroshi Fujimoto[1], Minoru Etoh[2], Akira Kinno[1], and Yoshikazu Akinaga[1]

[1] NTT DOCOMO R&D Center, 3-6, Hikarino-oka, Yokosuka-shi, Kanagawa, Japan
{fujimotoh,kinno,akinaga}@nttdocomo.co.jp
[2] Osaka University Cybermedia Center, 1-32 Machikaneyama,
Toyonaka, Osaka, Japan
etoh@ieee.org

Abstract. We propose a web user profiling and clustering framework based on LDA-based topic modeling with an analogy to document analysis in which documents and words represent users and their actions. The main technical challenge addressed here is how to symbolize web access actions, by words, that are monitored through a web proxy. We develop a hierarchical URL dictionary generated from Yahoo! Directory and a cross-hierarchical matching method that provides the function of automatic abstraction. We apply the proposed framework to 7500 students in Osaka University. The framework is used to analyze their 40GB click streams over a 4 month period. We evaluate clustering-based recommendation effectiveness to confirm the optimality of the framework. The results show high hit precision compared with existing methods.

Keywords: Web user clustering, Latent Dirichlet Allocation, topic modeling, Proxy logs based analysis.

1 Introduction

Web access user behavior analysis is, in general, the first crucial step in personalizing web applications such as advertizing, recommendation, and web search. A survey by Guandong Xu et al. [10] indicated that those applications include personalization and recommendation systems [12,18,7,3], web site modification or redesign [16] and business intelligence and e-commerce [1].

To realize the analysis needed, the application system monitors web access behavior at sites, which are categorized into clients, servers and proxies. Depending on the application, the monitoring site category and modeling of user web access may differ. This paper focuses on "topic modeling" which means that documents (i.e., users) are represented as mixtures of topics (i.e., abstracted user profile components), where a topic is a probability distribution over words (i.e., user web access actions). There have been comprehensive contributions regarding the topic modeling of user web access behavior as summarized in Table 1.

Most successful topic modeling techniques target domain-specific and application-oriented web analysis. By narrowing user actions to viewed contents,

X. Du et al. (Eds.): APWeb 2011, LNCS 6612, pp. 107–118, 2011.

Table 1. Topic Modeling of User Web Access

Monitoring Site	Web Action Modeling	Applications	Probabilistic Models	Contributions
Application Server	Viewed Contents within the Application	Content Recommendation, Site Optimization	pLSA	[2,20,9]
Search Engine	Search Words, Clicked URL	Target Advertisement	pLSA	[5,21,22]
Proxy / Tool bar	Viewed Page	Content Recommendation, Site Optimization	LDA	[11]
SNS server	Words in tweets	Collaborative Filtering	LDA	[13]

it offers excellent performance for recommendation and targeted advertisement [2,21,9,5,22,7]. The extracted topics, in other words, abstracted user intentions, enable the system to infer the user's next action. Please note that they used SVD (singular value decomposition) [8], LSI (Latent Semantic Indexing) [17] or pLSA (probabilistic Latent Semantic Analysis) [19] as the probabilistic models, since their contributions appeared in the early 2000's. As an update, LDA (Latent Dirichlet Allocation) [6] or more sophisticated models could be used instead.

The motivation for this paper lies in the authors' belief that combining proxy data with a better topic and action model will yield deeper user analysis, whose results are not domain-specific nor application-oriented, but rather broadened to represent social group descriptions. The research scope of this paper seems to be similar to [11], which compared LDA to pLSA for probabilistic modeling, and associated user sessions with multiple topics to describe the user sessions in terms of viewed web pages. This paper, however, focuses on the association between words (i.e., user web accesses) and the observed click streams rather than probabilistic modeling. We also use an LDA model for topic modeling though, simply taking viewed pages as words doesn't work, since a click stream contains many meaningless pages. Given a lot of proxy data, the key issue is how to select the appropriate words so as to symbolize sessions.

Our original contributions consist of

1. a word association scheme: we call it the "cross-hierarchical directory matching method". It extracts multiple words from each user session by matching against a directory database. We use Yahoo! Directory for cross-hierarchical directory matching since it resolves the ambiguity caused by multiple matches in the same domain by choosing most the abstract URL (i.e., the uppermost URL in the directory tree). Its benefits are shown in the following section.
2. an empirical study at Osaka University of proxy log analysis. The log contains 40GB click streams of 7500 students collected over several months. The study indicate optimality of the proposed framework in context of clustering-based recommendation.

1.1 LDA Formulation

We assume topic modeling where the user accesses Web pages under certain topics (i.e., abstracted user intentions or tasks). For example, the user accesses a certain SNS site under his latent topic "SNS-addict", or accesses a certain job site under her latent topic "Job Hunting". In this case, by applying the concepts of LDA, a Web user should correspond to a document, accessed web contents correspond to words, and their latent topics correspond to topics of documents. The observed accesses of each user are input to the LDA model, which then outputs the association between users and topics.

1.2 Cross-Hierarchical Directory Matching

Given a session with time duration t, the task is to label the session with multiple words. In the text mining domain, dictionaries and abstraction are being used with promising results [14,24]. A dictionary should cover a broad set of comprehensive concepts and words.

We use Yahoo! Directory [25] for the dictionary as it has a simple ontology structure, a category hierarchy containing paths of abstraction. For example, we extract a specific site 'The New York Times' from a session from web accesses it contains.

We may have more specific sites such as 'China - The New York Times' as a sub-category of Newspapers. In this example, we abstract those specific subcategories to the uppermost sites that appear in the session. The results mirror breadth-first search (BFS) with multiple outputs. **Our underlying assumption is that the most abstract URL that appears in the session best represents the user's intention.** Those abstracted URLs are identified from bookmarks and the landing URLs of search results. Thus, along with the directory structure, we can apply automatically adjusted abstraction to the found URLs. We call our proposal 'Cross-Hierarchical Directory matching.

2 LDA-Based Topic Modeling

User topic modeling is the action of identifying topics that web users are interested in based on their web actions. To realize it, we employ the LDA model, which was originally proposed as a probabilistic document-topic model in the document categorization domain. LDA assumes a "bag of words", i.e. each document is thought of as a vector of word counts. Each document is represented as a probability distribution of some topics, while each topic is represented as a probability distribution over a number of words. More formally, by assuming the Dirichlet distribution for per-topic word multinomial as shown in [20], the document-topic distribution $p(z|d)$ is denoted as θ and the topic-word distribution $p(v|z)$ is denoted as ϕ, where z represents a topic, d represents a document, v represents a word, α and β represent hyper-parameters , N_d is the total number of words in document d under the graphical model shown in Figure 1.

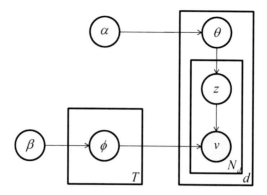

Fig. 1. Graphical model represenation of LDA

Table 2. Notations for LDA model

D	A set of documents : $D=\{d_m\|1 \leq m \leq \|D\|\}$ where d_m represents each user.
V	A set of words : $V = \{v_w\|1 \leq w \leq \|V\|\}$ where v_w represents each accessed URLs.
Z	A set of topics : $Z = \{z_k\|1 \leq k \leq \|Z\|\}$ where z_k represents each topic.
$\nu(d_m, v_w)$	An element of matrix N.
$\theta(d_m, z_k)$	An element of matrix Θ.
$\phi(z_k, v_w)$	An element of matrix Φ.

We assume topic modeling where the user accesses Web pages under certain topics (i.e., abstracted user intentions or tasks). For example, the user accesses a certain SNS site under his latent topic "SNS-addict", or accesses a certain job site under her latent topic "Job Hunting". In this case, by applying concepts of LDA, a Web user should correspond to a document, accessed web pages correspond to words, and their latent topics correspond to topics of documents. The observed accesses of each user are input to the LDA model, which then outputs the association between users and topics. In detail, under the notation shown in Table 2, the input and the outputs are as follows:

Inputs: matrix N where each line denotes the counts of words each user accessed.

Output1: matrix Θ where each line denotes the topic distribution of each user.

Output2: matrix Φ where each line denotes the word distributions of each topic.

The goal of topic modeling is to derive the optimal outputs Θ and Φ, where the topics of each user are represented by Θ and each topic is represented by Φ. To

Table 3. Notations for Proxy log for user u_m

$S_i^{(m)}$	The $i - th$ user session : $S_i^{(m)} = \{s_{ij}^{(m)}|1 \leq j \leq |S_i^{(m)}|\}$ where $s_{ij}^{(m)}$ represents $j - th$ record of the session recorded at $t_{ij}^{(m)}$.
$L_i^{(m)}$	A set of URLs accessed in session $S_i^{(m)}$: $L_i^{(m)} = \{l_{ij}^{(m)}|1 \leq j \leq |S_i^{(m)}|\}$ where $l_{ij}^{(m)}$ represents accessed URL of $s_{ij}^{(m)}$.
$V_i^{(m)}$	A word set labeled to session $S_i^{(m)}$: $V_i^{(m)} = \{v_w|v_w \in V \cap v_w$ is labeled to session $S_i^{(m)}\}$

realize this, optimal input N is needed. The simplest approach, which takes all the accessed URLs as words (i.e. the approach of [11]) doesn't work, since many URLs are not related to the users' intention. Moreover, it is said in the text mining domain that word sets should be abstracted by dictionaries if a proper model is desired.

3   Symbolizing URLs from Proxy Log

Our goal is to model user-topic association. This can be realized by deriving the optimal input matrix N for the LDA model by using dictionaries in the abstraction of the original web accesses. In this section, we will show an approach based on the use of proxy logs.

3.1   Description of Proxy Log

To reach our goals, we require that the proxy log for each user d_m satisfies the following conditions; each record has, at least, access time and accessed URL. The records are sorted in chronological order. A user session is also defined as a series of continuous records for each user. Each session has a time out interval δ, so the session ends when the user does not access any web page in interval δ. Formally, under the notation shown in Table 3, for each user d_m, each session $S_i^{(m)}$ consists of a series of records and each record $s_{ij}^{(m)}$ consists of access time $t_{ij}^{(m)}$ and accessed URL $l_{ij}^{(m)}$.

3.2   Basic Idea of Labeling Words to User Session

We define word set $V_i^{(m)}$ is the abstraction of URLs from $L_i^{(m)}$, a series of URLs in session $S_i^{(m)}$. For example, when user d_m accesses a certain SNS community site, the abstracted URL is v_w, so word v_w is assigned as the session label. Details of the abstraction process are explained in the next subsection.

An example of the relationships between sessions and words is shown in Figure 2. Each session is labeled by one or more words. For example in Session1, both URLs are abstracted to v_1 and v_1 is assigned to the session. Note

UID	Time	URL	Abstracted URL	Assigned Word Set
u1	t1	http://x.y.z/a.html	v1	v1
u1	t2	http://x1.y.z/a.html	v1	
u1	t3	http://x4.y.z/	v1	v1,v2
u1	t4	http://x4.y.z/b.gif	v1	
u1	t5	http://x.y.z/w5/c.html	v2	
u1	t6	http://x.y.z/a.html	v3	v3,v4,v5
u1	t7	http://x1.y.z/a.html	v4	
u1	t8	http://x4.y.z/	v5	

Session 1

$t_3 - t_2 > \delta$

Session 2

$t_6 - t_5 > \delta$

Session 3

Fig. 2. The relationships between sessions and actions

that we allow multiple words to be labeled to a single session shown by Session 2 and 3 in the figure.

After all sessions of all users are labeled, a set of words V can be derived as the union of words in all sessions of all users, while the number of words accessed by each user $\nu(u_m, c_w)$ can be derived as the number of sessions labeled v_w for each user u_m. This is formally represented as follows:

$$V = \cup_{m=1}^{M} \cup_{i=1}^{|S^{(m)}|} V_i^{(m)}, N(d_m, v_w) = |\{S_i^{(m)} | \exists i : v_w \in V_i^{(m)}\}| \qquad (1)$$

3.3   Cross-Hierarchical Directory Matching

Cross-Hierarchical Directory matching (CHDM) is a method that uses a hierarchical dictionary to get a set of abstracted URLs that are broader in concept than the originally accessed URLs. Dictionary $|C|$ should have an ontology structure, a category hierarchy that supports path abstraction. Categories c_h are numbered $\{1, 2, ..., |C|\}$ in order of breadth-first search, so h is smaller than h' if c_h is an ancestor and broader in concept than $c_{h'}$. For example, if c_h is "newspaper" and $c_{h'}$ is "local newspaper", $c_{h'}$ is subordinate to c_h.

Moreover one or more URLs of Web site $l_n^{(c_h)}$ can be registered to each category c_h. (To distinguish these URLs from proxy log entries, we call the former SURL.) If two SURLs are registered to two different categories and one category is subordinate to the other, the two sites have the same relationship with regard to conceptual hierarchy. For example, 'The New York Times' registered to c_h is broader in concept than 'China - The New York Times' which is registered to $c_{h'}$.

A basic idea of CHDM is to get a set of abstracted URLs by getting the hierarchical relationships of all URLs and abstracting URLs of subordinate concepts. To know the hierarchy of URLs, we get a set of SURLs that the URLs belong to (matching step). Since we know their hierarchical relationship, we can abstract all the SURLs of the subordinate concepts and so create set of abstracted SURLS (abstraction step). This is the word set assigned to the session.

The matching process searches for the associated category in the dictionary for each URL by string matching against the corresponding SURL, while the

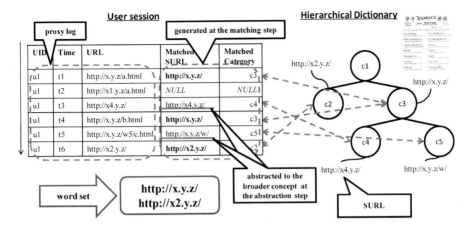

Fig. 3. Example of Cross-Hierarchical Directory matching

abstraction step discards URLs of subordinate concepts in the session by comparing their associated categories. More details are shown in the Appendix.

Simple examples of both steps are shown in Figure 3. The figure places a user session on the left, and the hierarchical dictionary on the right. 6 URLs are accessed in the session, and there are 5 categories (c1-c5) and 4 SURLs are found in the dictionary.

At the matching step, URLs accessed at t1, t2, t3, t4, and t5 belong to the respective SURLs in the dictionary as shown in the column 'Matched SURL'. Corresponding categories of the matched SURLs are also obtained straightforwardly in the column 'Matched Category'. This yields pairs (c2, 'http://x2.y.z/'), (c3, 'http://x.z.y/'), (c4, http://x4.y.z/), and (c5, 'http://x.y.z/w/') which are assigned to the session.

In the abstraction step, both 'http://x4.y.z/' and 'http://x.y.z/w/' are abstracted to 'http://x.y.z/' since corresponding categories (c4 and c5) are subordinate concepts of c3.

As a result, the set of remaining SURLS, i.e. ('http://x.y.z/', and 'http://x2.y.z') is the abstracted set of accessed web URLs in the session, and so is assigned as the word set.

4   Experiments and Results

In this section, we show results of an experiment on a real proxy log. The result shows the optimality of CHDM in context of clustering-based recommendation.

4.1   Data Sets and Evaluation Settings

We captured a set of proxy log accesses from over 7500 students in Osaka University; data occupied 40 GB. The log covered the four month period from April to July 2010. We divided the records into sessions for each user where session timeout δ was set to 1800 [sec]. This yielded 175831 sessions for 7537 users. We prepared a dictionary by crawling Yahoo! JAPAN Directory [25] in July 2010. This yielded a hierarchical dictionary with about 570 thousand distinct SURLs.

We matched the log entries against the dictionary in the manner of CHDM. This yielded, as the first result, over 20 thousand distinct words including many very minor words. We eliminated minor words (those with fewer than 4 users) to obtain 1150 test words.

We run LDA following [23] which describes the parallel implementation of LDA and uses Gibbs sampling as the inference algorithm. We set hyper-parameters α and beta to $|Z|/50$ and 0.01 respectively as recommended by the authors.

4.2   Evaluation Metrics

To show the optimality of CHDM, we evaluated clustering-based recommendation effectiveness with the following settings. We first assigned each user to a single topic by the LDA model using the data collected in the first 3 months. The topic was determined by choosing the topic with maximum probability on matrix Θ. Note that we must determine the number of topics before running LDA, so we evaluated various numbers of topics and empirically determined that 24 yielded the best model. Details are shown in the next subsection.

After clustering, we prepared a proxy log of the data gathered over the last 1 month for evaluation. We choose 1000 users randomly as the test-set, and predicted the web pages they would access in each session by generating a recommendation set from the accesses of the remaining 6537 users (learning-set users). For each session, assigning a test-set user to topic z_k, the recommendation set was generated from the top-N accessed pages by the learning-set users assigned to the same topic.

Moreover we introduce the notion of serendipity [4] such that recommendation of minor or surprising content is worth more than that of major or general content. We thought that web page accesses should be weighted by popularity. Accordingly, we used IDF (Inverse Document Frequency) as the weight since it is widely used in the text categorization domain for filtering general words. Formally, the weight of web page l is defined as follows:

$$weight(l) = \log \frac{total \quad number \quad of \quad test \quad sessions}{number \quad of \quad sessions \quad in \quad which \quad l \quad was \quad accessed} \quad (2)$$

The weight is large if the web page is accessed in few sessions. Please note that, although we heuristically selected IDF, there may be more appropriate criteria for weighting minor content. Under the settings mentioned above, we derived the hit precision of the recommendation for each session. Formally the hit precision is:

$$hit \quad precision(S_i) = \frac{\Sigma_j h_{ij} weight(l_{ij})}{\Sigma_j weight(l_{ij})} \quad (3)$$

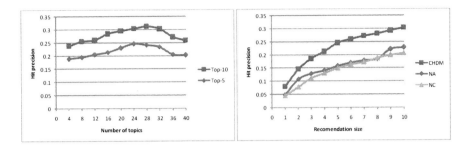

Fig. 4. Optimality Analysis of the model

where S_i is a test session, l_{ij} is the $j - th$ page accessed in session S_i, and if l_{ij} is in the recommendation set h_{ij} is 1 else 0. The hit precision is the biased hit-rate of the recommendation such that recommendation of minor content yields higher value than major content. The hit precision is increased by increasing recommendation size, i.e. the top-10 recommendations yield higher hit precision than the top-5. Recommendation size is, however, generally restricted by recommendation system, application, user experience and so on.

4.3   Evaluation Results

We first evaluate the number of topics in terms of optimizing the LDA model. The results are shown in Figure 4 (left). The figure represents changes in average hit precision for all test sessions with changes in the number of topics where the recommendation set is top-5 or top-10. The results show that 24 topics is a good choice and yields a better LDA model than the other values.

Next we show an optimality analysis of CHDM with other two evaluation sets. The first is non-clustering (NC) in which all test-set users are assigned to a single cluster. In this case, the recommendation set is simply chosen from all the learning-set users. The second is non-abstraction (NA) in which the word set of the LDA model is a simply URLs accesses in the session, i.e. $L_i^{(m)}$. This is the same approach of [11].

The results are shown in Figure 4 (right). The figure plots changes in average hit precision for all test session versus recommendation size from top-1 to top-10. CHDM yields higher hit precision than the others. In particular, the hit precision of CHDM is 1.5 times larger than that of NC, so our abstraction approach is quite effective even when the same data set is used.

Please note that our approach is based on the heuristic assumption that the most abstract URLs appearing in the session represent the user intension. Although the assumption supported by the empirical evaluation provided by our data sets, there is no assurance that the same results would be derived from other data.

5 Conclusion

Clustering Web users by their interests is a key technique for many Web applications such as recommendation, site optimization, and collaborative filtering. In this paper, we proposed a user clustering method that uses the LDA model to assess Web access patterns and their latent topics. To derive an optimal model, our method employs a hierarchical URL dictionary to abstract Web accesses into broader concept words. Experiments on real proxy log data showed the optimality of our method in context of clustering-based recommendation. In future, we intend to apply our model to a Web recommendation system based on collaborative filtering and evaluate its effectiveness in real applications.

References

1. Buchner, A.G., Mulvenna, M.D.: Discovering Internet Marketing Intelligence through Online Analytical Web Usage Mining. The ACM SIGMOD Record 27(4), 54–61 (1998)
2. Das, A.S., Datar, M., Garg, A., Rajaram, S.: Google News Personalization: Scalable Online@Collaborative Filtering. In: Proc. of The 16th International Conference on World Wide Web, Alberta, Canada (2007)
3. Mobasher, B., Cooley, R., Srivastava, J.: Creating Adaptive Web Sites through Usage-based Clustering of URLs. In: Proc. of the 1999 Workshop on Knowledge and Data Engineering Exchange (1999)
4. Mobasher, B., Dai, H., Luo, T., Nakagawa, M.: Discovery and Evaluation of Aggregate Usage Profiles for Web Personalization. Data Mining and Knowledge Discovery 6(1), 61–82
5. Lin, C., Xue, G.R., Zeng, H.J., Yu, Y.: Using Probabilistic Latent Semantic Analysis for Personalized Web Search. In: Zhang, Y., Tanaka, K., Yu, J.X., Wang, S., Li, M. (eds.) APWeb 2005. LNCS, vol. 3399, pp. 707–717. Springer, Heidelberg (2005)
6. Blei, D.M., Ng, A.Y., Jordan, M.I.: Latent dirichlet allocation. The Journal of Machine Learning Research archive 3, 993–1022 (2003)
7. Weng Ngu, D.S., Wu, X.: Sitehelper: A Localized Agent That Helps Incremental Exploration of the World Wide Web. In: Proc. of the 6th International World Wide Web Conference, Santa Clara (1997)
8. Golub, G.H., Reinsch, C.: Singular value decomposition and least squares solutions. Numerische Mathematik 14(5), 403–420
9. Xu, G., Zhang, Y., Zhou, X.: Using probabilistic latent semantic analysis for Web page grouping. In: Proc. of the Research Issues in Data Engineering: Stream Data Mining and Applications (2005)
10. Xu, G., Zhang, Y., Zhou, X.: A web recommendation technique based on probabilistic latent semantic analysis. In: Ngu, A.H.H., Kitsuregawa, M., Neuhold, E.J., Chung, J.-Y., Sheng, Q.Z. (eds.) WISE 2005. LNCS, vol. 3806, pp. 15–28. Springer, Heidelberg (2005)
11. Xu, G., Zhang, Y., Yi, X.: Modeling User Behavior for Web Recommendation Using LDA Model. In: Proc. of the 2008 IEEE/WIC/ACM International Conference on Web Intelligence and Intelligent Agent Technol, Melbourne (2008)
12. Lieberman, H.: An Agent That Assists Web Browsing. In: Proc. of the 13th International Joint Conference on Artificial Intelligence, Montreal, Canada (1995)

13. Weng, J., Lim, E.P., Jiang, J., He, Q.: TwitterRank: finding topic-sensitive influential twitterers. In: Proc. of the 3rd ACM International Conference on Web Search and Data Mining (2010)
14. Bessho, K.: Text Segmentation Using Word Conceptual Vectors. The Transactions of Information Processing Society of Japan 42(11), 2650–2662 (2001)
15. Tipping, M.E.: Sparse bayesian learning and the relevance vector machine. The Journal of Machine Learning Research archive 1, 211–244 (2001)
16. Perkowitz, M., Etzioni, O.: Adaptive Web Sites: Automatically Synthesizing Web Pages. In: Proc. of the 15th National Conference on Artificial Intelligence, Madison (1998)
17. Deerwester, S., Dumais, S.T., Furnas, G.W., Landauer, T.K., Harshman, R.: Indexing by latent semantic analysis. The Journal of the American Society for Information Science 41(6), 391–407 (1990)
18. Joachims, T., Freitag, D., Mitchell, T.: Webwatcher: A Tour Guide or the World Wide Web. In: Proc. of the 15th International Joint Conference on Artificial Intelligence, Nagoya, Japan (1995)
19. Hofmann, T.: Probabilistic Latent Semantic Analysis. In: Proc. of the 22nd Annual ACM Conference on Research and Development in Information Retrieval, California (1999)
20. Griffiths, T.L., Steyvers, M.: Finding scientific topics. Proc. of the National Academy of Sciences of the United States of America 101, 5228–5235 (2004)
21. Jin, X., Zhou, Y., Mobasher, B.: Web usage mining based on probabilistic latent semantic analysis. In: Proc. of the 10th ACM SIGKDD International Conference on Knowledge Discovery and Data Mining, Seattle (2004)
22. Wu, X., Yan, J., Liu, N., Yan, S., Chen, Y., et al.: Probabilistic latent semantic user segmentation for behavioral targeted advertising. In: The Proc of the 3rd International Workshop on Data Mining and Audience Intelligence for Advertising, Paris (2009)
23. Wang, Y., Bai, H., Stanton, M., Chen, W., Chang, E.Y.: PLDA: Parallel Latent Dirichlet Allocation for Large-Scale Applications. In: Goldberg, A.V., Zhou, Y. (eds.) AAIM 2009. LNCS, vol. 5564, pp. 301–314. Springer, Heidelberg (2009)
24. Elberrichi, Z., Rahmoun, A., Bentaalah, M.A.: Using WordNet for Text Categorization. The International Arab Journal of Information Technology 5(1) (January 2008)
25. Yahoo! Directory, http://dir.yahoo.com/, http://dir.yahoo.com/

A   Appendix

Algorithm of the both steps is as follows under the notation given in Table 4:

Table 4. Notations for Cross-Hierarchical Directory matching

C	A hierarchical dictionary: $\{(c_h, L^{c_h})\|1 \leq h \leq |C|\}$ where c_h represents a category.
$L^{(c_h)}$	A set of SURLs: $\{l_n^{(c_h)}\|1 \leq n \leq |L^{c_h}|\}$ where $l_n^{(c_h)}$ represents a SURL registered to a category c_h.
$P_i^{(m)}$	A pair of a set of categories and a set of SURLs: $\{(c_{ij}^{(m)}, l_{ij}'^{(m)})\|c_{ij}^{(m)} \in C_i^{(m)}, l_{ij}'^{(m)} \in L_i'^{(m)}, 1 \leq \hat{j} \leq |P_i^{(m)}|\}$ where $C_i^{(m)}$ is a set of categories and $L_i'^{(m)}$ is a set of SURLs assigned to the session $S_i^{(m)}$.

Process 1: the matching step

1 $P_i^{(m)} := \phi$

*2 **for** j **from** 1 to $S_i^{(m)}$*

3   ***for** h **from** $|C|$ to 1*

4   ***for** n **from** 1 to $|L^{(c_h)}|$*

5   ***if** $l_n^{c_h}$ is substring of $l_{ij}^{(m)}$ **then***

6 $P_i^{(m)} := P_i^{(m)} \cup (c_h, l_n^{(c_h)})$

*7 **return** $P_i^{(m)}$*

Process 2: the abstraction step

1 $V_i^{(m)} := L_i'^{(m)}$

*2 **for** $\hat{j}$ **from** 1 to $|P_i^{(m)}| - 1$*

3   ***for** $\check{j}$ **from** $\hat{j} + 1$ to $|P_i^{(m)}|$*

4   ***if** $c_{ij}^{(m)}$ is found by BFS (breadth-first search)*

5   *from the node $c_{i\hat{j}}^{(m)}$ to the bottom of the tree*

6   ***then** $V_i^{(m)} := V_i^{(m)} - l_{ij}^{(m)}$*

*7 **return** $V_i^{(m)}$*

Exploring Folksonomy and Cooking Procedures to Boost Cooking Recipe Recommendation

Lijuan Yu, Qing Li, Haoran Xie, and Yi Cai

Department of Computer Science, City University of Hong Kong,
83 Tat Chee Avenue, Kowloon, Hong Kong SAR, China
{lijuanyu2,hrxie2}@student.cityu.edu.hk,
{itqli,yicai3}@cityu.edu.hk

Abstract. Recommender systems have gained great popularity in Internet applications in recent years, due to that they facilitate users greatly in information retrieval despite the explosive data growth. Similar to other popular domains such as the movie-, music-, and book- recommendations, cooking recipe selection is also a daily activity in which user experiences can be greatly improved by adopting appropriate recommendation strategies. Based on content-based and collaborative filtering approaches, we present in this paper a comprehensive recipe recommendation framework encompassing the modeling of the recipe cooking procedures and adoption of folksonomy to boost the recommendations. Empirical studies are conducted on a real data set to show that our method outperforms baselines in the recipe domain.

1 Introduction

The fast development of information techniques has resulted in an explosive growth of various resources on the Internet. Nevertheless, what the overwhelming information choices bring to people is not greater convenience, but more complexity in getting the desired ones efficiently. Information filtering techniques have become increasingly popular in the last two decades due to this reason. As an indispensable component of information filtering, recommender systems have become prevalent since mid-1990s, aiming to make personalized information recommendations through learning users' preference patterns from the historical interactions with the systems. So far, the recommendation techniques have got success mainly in the domains of entertainment resources sharing, such as movies, music, images, etc. Similar to these applications, choosing/deciding the 'right' cooking recipes for meals is also a daily activity in which user experiences can be greatly improved by adopting recommendation algorithms, so as to achieve better life qualities of human kinds. Different from other domains, however, recipe domain has its own characteristics. For example, the description of a recipe is often not a set of features but a procedure which introduces the cooking flow of a recipe. Even two recipes share similar `features', e.g., ingredients and cooking actions, but if the sequences of cooking actions on their ingredients are different, and then the tastes, smells and appearances of two recipes would be quite different.

X. Du et al. (Eds.): APWeb 2011, LNCS 6612, pp. 119–130, 2011.
© Springer-Verlag Berlin Heidelberg 2011

Although there have been some pieces of work done in the cooking and food recommendation topics (as reviewed in section 2), few of these have considered the characteristics of recipe data itself. This could be due to the difficulties in representing the complex recipe features, such as cooking process, taste and visual features, etc., in a computer-readable way. Traditional recommendation works can be categorized in the following categories, viz., content-based (CB), collaborative filtering (CF) and the hybrid of the first two. However, due to the differences between recipe domain and other domains, the traditional methods have limitations and become inadequate in making recommendation of recipes to users. In particular, the content-based recommendation methods do not provide a way to model and describe cooking procedures of recipes (but only a flat feature vector). For collaborative filtering methods, they can apply the social or collaborative techniques to recommend the 'most popular' recipes to users by only taking resource ratings of similar users into consideration, and ignore the content analysis of recipes. Such CF methods fail to measure user similarity from the perspective of individual users' own preferences over the recipe characteristics.

In our viewpoint, an ideal recipe recommender system should be aware of users' specific interests from a more comprehensive perspective, like the cooking skills, taste features, etc. Recently, folksonomy (also named collaborative tagging) provides a way to let user annotate resources, and users can express their subjective feeling, perception and perspectives on resources. To achieve effective recommendation in such a specific domain, we present in this paper a data model to represent recipes by exploiting cooking procedures and combining the advantages of folksonomy. Based on the recipe data model, we then propose a recipe recommendation framework. The characteristics of our proposed framework and the contributions of our work are as follows.

1. A cooking recipe model is devised from the perspectives of cooking procedures and users' social tags, representing cooking oriented and user perceptual features;
2. We design a folksonomy-driven, hybrid recipe recommendation strategy, by taking both content features and social factors into consideration. Such a strategy leverages the folksonomy mechanism and cooking procedures to discover the taste-similar users from their common favorites and tagging histories;
3. Experiments are conducted on a real data set to evaluate our method. The results show that our method outperforms baseline methods on recipe recommendation.

The organization of this paper is as follows. Section 2 provides a review of some related works. Section 3 introduces our recipe data model. In section 4, we present our recommendation strategy in detail. Empirical studies and experiment results are given in section 5. We conclude the paper in section 6.

2   Related Work

In this section, we review some existing works related to the cooking and food topics, as well as some recommendation and folksonomy techniques relevant to our work.

Cooking & Food Related Research. Research on the cooking-related topics by computer science researchers has not started for long, and there are only a small number of works published. One of the early efforts is a social recipe navigation system designed by Svensson et al. [12], where users can browse recipes in a social context that supports viewing other users' actions and choices, chatting with them, getting collaborative-filtering based recommendations, and giving feedbacks. Wang et al. [13] proposed a graphical representation of the cooking procedure of recipes, named as cooking graph, where the cooking materials, actions are modeled as nodes, connected by directed edges indicating the ingredient and cooking flows; a sub-structure similarity metric between the cooking-graphs was also brought up in that work. Sobecki et al. [7] presented a hybrid recipe recommendation strategy based on demographic, content-based, and collaborative filtering methods, as well as the application of fuzzy inference. Another recent work was done by Freyne et al. [5] compared the recommendation strategies from both food and recipe levels, using the content-based, collaborative filtering and hybrid approaches. Other than applying the general social-based recommendation strategies to the cooking and food domain, none of the above works have provided an explicit and comprehensive content model for the recipe data.. One of the major objectives of our work is to overcome this problem, so as to provide an efficient content-based approach.

Recommender Systems. The recommender systems can be generally classified into four categories based on their different recommendation strategies: content-based, knowledge-based, collaborative filtering and hybrid ones [1].

The *content-based recommendations* concentrate on learning user preference patterns from analyzing the features of his/her selected items, and recommend items with similar content to the user's favorite ones. Pazzani et al. did a survey to content-based recommender systems in [8], and introduced methods of item representations, user profiling methods, and the comparison metrics of user-item profile relevance. *Collaborative recommendations* focus on identifying neighborhoods with similar preferences of a particular user, and recommend to him/her the items with the highest ratings among the neighbors. Herlocker et al. evaluated research on collaborative recommender systems comprehensively in [6] from aspects like user tasks, data sets, accuracy metrics, etc. The collaborative filtering techniques have got great popularity and success in systems like *MovieLens* [1] . There are also *knowledge-based recommendations*, which make advices based on predicting how a particular item would match a user's needs/requirements [2]. Since each recommendation technique has both strengths and limitations, there are also systems trying to use *hybrid approaches* to make better recommendations. The combination types and example systems were introduced in [2].

Folksonomy. Folksonomy, also named collaborative tagging, has gained great popularity in the web2.0 services community since 2004 or so, for its attraction in allowing user to annotate resources with personalized tags. In the multimedia resource sharing systems like *Del.icio.us*[2], *Flickr*[3], and *Last.fm*[4], tagging has become a popular tool to share users' comments on their interested resources.

---

[1] http://www.movielens.org
[2] http://delicious.com
[3] http://www.flickr.com
[4] http://www.last.fm

In the research field, folksonomy work mainly focuses on investigating and analyzing the features of user-generated free tags, and attempting to use them in applications such as personalized search, item- and topic- recommendations, etc. An overview of the tag usage patterns, and user activities in collaborative tagging systems has been done in [9]. Tags can be used to construct user profiles to facilitate personalized search in [17] [18]. In [4], three approaches of this as well as their comparisons were presented by Michlmayr et al. In [11], the authors proposed a hierarchical clustering method to get the highly relevant tag clusters, and use them to make personalized recommendations. In [14], Zhen et al. suggested to utilize the tags in the collaborative filtering procedure, so as to improve the recommendation accuracy by calculating user similarity based on their tag-based profiles. To measure the semantic similarity between tag-based profiles, multiple metrics were surveyed in [3], and empirical experiments were also conducted to evaluate these metrics.

3   Cooking Recipe Data Modeling

In this section, we firstly analyze the important features of recipes that we aim to model, and then introduce a cooking recipe data model which represents the recipe features into cooking oriented and user perspective features. As exemplified by Figure 1, a typical cooking recipe usually appears as a (multimedia) document describing the cooking time, the food ingredients and the cooking directions [13]. In this section, we categorize such information into two aspects, namely, the cooking oriented features, and user perceptual features, to capture and reflect cooking procedures and user's subjective perception.

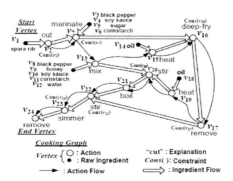

Fig. 1. Cooking graph of 'Chinese Braised Spare Ribs'

3.1   Cooking Oriented Features

Different from other domains, in the recipe domain, there are procedures telling users how to cook dishes out. A typical cooking process may include a preparation step to get the cooking materials in required condition, and a boiling step to heat the ingredients following a particular action sequence, until achieving the desired final dish. This process involves elements like food ingredients, seasonings, cooking actions, cooking constraints namely cooking temperature, cooking time, etc. We name these most basic but important information of food recipes as *cooking oriented features (COFs)*. Besides, the relationships between these elements are complicated, for example, 'cut' the 'spareribs' 'into pieces', 'cook' and 'stir' the 'spareribs' for '3 to 5 minutes', 'cook' and 'stir' the 'vegetable' 'until fragrant', etc. There is also temporal information indicating when to perform which steps. Such complexities determine the difficulty in modeling the cooking oriented features.

To overcome the inconvenience of viewing cooking oriented features by reading the long textual directions and make it computer readable, we adopt here the cooking graph model previously proposed by our group [13], for representing the cooking flows of various recipes. More specifically, the cooking flow of a recipe is formulated by a graph composed of four elements:

$$R^{CG} = (V, E, Cons, Ingr) \tag{1}$$

where V is the set of ingredient or action nodes, E is the set of edges between two nodes, *Cons* is the set of constraints associated with either action nodes or edges, and *Ingr* is the set of ingredients needed to cook the dish.

Figure 1 shows an example of graphical presentation of the recipe `Chinese Braised Spare Ribs'. From Figure 1 we can see that the cooking process of the recipe is divided into 24 actions. Each action is associated with some ingredient nodes and cooking constraints, and connected by the so-called 'Action Flow' edges. The 'Ingredient Flow' indicates the state changes of ingredients by connecting the relevant ingredient- and action- nodes. One benefit of this graphical representation is that users can view the cooking process in an easy and intuitive way, which has not been considered by other previous works.

3.2 User Perceptual Features

In the recipe domain, there are some features mainly involving the taste, smell, texture (whether the food is hard or tender), temperature, and all the other properties that contribute to people's perceptions of foods. We name these as *user perspective features (UPFs)*. For example, the UPFs of cheese cakes are usually described as 'sweet', 'smooth', 'soft', 'creamy', etc. The taste of the final dish could be represented by properties that users care most about a recipe. Such a kind of information is subjective perception of users and often not shown in traditional recipe cooking procedure description.

Depicting the UPFs of foods is not an easy task, due to people's subjective perceptions of them. This can be easily understood if we recall some common scenarios in our daily life. Taking the Japanese Sushi as an example, people who love fresh seafood would regard it as very delicious food, while those who cannot get used to raw foods may feel difficult to enjoy it. For such a reason, it is difficult (or too brutal) to prescribe statically user subjective perceptions on recipes.

Inspired by the prosperous collaborative tagging applications, we observe that the Internet users are quite active in contributing content to the online systems, especially for sharing their comments on resources. Collaborative tagging provides a way to collect the common point of view on a resource. According to the observation on our prototype system, we find that users like to tag their perceptual feeling. Motivated by this, we adopt the folksonomy mechanism into our system for deriving the UPFs of recipes under the help of users' collaborative editing. Specifically, the UPFs of a recipe are formulated by a tag vector as follows:

$$R^{UPF} = (\ t_1{:}w_1,\ t_2{:}w_2,\ t_3{:}w_3,\ ...,\ t_n{:}w_n) \tag{2}$$

where t_i is a tag, and w_i is the frequency that tag t_i has been annotated to a recipe R by all users.

Considering collaborative tagging would become effective only if the users are instructed in a correct way, in our prototype system, the objective to incorporate tagging mechanism is clear, i.e., to collect useful tags conveying additional knowledge besides those which can be directly read from the resources. For example, if a user tags the recipe 'Chinese Braised Spare Ribs' with 'pork spareribs', 'green onion', then it would be meaningless for the sharing purpose as the ingredient information has already been described in the resource.

4   Recommendation Framework

In this section, by taking both content features and social factors into consideration, we propose a hybrid recipe recommendation method which leverages the folksonomy mechanism and cooking procedures to predict users' preferences on recipes.

4.1   Content-Based Relevance

COF Similarity. The *cooking oriented feature (COF)* similarity is to measure how a recipe is similar to another via the cooking flow perspective. This factor is important when making recommendations as the cooking skills of users usually vary from one to another, and recommending a delicious but difficult-to-cook dish to an amateur cooking learner would not make sense. Since the cooking flow in our work is modelled by the cooking graph, we compute the similarity based on the sub-structure similarity between two cooking graphs, as presented in [14]:

$$sim(R_i^{CG}, R_j^{CG}) = \left[\left(\sum_{x=1}^{m} \left|E_{S_x}\right| \left(\mu \left|E_{SA_x}\right| + \gamma \left|E_{SI_x}\right| \right) \right) \cdot Per(R_i^{CG}, R_j^{CG}) \right]^{\frac{1}{2}} \tag{3}$$

where R_i^{CG}, R_j^{CG} are the cooking graphs of recipe R_i and R_j,   and they share m sub-graphs S_x where $x = [1, ..., m]$; $\left|E_{S_x}\right|$ is the total number of edges of the sub-graph S_x; $\left|E_{SA_x}\right|$ and $\left|E_{SI_x}\right|$ are the numbers of action and ingredient edges in S_x, respectively; μ and γ are adjustable weights for action and ingredient edges; $Per(R_i^{CG}, R_j^{CG})$ is the shared percentage of common *ReciSet* (defined as two connected nodes with a directed edge in the graph). A larger value of $sim(R_i^{CG}, R_j^{CG})$ indicates that the two recipes share more similar cooking flows.

UPF Similarity. The purpose to evaluate the *user perceptual feature (UPF)* similarity is to learn how similar the final dishes of two recipes are in terms of taste, smell, textures, etc., from a folksonomy perspective. No matter cooking at home or eating at restaurant, such features are always the factors which users would care most. As the UPFs are represented by tag vectors in our work, we calculate their similarity using the cosine similarity:

$$sim(R_i^{UPF}, R_j^{UPF}) = \frac{R_i^{UPF} \cdot R_j^{UPF}}{\left\| R_i^{UPF} \right\| \left\| R_j^{UPF} \right\|} \tag{4}$$

where R_i^{UPF} and R_j^{UPF} are the sets of tags describing the user perceptual features of recipes i and j, respectively. The larger value of $sim(R_i^{UPF}, R_j^{UPF})$ indicates that the two recipes are more similar from the users' perceptual point of view.

Recipe Content Relevance for Users. Suppose each user has a set of recipes in his/her favourite list, and analyzing the features of such recipes can help understand a user's preference. Based on the traditional content-based techniques, we predict the recipe content relevance for a user by comparing a candidate recipe with the ones which are the user's favours according to his/her past history. The overall content relevance is a weighted combination of two similarities:

$$RScore(U_i, R_j) = \alpha \cdot \max_{\forall R_x \in R_{Ui}} (sim(R_x^{CG}, R_j^{CG})) + (1 - \alpha) \cdot \max_{\forall R_x \in R_{Ui}} (sim(R_x^{UPF}, R_j^{UPF})) \tag{5}$$

where R_{Ui} is the set of recipes in user i's favourite list, R_j is a candidate recipe to be compared, $sim(R_x^{CG}, R_j^{CG})$ is the COF similarity between R_x and R_j, and $sim(R_x^{UPF}, R_j^{UPF})$ denotes the UPF similarity between R_x and R_j. The larger value of $RScore(U_i, R_j)$ is, the more relevant recipe j is to user i's preference from the content feature perspective.

4.2 Measuring User (Interests) Similarity

In our real lives, advices from trusted friends usually play an important role in our decision making. This factor could also be significant for the recommender systems, as it can contribute to the success of collaborative filtering (also known as social recommendation) techniques. According to the characteristics of recipe domain, which are different from other domains, we consider the `friends' (i.e., similar users) of a user a as a set of users not only sharing similar tastes but also having common preferences on structurally similar recipes for a.

In cooking recipe domain, it is important and indispensable to measure the structural similarity between two recipes to find recipes which are similar in terms of their cooking procedures to help a cook. Besides, the subjective features of recipes are derived from users' tagging in our system, whose values are often user-dependent. Thus by analyzing the users' personomies (tagging profiles), as well as their similar favorite recipes and their structure, it is helpful to predict and measure whether two users share the similar taste and common preference in cooking procedure.

The interest similarity between two users is defined as a combination of average structurally similarity and personomy similarity of their favorite recipes, i.e.:

$$sim(U_a, U_b) = \beta \cdot \frac{T_a \cdot T_b}{\|T_a\| \|T_a\|} + (1-\beta) \underset{\forall Ry \in R_{Ua}, \forall Rz \in R_{Ub}}{avg} (sim(R_y^{CG}, R_z^{CG})) \quad (6)$$

where T_a and T_b are the tag vectors that have been used by users a and b to annotate recipes, R_{Ua} and R_{Ub} are the set of recipes in the favourite lists of user a and correspondingly, $avg(sim(R_y^{CG}, R_z^{CG}))$ is the average recipe graph similarity of favorite recipes of users a and b. The larger value of $sim(U_a, U_b)$ indicates that users a and b share similar tastes .

4.3 Folksonomy Boosted Recommendation

For the reason that a user's choice on items is usually influenced by multiple factors, among which his/her own interest (preferences) and the friends' recommendations/comments are more dominant, thus we incorporate both factors into our recommendation strategy. Our overall prediction of users' preference on recipes is obtained by combining their content-based relevance and the social-based recommendations from the discovered 'friends'. Specifically, , the predicted rating of a user a on a recipe j can be calculated as follows:

$$p(\hat{r}_{a,j}) = \frac{1}{n} \cdot RScore(U_i, R_j) \cdot \sum_{\forall U_x \in F_{Ua}} sim(U_a, U_x) \cdot r_{x,j} \quad (7)$$

where $p(\hat{r}_{a,j})$ is the final predicted rating for user a on recipe j, F_{Ua} is a set of friends of user a, $r_{x,j}$ is user x's actual rating on recipe j, and n is the total number of 'friends' of user a. The final output of our recommendation framework is a list of recipes ranked by their $p(\hat{r}_{a,j})$ values in descending order.

5 Experiment

As a part of our research, we have developed a prototype system implementing the proposed recommendation strategy, based on which some experimental studies are conducted. In this section, we introduce the experiment setup and methodologies to evaluate our proposed strategy and discuss the experiment results.

5.1 Experiment Setup

Data Set. In order to collect real data from users, a prototype is developed and our experiment is based on a data set collected from 203 users with various backgrounds in our prototype system. Each of them was asked to select at least 10 favorite recipes out of 300, and give ratings and tags to the selected recipes. There are totally 3045 preferences and 7889 tags in the data set, with each user on average selecting 15 recipes and providing 39 tags.

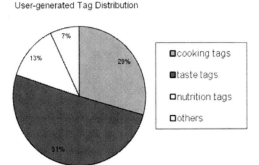

User-generated Tag Distribution

- cooking tags
- taste tags
- nutrition tags
- others

Fig. 2. The distribution of user generated tags

By analyzing on the user-generated tags, we found that 51% of these are about the taste descriptions, 29% on the cooking tips, 13% for other features like nutrition aspects, and also 7% remaining tags talk about users' impressions, emotional feelings, etc. on the recipes. An overall distribution of the annotated tags is shown in Figure 2.

Metrics. The first metric we use is *Hit Rate* (**HR**). It is commonly used in recommender systems [11] to measure how many items in the recommendation lists have hit (match) users' interests, and it reveals the accuracy of a recommendation algorithm. The calculation of *HR* is defined as follows:

$$HR = \frac{1}{n} \cdot \sum_{a=1}^{n} \frac{|T_{U_a} \cap X_{U_a}|}{|T_{U_a}|} \tag{8}$$

where T_{U_a} are the recipes relevant to user a in the test set, and X_{U_a} is the result set of top-N returned recipes; the overall Hit Rate of the top-N results is computed as average personal Hit Rate for all users in the test data set.

In addition, we also use *Improvement* (***imp***) [8] as the second metric in our experiment. Its purpose is to evaluate how the ranks of the hit recommendations have been improved by a recommendation algorithm when compared with a baseline approach. The formal definition of *imp* metric is as follows.

$$imp = \frac{1}{n} \cdot \sum_{a=1}^{n} \sum_{i-1}^{m} \frac{1}{r_p(R_{i,U_a})} - \frac{1}{r_b(R_{i,U_a})} \tag{9}$$

where R_{i,U_a} denotes a recipe in the personal favorite list of user a, $r_p(R_{i,U_a})$ and $r_b(R_{i,U_a})$ refer to the ranks of R_{i,U_a} in the proposed and baseline approach, respectively; m is the total number of recipes in a user's favorite list, and n is the total number of users.

Baseline Methods. In order to evaluate our proposed approach, we compare our method with two baseline methods implemented in [5]:

1. Content-based approach (CN): denoted as *recipe_cn* in [5], where each recipe is represented by a set of component food items, the predicted ratings in this

approach are calculated based on the food-item similarity between any two recipes.

2. Collaborative filtering approach (CF): denoted as *recipe_cf* in [5], it is a standard collaborative filtering algorithm for making predictions based on the neighborhood's ratings, with the user-similarity serving as the weights.

We divide the whole data set randomly into 5 groups, and each time use 80% of them as the training data (e.g. for a user with 15 preferred recipes, 12 of them will be used for training and 3 for testing). This process is repeated for five times for cross validation to reduce random noises, and the final *HR* is calculated as an average of the five results.

5.2 Experiment Results

The comparisons between our approach (denoted as *FolkBo*) and the baseline approaches *CN* and *CF* are shown in Figures 3, where the horizontal coordinate @*N* means the top-N recipes to be recommended to users.

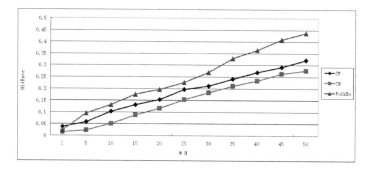

Fig. 3. Comparison of *HR@N* between our approach and baselines

We can observe from Figure 3 that the three methods have very tiny difference for *HR@1* (less than 1%), but for *HR@5*, it is clear that our approach outperforms CF and CN, reaching at around 10% (or 0.1). The improvement trend keeps increasing with the value of N, and at @*50* it achieves a hit rate of 43.7%, outperforming the CF method by 11.7% and the CN method by 16.1%. We consider that the better performance comes from both our refined data model of recipe features, and the folksonomy-based approach to predict users' interest relevance. Admittedly, the overall *HR* of our approach is still a bit low comparing to the recommendation accuracy in other applications, such as movie lens, due to that the preference data in our training set is a bit small (only 12 for each user on average) to make the algorithm converged.

Besides the hit rate, we also test the algorithms further to see how our approach improves the rankings of the hit recipes in the recommendation list. The result of *imp* with the recommendation list length as 50 is presented in Figure 4.

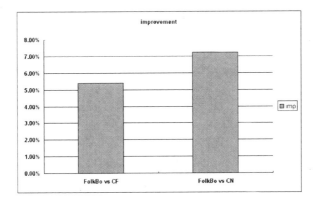

Fig. 4. Comparison of *imp* between our approach and baselines

According to Figure 4, our method improves the overall rankings of the target recipes by around 5.8% comparing to CF, which means the target recipes on average ascend 2.8 positions higher in the recommendation list. And comparing to the CN method, the improvement is more significant, reaching around 9.5%, indicating on average 4.75 positions' ascending in the result ranking. This improvement is important in the real-life applications, as we know that users would seldom go to the bottom of a result list, or even unlikely to the second page to look for their desired resources. Thus, improving the ranks of user's interested recipes in the top-K of the first page is more useful to achieve better recommendation quality.

6   Conclusions

In this paper we have presented a multi-faceted solution for the problem of cooking recipe recommendation. Firstly, we analyze the domain-specific characteristics of recipes from both objective and subjective features, and present a recipe data model to represent these features. Based on the proposed recipe data model, we have presented a recipe recommendation framework. Experiments are conducted on a real data set to evaluate our method and the results show that our method outperforms baseline methods on recipe recommendation.

Currently, we are working on further issues related recipe recommendation and management, including context-aware recipe recommendation based on user community information, and personalized recipe retrieval and adaptation based on (dynamic) user profiles.

Acknowledgement

The research work presented in this paper has been supported, substantially, by a grant from the Research Grants Council, the Government of the Hong Kong SAR, China [Project No. CityU 117608].

References

1. Adomavicius, G., Tuzhilin, A.: Toward the next generation of recommender systems: a survey of the state-of-the-art and possible extensions. IEEE Transactions on Knowledge and Data Engineering 17(6), 734–749 (2005)
2. Burke, R.: Knowledge-based Recommender Systems. In: Kent, A. (ed.) Encyclopedia of Library and Information Systems. Supplement 32, vol. 69, Marcel Dekker, New York (2000)
3. Markines, B., Cattuto, C., Menczer, F., Benz, D., Hotho, A., Stumme, G.: Evaluating similarity measures for emergent semantics of social tagging. In: Proceedings of the 18th International Conference on World Wide Web, Madrid, Spain, April 20-24 (2009)
4. Michlmayr, E., Cayzer, S.: Learning User Profiles from Tagging Data and Leveraging them for Personal(ized) Information Access. In: Proc. of the Workshop on Tagging and Metadata for Social Information Organization, Int. WWW Conf. (2007)
5. Freyne, J., Berkovsky, S.: Intelligent food planning: personalized recipe recommendation. In: Proceeding of the 14th International Conference on Intelligent User Interfaces. IUI 2010, pp. 321–324. ACM, New York (2010)
6. Herlocker, J.L., Konstan, J.A., Terveen, L.G., Riedl, J.T.: Evaluating collaborative filtering recommender systems. ACM Trans. Inf. Syst. 22(1), 5–53 (2004)
7. Sobecki, J., Babiak, E., Słanina, M.: Application of hybrid recommendation in web-based cooking assistant. In: Gabrys, B., Howlett, R.J., Jain, L.C. (eds.) KES 2006. LNCS (LNAI), vol. 4253, pp. 797–804. Springer, Heidelberg (2006)
8. Pazzani, M.J., Billsus, D.: Content-Based Recommendation Systems. In: Brusilovsky, P., Kobsa, A., Nejdl, W. (eds.) Adaptive Web 2007. LNCS, vol. 4321, pp. 325–341. Springer, Heidelberg (2007)
9. Golder, S.A., Huberman, B.A.: Usage patterns of collaborative tagging systems. Journal of Information Science 32(2), 198–208 (2006)
10. Burke, R.: Hybrid Recommender Systems: Survey and Experiments. User Modeling and User-Adapted Interaction 12, 331–370 (2002)
11. Shepitsen, A., Gemmell, J., Mobasher, B., Burke, R.: Personalized recommendation in social tagging systems using hierarchical clustering. In: Proceedings of the 2008 ACM Conference on Recommender Systems, RecSys 2008, pp. 259–266. ACM, New York (2008)
12. Svensson, M., Höök, K., Cöster, R.: Designing and evaluating kalas: A social navigation system for food recipes. ACM Trans. Comput.-Hum. Interact. 12(3), 374–400 (2005)
13. Wang, L., Li, Q., Li, N., Dong, G., Yang, Y.: Substructure similarity measurement in chinese recipes. In: Proceeding of the 17th international Conference on World Wide Web, WWW 2008, Beijing, China, April 21 - 25, pp. 979–988. ACM, New York (2008)
14. Zhen, Y., Li, W., Yeung, D.: TagiCoFi: tag informed collaborative filtering. In: Proceedings of the Third ACM Conference on Recommender Systems, RecSys 2009, October 23-25, pp. 69–76. ACM, New York (2009)
15. Wang L.: CookRecipe: towards a versatile and fully-fledged recipe analysis and learning system, Thesis (Ph.D.), City University of Hong Kong (2008)
16. Li, Y., Meng, X., Wang, L., Li, Q.: recipeCrawler: collecting recipe data from WWW incrementally. In: Yu, J.X., Kitsuregawa, M., Leong, H.-V. (eds.) WAIM 2006. LNCS, vol. 4016, pp. 263–274. Springer, Heidelberg (2006)
17. Cai, Y., Li, Q.: Personalized search by tag-based user profile and resource profile in collaborative tagging systems. In: CIKM, pp. 969–978 (2010)
18. Cai, Y., Li, Q., Xie, H., Yu, L.: Personalized Resource Search by Tag-Based User Profile and Resource Profile. In: Chen, L., Triantafillou, P., Suel, T. (eds.) WISE 2010. LNCS, vol. 6488, pp. 510–523. Springer, Heidelberg (2010)

Effective Hybrid Recommendation Combining Users-Searches Correlations Using Tensors

Rakesh Rawat, Richi Nayak, and Yuefeng Li

Faculty of Science and Technology, Queensland University of University
Brisbane Australia
{r.rawat,r.nayak,y2.li}@qut.edu.au

Abstract. Most recommendation methods employ item-item similarity measures or use ratings data to generate recommendations. These methods use traditional two dimensional models to find inter relationships between alike users and products. This paper proposes a novel recommendation method using the multi-dimensional model, tensor, to group similar users based on common search behaviour, and then finding associations within such groups for making effective inter group recommendations. Web log data is multi-dimensional data. Unlike vector based methods, tensors have the ability to highly correlate and find latent relationships between such similar instances, consisting of users and searches. Non redundant rules from such associations of user-searches are then used for making recommendations to the users.

Keywords: Tensor, clustering, association rule mining, web log data, recommendation.

1 Introduction

With the popularity of World Wide Web, use of recommenders to suggest relevant products and services to online users is gaining momentum. Collaborative filtering (CF) techniques of recommendation have been used by many websites like Amazon, ebay, CdNow, Netflix, Yahoo Answers and many more. CF techniques can be grouped into two general classes, namely the neighbourhood and model-based methods [1]. Neighbourhood (or memory-based or heuristic-based) methods use Item-to-Item or User-to-Item correlations to find the nearest neighbours and then subsequently use this information to make recommendations. The Item-to-Item correlation methods adopt a content-based approach where knowledge about the products (contents) is used for recommendation and only similar matching content/products are recommended [2], [4]. The User-to-Item correlation methods combine interests of a group of people to find the highest rated interests and then interests consisting of items/ products/people are recommended to the individuals in a group [4],[14].

A user's search generally consists of multiple attributes e.g. in the case of a car website, a user may search for a particular make, model, body type, cost, new or old car type. A user may have made n number of searches within a website. Modeling or comparing such users-items data having multi-dimensional properties is a complex

X. Du et al. (Eds.): APWeb 2011, LNCS 6612, pp. 131–142, 2011.

process. Traditional CF methods employed for finding similarity between users-items ignore this multi-dimensional nature of search log data and are unable to recommend unique items to different users [2]. These methods consider an item as an object, whereas, the item may be a combination of many features, represented as a vector itself. Finding the latent relationships between user's searches and item's features is often ignored by two dimensional data models such as vectors and matrix. Recommendation systems need to handle very high dimensional profiles of users-items, in order to find the correlations between users and items. A noteworthy consideration as also discussed by [3], is that distance measures used for clustering or comparisons may reflect strange properties in high dimensional space and might not be as useful as they seem.

In this paper, we propose a novel recommendation method which utilizes the implicit information about users by using the search log data. This methodology utilizes the search log data to infer the user rating about items. This is a collaborative approach which group users based on their common searches and then finds users–items correlations within a group. To find the correlations between users according to their usage of items, we employ tensor, the high dimensional data model. Once users are clustered using the proposed tensor based clustering method, the associations shared by a group of users represented as top 'n' items, are used for making recommendations within the group. Unlike most of previously adopted tensor models consisting of three dimensions, we have modeled users search log into more than three dimensions, and used the tensor factorization information for making recommendations. Empirical results on real car sales datasets show that the recommendation for all users suggested by the proposed tensor based recommendation method outperforms the recommendations given by the traditional collaborative based methods, which mostly employ vector/matrix methods to find similarity between users-items. Taking the average of recommendations done by CF methods and tensor based methods, on an average there was an improvement of about 40% in the precision, 52.78% in recall and 36.84% in F-score values.

2 Related Work

There are a myriad of work published, we present some of the recent related work employing CF and hybrid recommendations techniques. A collaborative filtering method to provide an enhanced quality of recommendations, derived from user-created tags is discussed by [4]. In this work collaborative methods of tagging item are employed to find users preferences for items. Data cubes consisting of three 2 dimensional matrices (User–item, User–tag and Tag–item frequency) which are transformed from three dimensional space for collaborative tagging are used. For recommendation Naïve Bayes classifier are used. The performance of such an approach was found out to be far superior than the plain collaborative recommendation approaches. A genetic algorithm that formulates purposeful association rules out of the transactions database of a transportation management system has been proposed by [5]. The constructed rules are recommended to the associated users. The recommendation process takes into account the constructed rules and techniques that are derived from collaborative filtering.

In another work [6] a novel hybrid recommendation method that combines the segmentation-based sequential rule method with the segmentation-based KNN-CF method is proposed. Here a sequential rule-based recommendation method analyses customers purchase behaviour over time to extract sequential rules. Sequential rules are extracted for each group from the purchase sequences to make recommendations. Consequently, the segmentation-based KNN-CF method provides recommendations based on the target customer's purchase data for the current period. The results of the two methods are combined to make final recommendations. A hybrid recommendation for an online retail store is proposed by [7]. The method adopts six steps for recommendation which are product taxonomy formation, grain specification, extracting product, category attributes, user (customer) profile creation, and user–user and user–product similarity calculation and recommendation generation. Experimental results show that proposed technique improves recommendation when compared to other similar CF based methods. Another recent work that proposes a hybrid approach that uses neural nets for recommendation is [8]. The proposed approach trains the artificial neural networks to group users into different clusters, and applies the well-established Kano's method for extracting the implicit needs of users in different clusters. The approach is applied on a tour and travel website to demonstrate the improvement for the problem of information overload.

Tensors have previously been used extensively in chemometrics and psychometric and some Web mining tasks like Web link analysis [9] and chat room analysis [10]. Recently some recommendation models, which have used three dimensional tensors for recommending music, tags and objects, have been proposed. A recommender model, using HOSVD for dimension reduction, have been proposed for recommending personalized music [11] and Tags [12]. Researchers [13] have used TSM based tag recommendation model which uses tensor factors by multiplying the three features matrices with core matrix each consisting of user, items and tags. Another collaborative filtering approach based on tensor factorization for making recommendations, where, the users, items and related contextual information are modeled as a three dimensional tensor is proposed by [20].

3 Model Construction, Decomposition and Clustering

Tensor notations and conventions used are similar to the notations used by previous authors [14]. Scalars are denoted by small letters a, b, vectors are denoted by boldface small letters like $\boldsymbol{a}$, $\boldsymbol{b}$. All subscript are shown by small letters starting from $l..n$. Matrices are shown using capital bold letters like $\mathbf{A}, \mathbf{B}$ and the element (i, j) of a matrix is shown by a_{ij}. All tensors are represented using calligraphic fonts $\mathcal{T}$, e.g. $\mathcal{T} \in \mathbb{R}^{M_1 \times M_2 \times M_3 \times .. M_n}$ and the entries are shown using a_{ijk} and the subscript (i, j, k) range from $1..$to I, J, K in each mode. A tensor is a multi-dimensional data array which has n dimensions (or modes). The order of a tensor is the number of dimensions. For example, the tensor $\mathcal{T} \in \mathbb{R}^{M_1 \times M_2 \times M_3 \times .. M_n}$ is of an order n with n dimensions. Each element of a tensor needs n indices to represent or reference its precise position in a tensor, for example, the element a_{ijklmn} is an entry value at the i, j, k, l, m and n modes. In various tensor decomposition techniques the dimensions are flattened to represent matrices of various sizes before the subsequent decomposition technique is applied. Matricizing, unfolding or flattening of a tensor is a useful operation for transforming a given multi-dimensional array into a matrix. A third order tensor $\mathcal{A} \in \mathbb{R}^{I \times J \times K}$ is able to form three matrices of $I \times JK, J \times IK, K \times IJ$. More details on tensors and their properties can be found out in [14],[15].

3.1 Model Construction

Prior to creating the tensor model, the data is preprocessed. Pre processing includes arranging searches in different sessions made by a user into records, removing unwanted attributes from such records. This data is then grouped for each user based on the various searched parameters, and frequency of such records as grouped is counted. The prominent attributes from the users' data are identified and such attributes are then represented as modes of the tensor model. A tensor is created with all such features and the users as one dimension. For example, the structure of a tensor created, consisting of 5 searched dimensions and the users are as follows:

$$\mathcal{T} \in \mathbb{R}^{Make \times Model \times Bodytype \times Search\, Type \times Cost\, Type \times Users} \tag{1}$$

For each user the term frequency of each similar search is counted. A similar search is a search whose all searched parameters are same. As an example for a given user, the different searched parameters like make, model, body type, search type and cost of a car may be same, and user may have searched them many times in different sessions. The term frequency value for all the searches of a user are found out. Next theses values are populated in the tensor. As an example, the term frequency t_{ijklmn} is an entry value at the i, j, k, l, m and n modes, where i represents the *Make*, j the *Model*, k the *Bodytype*, l the search type, m the cost ranges and n the user id.

3.2 Decomposition

In multi-dimensional data modeling, the decomposition process enables to find the most prominent components (i.e. tensor entries and modes) as well as the hidden relationships that may exist between different components. We have used the popular and widely used PARAFAC [16] tensor decomposition technique to decompose the constructed model. PARAFAC is a generalization of PCA (Principal Component Analysis) to higher order arrays. Given a tensor of rank 3 as $\mathcal{X} \in \mathbb{R}^{I \times J \times K}$, a R-component PARAFAC model can be represented as

$$x_{ijk} = \sum_{r=1}^{R} a_{ir} b_{jr} c_{kr} + E \tag{2}$$

where a_i, b_i, c_i are the ith column of component matrices $\mathbf{A} \in \mathbb{R}^{I \times R}, \mathbf{B} \in \mathbb{R}^{J \times R}$ and $\mathbf{C} \in \mathbb{R}^{K \times R}$ respectively and $E \in \mathbb{R}^{I \times J \times K}$ is the three way array containing residuals. x_{ijk} represents an entry of a three way array of $\mathcal{X}$ and in the ith row, j^{th} column and k^{th} tube. Thus in our case when the users tensor (equation 1) is decomposed using [17], the various matrices formed are as shown in the figure 1 below. In figure 1, $\mathbf{M_1}, \mathbf{M_2}..\mathbf{M_n}$ are the various component matrices formed after the decomposition of the tensor, and R is the desired best rank tensor approximation, which is set as 1, 2, and 3 in all the experiments.

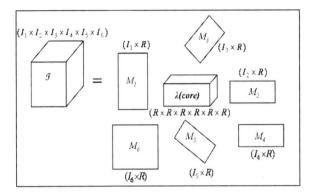

Fig. 1. PARAFAC Decomposed tensor of users-searches, gives component matrices as shown

3.3 Clustering

Clustering is done on the component matrices $\mathbf{M_1}, \mathbf{M_2}..\mathbf{M_n}$ obtained after PARAFAC decomposition and representing decomposed values of each mode, from $M_1, M_2,...M_n$ respectively. Each component matrix $\mathbf{M_1}, \mathbf{M_2}..\mathbf{M_n}$ is of dimension $\mathbf{M}_{i \times r}$ where i is the number of ways in a mode M_n, and r is the value of best rank approximation of the tensor. The nth matrix obtained after PARAFAC decomposition (Shown as $\mathbf{M_6}$ in fig. 1) represents the users' dimension. Clustering on row values of this matrix is achieved by using two clustering methods the EM (Expectation Maximization) and k means [18]. We have taken the last component matrix $\mathbf{M}_n$ as clustering input since it represents the users dimension in the proposed data model. Entries (or values) in the matrix $\mathbf{M}_n$ depicts the correlations between users based on the multiple factors of the tensor. Therefore, clustering on the matrix $\mathbf{M}_n$ results in grouping users according to similar search behaviour which is based on multiple search components as modeled in the tensor.

4 Discovering Association within Clusters

All searches made by users in a cluster are grouped and frequent associations based on desired query components (like make-model in our case) are mined. Thus each cluster contains the searched parameters, as searched by users of the respective cluster. Association rules are mined from each individual cluster. We have considered

Input: Processed search data users and searched parameter wise, n the number of recommendations desired.

Output: Top n recommendations.

Let k be the number of clusters (here $k=10, 20, 30$). Let $\text{Rule}_{rk} = X_r \Rightarrow Y_r$ be the rule mined for cluster k on item set X_m and Y_m such that $X_r, Y_r \in X_m, Y_m$ and $X_r \cap Y_r = \varnothing$, where m is the number of item sets in a cluster, on which association rules are mined and r be number of rules generated for each cluster k. Let j be a counter variable, initialized at zero.

Begin

    Step 1. Create users-searches tensor ($\mathcal{T}$).

    Step 2. Decompose the tensor ($\mathcal{T}$), and cluster the last component matrix $\mathbf{M}_n$.

    Step 3. // Find Association rules for each cluster from 1 to k.

    Step 4. //Order Association rules in decreasing order of confidence score.

       for $i=1$ *to* k //Evaluating distinct rules for each cluster from 1 to k

            $j=0; r=1;$

          Do While ($j!=n$)

              Extract (r); //Function retrieves top r^{th} rule with highest confidence from cluster k.

              If $(X_r \Rightarrow Y_r) = (Y_r \Rightarrow X_r)$ then

                  $Save(X_r \Rightarrow Y_r)$; //Save(r) function saves the distinct item sets from association rules for the cluster k.

                  $j=j+1; r=r+1;$

              Else

                  //Retrieve rules with next highest confidence score.

                  $Save(X_r \Rightarrow Y_r)$;

                  $j=j+1;$

                  $Save(Y_r \Rightarrow X_r)$;

                  $j=j+1; r=r+1;$

       End for

       // Cluster wise Top n recommendations

       $Top_k n = X_n \Rightarrow Y_n;$

End

Fig. 2. Complete Recommendation algorithm

associations of length two, as the occurrence of associations of length greater than 2 was very rare, especially when the number of users in a cluster is small. The whole recommender process is explained in the algorithm in figure 2.

5 Experimental Design

All experiments have been done on server log data collected from a live Australian car sales website[1]. Search log data of duration 1 month was used. Out of a total

---

[1] Due to privacy issues we are unable to specify the details about the website.

number of users, 949 users who had made searches in that period were used in experiments. These users were identified based on IP address (IP), web browser and Geo segmentation details like location and PIN number. These users had made leads or enquiries about a car of their interest, showing that these users were interested in buying the particular car, for which they had made the lead. All leads about a car were made by emailing the dealer through the e-mail feature (contact us) provided by the website. In all the experiments, the data used for making recommendations were taken prior to a user had made a lead. The major objective was to match leads with the correctly made recommendations by various traditional collaborative methods and collaborative methods based on tensor model. Some statistics for the user searches and leads are shown below in Table 1:

Table 1. Statistics of filtered data used for tensor modeling

No. of Sessions	No. of Users	Average searches per user	No. of Leads	Avg. Lead/User
2692	949	14	1649	1.74

Five parameters used for searching like make, model, body type, cost, search type (e.g. like new or used) plus a sixth dimension (users) were taken as dimensions of the tensor model (Table 2).

Table 2. Number and Sample of Dimensions Used in the Tensor Model

Dimension Name	No of unique ways or modes	Sample dimension values
Make	68	Toyota, Holden, Ford
Model	644	RAV4, Liberty.
Body Type	12	Sedan, Hatch
Search Type	5	New, Used, Ex Demo.
Cost Type	13	$1-2500
Users Id	949	1, 2,...949

Once the data was pre processed the users-items tensor model was created as $\mathcal{T} \in \mathbb{R}^{68 \times 644 \times 12 \times 5 \times 13 \times 949}$. Subsequently, decomposition was achieved using the PARAFAC model. In the absence of an ideal clustering solution, three cluster solutions consisting of 10, 20 and 30 clusters were used for evaluation. For each cluster, association rules with highest confidence score and make/model as associated items were taken. For each cluster the number of association rules generated was 3, 5, 10, and 15. Once frequent patterns in the form of association rules are mined cluster wise, only rules having highest support and distinct make-model were taken for recommendations from these rules. These top n make-models were given as recommendation to each of the users belonging to a same cluster. A sample of some association rules are shown as below.

Best rules found:
1. X-TRAIL=X-TRAIL 10 ==> NISSAN=NISSAN 10   conf:(1)
2. NISSAN=NISSAN 10 ==> X-TRAIL=X-TRAIL 10   conf:(1)
3. X-TRAIL=GRAND VITARA 2 ==> NISSAN=SUZUKI 2 conf:(0.89)
4. NISSAN=SUZUKI 2 ==> X-TRAIL=GRAND VITARA 2 conf:(0.75)

Fig. 3. Sample of associations found out for a cluster1

Example for the above case as shown in Figure 3 the top 3 cars recommended with make-model were: 1. NISSAN- X-TRAIL, 2 SUZUKI- GRAND VITARA and 3 SUBARU- OUTBACK. All users with similar interests belonging to a cluster (Evaluated based on users-searches using tensor) will be recommended these top n cars.

5.1 Evaluation Criteria

To evaluate the quality of top-m recommendations given by each method we used the following metrics. Let L_n be the number of leads made by a user U_n, and let R_n^m, be the top-m recommendations given by various methods to U_n, where $m \geq 3$ and $m \leq 15, m \in \{\{3\},\{5\},\{10\},\{15\}\}$. Precision (Pr_n) and recall (Re_n) for each user U_n is evaluated as

$$Pr_n = \frac{R_n^m \cap L_n}{R_n^m \cap L_n + (R_n^m - L_n)} \quad \text{And recall as } Re_n = \frac{R_n^m \cap L_n}{R_n^m \cap L_n + (L_n - R_n^m)} \quad (3)$$

5.2 Results

The average details of clustering results for each method are shown below in Table 3, where the acronyms used in the table 3 are U-ESM (Users-users Euclidian Similarity

Table 3. Average Recommendations for various methods

	Top 3 Recommendation			Top 5 Recommendation			Top 10 Recommendation			Top 15 Recommendation		
	Pr	Re	Fs	Pr	Re	Fs	Pr	Re	Fs	Pr	Re	Fs
U-ESM	0.08	0.10	0.06	0.06	0.12	0.06	0.07	0.16	0.06	0.07	0.16	0.06
U-CSM	0.24	0.46	0.31	0.23	0.56	0.31	0.22	0.59	0.30	0.22	0.68	0.29
P1-EM	**0.35**	**0.60**	**0.43**	**0.32**	**0.59**	**0.40**	0.22	0.67	0.30	0.21	0.69	0.28
P2-EM	0.25	0.38	0.25	0.24	0.4	0.25	0.17	0.55	0.18	0.16	0.61	0.18
P3-EM	0.34	0.56	0.40	0.29	0.57	0.36	**0.24**	**0.69**	**0.29**	**0.23**	**0.72**	**0.28**
P1-kM	0.15	0.25	0.17	0.11	0.26	0.15	0.13	0.53	0.21	0.11	0.57	0.17
P2-kM	0.22	0.47	0.29	0.21	0.51	0.27	0.14	0.65	0.22	0.12	0.73	0.19
P3-kM	0.22	0.44	0.28	0.19	0.48	0.25	0.16	0.63	0.24	0.14	0.67	0.21
Average	**0.23**	**0.41**	**0.27**	**0.21**	**0.44**	**0.26**	**0.17**	**0.56**	**0.23**	**0.16**	**0.60**	**0.21**

Measure) and U-CSM (Users-users Cosine Similarity Measure). U-ESM and U-CSM are evaluated based on users-items relationships, where such relationships are represented as vectors in two dimensional spaces. Clustering is done on these vectors to find users with similar interests. P1-EM, P2-EM and P3-EM are the PARAFAC best rank approximation of rank 1, 2, 3 respectively, with clustering achieved using Expectation Maximization (EM) [18]. Similarly P1-kM, P2-kM and P3-kM are the PARAFAC best rank approximation of rank 1, 2, 3 respectively, with clustering achieved using k means. The other values are Pr=precision, Re=recall and Fs=F Score.

6 Discussion

Due to very high number of dimensions of interest vector of users (742 dimensions excluding users and as shown in table2) clustering based on Euclidean (ESM) using k means and EM was unable to produce good clustering of users which ultimately resulted in not so high quality of recommendations. This can happen because as the number of dimensions grows significantly, ESM (Euclidian Similarity Measure) and CSM (Cosine Similarity Measure) eventually become less similar. In very high dimensional spaces as dimension gets higher (≥ 128) [19], the two similarity measures start having small variations between them. However, the rate of decrease of similarity is very slow. Similarity vectors of instances in such high dimensional spaces starts loosing inter component relationships, when traditional two dimensional distance based methods are employed. On the other hand cosine (CSM) measure produced average quality of results. This happens because cosine similarity is able to map the different dimensions, but due to the two dimensional model, latent relationships between users-items are lost.

In contrast the tensor based methods are able to extract hidden relationships between the datasets and give much improved similarity results for the users-items data. For making CF based recommendations, traditional k means algorithm with Euclidian and Manhattan similarity measures performed worse, whereas CSM methods performed average. On the other hand, EM based methods performed exceptionally well. These contrasting results confirm that distance based approaches using Euclidian, Manhattan or cosine base similarity measures used in high dimensional data mappings reflect strange properties [18] which include loss of inter component relationships and unable to map inter component latent relationships. EM is density based clustering

Table 4. Average Aggregate Recommendations for various methods

Methods	Precision	Recall	F-Score
CF-Distance Based	**0.15**	**0.36**	**0.19**
PARAFAC-kM	0.16	0.52	0.22
PARAFAC-EM	0.25	0.59	0.30
Avg. PARAFAC-EM+kM	**0.21**	**0.56**	**0.26**

algorithm, and rather than using a distance based clustering measure, it assigns a probability distribution to each instance, indicating the probability of it belonging to each of the other clusters.

From the results (table 3, figure 4) it is clear that, the number of best recommendations made to a user is around 3-5. The aggregate recommendation scores of precision, recall and F Score for each datasets using simple CF based methods and PARAFAC –EM and PARAFAC-kM based methods are shown below in table 4.

The other noteworthy observation was that in most associations with large lengths (> 2) , there was a reduction in number of frequent item sets discovered in the process, which had high support and confidence values. Association with length=1 had too many rules, and such rules were biased towards rules with highest frequency. Hence such rules were ignored for analysis in the experiments. An interesting observation which is shown in figure 5 shows relevant F-score when tensor best rank decomposition are considered. In the figure 5, 1EM, 2EM, 3EM refer to PARAFAC decomposition with best rank approximation of 1, 2 and 3 respectively, where clustering was achieved using EM clustering algorithm. Similarly 1KM, 2KM, 3KM refer to PARAFAC decomposition with clustering achieved using k-means clustering method. In both cases, using EM and KM performance starts decreasing with the increase in rank and number of recommendations. In case of KM clustering, rank 1 performs worst. KM clustering methods use distance measure for clustering. Due to the availability of singular values per instance for clustering, some useful relationships between instances may not be clearly distinguishable. When decomposing with higher ranks the extra factors available for clustering, which have the ability to preserve some information, KM's performance starts to increase. However decomposition at higher ranks may start loosing valuable inter component relationships, due to complete flattening of the tensor.

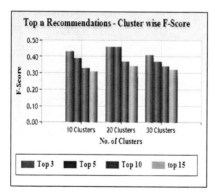

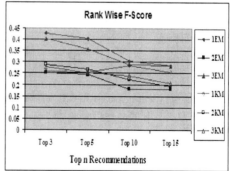

Fig. 4. F-Score for Top *n* recommendations cluster sizes

Fig. 5. F-Score Rank Wise, top n with various recommendations

The figure 5 clearly separates top3, top 5 recommendations from top10 and 15. Rank 1 approximation using EM clustering performed the best among all cases. In case of KM, rank 1 gave best results when number of clusters was 30 and for rank 3,

KM gave best when it was 30. These contrasting results in case of KM are clear indicator that distance measures may not work well when intra distance between instances is small. The other observation is that distance measures need larger cluster sizes to maximize the distance between instances and in hence improve overall performance.

7 Conclusion

This paper presented a novel method of recommendation based on tensor clustering and associations. Users-items similarity measures were evaluated using high dimensional data model tensors. Once the model was decomposed using PARAFAC, clustering was achieved on the users matrix. Association rules for users, clustered in a group were found out. Once such rules were found out, only unique rules with highest confidence were taken as top n recommendation for the users in a cluster. Experimental results show that tensor based recommendation method and unique association rule generation for making recommendations outperforms the traditional CF based methods, which perform user-items similarity measure using vector or matrix based methods. Since most of the processes for generating rules, creating and clustering users can be done offline, the system can effectively be used for generating high quality of online recommendations, thus limiting recommendation to top 3-5 recommendations per user. However as of now, since the process of identifying top n unique recommendations from association rules is not automated, a lot of time is needed to generate such top n recommendations for each group.

Acknowledgement

This research has been funded by CRC (Co-operative Research Centre), Australia and Queensland University of Technology, Brisbane Australia under the CRC Smart Services Web Personalization Project 2009-10.

References

1. Desrosiers, C., Karypis, G.: A comprehensive survey of neighborhood-based recommendation methods. In: Handbook on Recommender Systems. Springer, Heidelberg (2009)
2. Park, Y.J., Chang, K.N.: Individual and group behavior-based customer profile model for personalized product recommendation 36, 1932–1939 (2009)
3. Skillicorn, D.: Understanding complex datasets: data mining with matrix decompositions. Chapman & Hall/CRC (2007)
4. Kim, H.N., Ji, A.T., Ha, I., Jo, G.S.: Collaborative filtering based on collaborative tagging for enhancing the quality of recommendation. Electronic Commerce Research and Applications 9, 73–83 (2009)
5. Lazanas, A., Karacapilidis, N.: Enhancing Recommendations through a Data Mining Algorithm. In: Lovrek, I., Howlett, R.J., Jain, L.C. (eds.) KES 2008, Part I. LNCS (LNAI), vol. 5177, pp. 525–532. Springer, Heidelberg (2008)

6. Liu, D.R., Lai, C.H., Lee, W.J.: A hybrid of sequential rules and collaborative filtering for product recommendation. Information Sciences 179, 3505–3519 (2009)
7. Albadvi, A., Shahbazi, M.: A hybrid recommendation technique based on product category attributes. Expert Systems with Applications 36(9), 11480–11488 (2009)
8. Chang, C.C., Chen, P.L., Chiu, F.R., Chen, Y.K.: Application of neural networks and Kano's method to content recommendation in web personalization. Expert Systems with Applications 36(3), 5310–5316 (2009)
9. Kolda, T.G., Bader, B.W., Kenny, J.P., Livermore, C.A., Albuquerque, N.M.: Higher-Order Web Link Analysis Using Multilinear Algebra. In: Fifth IEEE International Conference on Data Mining (ICDM 2005), pp. 242–249 (2005)
10. Acar, E., Çamtepe, S.A., Krishnamoorthy, M.S., Yener, B.: Modeling and Multiway Analysis of Chatroom Tensors. In: Kantor, P., Muresan, G., Roberts, F., Zeng, D.D., Wang, F.-Y., Chen, H., Merkle, R.C. (eds.) ISI 2005. LNCS, vol. 3495, pp. 256–268. Springer, Heidelberg (2005)
11. Ruxanda, P.S.M., Manolopoulos, A.N.Y.: Ternary Semantic Analysis of Social Tags for Personalized Mucis Recommendation. In: Proceedings of the 9th International Conference on Music Information Retrieval, p. 219 (2008)
12. Symeonidis, P., Nanopoulos, A., Manolopoulos, Y.: Tag recommendations based on tensor dimensionality reduction. In: RecSys 2008, Lausanne, Switzerland, pp. 43–50 (2008)
13. Rendle, S., Marinho, B.: Learning optimal ranking with tensor factorization for tag recommendation. In: Proceedings of the 15th ACM SIGKDD International Conference on Knowledge Discovery and Data Mining, Paris, France, pp. 727–736 (2009)
14. Kolda, T.G., Bader, B.W.: Tensor decompositions and applications. Technical Report SAND2007-6702, Sandia National Laboratories, Albuquerque, NM and Livermore, CA, (November 2007)
15. Kolda, T.G., Sun, J.: Scalable Tensor Decompositions for Multi-aspect Data Mining. Time 18, 19
16. Harshman, R.A.: Foundations of the PARAFAC procedure: Models and conditions for an explanatory multi-modal factor analysis. UCLA working papers in phonetics 16, 1 (1970)
17. Bader, B.W., Kolda, T.G.: Efficient MATLAB computations with sparse and factored tensors. SIAM Journal on Scientific Computing 30, 205–231 (2007)
18. Ian, E.F., Witten, H.: Data Mining: Practical machine learning tools and techniques, 2nd edn. Morgan Kaufmann, San Francisco (2005)
19. Qian, G., Sural, S., Gu, Y., Pramanik, S.: Similarity between euclidean and cosine angle distance for nearest neighbor queries. In: ACM Special Interest Group on Applied Computing, pp. 1232–1237. ACM, New York (2004)
20. Karatzoglou, A., Amatriain, X., Baltrunas, L., Oliver, N.: Multiverse recommendation: n-dimensional tensor factorization for context-aware collaborative filtering. In: ACM Recommender Systems, Barcelona, Spain, pp. 79–86 (2010)

Maintaining Internal Consistency of Report for Real-Time OLAP with Layer-Based View

Ziyu Lin[1,*], Yongxuan Lai[2], Chen Lin[1], Yi Xie[1], and Quan Zou[1]

[1] Department of Computer Science, Xiamen University, Xiamen, China
[2] Software School, Xiamen University, Xiamen, China
{ziyulin,laiyx,chenlin,csyxie,zouquan}@xmu.edu.cn

Abstract. Maintaining internal consistency of report is an important aspect in the field of real-time data warehouses. OLAP and Query tools were initially designed to operate on top of unchanging, static historical data. In a real-time environment, however, the results they produce are usually negatively influenced by data changes concurrent to query execution, which may result in some internal report inconsistency. In this paper, we propose a new method, called layer-based view approach, to appropriately and effectively maintain report data consistency. The core idea is to prevent the data involved in an OLAP query from being changed through using lock mechanism, and avoid the confliction between read and write operations with the help of layer mechanism. Our approach can effectively deal with report consistency issue, while at the same time avoiding the query contention between read and write operations under real-time OLAP environment.

Keywords: OLAP; report consistency.

1 Introduction

Real-time data warehouses have been receiving more and more attention (e.g. [1,2,3,4, 5,6]) during the past few years, which is updated in as close to real time as possible [7]. However, OLAP and Query tools were initially designed to operate on top of static data, and they do not ensure that the data involved is protected from being modified. Therefore, the report result may be negatively influenced by the underlying changing data. In real-time data warehouse environment, this can lead to inconsistent and confusing query results, which is called *internal inconsistency of report* [8].

Take the simple report in Fig. 1 for example. It includes a multi-pass SQL statement made up of many smaller SQL statements. All these SQL statements will sequentially operate on a set of temporary tables. There will be no problem when the data is static, but it is not the case when the underlying data changes while the first temp table is being created. Most database systems (including multi-version databases [9]) will return the data that was current at the point that the query started to run [8]. At 0:01, the INSERT statement into TEMP1 started to run and lasted for four seconds. Then the query to load

---

* Supported by the Natural Science Foundation of China under Grant No. 61001013 and 61001143, and the Fundamental Research Funds for the Central Universities under Grant No. 2010121066.

X. Du et al. (Eds.): APWeb 2011, LNCS 6612, pp. 143–154, 2011.

```
0 : 00    create table TEMP1{Category_Id LONG, DOLLARSALES DOUBLE}
0 : 01    insert into TEMP1
              select all.[Category_Id] AS Category_Id
                sum (all.[Tot_Dollar_Sales]) AS DOLLARSALES
              from [YR_CATEGORY_SLS] all
              group by all.[Category_Id]
0 : 05    create table TEMP2 (ALLPRODUCTSD DOUBLE)
0 : 06    insert into TEMP2
              select sum((all.[Tot_Dollar_Sales]) AS ALLPRODUCTSD
              from [YR CATEGORY_SLS] all
0 : 08    select distinct pa1.[Category_Id] AS Category_Id,
              all.[Category_Desc] AS Category_Desc,
              all.[DOLLARSALES] AS DOLLARSALES,
              pa1.[DOLLARSALES]/pa2.[ALLPRODUCTSD]) AS DOLLARSALESC
          from [TEMP1] pa1,
              [TEMP2] pa2,
              [LU CATEGORY] all
          where pa1.[Category_Id]=all.[Category_Id]
0 : 09    drop table TEMP1
0 : 10    drop table TEMP2
```

Fig. 1. Sales by category with percent contribution

data into TEMP2 began to run at second 6. This means that TEMP1 will contain data current as of 0:01, but for TEMP2, it will contain data current as of 0:06. Suppose that during those five seconds, a few large sales were registered, they will be included in the total dollar amount contained in TEMP2, but won't be represented in the category-level data that is in TEMP1. So, when the data is brought together in the final SELECT statement, the total in TEMP2 will be larger than the sum of the categories in TEMP1, and then the total percentage number will be less than 100%. Obviously, this will lead to an incorrect report.

Multi-version database is a desirable approach to ensuring read consistency. However, read consistency in multi-version database is only achieved on the level of single-pass SQL statement [9], which means that it can not be used to deal with the internal consistency of report that contains multi-pass SQL statement made up of many smaller SQL statements (see Fig.1). Moreover, multi-version database is not good at dealing with the query contention issue resulting from the real-time update and query. Temporal model (e.g. [10]) is another one of the methods that can be used to solve the report inconsistency issue. However, keeping temporal data warehouses up-to-date is complex [11], and in some cases, the data warehouse may even become blocked due to the query contention issue resulting from performing queries on changing data.

We here propose a new layer-based view approach for appropriately and effectively maintaining report's internal consistency, and at the same time avoiding query contention issue which is a hard work for other available methods. The concepts of layer, view and lock are introduced to effectively control read and write operations upon fact tables. The core idea is that, all the data involved in an OLAP query is read-locked and is not allowed to be updated until the query finishes its reading job, so as to maintain the internal consistency of report. When the data is read-locked, if there is confliction between read and write operations on certain layer, the write operation will be redirected to another layer to continue its work so as to avoid waiting time and maintain the consistency as well. To achieve this target, we will present in detail the mechanisms for layer generating and deleting, view generating and deleting, and lock applying and confliction resolving. Also the algorithm and an example of our approach will be given here. Compared with the other available methods, the advantage of our method is the avoiding of confliction between read and write operations upon fact tables, which means that there will be no waiting time any more and therefore desirable system performance can be achieved. Also, our method can be easily used in memory database. Unlike most of the multi-version databases, in which multi-version method is deeply integrated with the database systems, our method is completely independent of the type of database, which means that it can be used together with any database product.

The remainder of this paper is organized as follows: Section 2 gives the detailed description of layer-based view approach. Then we present experimental results in Section 3, followed by the discussion of the related work (Section 4). Finally, we give the discussion and conclusion in Section 5.

2   Maintaining Report's Internal Consistency with Layer-Based View Approach

In this section, the frequently used concepts will be defined first, followed by the description of the mechanisms of our approach in detail. Finally, we will give the algorithm and an example of our approach.

2.1   Term and Definition

Definition 1. *Layer: A layer, denoted by Δ , is a table to store a set of records $\{r_0, r_1, ..., r_n\}$. Layer can be classified into root layer and non-root layer. If, for two layers Δ_A and Δ_B, Δ_B is generated from Δ_A, we say Δ_B is the child of Δ_A, which is denoted by $\Delta_B \succ \Delta_A$, and Δ_A is the parent of Δ_B, which is denoted by $\Delta_A \prec \Delta_B$.*

In real-time data cache, every fact table is corresponded to one root layer. Every non-root layer, which is initially an empty table without any data when generated, has the same table structure as its parent layer.

Definition 2. *View: Let $L = \{\Delta_0, \Delta_1, ..., \Delta_{m-1}\}$ be a set of layers, and $R = \{r_0, r_1, ..., r_{n-1}\}$ be a set of records, where $r_i \in \Delta_j, 0 \leq i \leq n-1$ and $0 \leq j \leq m-1$. A view is defined as $\Gamma = (V, \phi)$, where $V = \{v_0, v_1, ..., v_{n-1}\}$ is a set of records, $\phi(v_p) = r_q, 0 \leq p \leq n-1$ and $0 \leq q \leq n-1$. Here we say Γ is composed of $\Delta_0, \Delta_1, ..., \Delta_{n-1}$, and Δ_j is a composing layer of Γ.*

A view defines the mapping between every record of itself and the records of its composing layers, from which OLAP tools get to know where the required data actually locate. There may be many views in the system, but the only view that can be seen by newly arrived OLAP queries is the "current view", which is denoted by $\Gamma_{current}$. After OLAP queries get $\Gamma_{current}$, they will use it during the whole reading process, even though the "current view" now may become "old view" in the future.

Definition 3. Area: *Let* $R = \{r_0, r_1, ..., r_n\}$ *be the set of records that a layer* Δ *(or a view* Γ *) contains. An area, denoted by* δ, *is a subset of* R. *Here we say the area* δ *is in the layer* Δ *(or in the view* Γ *), which is denoted by* $\delta \subseteq \Delta$ *(or* $\delta \subseteq \Gamma$ *). Suppose there are* $\delta, \delta_0, \delta_1, ..., \delta_{n-1}$, *where* $\delta \subseteq \Gamma$, $\delta_i \subseteq \Delta_j$, $0 \le i \le n-1$, $0 \le j \le m-1$ *and* m *is the number of layers. If, by the function* ϕ *of* Γ, *the records in* δ *are mapped to the records in* $\delta_0, \delta_1, ..., \delta_{n-1}$, *we say that* δ *is composed of* $\delta_0, \delta_1, ..., \delta_{n-1}$, *and* δ_j *is a composing area of* δ.

In order to better understand the concepts of layer, area and view, we give an example in Fig.2. As can be seen in Fig.2, View1 is composed of three layers, i.e. Layer1, Layer2 and Layer3. The three records in View1, which are what OLAP tools can see, are actually located in the three different layers. This is similar to the layer technology used in painting software (e.g. PhotoShop), where a photo is composed of many layers, and what we can see is the result of combining the objects in different layers together. This also explains why a table is called a layer in our paper. In Fig.2, there are four areas, where Area1 $\subseteq$ Layer1, Area2 $\subseteq$ Layer2, Area3 $\subseteq$ Layer3 and Area4 $\subseteq$ View1. Area1 contains one record, Area2 contains three records, Area3 contains two records and Area4 contains three records.

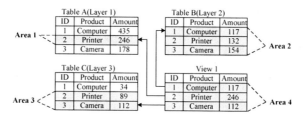

Fig. 2. An example of layer, area and view

Definition 4. Lock: *A lock* Ω *is used to lock certain area* δ *so as to control the read and write operations upon* δ, *where* $\delta \subseteq \Gamma$ *or* $\delta \subseteq \Delta$. $\Omega_{on}(\delta)$ *and* $\Omega_{off}(\delta)$ *mean placing locks on and removing locks from* δ *respectively.*

Lock includes "read lock" and "write lock" (see Table 1), and read lock has higher priority than write lock. The former is assigned by system to the OLAP tools to protect the target area being read from being updated, and the latter is used to inform the other operations that the target area is being updated by write operation, or else it will probably lead to the occurrence of inconsistency. Table 2 gives the compatibility relationship between read lock and write lock. Also a lock can be a virtual lock or an actual lock. A virtual lock is put on an area in a view, while the target object of an actual lock is an area in a layer.

2.2   Lock Mechanism

Lock mechanism is responsible for the jobs such as lock applying, lock translating and lock confliction resolving, so as to effectively control the read and write operations upon the changing data to maintain the internal consistency of report.

Table 1. Different types of locks

lock	virtual lock	read lock
		write lock
	actual lock	read lock
		write lock

Table 2. Lock compatibility matrix

	read lock	write lock
read lock	compatible	incompatible
write lock	incompatible	incompatible

Lock applying. Only virtual lock can be applied by OLAP tools and data loading and updating tools, for what these tools can see in the first place are views instead of layers.

– *Read lock applying.* The purpose of read lock is to declare the "occupation" of specific area by OLAP tools, which means the loading tools have no right to updating the "occupied" area. Since read lock has higher priority than write lock, read lock applying process never fails. For a multi-pass SQL statement, its read lock application is submitted as a whole to the system.
– *Write lock applying.* Write lock is used by updating operation to express the intention of updating certain area. Write lock has lower priority than read lock, and therefore its request may not be satisfied all the time. For a transactional updating statement, its write lock application is submitted as a whole to the system.

Definition 5. *Lock transforming: Suppose δ is composed of $\delta_0, \delta_1, ..., \delta_{n-1}$, where $\delta \subseteq \Gamma, \delta_i \subseteq \Delta_j, 0 \leq i \leq n-1, 0 \leq j \leq m-1$ and m is the number of layers. Lock transforming is the process of transforming the virtual lock on δ into actual locks on $\delta_0, \delta_1, ..., \delta_{n-1}$, which is denoted by $\Omega_{on}(\delta) \rightarrow \Omega_{on}(\delta_0) \cup \Omega_{on}(\delta_1) \cup ... \cup \Omega_{on}(\delta_{n-1})$.*

A view is composed of one or more layers, and therefore there is a need to transform the lock on it into one or more locks on layers. The following is an example to explain the transformation mechanism between virtual lock and actual lock.

In Fig. 3, View1 is composed of three layers, i.e. Layer1, Layer2 and Layer3. While what OLAP tools can see is View1, with the help of view definition, they will finally be "guided" to the three layers where the data actually locate. This also takes place for write operations. Now suppose that Lock0 is imposed on View1, which can be either a read lock or a write lock. Through the analysis of view definition, Lock0 is finally translated into three locks with Lock1 on the second record of Layer1, Lock2 on the first record of Layer2 and Lock3 on the third record of Layer3.

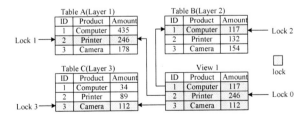

Fig. 3. The transformation of virtual lock into actual lock

Definition 6. *Lock confliction* *If, according to the lock compatibility matrix in Table 2, two locks* $\Omega_{on}(\delta)$ *and* $\Omega'_{on}(\delta)$ *are incompatible, we say that there is lock confliction on* δ.

Whenever there is lock confliction on δ ($\delta \subseteq \Delta$), the write operation has to be redirected to another layer Δ_{child}, which is a child layer of Δ. If Δ_{child} does not exist, it will be generated automatically by system through the mechanism of layer generating. This process is called *lock confliction resolving*, during which, one important aspect is to maintain the transactional consistency of the undergoing write operation.

2.3   Layer and View Mechanism

Layer generating: The layer generating process is activated whenever the write operation needs to be completed in another layer due to the lock confliction, and at the same time there is not one layer available for it. The newly generated layer has the same table structure as its parent layer, but is initially empty.

Layer merging and deleting: In order to achieve better system performance, there is also a need to merge and delete layers under certain condition. Whenever the preset threshold is met, the layer merging process begins.

View generating: Whenever there is updating against the definition of a view, a new view containing the newest definition will be generated above the old one with the latter unchanged, and then the new view becomes $\Gamma_{current}$. The reason for the generation of a new view when the definition of the current view is changed, is out of the consideration of avoiding the confliction between the read operation of OLAP tools and the write operation of updating against $\Gamma_{current}$.

View deleting: The process of view deleting begins when certain view is no longer used by OLAP tools, or it will lead to the depletion of system resources, because new views are generated constantly along the time. Sometimes the deleting process of views is also accompanied with the merging of layers.

2.4   The Algorithm for Layer-Based View Approach

Based on the mechanisms described above, we can now implement the layer-based view approach. Fig. 4 shows the main algorithm for this method. The 5th and 12th lines are executed concurrently instead of sequentially with the help of multi-thread technology. In other words, the read process and the confliction resolving process go at the same time. And for every layer, the confliction resolving process can also execute concurrently. The read operation will read the right now available data first so as to avoid the waiting time, and then read the previously write-locked but currently available data. Even though the transactional updating has to be assured to be finished when the undergoing write operation needs to be redirected to another layer during the confliction resolving process, the overall waiting time of read operation is still very little and usually can be neglected due to the adoption of multi-thread technology as is described above.

In the 7th line, if the undergoing write operation is a part of transactional updating, it can not be stopped until the transaction is finished. In the 8th and 21th lines, as far as

---

Input: an SQL statement S
        the current view $\Gamma_{current}$
Output: execution result of S

1.   scan the statement to get the value of δ;
     /* δ is the area in $\Gamma_{current}$ that S is requried to lock;*/
2.   **if** S is a read operation
3. $\Omega_{on}(\delta)$;
4. $\Omega_{on}(\delta) \rightarrow \Omega_{on}(\delta_0) \cup \Omega_{on}(\delta_1) \cup ... \cup \Omega_{on}(\delta_{n-1})$;
5.   **for** $(i = 0; i < n; i + +)$
6.   **if** (there is lock confliction on δ_i) /*$\delta_i \subseteq \Delta$*/
7.   stop the undergoing write operation rightly;
8.   find another right layer Δ_{child} for the undergoing write operation;
            /* Δ_{child} is a child layer of Δ; */
9.   continue to do the suspended write operation on Δ_{child};
10.   generate a new view Γ_{new} to record the layer information;
11. $\Gamma_{current} = \Gamma_{new}$;
12.   read the data on the target layers into a temporary table T;
13.   **for** $(i = 0; i < n; i + +)$
14. $\Omega_{off}(\delta_i)$;
15.   **return** T;
16. **else** /* S is a write operation*/
17. $\Omega_{on}(\delta)$;
18. $\Omega_{on}(\delta) \rightarrow \Omega_{on}(\delta_0) \cup \Omega_{on}(\delta_1) \cup ... \cup \Omega_{on}(\delta_{n-1})$;
19.   **for** $(i = 0; i < n; i + +)$
20.   **if** (there is lock confliction on δ_i) /*$\delta_i \subseteq \Delta$*/
21.   find another right layer Δ_{child} for the write operation of S;
22.   do the write operation of S on Δ_{child};
23.   generate a new view Γ_{new} to record the layer information;
24. $\Gamma_{current} = \Gamma_{new}$;
25.   **else**
26.   do the write operation of S on Δ ;
27.   **for** $(i = 0; i < n; i + +)$
28. $\Omega_{off}(\delta_i)$;
29.   **return** success information;

Fig. 4. The algorithm for layer-based view approach

Δ_{child} is concerned, it may be the existing child layer of Δ or a newly generated child layer of Δ according to the different conditions, and it has the same table structure as Δ.

2.5   An Example of Layer-Based View Approach

In order to better understand how our layer-based view approach works, we here give an example.

As can be seen in Fig.5, at time T_1, there is only one layer, i.e. Layer1 in the system. View1 ($\Gamma_{current}$) is composed of Layer1, and is also what OLAP tools can see at time T_1. We suppose that a query Q_1 has already read-locked all the records (i.e. v_1, v_2 and v_3) of View1 before time T_1, and it will not release its read locks until time T_3.

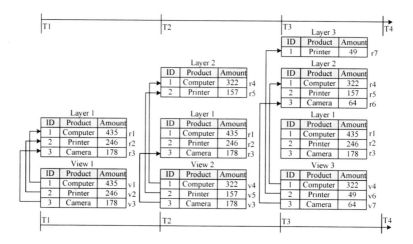

Fig. 5. The generating process of layers and views

According to the knowledge about the lock transformation, the virtual read locks on v_1, v_2 and v_3 of View1, are transformed into the actual read locks on r_1, r_2 and r_3 of Layer1. So r_1, r_2 and r_3 of Layer1 are all read-locked. Then, at time T_1, an update operation U_1 arrives at the system, expecting to update the records v_1 and v_2 of View1. Before starting the update work, U_1 must apply to the system for the virtual write locks on v_1 and v_2. The virtual write locks will then be transformed into the actual write locks on r_1 and r_2. However, r_1 and r_2 are already read-locked by Q_1, and also in our method, read lock has higher priority to write lock, so the write lock application of U_1 fails due to the lock confliction. But U_1 will not wait for the read locks to be released, and it will be redirected to Layer2, the child layer of Layer1, to continue its job. In other words, U_1 will write the update results (r_4 and r_5) into Layer2. Layer2 is automatically generated by the system to accommodate the write operation U_1. After U_1 completes its work, the system will generate a new view, i.e. View2, to reflect the most recent data, and then View2 becomes $\Gamma_{current}$. From now on, View2 is what OLAP tools can see, but View1 will not be deleted until Q_1 finishes its work.

Then, at time T_2, another query Q_2 arrives at the system, and needs to read v_4, v_5 and v_3. Q_2 first applies virtual read locks on v_4, v_5 and v_3 of View 2, which will be transformed into three actual locks, i.e. the actual read locks on r_4 and r_5 of Layer2, and the actual read lock on r_3 of Layer1. Read lock application will never fail, so Q_2 starts to read the locked records, and here we suppose that it will not release its read locks until the time T_3. When Q_2 is undergoing its read work, another update operation U_2 arrives at the system, and expects to update the records v_5 and v_3 of View2. Its virtual write locks on v_5 and v_3 of View2 will be transformed into actual write locks, i.e. the lock on r_5 of Layer2 and the lock on r_3 of Layer1. Because r_3 of Layer1 is still read-locked, lock confliction occurs, and the system will redirect U_2 to Layer2, the child layer of Layer1, to do the update against r_3. Similarly, update operation against r_5 will also be redirected to Layer3, the child layer of Layer2. After U_2 finishes its work, the system will generate a new view, i.e. View3, to present the most recent data, and then View3 becomes $\Gamma_{current}$. From now on, View 3 is what OLAP tools can see, but View2 will not be deleted until Q_2 completes its work.

At time T_3, both Q_1 and Q_2 finish their reading job, and release their read locks. During the whole reading process, data involved in Q_1 (r_1,r_2 and r_3) and that involved in Q_2 (r_4,r_5 and r_3) are never changed, so the internal consistency of report is well maintained. Also, when the reading work is undergoing, the update operations of U_1 and U_2 are not blocked, instead they perform their jobs smoothly and successfully.

3   Empirical Study

Now we report the performance evaluation of our method. The algorithms were implemented with C++. All the experiments are conducted on 4*2.4GHz CPU (double core), 32G memory HP Proliant DL585 Server running Windows Server 2003 and Oracle 10g (for operational system and data warehouse) and Oracle TimesTen In-Memory Database (for real-time data cache).

We use TPC benchmark TPC-H to get the required datasets in our experiment. DB-GEN, a tool provided by TPC, is used here to generate the required datasets to populate the database in the data source. We take real-time data cache [8] running Oracle TimesTen In-Memory Database to store all the real-time data. The external data cache database is generally modeled identically to the data warehouse, but typically contains only the tables that are real-time. Also through JIM or RJIM system [8], we can seamlessly combine the real-time data in the data cache and historical data in the data warehouse. With the help of Streams Components provided by Oracle 10g, it is easy to capture the change data in the data source and send them to the destination queue, from which they are dequeued to be integrated into the data caches.

Performance ratio. In this experiment, we will show that our method can not only maintain report internal consistency, but also effectively avoid the contention between read and write operations so as to achieve desirable performance for both update and query. In order to show the influence of read and write operations upon system performance, we change the contention ratio (denoted by r) between these two kinds of

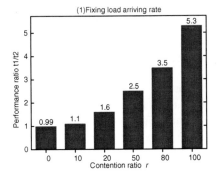 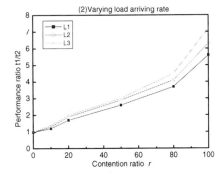

Fig. 6. Performance ratio. (1) Fix the load arriving rate, and change the value of contention ratio r from 0 to 100%. (2) Change the load arriving rate, and for each type of load L_1, L_2 and L_3, change the value of contention ratio r from 0 to 100%.

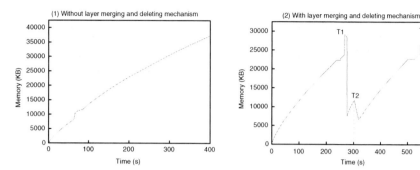

Fig. 7. Memory usage. (1) When there is no layer merging and deleting mechanism, the memory space is to be exhausted. (2) When there exists layer merging and deleting mechanism, the memory can be recycled.

operations in the load. Fig. 6 (1) shows how the performance ratio t_1/t_2 changes when varying the value of r from 0 to 100%, where t_1 denotes the total running time for the given update and query load L when not taking layer based view approach, and t_2 the total running time for L when taking our method. In Fig. 6 (1), we can see that when r equals 0, which means there is no read and write contention, the value of t_1/t_2 is 0.99. In another word, under such circumstance, it will bring negative benefits, though only a few, when taking layer based view approach. This is due to the additional cost for layer management. However, benefits from our method will become more and more evident when the contention ratio r increases. As can be seen in Fig. 6 (1), the performance ratio is 5.3 when r equals 100%.

We also test our method under three different kinds of workloads, i.e., L_1, L_2 and L_3. These workloads contain the same sequence of update and query operations with the same contention ratio, but feeding the system at different rates. The update and query operations in L_1 arrive at the slowest rate, and those in L_3 at the fastest rate. Fig. 6 (2) shows that our method can achieve much better performance ratio (the same as the definition above, i.e., t_1/t_2) when arriving rate is faster. Since there will be much more contention between read and write operations within a given time window when increasing the arriving rate, it can be concluded that our method can effectively deal with the query contention issue besides maintaining report internal consistency.

Memory usage. In this experiment, we will show that our method can effectively manage memory usage through the mechanism of layer merging and layer deleting. Fig. 7 (1) shows the memory usage when there is no layer merging and deleting mechanism. In such case, more and more memory is used to support the continuously generated layers. Since there is no layer dropping mechanism, those layers without any use in the future still reside in memory, which wastes much memory space and is to exhaust the limited memory resource in the end. In contrast, the memory resource can be recycled (see Fig. 7 (2)) when there exists layer merging and deleting mechanism. In Fig. 7 (2), there are three turning point T_1, T_2 and T_3 in the memory usage curve. At T_1 and T_3, the memory usage amounts to a predefined value M, layer merging process starts, which merges different layers into one if condition is met and drops many useless layers to release the memory. At T_2, a lot of DELETE operations occurs in the system, which leads to the start of layer dropping process and much memory is recycled.

4   Related Work

Report's internal consistency is an important issue for real-time OLAP. Temporal model (e.g. [10]) can be used to solve the problem, which enables analytical processing of detailed data based on a stable state at a specific time [11]. While research in temporal databases has grown immensely in recent years, only a few DWH research projects paid attention to such problem as temporal model. Until the work done by Bobert et al. in [10], most of the previous research has been concentrated on performance issues, rather than higher-level issues, such as conceptual modeling [12]. In [10], an approach is presented to model conceptual time consistency problems, in which all states that were not yet known by the system at specific point in time are consistently ignored, and thus enables timely consistent analyses.

Temporal model addresses the issue of supporting temporal information efficiently in data warehousing systems, however, keeping temporal data warehouses up-to-date is complex [11]. Usually, it is more appropriate to be used to deal with the temporal consistency problem brought by late-arriving data, than to be used to solve the problem of report's internal consistency resulting from the continuous data integrating in the real-time data warehouse environment.

Another desirable way is to use an external real-time data cache, and at the same time without compromising report's internal consistency, data latency, or the user experience [8]. However, there are problems with the method even with the help of JIM system. The most obvious one is that, in the real-time data warehouses environment, real-time data cache is continuously updated, and therefore the "read" operation of getting a snapshot for OLAP tools will undoubtedly conflict with the "write" operation of data updating. If there are many concurrent users in the system, such confliction is to deteriorate system performance greatly. Such problem can not get resolved simply through adding more hardware to the system.

Multi-version database is a desirable method to maintain read consistency [9]. To some extent, our method is similar to multi-version database. However, there are still great differences between them. For example, our method can provide read consistency for multi-pass SQL statement, but it is hard for multi-version database. Also, our method is independent of DBMS and can be used in memory database. However, multi-version method is usually integrated into DBMS products, and can not be used in memory databases in some cases. Furthermore, our method can effectively deal with query contention issue besides maintaining report internal consistency, while multi-version database can not.

There are also some other methods available now, such as using a near real-time approach and risk mitigation for true real-time [8]. In [13], a new method is presented, which raises the level of abstraction for the use of replicated and cached data by allowing applications to state their data currency and consistency requirements explicitly and having the system take responsibility for producing results that meet those requirements.

5 Discussion and Conclusion

In this paper, we have revisited the issue of maintaining internal consistency of report for real-time OLAP. We propose a new method, called layer-based view, to effectively maintain report's internal consistency in real-time data warehouses environment. Important concepts such as layer, view and lock are defined, and the related mechanism, especially the layer generating mechanism, are discussed in detail. The advantages provided by layer-based view approach include no confliction between read and write operations, achievement of report's internal consistency and faster response time for OLAP queries.

Important future research directions in this field will be the appropriate definition of threshold for layer merging process, and the application of the theory into the field of read-time data warehouses in business environment.

References

1. Tho, M.N., Tjoa, A.M.: Zero-Latency Data Warehousing for Heterogeneous Data Sources and Continuous Data Streams. In: 5th International Conference on Information Integrationand Web-based Applications Services, pp. 55–64. Austrian Computer Society, Vienna (2003)
2. Thalhammer, T., Schrefl, M.: Realizing active data warehouses with off-the-shelf database technology. Software-Practice & Experience 32(12), 1193–1222 (2002)
3. Chen, L., Rahayu, J.W., Taniar, D.: Towards Near Real-Time Data Warehousing. In: 24th IEEE International Conference on Advanced Information Networking and Applications, pp. 1150–1157. IEEE Computer Society, New York (2010)
4. Santos, R.J., Bernardino, J.: Real-time Data Warehouse Loading Methodology. In: 12th International Database Engineering and Applications Symposium, pp. 49–58. ACM, New York (2008)
5. Bateni, M., Golab, L., Hajiaghayi, M.T., Karloff, H.J.: Scheduling to Minimize Staleness and Stretch in Real-time Data Warehouses. In: 21st Annual ACM Symposium on Parallel Algorithms and Architectures, pp. 29–38. ACM, New York (2009)
6. Polyzotis, N., Skiadopoulos, S., Vassiliadis, P., Simitsis, A., Frantzell, N.: Supporting Streaming Updates in an Active Data Warehouse. In: 23rd International Conference on Data Engineering, pp. 476–485. IEEE Computer Society, New York (2007)
7. Conn, S.S.: OLTP and OLAP Data Integration: a Review of Feasible Implementation Methods and Architectures for Real Time Data Analysis. In: Proceedings of SoutheastCon 2005, pp. 515–520. IEEE Computer Society, New York (2005)
8. Langseth, J.: Real-Time Data Warehousing: Challenges and Solutions, http://www.dssresources.com
9. Korth, H.F.: Database System Concepts, 3rd edn. McGraw-Hill, New York (1999)
10. Bruckner, R.M., Tjoa, A.M.: Capturing Delays and Valid Times in Data Warehouses-Towards Timely Consistent Analyses. Journal of Intelligent Information Systems 19(2), 169–190 (2002)
11. Bruckner, R.M., Tjoa, A.M.: Managing Time Consistency for Active Data Warehouse Environments. In: Kambayashi, Y., Winiwarter, W., Arikawa, M. (eds.) DaWaK 2001. LNCS, vol. 2114, pp. 254–263. Springer, Heidelberg (2001)
12. Pedersen, T.B., Jensen, C.S., Dyreson, C.E.: A Foundation for Capturing and Querying Complex Multidimensional Data. Information Systems 26(5), 383–423 (2001)
13. Guo, H., Larson, P.A., Ramakrishnan, R., Goldstein, J.: Relaxed currency and consistency: how to say "good enough" in SQL. In: ACM SIGMOD 2004 International Conference on Management of Data, pp. 815–826. ACM, New York (2004)

Efficient Approximate Top-k Query Algorithm Using Cube Index

Dongqu Chen, Guang-Zhong Sun[*], and Neil Zhenqiang Gong[**]

Key Laboratory on High Performance Computing, Anhui Province
School of Computer Science and Technology
University of Science and Technology of China
cdq2012@mail.ustc.edu.cn, gzsun@ustc.edu.cn,
neilz.gong@berkeley.edu

Abstract. Exact top-k query processing has attracted much attention recently because of its wide use in many research areas. Since missing the truly best answers is inherent and unavoidable due to the user's subjective judgment, and the cost of processing exact top-k queries is highly expensive for datasets with huge volume, it is intriguing to answer approximate top-k query instead. In this paper, we first define a novel kind of approximate top-k query, called *μ-approximate* top-k query. Then we introduce an efficient index structure, i.e. *cube index*, based on which, we propose our novel Cube Index Algorithm (CIA). We analyze the complexity of both constructing cube index and CIA algorithm. Moreover, extensive experiments show that CIA performs much better than the well-known approximate TA$_\theta$ algorithm [3].

Keywords: Top-k Query Processing, Algorithms, Index.

1   Introduction

Exact top-k query processing has gained more and more attention recently because of its wide use in many fields, such as information retrieval [16][17], multimedia databases [20][21], P2P and sensor networks [18][19], etc. The main reason for such attention is that top-k queries avoid overwhelming the user with a large number of uninteresting answers that are resource-consuming.

However, two main reasons convince us to abandon exact top-k query processing. First, the top-k query concept is heuristic anyway. Hardly any user is interested in all the exact k answers of a top-k query. Instead, they may be only interested in one or several relevant objects in the top-k (e.g. 500 or 2000) answers. So, due to the subjective judgment of the user, missing the truly best answers is inherent and unavoidable. This argument enlightens us to relax exact top-k query to approximate top-k query. Second, the cost of processing exact top-k queries is highly expensive for datasets

---

[*]   Corresponding author.
[**] Neil Z. Gong is now a graduate student in EECS Department, UC Berkeley. This work was completed when he was an undergraduate student of USTC.

X. Du et al. (Eds.): APWeb 2011, LNCS 6612, pp. 155–167, 2011.
© Springer-Verlag Berlin Heidelberg 2011

with huge volume, and the size of datasets in practice is always quite huge. So it's intriguing to answer approximate top-k query instead of exact top-k query.

To answer approximate top-k queries, Fagin et al. propose the TA$_\theta$ algorithm in [3], which is based on the TA algorithm. Based on TA$_\theta$, Theobald et al. [6] introduced a scheme to associate probabilistic guarantees with approximate top-k answers. In [8], Amato used a *proximity* measure to decide if a data region should be inspected or not. Only data regions whose *proximity* to the query region is greater than a specified threshold are accessed. This method is used to rank the nearest neighbors to some target data object in an approximate manner. Approximate top-k query processing has been also studied in peer-to-peer environments. The KLEE system (Michel et al. [2]) addressed this problem, where distributed aggregation queries are processed based on index lists located at isolated sites. KLEE assumes no random accesses are made to index lists located at each peer.

In this paper, we first define a novel approximate top-k query, called μ-*approximate* top-k query. Then we introduce an efficient index structure, i.e. *cube index*, based on which, we propose our new Cube Index Algorithm (CIA). We analyze the complexity of both constructing cube index and CIA algorithm. Moreover, extensive experiments show that CIA performs much better than the well-known approximate TA$_\theta$ algorithm.

The rest of this paper is organized as follows: First, we define the computation model formally and review the TA$_\theta$ algorithm in Section 2. In Section 3, we describe our method on setting up the cube index and then analyze its time complexity. Based on these, we show our algorithm CIA and analyze its cost in Section 4. Thereafter, we show the experimental results in Section 5. Finally, in Section 6, we conclude this paper and introduce our future work.

2 Computation Model and TA$_\theta$ Algorithm

In this section, we describe the model of top-k problem and review the TA$_\theta$ algorithm [3].

Our model of the dataset can be described as follows: assume the database D consists of n objects, which are denoted as $x_1, x_2 \ldots x_n$. Each object x is an m-dimensional vector $(s_1(x), s_2(x) \ldots s_m(x))$, where $s_i(x)$ is the ith local score of x as a real number in the interval [0, 1]. For a given object x, x has a total score of $f(x)= f(s_1(x), s_2(x) \ldots s_m(x))$, where the m-dimensional aggregate function f is supposed to be increasingly monotonic:

Definition 2.1. *Monotonic Function.* An aggregate function f is *increasingly monotonic* if $f(a_1, a_2 \ldots a_m) \leq f(a_1', a_2' \ldots a_m')$, whenever $a_i \leq a_i'$ for every i.

In this paper, we assume the aggregate function is weighted summation function, $f(x) = \sum_{i=1}^{m} w_i s_i(x)$, where $s_i(x) \in [0, 1]$ and $\sum w_i = 1$ ($w_i \neq 0$). We can easily verify that weighted summation function is increasingly monotonic. Exact top-k query is to find k objects with the highest total scores. For approximate top-k query, Fagin et al. [3] defined a θ-*approximation* to the top-k answers:

Definition 2.2. *θ-Approximation* [3]. Let *Y* be a collection of *k* objects such that for each *y* among *Y* and each *z* not among *Y*, there are $\theta f(y) \geq f(z)$, where $\theta > 1$. Then *Y* is one of the top-*k* answers with *θ-approximation* and *θ* is the *relative approximation coefficient*.

To solve the *θ-approximation* top-*k* query, Fagin et al. [3] proposed the TA_θ algorithm, based on the threshold algorithm (i.e. TA). TA_θ is described in Fig. 1.

Threshold Algorithm with *θ-Approximation* (TA_θ)

Pre-computing Phase:
For each attribute $i \in \{1, 2 \dots m\}$, get every $s_i(x_j)$ where $j \in \{1, 2 \dots n\}$ and insert them into a sorted list L_i. Sorted list means that objects in each list are sorted in descending order by the $s_i(x_j)$ value.

Computing Phase:
1: Do sorted access in parallel to each of the *m* lists. As an object is seen through sorted access in some list, do random access to the other lists to find all its remaining local scores, and compute its overall score. Maintain a set *Y* containing the *k* objects whose overall scores are the highest among all the objects seen so far.
2: For each list L_i, let $\underline{s_i}$ be the last local score seen under sorted access in L_i. Define the *threshold value* τ to be $\tau = f(\underline{s_1}, \underline{s_2} \dots \underline{s_m})$..
3: Halt when $\theta \cdot M_k \geq \tau$, where $M_k = \min\{f(x) \mid x \in Y\}$.

Fig. 1. Threshold Algorithm with *θ-Approximation*

3   Cube Index

Before proposing our algorithm, we first introduce an efficient indexing structure called cube index to support such *μ-approximation* top-*k* query processing.

3.1 Description of Cube Index

We map the database to an *m*-dimensional hyperspace $[0, 1]^m$; each object x_j with scores $(s_1(x_j), s_2(x_j) \dots s_m(x_j))$ in the database is mapped to an *m*-dimensional point $p_j = (s_1(x_j), s_2(x_j) \dots s_m(x_j))$ in $[0, 1]^m$. We will not distinguish between object *x* and its corresponding point *p* discussed below. Similarly, $s_i(p)$ is the value of *p*'s ith dimension and $f(p)$ is *p*'s total score.

Now we define a *μ-approximation* to the top-*k* answers.

Definition 3.1. *μ-Approximation.* Let *Y* be a collection of *k* objects such that for each *y* among *Y* and each *z* not among *Y*, there are $f(y) + \mu \geq f(z)$, where $0 < \mu \leq 1$. Then *Y* is one of the top-*k* answers with *μ-approximation* and *μ* is the *proportional approximation coefficient*.

Definition 3.2. *Dominate* [7]. Point p_1 dominates point p_2 if and only if for each $i \in \{1, 2 \dots m\}$, $s_i(p_1) \geq s_i(p_2)$ and there exists at least one member *j* of $\{1, 2 \dots m\}$ satisfying $s_j(p_1) > s_j(p_2)$.

Observation 3.1. If point p_1 dominates point p_2, then $f(p_1) > f(p_2)$, where f is an *aggregate monotone function*.

Proof. We can easily get the correctness of Observation 3.1 according to the definitions of *aggregate monotone function* and *dominate*. □

Definition 3.3. *Skyline* [7]. The *skyline* of a dataset D is the set of points that are not *dominated* by any point in D.

Definition 3.4. *Bottom Point*. The *bottom point* of a hypercube is the vertex whose values of every dimension are all lowest in the hypercube.

For example, the *bottom point* of the 3-dimensional cube $[0.2, 0.3] \times [0.1, 0.2] \times [0.5, 0.6]$ is $(0.2, 0.1, 0.5)$.

Observation 3.2. All other points in a hypercube *dominate* the *bottom point*.

Proof. We can easily get the correctness of Observation 3.2 according to the definitions of *dominate* and *bottom point*. □

Now we show the cube partition method on the m-dimensional hyperspace $[0, 1]^m$, which is described as follows:

Firstly, we set the length of the edge of each hypercube as μ, where $\mu \in [0, 1]$. Then we divide the interval $[0, 1]$ into several μ-segments from 1 to 0 until the rest is shorter than μ. Each dimension is divided in this way so that the m-dimensional hyperspace $[0, 1]^m$ is partitioned into several hypercubes or sub-hyperspaces. Thereafter, we classify all the points in database into several sets: Point p_i belongs to bp_i's associated point set S_i if and only if p_i is in the hypercube whose *bottom point* is bp_i.

We call this partition method the *μ-cube partition*.

Definition 3.5. *Sky Point*. For a *μ-cube partition*, the *sky point* is the point whose values in every dimension are all $1 - \mu$, that is, the point $(1 - \mu, 1 - \mu \ldots 1 - \mu)$.

Apparently, *sky point* is the very *bottom point* which *dominates* all the other *bottom points* and the set {*sky point*} is the *skyline* of the set of *bottom points*.

Definition 3.6. *Neighbor*. Bottom point bp_1 is a *neighbor* of bottom point bp_2 if and only if they satisfy $\sum\limits_{i=1}^{m} \left\lceil \dfrac{|s_i(bp_1) - s_i(bp_2)|}{\mu} \right\rceil = 1$.

Definition 3.7. *Superior*. Bottom point bp_1 is a *superior* of bottom point bp_2 if and only if bp_1 is a *neighbor* of bp_2 and bp_1 *dominates* bp_2.

Definition 3.8. *Inferior*. Bottom point bp_1 is an *inferior* of bottom point bp_2 if and only if bp_1 is a *neighbor* of bp_2 and bp_1 is *dominated* by bp_2.

Discussions on special cases:

1)   For the points in the hypercube whose *bottom point* is the *sky point* belong to the 0th set S_0.

2)   The points on the intersecting hyperplane of several neighboring hypercubes belong to the hypercube whose *bottom point* dominates the others' *bottom point*.

3)   The points coinciding with bp_i belong to set S_i.

4)   If $S_i.size = 0$ and $i \neq 0$, then remove bp_i from the set of *bottom points*. Meanwhile, for each *inferior inf* of bp_i, regard all the *superiors* of bp_i as *inf*'s *superiors* too; for each *superior sup* of bp_i, regard all the *inferiors* of bp_i as *sup*'s *inferiors* too.

Definition 3.9. *μ-Cube Index*. For a *μ-cube partition*, the *μ-cube index* is an index list or array whose entries are ids of the *bottom points*. Each *bottom point* bp_i has its associated point set S_i as well as its *superiors*' ids and *inferiors*' ids.

3.2   Complexity Analysis of μ-Cube Indexing Method

Now we analyze the time complexity of the method on setting up the cube index, which is done in the pre-computing phase.

According to the description, the most time-consuming calculations in a *μ-cube partition* are to find the *superiors* and *inferiors* of each *bottom point* and to classify all the points in database into their corresponding sets.

Actually, the *superiors* and *inferiors* of each *bottom point* bp can be determined by the following two simple formulas:

1.   For each $i \in \{1, 2 \ldots m\}$ and $s_i(bp) \neq 0$, bottom point bp' is one *inferior* of bp, satisfying

  a. $s_i(bp') = (s_i(bp) \ \mu) \cdot H(s_i(bp) \ \mu)$, where $H(x)$ is the Heaviside step function;

  b. $s_j(bp') = s_j(bp)$ for each $j \in \{1, 2 \ldots m\}$ and $j \neq i$.

2.   *Bottom point* bp' is one *inferior* of bp if and only if bp is one *superior* of bp'.

There are $\left\lceil \dfrac{1}{\mu} \right\rceil^m$ *bottom points* in total, so the time complexity to find the *superiors*

and *inferiors* of each *bottom point* is $O\left(\left\lceil \dfrac{1}{\mu} \right\rceil^m \times m\right)$.

On the other hand, each point p in database belongs to set S_i if and only if set S_i's corresponding *bottom point* bp_i satisfies that for each $i \in \{1, 2 \ldots m\}$,

  a. $s_i(bp) = (1 - \left\lceil \dfrac{1 - s_i(p)}{\mu} \right\rceil \times \mu) \times H(1 - \left\lceil \dfrac{1 - s_i(p)}{\mu} \right\rceil \times \mu)$ if $s_i(p) \neq 1$, where $H(x)$ is the Heaviside step function;

  b. $s_i(bp) = 1 - \mu$ when $s_i(p) = 1$.

Similarly, there are n points in database, so the time complexity to classify all the points in database into their corresponding sets is $O(mn)$.

Therefore, the total time complexity in the pre-computing phase is $O\left(m\left\lceil \dfrac{1}{\mu} \right\rceil^m + mn\right)$.

4   The Cube Index Algorithm

4.1   Description of Cube Index Algorithm

Based on the *μ-cube index*, we now propose a novel algorithm to answer the *μ-approximation* top-k query: the Cube Index Algorithm (i.e. CIA), which is described by the pseudo-code in Fig. 2.

Cube Index Algorithm (CIA)

Pre-computing Phase:

    Execute the normalization then set up the μ-cube index on the database.

Computing Phase:

1: $Y = \emptyset$, $CL = \emptyset$, $T = \{sky\ point\}$, where Y is the result set while CL is the sorted candidate list according to the total scores and T is a temp set.

2: **if** $S_0.size \leq k$ **then**

3:    add all points in S_0 into Y

4: **else**

5:    *Selectively Add* k points in S_0 into Y.

6: $bp_i = sky\ point$.

7: **while** ($Y.size < k$) **do**

8:    **for** each *inferior inf* of bp_i **do**

9:      **if** *inf* has not been accessed before and all *superiors* of *inf* is among T **then**

10:          Access *inf* and insert it into CL

11:    **else**

12:          Continue.

13:    **if** $CL.size > k - Y.size$ **then**

14:    Only keep the first $k - Y.size$ points in CL.

15:    Let bp_i be the *bottom point* with the highest score in CL and move it into T.

16:    **if** $S_i.size \leq k$ $Y.size$ **then**

17:    add all points in S_i into Y

18:    **else**

19:    *Selectively Add* k $Y.size$ points in S_i into Y.

20: **Return** Y.

Fig. 2. Cube Index Algorithm

Here *Selectively Add* in the pseudo-code is a sub-method to improve the precision of the algorithm qualitatively. It can be to add the points at random, or to add them from the points in *skyline* of S_i or others ways.

4.2 μ-Approximation of Cube Index Algorithm

To proof the μ-approximation of CIA, we first introduce three lemmas and a corollary as follows.

Lemma 4.1. Set T is always the top-($T.size$) answers to the set of *bottom points*.

Proof. (By mathematical induction) *Basis:* Set $T = \{sky\ point\}$ is the top-1 answers to the set of *bottom points*. Actually, *sky point dominates* all the other *bottom points* for the formula of μ-cube index and the definition of *sky point*. According to Observation 3.1, the *sky point* is the top-1 in the set of *bottom points*.

    Inductive step: Assume that set T is the top-j answers to the set of *bottom points* now, then the *bottom point* bp_i with the highest score in CL is the top-($j + 1$) in the set of *bottom points* and is supposed to be moved to set T from CL.

Actually, only the points in the *CL* now have the chance to be the top-$(j + 1)$. Otherwise, for a point *bp* which is not in *CL* or set *T*, either *bp* has been accessed before or *bp* has at least one *superior* that is not in set *T*. In the first case, according to the algorithm, CIA halts if and only if *Y.size* = *k*, so *Y.size* < *k* before the algorithm halts. If *bp* has been accessed before and be removed from *CL*, then there exist at least $T.size+(k-Y.size) \geq T.size+1 = j + 1$ points whose total scores are higher than *bp* so that *bp* even has no chance to be one of the top-$(j + 1)$ answers. In the other case, according to the definition of *superior* and Observation 3.1, every *superior* *sup* of *bp* satisfies $f(sup) > f(bp)$, so once *sup* is not in the top-*j* answers, or set *T*, *bp* has no chance to be one of the top-$(j + 1)$ answers. Furthermore, for each point *bp* in *CL*, where $bp \neq bp_i$, *bp* is impossible to be one of the top-$(j + 1)$ answers because even bp_i is not in the top-*j* answers. Therefore, bp_i is the top-$(j + 1)$ in the set of *bottom points*.

Conclusion: When CIA halts, set *T* is the top-(*T.size*) answers to the set of *bottom points*. □

Corollary 4.1. *Bottom points* are moved into set *T* in descending order of total score.
Proof. From the proof of Lemma 4.1, we easily conclude that *bottom points* are moved into set *T* in descending order of total score. □

Lemma 4.2. When CIA halts, there is at most one *bottom point* bp_j in set *T* satisfying $S_j \nsubseteq Y$, where bp_j is the one with the lowest score in set *T* and for each $bp_i \in T$ and $bp_i \neq bp_j$, $S_i \subseteq Y$.
Proof. According to the algorithm, the sub-method *Selectively Add* is executed if and only if $S_j.size > k$ $Y.size$. In this case, we *Selectively Add* k $Y.size$ points in S_j into *Y* so that $S_j \nsubseteq Y$. Thus there would be *Y.size* = *k* once the *Selectively Add* has been executed, where CIA halts. So the sub-method *Selectively Add* can be executed at most once. For Corollary 4.1, bp_j is the one with the lowest score in set *T*. However, in the case that $S_i.size \leq k$ $Y.size$, we add the whole S_i into set *Y* so that $S_i \subseteq Y$.

Therefore, when CIA halts, there is at most one *bottom point* bp_j in set *T* satisfying $S_j \nsubseteq Y$, where bp_j is the one with the lowest score in set *T* and for each $bp_i \in T$ and $bp_i \neq bp_j$, $S_i \subseteq Y$. □

Lemma 4.3. For point $p_i \in S_i$ and point $p_j \in S_j$, if $f(bp_i) \geq f(bp_j)$, then $f(p_i)+\mu \geq f(p_j)$.
Proof. According to the formula of μ-cube index and the definition of *bottom point*, for each $l \in \{1, 2 \dots m\}$, there is $s_l(bp_j) \leq s_l(p_j) \leq s_l(bp_j) + \mu$. Considering $f(x) = \sum_{l=1}^{m} w_l s_l(x)$, where $s_l(x) \in [0, 1]$ and $\sum_{l=1}^{m} w_l = 1$, we have

$$f(bp_j) \leq f(p_j) \leq \sum_{l=1}^{m} w_l \left[s_l(p_j) + \mu \right] = \sum_{l=1}^{m} w_l s_l(p_j) + \sum_{l=1}^{m} w_l \mu = f(bp_j) + \mu$$

for Observation 3.1 and Observation 3.2. We can also get $f(bp_i) \leq f(p_i)$ in the same way.

Therefore, $f(p_i)+\mu \geq f(bp_i) +\mu \geq f(bp_j) +\mu \geq f(p_j)$. □

Theorem 4.1. CIA based on μ-cube index finds the top-*k* answers with μ-approximation.
Proof. According to the algorithm, if $bp_i \notin T$, any member of S_i has no chance to be

added into set Y. That is, for each $y \in Y$ and $y \in S_y$, there must be $bp_y \in T$. And from Lemma 4.1, we know that set T is the top-$(T.size)$ answers to the set of *bottom points*. For each point $z \notin Y$ and $z \in S_z$ and for each $y \in Y$ and $y \in S_y$, if $bp_z \notin T$, then $f(bp_y) \geq f(bp_z)$, so $f(y)+\mu \geq f(z)$ for Lemma 4.3. In the other case, if $bp_z \in T$, since $z \notin Y$, meaning $S_z \not\subseteq Y$, bp_z is the one with the lowest score in set T according to Lemma 4.2. So we also have $f(bp_y) \geq f(bp_z)$ and $f(y)+\mu \geq f(z)$.

Therefore, for each y among Y and each z not among Y, there is $f(y)+\mu \geq f(z)$. That is, CIA based on μ-*cube index* finds the top-k answers with μ-*approximation*. $\square$

4.3   Cost Analysis of Cube Index Algorithm

According to Fagin et al. [3], the cost of the top-k query is proportional to the times of accessing or aggregating the objects. For the CIA, the cost is the number of *bottom points* accessed in the query.

First, let $\underline{bp}$ be the last *bottom point* added into set T. Denote $B_1 = \{sky\ point\} + \{bp \mid bp$ is a *bottom point* which is accessed in the query$\}$ and $B_2 = T - \{\underline{bp}\}$. According to Lemma 4.1 and Corollary 4.1, B_2 is the top-$(T.size\ -1)$ answers to the set of *bottom points*.

Theorem 4.2. The cost of the CIA is $T.size - 2 + skyline(\overline{B_2}).size$, where $\overline{B_2}$ is the complementary set of B_2.

Proof. We only need to show that $B_1 = B_2 + skyline(\overline{B_2})$. Actually, it can be proved by apagoge.

Case 1: If there exists $bp \in B_2 + skyline(\overline{B_2})$ but $bp \notin B_1$, then we know bp is not the *sky point*.

sub-case 1: If $bp \in B_2$, since $B_2 = T - \{\underline{bp}\} \Rightarrow B_2 \subset T$, according to the algorithm, bp has no chance to be added into set T if bp has not been accessed. So it will conflict with the algorithm.

sub-case 2: If $bp \in skyline(\overline{B_2})$, then all the *superiors* of bp is in B_2 because there is no any point in $\overline{B_2}$ dominating bp according to the definition of *skyline*. However, in CIA, all points whose all *superiors* are in T must be accessed before the CIA halts. As $B_2 \subset T$, bp must be accessed, which contradicts the assumption that $bp \notin B_1$.

Case 2: If there exists $bp \in B_1$ but $bp \notin B_2 + skyline(\overline{B_2})$, then bp belongs to neither B_2 nor $skyline(\overline{B_2})$. The fact that $bp \notin B_2$ indicates bp is not in the top-$(T.size\ -1)$ answers to the set of *bottom points* so bp is not the *sky point* because $\{sky\ point\}$ is the top-1 answers. Therefore, bp has chance to be accessed if and only if all the *superiors* of bp are in B_2 for the algorithm. However, $bp \notin skyline(\overline{B_2})$, meaning that bp has at least one *superior* that is not in T so bp cannot be accessed. Thus the assumption has no chance to be true.

Therefore, $B_1 = B_2 + skyline(\overline{B_2})$. Besides, since $skyline(\overline{B_2}) \subseteq \overline{B_2}$, $B_2 \cap skyline(\overline{B_2}) = \emptyset$. So $B_1.size = B_2.size + skyline(\overline{B_2}).size$. Moreover, $B_2.size = T.size - 1$ and the cost of the CIA is $B_1.size - 1$, considering that the *sky point* is not accessed in the algorithm.

Therefore, the cost of the CIA is $B_1.size - 1 = T.size - 1 + skyline(\overline{B_2}).size - 1 = T.size - 2 + skyline(\overline{B_2}).size$. $\square$

5 Experiments

In this section, we conduct extensive experiments to evaluate the performance of our algorithm. Our algorithm is implemented in C/C++ language. We perform our experiments on an 8-CPU server with 8GB shared memory and each CPU is 4-core Intel Xeon E5430 2.66GHz.

5.1 Turning μ-Approximation into θ- Approximation

According to the definitions of μ- *approximation* and θ-*approximation*, if set Y is the top-*k* answers with μ-*approximation*, for each y among Y and each z not among Y, there are $f(y)+\mu \geq f(z)$. So we have $(1+\frac{\mu}{f(y)})f(y) \geq f(z)$. Let $\underline{f}(y)$ be the *k*th highest total score in set Y so that $\frac{\mu}{f(y)} \leq \frac{\mu}{\underline{f}(y)}$. Therefore, the *relative approximation coefficient*

$$\theta = \frac{\mu}{\underline{f}(y)}, \text{ or } \mu = \underline{f}(y) \cdot \theta.$$

In our experiments, we run the CIA over the databases to find the value of $\underline{f}(y)$ and then the TA$_\theta$ runs on θ-approximation of $\theta = \frac{\mu}{\underline{f}(y)}$. We choose the μ-approximation as the criterion of approximation to run our tests.

5.2 Evaluation Metrics

In our tests, the following measures are collected for efficiency comparison [6]:

   accesses: the number of items accessed in the query without duplication;

   precision: the fraction of top-*k* results in an approximate result that belongs to the exact top-*k* result;

   recall: the fraction of top-*k* results in the exact result that were returned by the approximate top-*k* query;

   rank distance: the *footrule distance* [14] between the ranks of the approximate top-*k* results and their true ranks in the exact top-*k* result, i.e., $\frac{1}{k}\sum_{i=1}^{k}|i-truerank(i)|$, where $truerank(i)$ is the *i*th returned object's true rank in the exact top-*k* result.

   score error: the absolute error between approximate and exact top-*k* scores, i.e.,

$$\frac{1}{k}\sum_{i=1}^{k}|totalscore_i^{(approx)} - totalscore_i^{(exact)}|,$$

where $score_i^{(approx)}$ is the total score of the *i*th object in the approximate top-*k* result while $score_i^{(exact)}$ is the total score of the *i*th object in the exact top-*k* result.

   Because the *precision* and the *recall* have the same denominator *k*, they have identical values in our setup. We regard the *recall* as a formal measure in our tests, instead of *precision*.

5.3 Description of Datasets

We do experiments on two synthetic datasets. All generated local scores belong to the interval [0, 1]. The two synthetic datasets are produced to model different input scenarios. They are UI and NI respectively. UI contains datasets in which objects' local scores are uniformly and independently generated for the different lists. NI contains datasets in which objects' local scores are normally and independently generated for the different lists. For synthetic datasets, our default settings for different parameters are shown in Table 1. As mentioned above, approximate top-k queries are usually applied in the cases that the values of n is fairly large, which could cause considerable cost and delays to return the exact query answers. Therefore, in our tests, the default number of data items in each list is 1,000,000, i.e. n=1,000,000. Typically, users are interested in a small number of top answers, thus we set $k = 500$ as the default value of k, which is a tiny value compared with n. We set m as 3 since most previous works evaluate their algorithms on datasets with 3 lists like [4]. Finally, we set 0.05 as the default value of μ.

We run our tests with default precision ($\mu = 0.05$) and high precision ($\mu = 0.005$) over each dataset respectively. Furthermore, we run the algorithms on the datasets with large value of k (2000) to observe the effect of k on the performance.

Table 1. Default values of experimental parameters

Parameters	Default Values
The number of objects, i.e. n	1,000,000
The number of lists, i.e. m	3
The number of results returned, i.e. k	500
The precision of results returned, i.e. μ	0.05
Aggregate function	$0.2s_1+0.3s_2+0.5s_3$

For real datasets, we choose El Nino dataset[1] and Forest Cover (FC) dataset[2]. El Nino dataset contains 93935 objects and FC dataset contains 581012 objects. El Nino contains oceanographic and surface meteorological readings taken from a series of buoys positioned throughout the equatorial Pacific. The data is expected to aid in the understanding and prediction of El Nino/Southern Oscillation (ENSO) cycles. FC contains 581012 forest land cells (i.e. objects), having four attributes (i.e. lists): horizontal distance to nearest surface water features, vertical distance to nearest surface water features, horizontal distance to nearest roadways, and horizontal distance to nearest wildfire ignition points. For both real datasets, we choose 3 lists and normalize the dataset with the formula: $\frac{s_i(t) - Min}{Max - Min}$, where $s_i(t)$ is t's ith local score.

5.4   Experimental Results

Fig. 3 illustrates the experimental results where all the parameters are set as default values. Apparently, CIA has significant reduction on the number of accesses over every dataset. Compared with the TA$_\theta$, CIA reduces more than 99% accesses during

---

[1] From UCI KDD. http://kdd.ics.uci.edu/databases/el_nino/el_nino.html
[2] From UCI KDD. http://kdd.ics.uci.edu/databases/covertype/covertype.html

the query process. Apart from this, CIA is also dominant on other evaluation metrics, namely *recall*, *rank distance* and *score error* over every dataset but FC, where CIA is a little inferior to TA$_\theta$ on these aspects.

The experimental results shown in Fig. 4 when $k = 2000$ on each dataset are similar to the results when all the parameters are set as default values. From the results, we can see that CIA also has great reduction on the number of accesses compared with the TA$_\theta$. In terms of the other aspects, CIA performs much better than TA$_\theta$ over every dataset except FC.

Fig. 5 shows us the experimental results where the parameters are set as default values except that μ, the precision of results returned is 0.005. Obviously, CIA is more efficient than TA$_\theta$ considerably but is transcended in other measures. Therefore, CIA has lower accuracy compared with TA$_\theta$ but still keeps its efficiency in the queries with high precision.

Results for UI	accesses	recall	rank distance	score error
TA$_\theta$	10527	0.50200	281.78800	0.008390
CIA	7	0.75600	88.404000	0.002428
Results for NI	accesses	recall	rank distance	score error
TA$_\theta$	10703	0.52600	242.084000	0.007883
CIA	7	0.76800	88.180000	0.002601
Results for EI	accesses	recall	rank distance	score error
TA$_\theta$	1890	0.29200	702.208000	0.006722
CIA	2	0.66600	124.584000	0.001354
Results for FC	accesses	recall	rank distance	score error
TA$_\theta$	5031	0.99200	0.506000	0.000017
CIA	61	0.83800	35.214000	0.001281

Fig. 3. Performance of CIA vs. TA$_\theta$ when $k = 500$ and $\mu = 0.05$

Results for UI	accesses	recall	rank distance	score error
TA$_\theta$	28778	0.77900	232.320500	0.003214
CIA	24	0.83700	158.745000	0.001843
Results for NI	accesses	recall	rank distance	score error
TA$_\theta$	29375	0.80200	194.234000	0.002834
CIA	26	0.85100	136.511000	0.001665
Results for EI	accesses	recall	rank distance	score error
TA$_\theta$	4519	0.70300	463.897000	0.006876
CIA	4	0.94750	17.667000	0.000258
Results for FC	accesses	recall	rank distance	score error
TA$_\theta$	10084	0.94150	23.150500	0.000418
CIA	138	0.89650	75.043000	0.001175

Fig. 4. Performance of CIA vs. TA$_\theta$ when $k = 2000$ and $\mu = 0.05$

Results for UI	*accesses*	*recall*	*rank distance*	*score error*
TA$_\theta$	40683	0.99800	0.030000	0.000001
CIA	532	0.97400	1.112000	0.000031

Results for NI	*accesses*	*recall*	*rank distance*	*score error*
TA$_\theta$	40371	0.99999	0.000001	0.000001
CIA	539	0.97800	0.678000	0.000023

Results for EI	*accesses*	*recall*	*rank distance*	*score error*
TA$_\theta$	8941	0.99999	0.000001	0.000001
CIA	22	0.96200	2.200000	0.000023

Results forFC	*accesses*	*recall*	*rank distance*	*score error*
TA$_\theta$	10482	0.99999	0.000001	0.000001
CIA	552	0.97400	0.840000	0.000027

Fig. 5. Performance of CIA vs. TA$_\theta$ when $k = 500$ and $\mu = 0.005$

Summary: From all the experimental results, we know that CIA improves significantly not only on the number of accesses, but also on other evaluation metrics in the queries with default precision. In addition, we can also learn the fact that CIA still keeps its efficiency and accuracy when the value of k is considerable large. However, CIA is not dominant on all the evaluation metrics over some datasets, like FC in our tests. Finally, in the queries with high precision, our algorithm is considerably superior to TA$_\theta$ on the number of accesses but have little advantage on other respects.

6   Conclusions and Future Work

In this paper, we analyzed the model of the top-k queries and gave some observations. To measure the approximation of the top-k answers, we defined a novel approximation, μ-*approximation* to the top-k answers. Then we introduce an efficient indexing structure called μ-*cube index* to support this kind of approximate query. Based on the μ-*cube index* on the dataset, we proposed our algorithm, the Cube Index Algorithm to answer the μ-*approximation* top-k queries. The main advantage of CIA is that we choose the *bottom point* of a hypercube to approximately represent the points in the hypercube and run the algorithm to find the top-T.*size* in the set of *bottom points* so that the number of accesses can be reduced significantly. Extensive experimental results on both generated and real-world datasets show that our algorithm owns higher accuracy with less cost, compared with TA$_\theta$.

In the future work, we plan to turn our algorithm into exact algorithm based on the cube index ideas.

Acknowledgments. This work is supported by the National Science Foundation of China under the grant [No. 60873210].

References

1. Ilyas, I., Beskales, G., Soliman, M.A.: A Survey of Top-k Query Processing Techniques in Relational Database Systems. In: ACM Computing Surveys, New York (2008)
2. Michel, S., Triantafillou, P., Weikum, G.: KLEE: A frame work for distributed top-k query algorithms. In: VLDB (2005)
3. Fagin, R., Lotem, A., Naor, M.: Optimal aggregation algorithms for middleware. In: PODS (2001)
4. Gong, N.Z., Sun, G.Z.: Parallel Algorithms for Top-k Query Processing. ACM SIGMOD (2010)
5. Cormen, T.H., Leiserson, C.E., Rivest, R.L., Stein, C.: Introduction to Algorithms. MIT Press, Cambridge (2001)
6. Theobald, M., Weikum, G., Schenkel, R.: Top-k Query Evaluation with Probabilistic Guarantees. In: VLDB (2004)
7. Zou, L., Chen, L.: Dominant Graph An Efficient Indexing Structure to Answer Top-K Queries. In: ICDE (2008)
8. Amato, G., Rabitti, F., Savino, P., Zezula, P.: Region Proximity in Metric Spaces and Its Use For Approximate Similarity Search. ACM Trans. Inform. Syst. (2003)
9. Xin, D., Han, J., Cheng, H., Li, X.: Answering Top-k Queries with Multi-Dimensional Selections: The Ranking Cube Approach. In: VLDB (2006)
10. Fagin, R., Kumar, R., Sivakumar, D.: Comparing Top K Lists. ACM-SIAM SODA (2003)
11. Donjerkovic, D., Ramakrishnan, R.: Probabilistic Optimization of Top N Queries. In: VLDB (1999)
12. Hellerstein, J., Haas, P., Wang, H.: Online Aggregation. ACM SIGMOD (1997)
13. Ilyas, I., Aref, W., Elmagarmid, A.: Supporting Top-K Join Queries in Relational Databases. In: VLDB (2004)
14. Kendall, M., Gibbons, J.D.: Rank Correlation Methods. Oxford University Press, Oxford (1990)
15. Re, C., Dalvi, N., Suciu, D.: Efficient Top-K Query Evaluation on Probabilistic Data. In: ICDE (2007)
16. Balke, W.-T., Nejdl, W., Siberski, W., Thaden, U.: Progressive distributed top-k retrieval in peer-to-peer networks. In: ICDE Conf. (2005)
17. Kimelfeld, B., Sagiv, Y.: Finding and approximating top-k answers in keyword proximity search. In: PODS Conf. (2006)
18. Akbarinia, R., Pacitti, E., Valduriez, P.: Reducing network traffic in unstructured P2P systems using Top-k queries. Distributed and Parallel Databases 19(2) (2006)
19. Akbarinia, R., Pacitti, F., Valduriez, P.: Processing top-k queries in distributed hash tables. In: Kermarrec, A.-M., Bougé, L., Priol, T. (eds.) Euro-Par 2007. LNCS, vol. 4641, pp. 489–502. Springer, Heidelberg (2007)
20. Chaudhuri, S., Gravano, L., Marian, A.: Optimizing top-k selection queries over multimedia repositories. IEEE Trans. on Knowledge and Data Engineering 16(8) (2004)
21. Nepal, S., Ramakrishna, M.V.: Query processing issues in image (multimedia) databases. In: ICDE Conf. (1999)

DVD: A Model for Event Diversified Versions Discovery*

Liang Kong[1], Rui Yan[2], Yijun He[2], Yan Zhang[1,**],
Zhenwei Zhang[3], and Li Fu[3]

[1] Department of Machine Intelligence, Peking University, Beijing 100871, China
Key Laboratory on Machine Perception, Ministry of Education, Beijing 100871, China
[2] Department of Computer Science, Peking University, Beijing 100871, China
[3] Service Software Chongqing Institute of ZTE Corporation, China
{kongliang,r.yan,withleave}@pku.edu.cn, zhy@cis.pku.edu.cn,
{zhang.zhenwei,fu.li3}@zte.com.cn

Abstract. With the development of the techniques of Event Detection
and Tracking, it is feasible to gather text information from many sources
and structure it into events which are constructed online automatically
and updated temporally. There are always diversified versions to describe
an event and users usually are eager to know all the versions. With the
huge quantity of documents, it is almost impossible for users to read all of
them. In this paper, we formally define the problem of event diversified
versions discovery. We introduce a novel and principled model (called
DVD) for discovering diversified versions for events. Unlike traditional
clustering methods, we apply an iterative algorithm on a bipartite graph
integrating co-occurrence and semantics to select the popular words and
filter them to reduce the tight correlation between documents in a specific event. Hybrid link structures between words are utilized to find the
hierarchical relationships. We employ a web communities discovery algorithm to construct virtual-documents which consist of a bag of words
indicating one of the diversified versions. Under Rocchio Classification
framework, we can classify the documents to diversified versions. With
our novel evaluation method, empirical experiments on two real datasets
show that DVD is effective and outperforms various related algorithms,
including classic K-means and LDA.

Keywords: Diversified Versions Discovery, popular words selection.

1 Introduction

With the fast development of modern technologies, every day there is a large
number of new information produced. Currently document flood which comes

---

* Supported by NSFC with Grant No. 61073081, National Key Technology R&D Pillar
Program in the 11th Five-year Plan of China with Research No. 2009BAH47B00,
ZTE University Partnership Fund.
** Corresponding author.

X. Du et al. (Eds.): APWeb 2011, LNCS 6612, pp. 168–180, 2011.

from different websites spreads throughout the web and users can be easily trapped in news sea. In some news websites, the mess-up information could be gathered together, turned into event, issue, topic or special manually by editors, and displayed to users, for instance, *Yahoo! News Topics*(http://news.yahoo.com/topics) and *CNN Special Coverage & Hot Topics*(http://edition.cnn.com/SPECIALS). With the development of the techniques of Topic Detection and Tracking (TDT)[1], some practical web service can gather news information from many websites and structure it into news topics which are constructed online automatically and updated temporally, such as *Google News*.

Owing to the techniques of TDT, it is feasible for a user to know "what's happening" and "what's new". However, there are always diversified versions to describe an event. With the huge quantity of documents constructed and updated at all times, it is almost impossible for users to read all of them. In many scenarios, it is appealing to have a technique to automatic discover diversified versions of a popular event. Hence, diversified versions discovery is a challenging and valuable research subject.

Fig. 1 shows a diversified versions example about the disputable truth of *Cheonan sink*. Although it is almost impossible to give the truth from those versions, we can try to show all the mainstream versions to users. As another example, after watching *Inception* which is a movie event, users will express their diversified views on how many levels of the dream in the movie is and whether the spanning top drops in ending. Although we cannot give what it really is, we can show fans the existing different versions.

Fig. 1. A diversified versions example: the controversial truth of *Cheonan sink*

In this paper, we formally define the problem of event diversified versions discovery. Intuitively, the task can be regarded as a problem of clustering. However,these documents can be divided in one specific event, just because of existing of the tight correlations between them. Therefore the traditional clustering methods have poor performance. Motivated by above problems of previous methods, we introduce a novel and principled model (called DVD) for discovering diversified versions of a specific event that consists of a quantity of documents information. DVD is composed of three major phases and effective to achieve the task of diversified versions discovery.

Discovering event diversified versions is important and challenging in many ways. The similarity between documents, which are influenced by the popular words, is changed after filtering the popular words. Indeed, our approach can be combined with any existing event detection algorithms and present the diversified versions for an event. We show that DVD is motivated by and well reflects the existing observations and findings. Empirical experiments on two real datasets show that DVD is effective and outperforms various related algorithms.

The rest of this paper is organized as follows. In Section 2, we revisit the related work. Section 3 gives the problem formulation. Section 4 introduces the model for event diversified versions discovery. Experiments in Section 5 show the performance of DVD. Section 6 summarizes contributions of this paper.

2   Related Work

Topic detection and tracking (TDT) research has been extensively studied in previous work. Pilot experiments in retrospectively and incrementally clustering of text documents have been done as a part of event detection task initiative [2] and query document like retrieval [3]. Others follow their work on event detection [4], improving clustering techniques [5] and novelty detection [6]. Topic tracking also stimulates many researches because events are dynamic and evolutionary character is studied in [7]. Tracking captured dependencies in news topics, either casual or temporal in [8] and threads topics. Our work is related with above-mentioned work. Finding a new event which have diversified versions requires technology of event detection. Event tracking helps us to access the documents stream. Nevertheless, our work is distinct from these work before. We focus on discovering diversified versions rather than whether events are found.

PageRank [9][10] and HITS[11] are both famous ranking algorithms. These algorithms use linkage analysis for ranking based on graph. Graph is used and acts well in many other aspects besides ranking. In [12], a method based on graph mining is used to discover topics about scientific papers, however it is crippled because of low efficiency. In[13], the researchers use a new way based on graph for multiple documents summarization. The thought of using graphs is employed in our paper, and we use the method to select popular words and compute the similarity between documents. Either Maximum Flow[14] or Community Trawling[15] is one of the most classical web communities discovery algorithm, in our paper, we employ Maximum Flow to discover the word communities.

To the best of our knowledge, our approach has different characteristics from all previous work. In this paper, we formally define the problem of event diversified versions discovery and present a novel model DVD to solve the problem.

3   Problem Formulation

In this section, we formally define the related concepts and the task of diversified versions discovery. Let us begin with defining a few key concepts as follows.

Document. Let $D^\varepsilon = \{d_1^\varepsilon, d_2^\varepsilon, ..., d_n^\varepsilon\}$ be the set of documents published about a specific event ε, where d_i^ε is the text document. d_i^ε is represented by a bag of words from a fixed vocabulary $W^\varepsilon = \{w_1^\varepsilon, w_2^\varepsilon, ..., w_m^\varepsilon\}$. That is, $d_i^\varepsilon = \{c(d_i^\varepsilon, w_1^\varepsilon),$ $c(d_i^\varepsilon, w_2^\varepsilon), ..., c(d_i^\varepsilon, w_m^\varepsilon)\}$, where $c(d_i^\varepsilon, w^\varepsilon)$ denotes the number of occurrences of word w^ε in d_i^ε.

Version. Let $V^\varepsilon = \{v_1^\varepsilon, v_2^\varepsilon, ..., v_k^\varepsilon\}$ be the set of versions describing the truth of event ε, where v_i^ε represents a version. v_i^ε is represented by a word distribution.

With the definitions of key concepts, we can now formally define the major task of diversified versions discovery. Given the input of a set of documents D^ε, the task is:

Diversified Versions Discovery. Formally, we want to infer the set of versions V^ε about the event ε. An event may consist of diversified versions. By inferring the set of versions, we expect to keep track of different aspects about an event, give users multilateral understanding of an event and classify documents to the diversified versions of an event, etc.

4   Diversified Versions Discovery

4.1   System Framework for DVD

Intuitively, the task can be regarded as clustering problem. However, in a specific event, there are the tight correlations between documents. Therefore the classic clustering methods have poor performance. For solving the problem, we propose DVD. As shown in Fig. 2, the system framework of DVD is composed of three major phases. It works as follows:

1) Hybrid Link Words Graph Construction. Words can be considered as *foundations* for supporting many documents. We apply an iterative algorithm on a bipartite graph to select the popular words and filter them to reduce the tight correlation between documents. Afterwards, in order to find the hierarchical relationships between words, we take efforts to construct the words graph to capture hybrid link structures.

2) Word Communities Discovery. Based on basic words graph, we employ a communities discovery algorithm to obtain word communities. Based on the word distribution of communities, we can build the virtual-documents. Each virtual-document consists of a bag of words, which can describe the word distribution of a version.

3) Versions Classification. We utilize a vector to indicate the centroid of every version represented by the virtual-document built in phase 2). The weight of a vector is computed based on the features of documents. Afterwards, we employ

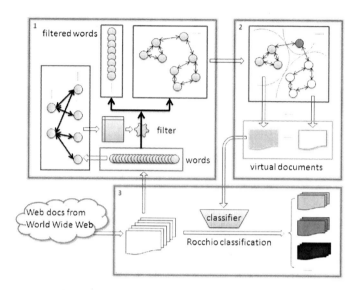

Fig. 2. The system framework for the model of DVD

the Rocchio classification framework to classify the documents to diversified versions.

In the following subsections, phases 1), 2), and 3) are formally expressed.

4.2   Hybrid Link Words Graph Construction

The words are the most elementary units to express the context of a document, we can regard them as foundations for supporting documents. Hence we can discover affiliations between documents by mining the relationships between words. The words graph is the most intuitive structure to illustrate the relationships between words. In this section, we present a novel hybrid link structure to find the hierarchical relationships between words.

The popular words play a important part in building the tight correlations between documents. Because the popular words exist, the classic clustering methods have poor performance on diversified versions discovery. Therefore, we will select the popular words and filter them. Note that, before we start, we have removed the stop words and stemmed all the other words.

Popular Words Selection. We employ an iterative algorithm to select popular words. Given the set of documents D^ε, we can construct a bipartite graph $G = (V_m \cup V_l, E)$. In G, each vertex in V_m corresponds to a word whose document frequency is more than average document frequency and each vertex in V_l corresponds to a word whose document frequency is less than average document frequency. An edge in E connects two vertexes in V_m and V_l respectively. Note that there is no intra-set edges linking two vertexes in V_m or V_l. Furthermore, each edge $e_{i,j} \in E$ is associated with a non-negative weight $m_{i,j}$. The score of

$m_{i,j}$ measures the relationship between word w_i^ε and w_j^ε. Hence we can utilize the bipartite graph to model the intrinsic relationship between V_m and V_l.

Besides using the co-occurrence frequency of words within sentences, we can also adopt sematic similarity to estimate m_{ij}. An obvious way to measure the semantic similarity between two words w_i and w_j is to measure the distance between them in WordNet[1]. This can be done by finding the paths from each sense of w_i to each sense of w_j and then selecting the shortest such path. we compute path length similarity between w_i and w_j using the formula:

$$sim(w_i, w_j) = -\log \frac{len(w_i, w_j)}{2 \cdot \max\{depth(w_i), depth(w_j)\}}. \qquad (1)$$

where $len(w_i, w_j)$ is the number of nodes in path from w_i to w_j and $depth(w)$ is the length of the path to word w from the global root. Therefore the m_{ij} can be defined as follow:

$$m_{ij} = \lambda \cdot CoSentence(w_i, w_j) + (1 - \lambda) \cdot sim(w_i, w_j). \qquad (2)$$

where $CoSentence(w_i, w_j)$ is co-occurrence frequency of words within sentences and λ is an application parameter. We want to show that by constructing a simple bipartite graph and adapting an iterative algorithm on it, we can select popular words effectively.

Inspired by HITS algorithm[11], we give an iterative algorithm to select the popular words. In our case, we assume that 1) If a word in V_m is connected to more words in V_l, the word is more popular. 2) If there is a tight sematic similarity between two words, then they can transform more popular value to each other. With the above assumptions, we present an iterative formula as follows:

$$Popular(w_p) = \sum_{q_i:q_i \to p} \frac{m_{pq_i}}{\sum_{p_j \in out(q_i)} m_{p_j q_i}} Popular(w_{q_i}). \qquad (3)$$

where $out(q_i)$ indicates the set of words connected by the word w_{q_i}. $Popular(w)$ is the current popular value of word w. Based on Equation (3), we can iterate to compute every word popular value and select the top k popular words to filter.

Words Graph Construction. In order to discover the relationships between words, we can construct the words linkage graph $G(V, E)$. Words are considered as vertexes V and links as edges E. Each edge e_{ij} is associated with an affinity weight $f(w_i \to w_j)$ between w_i and w_j ($i \neq j$). Two vertices are connected if their weight is larger than 0 and we let $f(w_i \to w_i) = 0$ to avoid self transition.

Except document and sentence, in the hybrid link structures, we take another structure called event snippet into consideration. [16] firstly defines the concept of event snippet which are the different Latent Ingredients(LIs) in one document. They present a novel event snippet recognition models based on LIs extraction and exploit a set of useful features consisting of context similarity, distance restriction, entity influence from thesaurus and temporal proximity.

---

[1] http://wordnet.princeton.edu

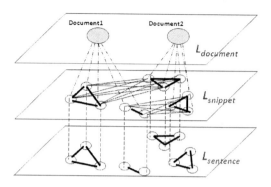

Fig. 3. Hierarchical hybrid linkage layer

Now we can take efforts to capture hybrid link structures to discover the hierarchical relationships between words. Note that in our proposed words graph, the linkage between words can be categorized into three different layers. The first layer is the intra-sentence links between words from the same sentences, denoted by $L_{sentence}$. These words are closely connected. The second layer of intra-snippet links is between the words from the same snippets, namely $L_{snippet}$. Although from different sentences, they present the "citation" relationship within the same snippets. The last layer $L_{document}$ represents the intra-doc from same documents connections. All the hierarchical relationships between words can be described in Fig. 3. Utilizing the hierarchical linkages, $f(w_i \rightarrow w_j)$ is expressed as follow:

$$f(w_i \rightarrow w_j) = \alpha \cdot CoSentence(w_i, w_j) + \beta \cdot CoSnippet(w_i, w_j)$$
$$+ (1 - \alpha - \beta) \cdot CoDocument(w_i, w_j). \tag{4}$$

where $CoSentence(w_i, w_j)$, $CoSnippet(w_i, w_j)$ and $CoDocument(w_i, w_j)$ are the co-occurrence frequency of words within sentences, snippets and documents respectively. α and β are two application parameters, which are restricted between 0 and 1. Computing the hybrid affinity weight by Equation (4), we can construct a words graph to build the hierarchical relationships between words.

4.3   Word Communities Discovery and Versions Classification

After hybrid link words graph constructed in previous subsection, we can begin to mine the relationships between words. In our case, we assume that the words in one version should have tighter relationships to each other than to the words not in the version. With the above assumption, versions can be regarded as the communities in the hybrid link words graph.

We employ maximum flow[14] to identify the communities. The seed words are selected based on the popular value computed in Section 4.2, excluding the words already filtered. Afterwards, each version can be described by the word set within a community. Under the bag of words concept, a word set can be regarded as a document, called virtual-document. Therefore, the set of versions V^{ε} can be indicated by a virtual-document set $VD^\varepsilon = \{VD_1^\varepsilon, VD_2^\varepsilon, ..., VD_k^\varepsilon\}$, where VD_j^ε is one of virtual-documents. Each VD_j^ε is represented by a bag of words from a word community. That is $VD_j^\varepsilon = \{w_{1,vd_j^\varepsilon}^\varepsilon, w_{2,vd_j^\varepsilon}^\varepsilon, ..., w_{k_j,vd_j^\varepsilon}^\varepsilon\}$. Based on the virtual-document set, we can start to classify the real documents to versions.

Given the virtual-document set VD^ε, we try to build document-to-version classifier, which only depends on the virtual-document set VD^ε. We considered every virtual-document VD_j^ε as the centroid of a version for a nearest-neighbor classifier based on Rocchio classification framework[17] with only positive examples and no relevance feedback. Each centroid is defined as a vector $\boldsymbol{v}_j^\varepsilon = \{v^i | w_{v^i}^\varepsilon \in W^\varepsilon \cap VD_j^\varepsilon\}$, where v^i is the weight of word $w_{v^i}^\varepsilon$. The weight v^i is the sum of its tf-idf values in document set D^ε, normalized by RSS (Root Sum Square) of weights of the words in the virtual-document set VD_j^ε, that is:

$$v^i = \frac{1}{\|\boldsymbol{v}_j^\varepsilon\|} \sum_{d_k^\varepsilon} tfidf(w_{v^i}^\varepsilon, d_k^\varepsilon). \tag{5}$$

where $tfidf(w_{v^i}^\varepsilon, d_k^\varepsilon)$ is the tf-idf value of the word $w_{v^i}^\varepsilon$ on the document d_k^ε.

The classification is based on the cosine of the angle between the document and the virtual-document, that is:

$$\begin{aligned} v_{max}^\varepsilon &= \arg\max_{v_j^\varepsilon \in V^\varepsilon} \frac{\boldsymbol{v}_j}{\|\boldsymbol{v}_j\|} \cdot \frac{\boldsymbol{d}}{\|\boldsymbol{d}\|} \\ &= \arg\max_{v_j^\varepsilon \in V^\varepsilon} \frac{\sum_{i \in |F|} v^i \cdot d^i}{\sqrt{\sum_{i \in |F|} (v^i)^2} \sqrt{\sum_{i \in |F|} (d^i)^2}}. \end{aligned} \tag{6}$$

where F is the union set of words of document and virtual-document. The score is normalized by the document and virtual-document length to produce comparable score. The terms v^i and d^i represent the weight of the ith word in the virtual-document and the document respectively. The weights in document side are based on the standard tf-idf formula. As the score of the max version is normalized with regard to document length, the scores for different documents are comparable. Experiment shows that the Rocchio classifier performs well on judges on the document versions assignment.

5 Experiments

In this section, we describe the experimental designs and evaluation of event diversified version discovery. We show the effectiveness of our model DVD with experiments both on a smaller dataset *Cheonan sink* (including 533 documents, called CS) and on a larger dataset *Michael Jackson death* (including 812 documents, called MJD). CS is an event about the sinking of South Korean warship Cheonan and causes the discussion on who sank it. Analogously, MJD causes the discussion on truth of Michael Jackson's death. We crawl documents from the mainstream news web portals, such as *BBC, Reuters, MSNBC, NYTimes,*

People, to build the dataset. Since there is no existing standard test set for diversified versions discovery, we opt to construct our own test sets. We first present the evaluation criterion for DVD. Then we illustrate and analyze the experimental results.

5.1   Evaluation

With regard to an event, it is almost impossible to manually observe all the documents and classify them into different versions. Hence, we evaluate the effectiveness of DVD in a pairwise judge task.

In the pairwise judge task, we focus on judging whether a pair of documents belongs to the same version. Firstly we construct the pairwise standard test sets. We randomly sample 200 pairs of documents for CS and 300 pairs of documents for MJD. We make sure each document is different from others, which enhances the evaluation reliability. Afterwards, each pair of documents is shown to volunteers and whether the pair of documents belongs to the same version depends on the voting result. If a result is hard to judge, the pair of documents will be discarded and we will add a new pair of documents to test set, which guarantees effectiveness of the test set. Finally, we can obtain the pairwise test set, denoted as $T_\varepsilon = \{<< d_{i_1}^\varepsilon, d_{i_2}^\varepsilon >, v_i >| v_i \in \{0, 1\}, d_{i_1}^\varepsilon \in D^\varepsilon, d_{i_2}^\varepsilon \in D^\varepsilon\}$, where $v_i = 1$ indicates that $d_{i_1}^\varepsilon$ and $d_{i_2}^\varepsilon$ belong to the same version and $v_i = 0$ indicates otherwise situation. Specifically, we construct the pairwise test sets of CS and MJD, which are abbreviated to T_{cs} and T_{mj} respectively.

In our evaluation, we use the *Precision* criterion in the pairwise test task. Given the pairwise documents $< d_{i_1}^\varepsilon, d_{i_2}^\varepsilon >$, each method can give a judge v_i'. Hence, we can define *Precision* in the pairwise test task as P_{score}, that is:

$$P_{score} = \frac{\sum_{<d_{i_1},d_{i_2}>} v_i \odot v_i'}{|T_\varepsilon|}. \tag{7}$$

where $|T_\varepsilon|$ represents the size of the pairwise test set for event ε. $\odot$ is XNOR.

5.2   Parameter Tuning

In order to set application parameters, we start our experiment from examining their influence. There are four application parameters, α, β, λ and k. We discuss them respectively in this section.

Firstly, we examine the influence of factor α and β under specific λ, k. During every "training and testing" process, we vary α from 0 to 1 with the step of 0.1 and make the same move to β. We check the P_{score} when these two parameters change in Fig. 4 and get a best α and β value pair (α_{best} and β_{best}). The best P_{score} is achieved when $\alpha = 0.6$ and $\beta = 0.3$ in CS and $\alpha = 0.7$ and $\beta = 0.2$ in MJD. Furthermore, we can see that α can more significantly effect P_{score} than β and when α or β increases the P_{score} will increase and then decrease.

The parameter λ and k tuning procedure is listed in Table 1. We can adjust every parameter by fixing others. λ can balance the influence of co-sentence

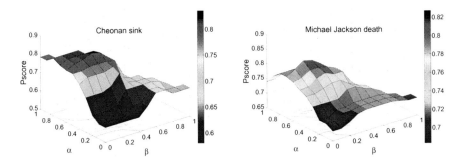

Fig. 4. α, β tuning under specific λ, k in CS and MJD

and semantics in the process of computing the popular words. When λ equals 0.2, P_{score} can get the optimum value. However, when λ gets larger the P_{score} will decay and when λ is 1 or 0 the P_{score} seems not too bad, which indicates the effect of semantics is weak. k can decide the granularity of filtering popular words. Table 1 illustrates that P_{score} is greatly influenced by filtering factor. Both high and low k significantly reduce the P_{score}. From Table 1, we can see that P_{score} can obtain the optimum value when $k = 100$ and $\lambda = 0.2$.

Table 1. Parameter tuning of λ, k in CS and MJD

	k	100						λ	0.2					
	λ	0.0	0.2	0.4	0.6	0.8	1.0	k	0	50	100	500	1000	5000
CS	α_{best}	0.6	0.6	0.6	0.6	0.5	0.5	α_{best}	0.9	0.7	0.6	0.1	0.0	0.0
	β_{best}	0.3	0.3	0.3	0.3	0.3	0.3	β_{best}	0.1	0.3	0.3	0.4	0.0	0.0
	P_{score}	0.800	**0.835**	0.795	0.745	0.715	0.715	P_{score}	0.615	0.785	**0.835**	0.645	0.555	0.505
MJD	α_{best}	0.7	0.7	0.7	0.6	0.6	0.6	α_{best}	1.0	0.7	0.7	0.2	0.0	0.0
	β_{best}	0.2	0.2	0.2	0.3	0.3	0.3	β_{best}	0.0	0.2	0.2	0.2	0.2	0.2
	P_{score}	0.787	**0.827**	0.807	0.767	0.757	0.757	P_{score}	0.657	0.767	**0.827**	0.703	0.623	0.477

5.3 Performace and Discussion

In this section, we study quantitatively the effectiveness of the proposed DVD. Comparisons against related algorithms are also conducted. The related algorithms studied include:

- **K-means**, which clusters the documents based on the text similarity.
- **LDA**, which clusters the documents based on the word distribution.
- **DVD-rf**, which implements a special version of DVD without removing any popular words.
- **DVD-s**, which implements a special version of DVD only using co-occurrence within sentence to construct words graph.
- **DVD-ss**, which implements a special version of DVD using co-occurrence within sentence and snippet to construct words graph.

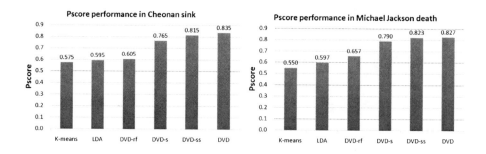

Fig. 5. P_{score} performance comparison in the pairwise test task of CS and MJD

We compare the performance of DVD with related algorithms in the two built pairwise test sets, T_{cs} and T_{mj}. The P_{score} performance of the pairwise judge task in T_{cs} and T_{mj} is demonstrated in Fig. 5. From Fig. 5 we can see that the classic K-means and LDA are not satisfactory. Since DVD-rf only using words cluster which is similar to K-means, it also dosen't perform well. DVD-s, DVD-ss and DVD outperform other related algorithms, which demonstrates that the popular words filtering is important and effective. Moreover, DVD performs little better than DVD-ss which performs better than DVD-s. It illustrates that the hybrid link can improve the performance. Besides, the snippet is more useful than document for improving the results.

Tangible benefit can be realized when applying our model to a demo system. We combine DVD with a graph-based multi-document summarization algorithm (GMS) proposed by Wan et al[18]. We apply GMS on the document set of each version. A snapshot from our demo system is present in Fig. 6. Each row shows the summarization of a version. Users can click button "article" to browse the document set. The versions proposed by our model perform well, for example, the upper three results talk *homicide*, *heart attack* and *propofol* respectively.

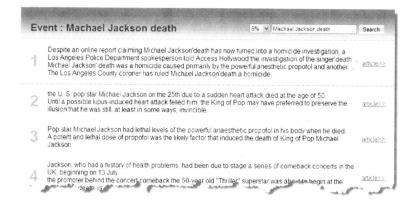

Fig. 6. A snapshot of our demo system

6   Conclusion

In this paper, we present an event diversified versions discovery model, which helps in quickly learning from multilateral description of a specific event. Within our innovative model, we take three phases to achieve this task. An iterative algorithm is applied on a bipartite graph integrating co-occurrence and semantics to select popular words. After filtering them and hence reducing the tight correlations between documents, we construct a hybrid link words graph to find the hierarchical relationships between words. With a communities discovery algorithm, we can build the virtual-documents to describe the centroid of versions. Under Rocchio Classification framework, we classify the documents to diversified versions. Experiments on real datasets show that our approach is effective.

While the work in this paper exhibits good performance, there is much more room for further improvement along this direction. In the future, combined with trustworthy analyzing techniques, we can present the most trustable version for users, which is very urgent and important in practical applications.

References

1. Allan, J., Carbonell, J., Doddington, G., Yamron, J., Yang, Y.: Topic detection and tracking pilot study: Final report, Tech. Rep. (1998)
2. Yang, Y., Pierce, T., Carbonell, J.: A study of retrospective and on-line event detection. In: SIGIR 1998, pp. 28–36 (1998)
3. Allan, J., Papka, R., Lavrenko, V.: On-line new event detection and tracking. In: SIGIR 1998, pp. 37–45 (1998)
4. Brants, T., Chen, F., Farahat, A.: A System for new event detection. In: SIGIR 2003, pp. 330–337 (2003)
5. Franz, M., Ward, T., McCarley, J.S., Zhu, W.: Unsupervised and supervised clustering for topic tracking. In: SIGIR 2001, pp. 310–317 (2001)
6. Yang, Y., Zhang, J., Carbonell, J., Jin, C.: Topic-conditioned novelty detection. In: KDD 2002, pp. 688–693 (2002)
7. Liu, S., Merhav, Y., Yee, W.G., Goharian, N., Frieder, O.: A sentence level probabilistic model for evolutionary theme pattern mining from news corpora. In: SAC 2009, pp. 1742–1747 (2009)
8. Nallapati, R., Feng, A., Peng, F., Allan, J.: Event threading within news topics. In: CIKM 2004, pp. 446–453 (2004)
9. Page. R. M. T. W. L, Brin. S. The pagerank citation ranking: Bringing order to the web (in Manuscript in Progress)
10. Sergey Brin, L.P.: The anatomy of a large-scale hypertextual web search engine. In: WWW 1998, pp. 107–117 (1998)
11. Kleinberg, J.: Authoritative sources in a hyperlinked environment. Journal of the ACM 46(5), 604–632 (1999)
12. Jo, C.G.Y., Lagoze, C.: Detecting research topics via the correlation between graphs and texts. In: KDD 2007, pp. 370–379 (2007)
13. XWan, J.Y.: Collabsum: exploiting multiple document clustering for collaborative single document summarizations. In: SIGIR 2007, pp. 143–150 (2007)
14. Flake, G.W., Lawrence, S., Giles, C.L.: Efficient Identification of Web Communities. In: KDD 2000, pp. 160–169 (2000)

15. Kumar, R., Raghavan, P., Rajagopalan, S.: Trawling the Web for emerging cyber-communities. Journal of Computer networks, 1481–1493 (1999)
16. Yan, R., Li, Y., Zhang, Y., Li, X.: Event Recognition from News Webpages through Latent Ingredients Extraction. In: Cheng, P.-J., Kan, M.-Y., Lam, W., Nakov, P. (eds.) AIRS 2010. LNCS, vol. 6458, pp. 490–501. Springer, Heidelberg (2010)
17. Rocchio, J.: Relevance feedback in information retrieval. In: The SMART Retrieval System: Experiments in Automatic Document Processing, pp. 313–323 (1971)
18. Wan, X., Yang, J.: Multi-document summarization using cluster-based link analysis. In: SIGIR 2008, pp. 299–306 (2008)

Categorical Data Skyline Using Classification Tree

Wookey Lee[1], Justin JongSu Song[1], and Carson K.-S. Leung[2]

[1] Department of Industrial Engineering, Inha University, South Korea
[2] Department of Computer Science, The University of Manitoba, Canada
trinity@inha.ac.kr, jaegal83@inha.edu, kleung@cs.umanitoba.ca

Abstract. Skyline query is an effective method to process large-sized multi-dimensional data sets as it can pinpoint the target data so that dominated data (say, 95% of data) can be efficiently excluded as unnecessary data objects. However, most of the conventional skyline algorithms were developed to handle numerical data. Thus, most of the text data were excluded from being processed by the algorithms. In this paper, we pioneer an entirely new domain for skyline query—namely, the categorical data—with which the corresponding ranking measures for the skyline queries are developed. We tested our proposed algorithm using the ACM Computing Classification System.

Keywords: Skyline, categorical data, domination, classification tree.

1 Introduction

As skyline query effectively retrieves results by utilizing the concept of domination among data [1, 2, 3], it has been recognized as a crucial research subject centered on decision support system and visualization and has gained significance in data processing of massive storage [4, 5, 6, 7, 8, 9, 10]. In many applications, skyline query provides more meaningful results (than "nearest neighbour answers" returned by the usual top-k queries) and more semantically valuable results. Despite the performance improvement provided by a diversity of skyline query algorithms, text-based data have been ignored as these algorithms primarily focused on handling numerical data. In this paper, we present a skyline query method that handles tree-constructed *categorical data*. Noticing the differences between the characteristics of numerical data and categorical data [11] when data were used as skyline query objects, we quantify the distance [12] between categorical data. As such, the relative semantic distances from categorical data are extracted, which are then applied to the skyline query. To avoid distraction, we limit the scope of this paper to cases where categories are given as tree-formed metadata like structure information of XML or ACM Classification System [13].

Table 1 shows the position of our current paper (which deals with *skyline query for categorical data*) with respect to related works (which handle classification and/or numerical data). Among the most relevant works, Ref. [14] focused on skylines with partially-ordered domains, whereas we emphasized on distance functions. Although categorical data have been studied for a long time in various contexts, the corresponding similarity measures for categorical data were data-driven (i.e., different measures for different data sets) [15, 16, 17, 18, 19].

X. Du et al. (Eds.): APWeb 2011, LNCS 6612, pp. 181–187, 2011.
© Springer-Verlag Berlin Heidelberg 2011

Table 1. Categorization of our approach and its related works

	Numerical data	Categorical data
Classification	Many conventional classification methods	Poset [5], flexible XML semantics [7], topologically sorted [11], SDC [14], Wu & Palmer [21]
Skyline query	NN [1], BNL [2], D&C [2], BBS [6], distance-based skyline [8], object space partitioning [9], skyline for uncertain data [10]	Our approach: The CDSS algorithm

The rest of the paper is organized as follows. Next section reviews relative semantic distance functions for categorical data. Section 3 introduces our categorical data skyline search (CDSS) algorithm. Section 4 shows experiments results conducted on the ACM classification. Finally, conclusions are presented in Section 5.

2   Formal Representation of Distance for Categorical Data

2.1   Semantic Distance Function

Categorical data consisting of nodes N (that specify keywords) and edges E (that connect nodes $x,y \in N$) can be represented in a tree form $T(N,E)$ as a *categorical inclusion relation*. See definitions below.

Definition 1. The **level difference (n)** between two arbitrary nodes is defined as the largest number of abstracted levels from the two nodes to their lowest common abstract node.

Definition 2. The **n-level neighbour nodes** are the nodes that share a common abstract node with a level difference value of n.

Example 1. Fig. 1 depicts a tree representation of categorical data in the ACM Computing Classification System [13]. The level difference between "Arrays" and "Data encryption" is 2 because the abstracted levels from these two nodes to "Data" (their lowest common abstract node) are 2 and 1, respectively. "Arrays" and "Trees" are 1-level neighbour nodes (with the common abstract node "Data structures"), "Data structures" and "Data encryption" are also 1-level neighbour nodes (with the common abstract node "Data"), whereas "Arrays" and "Code breaking" are 2-level neighbour nodes (with the common abstract node "Data").   ∎

Based on the above definitions, we observe the following two properties of categorical data: (i) Distances between every pair of nodes can be derived. (ii) The smaller the level difference between the two nodes, the more are their semantic similarity.

Example 2. Based on the above properties, distances between any two arbitrary nodes in the ACM Computing Classification System [13] can be derived. Then, we observed that the distance between "Data structures" and "Arrays" (i.e., level difference of 1) is smaller than that between "Data structures" and "Code breaking" (i.e., level difference of 2), which means the former pair is semantically more similar than the latter one.   ∎

In this paper, we define the following *distance function for categorical data*.

Definition 3. Given a categorical tree $T(N,E)$, the **distance** between categorical nodes $x,y \in N$ can be computed as follows:

$$D_{WP}(x,y) = 1 - [2n_{zr} / (n_{xz} + n_{yz} + 2n_{zr})] \tag{1}$$

where z is the common ancestor node for $x,y \in N$, n_{zr} is the level difference between z & the root, n_{xz} is the level difference between x & z, and n_{yz} is the level difference between y & z.

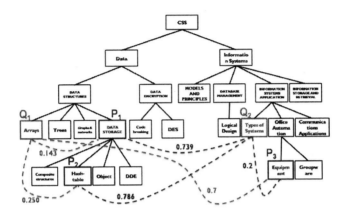

Fig. 1. ACM Computing Classification System (with EWP-based distance for skyline query)

Theorem 1. The distance defined in Definition 3 satisfies the following properties: (i) $D_{WP}(x,z) < D_{WP}(x,y) + D_{WP}(y,z)$ for all $x,y,z \in N$, (ii) $D_{WP}(x,x) = 0$ for all $x \in N$, (iii) $D_{WP}(x,y) = D_{WP}(y,x)$ for all $x,y \in N$, and (iv) $0 \leq D_{WP}(x,z) \leq 1$ for all $x,y \in N$.

Example 3. Given Fig. 1, the distance between "Arrays" and "Trees" can be computed using Equation (1): $D_{WP}(\text{"Arrays"}, \text{"Trees"}) = 1 - [2*3 / (1 + 1 + 2*3)] = 0.25$. Similarly, $D_{WP}(\text{"Code breaking"}, \text{"DES"}) = 1 - [2*3 / (1 + 1 + 2*3)] = 0.25$, $D_{WP}(\text{"Data structures"}, \text{"Data encryption"}) = 1 - [2*2 / (1 + 1 + 2*2)] = 1/3$, $D_{WP}(\text{"Data structures"}, \text{"DES"}) = 1 - [2*2 / (1 + 2 + 2*2)] = 3/7$, $D_{WP}(\text{"Data"}, \text{"Information systems"}) = 1 - [2*1 / (1 + 1 + 2*1)] = 0.5$, and $D_{WP}(\text{"Arrays"}, \text{"Arrays"}) = 1 - [2*4 / (0 + 0 + 2*4)] = 0$. ∎

With Equation (1), the distance between any two categorical nodes is between 0 and 1 inclusive (with 0 indicating two identical nodes and large value indicating distant nodes). The shorter the distance between two categorical nodes, the more is the relevance between them. This is congruous to other distance functions [20].

2.2 Improved Distance Function

The aforementioned distance is referred as *WP* because it utilized *Wu and Palmer* [21]. Despite its advantage of providing reference for categorical data, the WP distance suffers from several limitations such as the attribution of identical distance values for every node of the same dimension regardless of the number of their child nodes. For instance, although there are four child nodes (e.g., "Arrays", "Trees") for "Data structures" and only two child nodes (i.e., "Code breaking", "DES") for "Data encryption", D_{WP}("Arrays", "Trees") $= 0.25 = D_{WP}$("Code breaking", "DES"). In other words, the WP distance fails to capture information about the number of child or sibling nodes. To overcome such a limitation, we assign different weights based on the number of child or sibling nodes. The resulting distance function is referred as the *Extended-Wu and Palmer* (*EWP*). See the definition below.

Definition 4. Given a categorical tree $T(N,E)$, the **distance** between categorical nodes $x, y \in N$ can be computed as follows:

$$D_{EWP}(x,y) = 1 - [2w_z\, n_{zr} / (n_{xz} + w_y\, n_{yz} + 2w_z\, n_{zr})] \qquad (2)$$

where z is the common ancestor node for $x, y \in N$, $w_z = 1 +$ number of child nodes of z, n_{zr} is the level difference between z & the root, $w_y = 1 +$ the number of sibling nodes of y (including itself), n_{xz} is the level difference between x & z, and n_{yz} is the level difference between y & z.

Example 4. Given Fig. 1, the EWP distance between "Arrays" and "Trees" can be computed using Equation (2): D_{EWP}("Arrays", "Trees") $= 1 - [2*5*3 / (1 + 5*1 + 2*5*3)] = 1/6$. Similarly, D_{EWP}("Code breaking", "DES") $= 1 - [2*3*3 / (1 + 3 + 2*3*3)] = 2/11$. ∎

3 Our Skyline Query Algorithm

In this section, we propose an algorithm called *CDSS* for *categorical data skyline*, which uses the above distance functions. In abstract term, the algorithm finds a skyline set based on (i) the categorical tree T representing categorical data, (ii) a data set P containing multiple data objects (i.e., $P=\{P_1, \ldots, P_m\}$) and (iii) a user-specified keyword set K. Specifically, the algorithm first searches for top-k data objects relevant to the user keyword set K. Once these objects are found, the algorithm then computes the relative distance (e.g., using the WP or EWP distance) based on categorical keywords $\{Q_1, \ldots, Q_d\}$ and categorical properties of the top-k data objects that are in d-dimensional data set (where d is the total number of categorical keywords). Each dimension indicates a categorical keyword. After computing the distance, skyline query algorithms such as NN [1] and BBS [6] can then be applied to get the resulting skyline set. Note that objects corresponding to each categorical keyword Q_i in the categorical tree T ought to have a domination relation such that (i) $D(Q_i) = \min_j D_{EWP}(Q_i, P_j)$ and (ii) $\operatorname{argmin}_j D_{EWP}(Q_i, P_j)$ is the most semantically relevant object with respect to the categorical keyword Q_i.

Example 5. When a user submits a query with keyword set {"retrieval"}, the CDSS algorithm finds top-k data objects (say, top-k papers) relevant to the topic of retrieval. Once the data set P consisting of these top-k data objects are found, the algorithm computes the relative distance based on the two given categorical keywords {Q1="Arrays", Q2="Types of systems"} contained in the categorical tree T of categorical data. Then, these top-k data objects (e.g., P1="Data storage", P2="Hash table", and P3="Equipment") are in a two-dimensional data set. Using Equation (2), the CDSS algorithm computes DEWP(Q1,P1)=0.143, DEWP(Q1,P2)=0.25 and

DEWP(Q1,P3)=0.7. Similarly, the algorithm computes DEWP(Q2,P1)=0.739, DEWP(Q2,P2)=0.786 and DEWP(Q2,P3)=0.2. See Fig. 1. Based on these distance values, the algorithm computes the distance for each categorical keyword: D(Q1) = minj DEWP(Q1,Pj) = 0.143 and D(Q2) = minj DEWP(Q2,Pj) = 0.2, which means P1="Data storage" is the most relevant paper with respect to category Q1="Arrays" and P3="Equipment" is the most relevant paper with respect to category Q2="Types of systems". Afterwards, skyline query algorithms can be applied to these data.   ∎

4   Experimental Evaluation

The experiment was conducted on WWW 2007-2008 and *TODS* 2005-2008 papers in the ACM Digital Library [22]. Here, K_1={"TODS", "2005", "2006", "2007", "2008"} and K_2={"WWW conference", "2007", "2008"} are examples of input keyword sets used for selecting the initial data set of 165 and 300 top-k data objects; q1= {Q_{11}="Database applications", Q_{12}="Heterogeneous databases"}, q2={"Formal languages", "Knowledge acquisition"} and q3={"Relational databases", "Abstracting methods"} are examples of categorical keyword sets used in our experiments.

First, we compared the difference between WP and EWP distances. In Fig. 2, the skyline of data is arranged by the semantic distance computed by BBS [6]. In each experiment, data computed as the skyline were depicted by lines connecting the data points. The results showed that categorical data (which were previously excluded from computation of skyline) can now be applied to skyline query.

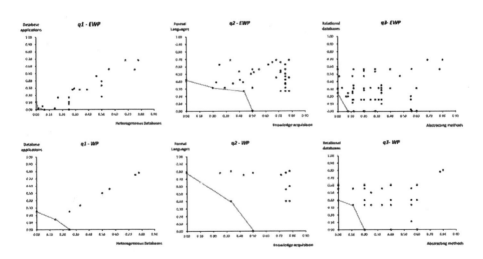

Fig. 2. Comparison between the WP and EWP distance functions

Second, we measured precision ratio (i.e., the ratio of the number of skyline result nodes to the total number of object nodes). Fig. 3 shows that WP returned about 10% of the total data as skyline results, whereas EWP returned about 0.5% to 4%. This indicates the capability of using EWP in retrieving outstandingly accurate results, i.e., effective processing of multi-dimensional data with massive storage. Specifically, a problem associated with the WP distance was related to the shallowness of categorical tree (e.g., maximum depth level is 5 for the ACM Classification System). The average depth level of the document set for K_1 was 4.62. As such, the utility of the skyline query was undermined. In contrast, EWP successfully improved the distance function.

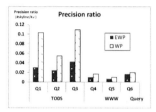 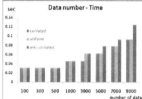 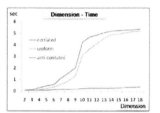

Fig. 3. Precision ratio   **Fig. 4.** Scalability: Runtime vs. amount of data   **Fig. 5.** Dimensionality: Runtime vs. #dimensions

Finally, we evaluated the scalability and dimensionality of the CDSS algorithm. Fig. 4 depicts the comparison associated with an increase in the amount of data, and Fig. 5 illustrates performance changes with respect to dimension growth. Here, we used three different data distribution—namely, uniform, correlated and anti-correlated distribution.

5 Conclusions

Previously, skyline queries were operated only for numerical data, and thus could not process queries for most of the text data. This paper suggested a new algorithm that *widens the applicability of skyline to categorical data*. The algorithm queries the skyline based on a function that calculates the distance between two nodes for categorical data, by which more semantically meaningful answers can be generated than the top-k results returned by conventional queries. Our experimental results on the ACM Computing Classification System showed that our two proposed distance functions—WP and EWP—led up to 30% and 60% enhancements, respectively.

Acknowledgments. This project is partially supported by (i) Basic Science Research Program through the NRF of Korea funded by the MEST(2010-0025366) and (ii) NSERC (Canada).

References

1. Kossmann, D., Ramsak, F., Rost, S.: Shooting stars in the sky: an online algorithm for skyline queries. In: VLDB 2002, 275–286 (2002)
2. Börzsönyi, S., Kossmann, D., Stocker, K.: The skyline operator. In: IEEE ICDE, pp. 421–430 (2001)
3. Tan, K., Eng, P., Ooi, B.C.: Efficient progressive skyline computation. In: VLDB 2001, pp. 301–310 (2001)

4. Chakrabarti, K., Chaudhuri, S., Hwang, S.: Automatic categorization of query results. In: ACM SIGMOD 2004, pp. 755–766 (2004)
5. Sarkas, N., Das, G., Koudas, N., Tung, A.K.H.: Categorical skylines for streaming data. In: ACM SIGMOD 2008, pp. 239–250 (2008)
6. Papadias, D., Tao, Y., Fu, G., Seeger, B.: An optimal and progressive algorithm for skyline queries. In: ACM SIGMOD 2003, 467–478 (2003)
7. Cohen, S., Shiloach, M.: Flexible XML querying using skyline semantics. In: IEEE ICDE 2009, pp. 553–564 (2009)
8. Tao, Y., Ding, L., Lin, X., Pei, J.: Distance-based representative skyline. In: IEEE ICDE 2009, pp. 892–903 (2009)
9. Zhang, S., Mamoulis, N., Cheung, D.W.: Scalable skyline computation using object-based space partitioning. In: ACM SIGMOD 2009, 483–494 (2009)
10. Atallah, M.J., Qi, Y.: Computing all skyline probabilities for uncertain data. In: PODS 2009, pp. 279–287 (2009)
11. Dimitris, S., Stavros, P., Dimitris, P.: Topologically sorted skylines for partially ordered domains. In: IEEE ICDE 2009, pp. 1072–1083 (2009)
12. Shin, M., Huh, S., Park, D., Lee, W.: Relaxing queries with hierarchical quantified data abstraction. J. Database Management 19(4), 47–61 (2008)
13. The 1998 ACM Computing Classification System (1998),
    http://www.acm.org/about/class/1998
14. Chan, C., Eng, P., Tan, K.: Stratified computation of skylines with partially-ordered domains. In: ACM SIGMOD 2005, 203–214 (2005)
15. Boriah, S., Chandola, V., Kumar, V.: Similarity measures for categorical data: a comparative evaluation. In: SIAM SDM 2008, pp. 243–254 (2008)
16. Burnaby, T.: On a method for character weighting a similarity coefficient, employing the concept of information. Mathematical Geology 2(1), 25–38 (1970)
17. Eskin, E., Arnold, A., Prerau, M., Portnoy, L., Stolfo, S.: A geometric framework for unsupervised anomaly detection. In: Applications of Data Mining in Computer Security, pp. 78–100. Springer, Heidelberg (2002)
18. Goodall, D.W.: A new similarity index based on probability. Biometrics 22(4), 882–907 (1966)
19. Lin, D.: An information-theoretic definition of similarity. In: ICML, pp. 296–304 (1998)
20. Hwang, S., Yu, H.: Mining and processing category ranking. In: ACM SAC 2007, pp. 441–442 (2007)
21. Wu, Z., Palmer, M.: Verbs semantics and lexical selection. In: ACL 1994, pp. 133–138 (1994)
22. ACM Digital Library, http://portal.acm.org

Quality Evaluation of Wikipedia Articles through Edit History and Editor Groups

Se Wang and Mizuho Iwaihara

Graduate School of Information, Production and Systems
Waseda University
Fukuoka 808-0135, Japan
wsarthur@fuji.waseda.jp, iwaihara@waseda.jp

Abstract. Wikipedia is well known as a free encyclopedia, which is a type of collaborative repository system that allows the viewer to create and edit articles directly in the web browser. The weakness of the Wikipedia system is the possibility of manipulation and vandalism cannot be ruled out, so that the quality of any given Wikipedia article is not guaranteed. It is an important work to establish a quality evaluation method to help users decide how much they should trust an article in Wikipedia. In this paper we investigate the edit history of Wikipedia articles and propose a model of network structure of editors. We propose an algorithm to calculate the network structural indicator restoreratio. We use the proposed indicator combined with existing metrics to predict the quality of Wikipedia articles through support vector machine technology. The experimental results show that the proposed indicator has better performance in quality evaluation than several existing metrics.

Keywords: Wikipedia, quality evaluation, web mining, edit network, web trust.

1 Introduction

Wikipedia is well known as a free, collaborative, multilingual encyclopedia. It has over 15 million articles in over 270 languages, as measured in June 2010. These articles have been written collaboratively by volunteers around the world, and almost all of its articles can be edited by anyone with access to the site. Wikipedia is currently the largest and most popular general reference work on the Internet, which is constantly listed in the top ten most visited websites worldwide.

Every editable page on Wikipedia has an associated edit history, which is accessed by clicking the "history" tab at the top of the page. The page history contains a list of the page's previous revisions, including the date and time of each edit, the username or IP address of the user who made it, and their edit summary.

Wikipedia grows in the span of a few years to become one of the most widely used sources of information on the web. It owes its growth and breadth of coverage to its ability to harness the contributions of millions of individuals, ranging from casual visitors, to domain experts, to dedicated editors. It allows viewers to edit the articles directly within the web browser. With Wikipedia the articles are contributed voluntarily by everyday web users, whereas with traditional encyclopedias, the

X. Du et al. (Eds.): APWeb 2011, LNCS 6612, pp. 188–199, 2011.

articles are written by experts. The openness of the system attracts many volunteers, who write, update and maintain the articles. According to a study of the scientific magazine, Nature, the quality of Wikipedia is comparable to that of the traditional Encyclopedia Britannica.

On the other hand, the open process that gives rise to Wikipedia content makes it difficult for visitors to form an idea of the reliability of the content. Wikipedia articles are constantly changing, and the contributors range from domain experts, to vandals, to dedicated editors, to superficial contributors not fully aware of the quality standards the Wikipedia aspires to attain. Wikipedia visitors are presented with the latest version of each article they visit: this latest version does not offer them any simple insight into how the article content has evolved into its most current form, nor does it offer a measure of how much the content can be relied upon. These considerations generated interest in algorithmic systems for estimating the quality of Wikipedia articles.

The open access has been known to cause quality problems. The possibility of manipulation and vandalism cannot be ruled out. For example, inaccurate information is occasionally published by opportunistic or inexperienced contributors. Additionally, when articles are not being focused on by the Wikipedia community and hence there is a lack of volunteers providing content, these articles can be incomplete or insufficient. As a consequence, the quality and accuracy of any given Wikipedia article cannot be guaranteed. To overcome this weakness Wikipedia has developed several user-driven approaches for evaluating the articles. High quality articles can be marked as "Good Articles" or "Featured Articles" whereas poor quality articles can be marked as "Articles for Deletion". However, these user-driven evaluations can only partially solve the problem of quality transparency since only a very small part of Wikipedia is evaluated by them. For example in June 2010 only about 0.5% of articles were evaluated in the Wikipedia. Another difficulty of the user-driven evaluations is that Wikipedia contents are by their nature highly dynamic and the evaluations often become obsolete rather quickly. Due to these conditions, recent research work involves automatic quality assessment that is being developed specifically for Wikipedia. In this paper we provide a new approach using network structural indicators to effectively measure the quality of the Wikipedia articles. The indicators are based on the structure of editors of a certain article through its edit history.

The rest of this paper is organized as follows. In Section 2, we survey previous research about Wikipedia, especially on the trustworthiness and quality measurement of articles. In Section 3, we propose a model of Wikipedia article network structures and an algorithm to calculate the network structural indicator. Then we explain how the indicator can affect the quality of articles. In Section 4, we use the proposed indicator combining with existing metrics for evaluating the quality of articles. We use the support vector machine technology to predict the quality of randomly picked articles. The experiment results show that our proposed method is more accurate than existing ones. Section 5 concludes this paper.

2   Related Work

The incredible success of Wikipedia has attracted a lot of researchers. So it is not surprising that numerous researches about Wikipedia have appeared in the last few years. There is a wide and interdisciplinary array of issues being discussed, such as visualization tools [16], motivations for participation [8], the effects of coordination and collaboration, vandalism analysis and detection [10, 15, 17, 19], reputation systems [13], quality assurance and automatic quality measurement [1, 3, 4, 6, 12, 13, 18]. Relating to quality assessment there are two divisions of research. The first group investigates the trustworthiness of the text of a Wikipedia article whereas the second one is involved in the assessment of the quality of the article.

2.1   Computing the Trustworthiness of Text

The methods in this category offer a means for predicting the accuracy of certain facts of an article. Cross [4] introduces an approach that calculates the trustworthiness throughout the life span of the text in the article and marks this by using different colors. Adler and de Alfaro calculate the reputation of the authors of the Wikipedia by using the survival time of their edits as the first step. Then they analyze exactly which text of an article was inserted by precisely which author. Finally, based on the reputation score of the respective authors, Adler and de Alfaro are able to compute the trustworthiness of each word [2]. Analog to Cross they illustrate the trustworthiness by using color-coding.

2.2   Assessing the Quality of Articles

A first work in this category was published by Lih [12], who discovered a correlation of the quality of an article with the number of editors as well as the number of article revisions. Lim et al. define three models for ranking Wikipedia articles according to their quality level[13]. The models are based on the length of the article, the total number of revisions and the reputation of the authors, which is measured by the total number of their previous edits. Zeng et al. propose to compute the quality of a particular article version with a Bayesian network from the reputation of its author, the number of words the author has changed and the quality score of the previous version[15]. Furthermore, on the basis of a statistical comparison of a sample of Featured and Non-Featured Articles in the English Wikipedia, Stvilia et al. constructed seven complex metrics and used a combination of them for quality measurement[18]. Dondio et al. derived ten metrics from research related to collaboration in order to predict quality[6]. Blumenstock[3] investigates over 100 partial simple metrics, for example the number of words, characters, sentences, internal and external links, etc. He evaluates the metrics by using them for classifications between Featured and Non-Featured Articles. Zeng et al., Stvilia et al. and Dondio et al. used a similar evaluation method which enables the evaluation results to be compared. Blumenstock demonstrates, with an accuracy of classification of 97%, that the number of words is the best current metric for distinguishing between Featured and Non-Featured Articles.

2.3   Evaluating Method Using Lifecycle

Wöhner et al. proposed a novel method to calculate quality of Wikipedia articles using lifecycle calculations[20]. Wikipedia includes a great number of articles $i=1..n$ that were edited by the Wikipedia authors during the life span. With every contribution, a new article version $v_{i,j}$ is created. The index i refers to the article identification number and the index j to the version. The versions are chronologically ordered, starting with $j=1$. The version $v_{i,0}$ is technically defined as an empty one, in other words it is the version before any content was added. To analyze the changes over time, the life span has been divided into periods. As the periods of analysis, we use months, since a shorter period causes overly high volatility of the metrics, whereas a longer period does not enable us to track the metrics precisely. The period in which a version was generated is called $p(v_{i,j})$. If an article i gets a Wikipedia evaluation, we call the period in which the article becomes a candidate for the respective Wikipedia evaluation $c(i)$.

For the calculation of the persistent and the transient contributions we have to parameterize the differences between two article versions. Therefore, we define the editing distance $dis(i,j,k)$ as that which shows the difference between the versions $v_{i,j}$ and $v_{i,k}$. It refers to the number of words which were deleted from the former version and the number of words which were inserted into the newer version.

To calculate the lifecycle of Wikipedia articles they constructed two metrics, the persistent contribution and the transient contribution. They presented a Wikipedia model for analysis, and based on this model they described the meanings and measurements of the persistent and transient contributions.

To compute the persistent contribution they measured the editing distance between the last article version in a given period and the last one in the previous period. The index of the last version of an article i in a period p was defined as

$$x(i,p) = \max(x \mid p(v_{i,x}) \leq p)$$

Accordingly the persistent contribution is defined as

$$C^{per}_{i,p} = dis(i, x(i,p-1), x(i,p))$$

They define the transient contribution as

$$C^{tran}_{i,p} = \begin{cases} C^{tran}_{i,p} = 0 & \text{if } x(i,p) = x(i,p-1) \\ C^{tran}_{i,p} = \sum_{j=x(i,p-1)}^{x(i,p)-1} dis(i,j,j+1) - C^{per}_{i,p} & \text{else} \end{cases}$$

They used these metrics to assess the quality of Wikipedia articles. The experimental result showed a good accuracy. In our experiment we use the metrics proposed by Wöhner as the features in the machine learning technology. We will have a comparison between the existing methods and our proposed method.

3   Indicators of Network Structure for Quality Evaluation

We give models and algorithms to describe and analyze the collaboration among authors of Wikipedia from a network analytical perspective. The edit network encodes who interacts how with whom when editing an article; it significantly extends previous network models that code author communities in Wikipedia. Several characteristics summarizing certain aspects of the organization process and allowing the analyst to identify certain types of authors can be obtained from the edit network. Moreover, we propose two indicators characterizing the network structure. We show that the structural network indicators are correlated with the quality of the associated Wikipedia articles.

3.1   Network Model

The edit network associated with a Wikipedia page p has as nodes the authors of p and encodes how authors contributed to p and how authors interacted with each other while editing p. This information is computed from the complete history of p, from the sequence of its revisions, by determining which part of the text has been added, has been deleted, or remained unchanged when going from one version of the page to the next.

The edit network associated with a Wikipedia page p is modeled as a graph $G = (V, E, A)$, whose components are defined as follows [12].

1.   The nodes V of the graph (V, E) correspond to the authors that have done at least one revision on p.
2.   The directed edges $E \in V \times V$ of the graph (V, E) encode the edit interaction among authors. A particular pair of authors $(u, v) \in V \times V$ is in E, if u performed one of the following three actions with respect to v.
     a) u deletes text that has been written by v;
     b) u undeletes text that has been deleted by v (and written by a potentially different author w);
     c) u restores text that has been written by v (and deleted by a potentially different author w).
          Since authors may as well revise text written by themselves, loops, i. e., edges connecting an author with herself, are allowed.
3.   A is a set of weighted attributes on nodes and edges.

3.2   Attributes on Nodes and Edges

The weighted attributes on nodes and edges encode how much text users add, delete, or restore. Furthermore, in case of deletion we keep track of who has previously written the text and in case of restoration we keep track of both, the original author and the deleter of the restored text. By combining these attributes, we can explore deep insight into the various roles that users play when editing, as well as into relations between users. The amount of text added, deleted, or restored is measured by the number of words.

We will also keep track of the timepoint when edit actions occur by indexing attributes with the revision number. In the following we assume that the history of a given page is a sequence of revisions $R = (r_1, ..., r_N)$, ordered by increasing timestamps $1, ..., N$.

For each timepoint $i \in \{1, ..., N\}$ and each pair of authors $(u, v) \in V \times V$,

- $delete_i(u, v)$ denotes the number of words deleted by u in revision r_i and written by v at earlier timepoints j $(j < i)$;
- $undelete_i(u, v)$ denotes the number of words restored by u in revision r_i, deleted by v at timepoints j $(j < i)$, and written by a potentially different author w at timepoints $l(l < j < i)$;
- $restore_i(u, v)$ denotes the number of words restored by u in revision r_i, written by v at timepoints j $(j < i)$, and deleted by a potentially different author w at timepoints $l(j < l < i)$.

For each timepoint $i \in \{1, ..., N\}$ and each author $u \in V$,

- $add_i(u)$ denotes the number of words that are added by u at time i;

- authorship$_i(u)$ denotes the number of words in revision r$_i$ that have been authored by u, all words that have been added to the text by u in a revision $j \leq i$ and that are still there in r$_i$.

Summing values over all timepoints yields three weight functions for edges $(u, v) \in$ E, that are given by

$$delete(u, v) = \sum_{i=1}^{N} delete_i(u, v)$$

$$undelete(u, v) = \sum_{i=1}^{N} undelete_i(u, v)$$

$$restore(u, v) = \sum_{i=1}^{N} restore_i(u, v)$$

The sum over the two negative relations, denoted by

$$revise(u, v) = delete(u, v) + undelete(u, v)$$

The revise(u, v) encodes how much u undoes $v's$ edits. It is interpreted as a measure of how strongly u disagrees with v. Similarly, restore(u, v) is interpreted as a measure of how strongly u agrees with v.

3.3   Network Structural Indicators

In contrast to the negative disagreement edges that are given by the attribute revise(u, v), high values of the attribute restore(u, v) indicate a positive relationship from u to v. If the two opposing groups (V_1, V_2) really represent contradicting opinions, we expect that positive edges are mostly between members of the same group. The following indicator estimates to what extent this property holds.
Let

$$c = \sum_{u,v \in V_1} restore(u, v) + \sum_{u,v \in V_2} restore(u, v)$$

be the aggregated weight of positive edges within the groups and

$$w = \sum_{u \in V_1, v \in V_2} [restore(u, v) + restore(v, u)]$$

be the aggregated weight of positive edges between the groups. Now we introduce the following new indicator called *restoreratio*:

$$restoreratio(G) = \frac{c - w}{w + c}$$

is normalized to the interval [−1, 1]. It equals 1 if all restore-edges are within the groups, i. e., if no author restores text from an author of the other group, indicating contradicting opinions. It equals 0 if the restore-edges are independent from the group membership, indicating that the two groups do not have contradicting opinions. It equals -1 if all restore-edges are between the two groups, indicating no controversy between the two groups.

The restoreratio indicator is higher for controversial articles (although not significantly), indicating that authors are more likely to restore text written by authors in their own group. This suggests that for controversial articles the two computed groups indeed represent contradicting opinions, while the opposition is less clear for featured articles.

We hypothesize that the network structural indicator restoreratio has correlation with the quality of Wikipedia articles, because of the following observations:

- The restoreratio indicator is higher for controversial articles.
- The contents of a controversial article can be hampered by an "edit fight" between the two opposing groups, damaging its integrity and objectivity. Thus the article will not gain support from general readers.
- The two opposing groups will not support the controversy article, unless they can eliminate contents contradicting to their views.
- Less support from both the editors and general readers will prevent the article from being regarded as having high quality.

In the subsequent experiments, we verify the above correlation between restoreratio and the quality of articles.

4   Quality Evaluation Using Machine Learning Technology

4.1   Support Vector Machine

We will use the machine learning technology which is called support vector machine(SVM) to deal with the features. SVM is a kind of related supervised learning methods used for classification and regression. Given a set of training examples, each marked as belonging to one of the other categories, an SVM training algorithm builds a model that predicts whether a new example falls into one category or the other. Intuitively, an SVM model is a representation of the examples as points in space, mapped so that the examples of the separate categories are divided by a clear gap that is as wide as possible. New examples are then mapped into that same space and predicted to belong to a category based on which side of the gap they fall on.

In this paper we use the SVM tool named winSVM, which is developed by Marc Block based on libsvm.

4.2   Sample Articles Using Original Wikipedia Evaluation

To increase the trustworthiness of the quality, Wikipedia introduced the voting-based quality evaluations *"Articles for Deletion"*, *"Good Articles"* and *"Featured Articles"*. For the rest of the paper, we refer to them as Wikipedia evaluations. As we investigate the Wikipedia in this study, we describe how the Wikipedia evaluations are used in the English Wikipedia. First, for all of the Wikipedia evaluations, any user can nominate an article by listing it on the respective nomination site (*Articles for Deletion, Candidate for Good Article* and *Candidate for Featured Article*). When an article is nominated, the article is flagged with a special tag. According to the type of evaluation, there are particular criteria that are used for the decision. Featured Articles have the highest quality standard. They have to be accurate, complete and well written. Good Articles are also high quality articles, however, slight inconsistencies in the quality are tolerated, such as a lack of illustrations or small weaknesses in the writing style. Articles for Deletion are articles of particularly low quality that have been tagged for deletion. Criteria are, for example, an unsuitable representation or a lack of relevance for an encyclopedia. However, even Articles for Deletion actually maintain a minimum standard of quality. The articles that are generally uncontroversial for deletion, such as those victimized by vandalism or other nonsense, are deleted quickly by using the speedy deletion procedure.

After the nomination of an article, the community decides via a vote as to whether or not the article complies with certain criteria. The voting period and the voting rule depend on the kind of evaluation. For example, in order to become a Featured Article, a voting period of 20 days and a slight modification of the two-third voting rule are necessary. After a successful election, the Featured and Good Articles are marked by special tags and are displayed in the respective sections of the Wikipedia portal, whereas Articles for Deletion are deleted by an administrator.

In the experiment we will randomly choose samples of featured articles and non-featured. The featured articles represent higher quality ones, while the non-featured represent lower quality ones.

4.3   Features of SVM

For the support vector machine, we will use several existing metrics which are proposed by previous research to be the features. Our proposed methods are combined with the existing ones. We will conduct several experiments to test the accuracy of our methods.

The metrics that will be used as SVM features are shown below

- A^e : average number of editors per month;
- M^e : maximum number of editors per month;
- E : overall number of editors;
- Q^3 : Quotient of the sum of the transient contributions and the sum of the persistent contributions within the last three months until nomination;
-

$$Q^3 = \frac{\sum_{p=c(i)-3}^{c(i)-1} C^{tran}{}_{i,p}}{\sum_{p=c(i)-3}^{c(i)-1} C^{per}{}_{i,p}}$$

- RE : restoreratio (proposed feature in Section 3).

4.4   Experimental Results

We randomly choose 16 featured articles and 16 non-featured articles to form the sample of training data. We randomly choose another 16 featured articles and 16 non-featured articles for prediction. We will do three experiments using SVM for comparison and analysis.

First we use three features, A^e, M^e, E. The result is shown in Table 1.

In the first experiment, we use three features, A^e, M^e, E. The result shows that 5 higher quality articles are predicted to be lower quality articles while 4 lower quality articles are predicted to be higher quality articles. The total precision is 0.719. It is not a surprising result. Then we use four features, A^e, M^e, E, Q^3. The result is shown in Table 2.

Table 1. Experimental result of 3-dimensional features

	Predicted to be higher	Predicted to be lower
Higher quality	11	5
Lower quality	4	12

Table 2. Experimental result of 4-dimensional

	Predicted to be higher	Predicted to be lower
Higher quality	12	4
Lower quality	3	13

In the second experiment, we use four features, A^e, M^e, E, Q^3. The result shows that 4 higher quality articles are predicted to be lower quality articles while 3 lower quality articles are predicted to be higher quality articles. The total precision is 0.780, which is higher than the result of first experiment. The metric Q^3 is a novel metric which is proposed by Wöhner[20]. The result shows it has a high correlation with of quality of Wikipedia articles.

Then we use four features, A^e, M^e, E, RE. The result is shown in Table 3.

Table 3. Experimental result on 4-dimensional

	Predicted to be higher	Predicted to be lower
Higher quality	13	3
Lower quality	2	14

In the third experiment, we also use four features, A^e, M^e, E, RE, including our proposed metric. The result shows that 2 higher quality articles are predicted to be lower quality articles while 3 lower quality articles are predicted to be higher quality articles. The total precision is 0.844, which is higher than the result of first and second experiment.

We exchange the training data for another sample of randomly picked featured and non-featured articles, and repeat the previous experiment for three more times using the same prediction data. The precision and recall curves of the four experiments are shown in Figure 1 and Figure 2, respectively. The blue curve represents the four experiment of features A^e, M^e, and E. The red curve represents the four experiment of features A^e, M^e, E, and Q^3. The green curve represents the four experiment of features A^e, M^e, E, and RE.

As shown in Figure 1, the precision curve of the features contain our proposed indicator RE is higher than the curve of the features contain metric Q^3, except the experiment 2. The recall curves of Figure 2 show that the features including RE is always higher than the others. Overall, our proposed metric restoreratio has a better performance in quality evaluation than the existing metrics.

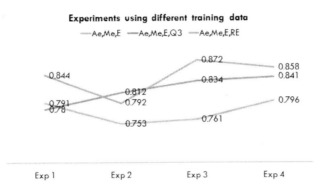

Fig. 1. Precision curve of four different training data

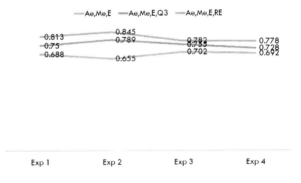

Fig. 2. Recall curve of four different training data

5 Conclusion

In this paper we investigated the quality evaluation method of Wikipedia articles. Wikipedia is an appropriate choice for applying quality measurement techniques. As compared to traditional websites and print media, Wikipedia offers within the edit history a vast array of information about the development process. This information includes implicit evidence about the quality and thus can be utilized explicitly for quality assessment.

In this paper we proposed a model of network structure and explained the algorithm to calculate the network structural indicator. Then we use the network structural indicator combining with several existing metrics for evaluating the quality of articles. The experimental results show that our proposed method has better performance than existing ones.

For further work, first we would like to find more efficient metrics. For example, the number of edits that a contribution survives could be used as an alternative measure. Furthermore we are interested in other reference articles instead of simply the articles judged via Wikipedia evaluations. For example, expert ratings, published in studies that compare Wikipedia with other traditional encyclopedias, could be used. It can be assumed that according to the voting procedure the most popular articles are elected for Good and Featured Articles but articles that truly maintain the highest quality standard may not be elected. Furthermore by using articles assessed by Wikipedia evaluations, the metrics for the time period in which an article is determined to be high quality, cannot be measured in that exact same period of time. The metrics are influenced by the attraction of editors after articles are listed on the respective Wikipedia pages. By using expert rated articles instead, this period of time can be analyzed too, which may provide more efficient metrics.

References

[1] Wikipedia: page history, http://en.wikipedia.org/wiki/Edit_history

[2] Adler, B.T., Chatterjee, K., de Alfaro, L., Faella, M., Pye, I., Raman, V.: Assigning Trust To Wikipedia Content. In: Proc. 2008 Int'l Symp. Wikis., Porto (September 2008)

[3] Blumenstock, J.E.: Size Matters: Word Count as a Measure of Quality on Wikipedia. In: Proc. 17th Int'l Conf. World Wide Web, Beijing, pp. 1095–1096 (April 2008)

[4] Cross, T.: Puppy smoothies: Improving the reliability of open, collaborative wikis. First Monday 11(9) (September 2006)

[5] Cunningham, W., Leuf, B.: The Wiki Way. Quick Collaboration on the Web. Addison-Wesley, Reading (2001)

[6] Dondio, P., Barrett, S.: Computational Trust in Web Content Quality: A Comparative Evalutation on the Wikipedia Project. Informatica – An International Journal of Computing and Informatics 31(2), 151–160 (2007)

[7] Giles, G.: Internet encyclopedias go head to head. Nature 438(7070), 900–901 (2005)

[8] Hoisl, B., Aigner, W., Miksch, S.: Social rewarding in wiki systems – motivating the community. In: Schuler, D. (ed.) HCII 2007 and OCSC 2007. LNCS, vol. 4564, pp. 362–371. Springer, Heidelberg (2007)

[9] Hunt, J., McIlroy, M.:: An algorithm for differential file comparison. Computer Science Technical Report 41. Bell Laboratories (1975)

[10] Kittur, A., Suh, B., Pendleton, B.A., Chi, E.H.: He says, she says: Conflict and coordination in Wikipedia. In: Proc. 25th Annual ACM Conf. Human Factors in Computing Systems (CHI 2007), San Jose, USA, pp. 453–462 (April/May 2007)

[11] Kramer, M., Gregorowicz, A., Iyer, B.: Wiki Trust Metrics based on Phrasal Analysis. In: Proc. 2008 Int'l Symp. Wikis, Porto, Portugal (September 2008)

[12] Lih, A.: Wikipedia as participatory journalism: Reliable sources? metrics for evaluating collaborative media as a news resource. In: Proc. 5th Int'l Symp. Online Journalism, Austin, USA (April 2004)

[13] Lim, E.P., Vuong, B.Q., Lauw, H.W., Sun, A.: Measuring Qualities of Articles Contributed by OnlineCommunities. In: Proc. 2006 IEEE/WIC/ACM Int'l Conf. Web Intelligence, Hong Kong, pp. 81–87 (December 2006)

[14] O'Reilly, T.: What is Web2.0, `http://www.oreillynet.com/pub/a/oreilly/tim/news/2005/09/30/what-is-web-20.html` (accessed 2005)

[15] Priedhorsky, R., Chen, J., Lam, S.K., Panciera, K., Terveen, L., Riedl, J.: Creating, Destroying, and Restoring Value in Wikipedia. In: Proc. Int'l ACM Conf. Supporting Group Work, Sanibel Island, USA, pp. 259–268 (November 2007)

[16] Sabel, M.: Structuring wiki revision history. In: Proc. 2007 Int'l Symp. On Wikis, Montreal, Canada, pp. 125–130 (October 2007)

[17] Smets, K., Goethals, B., Verdonk, B.: Automatic Vandalism Detection in Wikipedia: Towards a Machine Learning Approach. In: Proc. AAAI Workshop, Wikipedia and Artificial Intelligence: An Evolving Synergy (WikiAI 2008), Chicago, USA (July 2008)

[18] Stvilia, B., Twidale, M.B., Smith, L.C., Gasser, L.: Assessing information quality of a community-based encyclopedia. In: Proc. Int'l Conf. Information Quality, Cambridge, USA, pp. 442–454 (November 2005)

[19] Potthast, M., Stein, B., Gerling, R.: Automatic Vandalism Detection in Wikipedia. In: Macdonald, C., Ounis, I., Plachouras, V., Ruthven, I., White, R.W. (eds.) ECIR 2008. LNCS, vol. 4956, pp. 663–668. Springer, Heidelberg (2008)

[20] Wöhner, T., Peters, R.: Assessing the quality of Wikipedia Articles with Lifecycle Based Metrics. In: Proc. 2009 Int'l Symp. Wikis, Orlando (October 2009)

Towards Improving Wikipedia as an Image-Rich Encyclopaedia through Analyzing Appropriateness of Images for an Article

Xinpeng Zhang, Yasuhito Asano, and Masatoshi Yoshikawa

Kyoto University, Kyoto, Japan 606-8501
{xinpeng.zhang@db.soc.,asano@,yoshikawa@}i.kyoto-u.ac.jp

Abstract. In Wikipedia, knowledge related to an object is gathered into a single article that is mainly composed of text and images. We observed that some Wikipedia images are inappropriate for addition to some Wikipedia articles. In this paper, we propose a RCT model for analyzing the appropriateness of images for Wikipedia articles. The model assumes that an image appearing in article s indicates relations related to s. The model then examines the appropriateness of an image through analyzing relations indicated by the image. We also propose a system which examining Wikipedia images using the RCT model. The system also searches images on the Web which could serve as references for users to add appropriate images or text to edit high-quality and image-rich Wikipedia articles. Finally, we confirm through experiments that the RCT model could examine the appropriateness of Wikipedia images to a satisfactory degree.

Keywords: Wikipedia, Multimedia Information, relation.

1 Introduction

Wikipedia is a popular web-based encyclopaedia used for searching knowledge about objects. Wikipedia presents knowledge about objects of many categories such as people, science, geography, politic, and history. In Wikipedia, knowledge related to an object is gathered into a single article that is mainly composed of text and images. Wikipedia articles are edited cooperatively and constantly by volunteers. Therefore, Wikipedia always includes the most up-to-date information. Nevertheless, the quality of some Wikipedia articles is inferior [12]; some Wikipedia articles even include incorrect information [3]. Several works have been proposed for examining the quality or credibility of Wikipedia articles [1, 4]. However, previously presented works do not devote attention to images in Wikipedia articles despite the fact that images are interesting and helpful for understanding knowledge. To improve Wikipedia as an encyclopaedia that contains an abundance of appropriate images, we aim to analyze the appropriateness of images for Wikipedia articles.

In Wikipedia, every image has a caption describing it in a few words. The caption probably includes links to other Wikipedia articles. For example, Fig. 1 depicts an image appearing in Wikipedia article "Junichiro Koizumi," who was the Prime Minister of Japan. Two links to articles "Sea Island, Georgia" and "2004 G8 summit" appear

X. Du et al. (Eds.): APWeb 2011, LNCS 6612, pp. 200–212, 2011.

Koizumi meets children in Sea Island, Georgia, shortly before the 2004 G8 summit.

Tattooed Japanese men in 1870.

Fig. 1. Image appearing in English Wikipedia article "Junichiro Koizumi"

Fig. 2. Image appearing in Japanese Wikipedia article "Junichiro Koizumi"

Fig. 3. Images of Koizumi attending the 2004 G8 summit

in the caption of the image. The caption indicates to us that Koizumi met children in Sea Island, Georgia, shortly before he attended the 2004 G8 summit. The image shows a scene in which Koizumi meets chidlren. The image with its caption indicates the relation between "Junichiro Koizumi" and "Sea Island, Georgia," and that between "Junichiro Koizumi" and the "2004 G8 summit." Similarly, for an image in article s whose caption includes links to articles $t_0, t_1, ..., t_m$, we assume the image indicates the relation between s and each $t_j, 0 \leq j \leq m$. In this paper, based on the assumption, we propose an *RCT* model to examine the appropriateness of images for Wikipedia articles by the following three aspects through analyzing relations indicated by the image: relatedness, consistency and typicality.

Relatedness. For an image i existing in article s, we measure the relation indicated by i. If the relation is strong, then image i represents information that is strongly related to s; otherwise, i has low relatedness with s. Assuming that only a limited number of images can be added to an article, we then should add images that are strongly related to the article rather than images having low relatedness with the article. For instance, Fig.2 depicts an image of two tattooed Japanese people photographed in 1970. The image appears in the Japanese Wikipedia article "Junichiro Koizumi," whose text describes that Koizumi's grandfather was famous for sporting a full-body tattoo. The relation between Koizumi and 2004 G8 summit is expected to be stronger than that between Koizumi and Tattoo. That is, the 2004 G8 summit has higher relatedness with Koizumi than Tattoo has. Therefore, the image depicted in Fig.1 is expected to be more appropriate for the article "Junichiro Koizumi" than that depicted in Fig.2.

Consistency. If the text of an article includes no explanation or description about the relation between s and t_j, then we say that image i is inconsistent with the text. Images are usually difficult to interpret without description. Therefore, images that are inconsistent with the text of an article might be inappropriate for the article. For example, no description about the image depicted in Fig. 1 is included in the text of article "Junichiro Koizumi." It is difficult to understand why "Koizumi" met children in "Sea Island", or even why "Koizumi" attended the "2004 G8 summit," by reading the article. Users might be confused about why the image appears in the article. In Section 5.2, we propose a method to examine the consistency between an image i and text by investigating how many descriptions about the relations indicated by i appear in the text.

Typicality. As discussed above, for an image appearing in article s we assume that the image indicates the relation between s and an article t linked from the caption. Therefore, we also examine whether an image is visually typical for representing a relation. For instance, the image depicted in Fig. 1 indicates a relation between "Junichiro Koizumi" and the "2004 G8 summit." Fig. 3 depicts four visually similar images showing the relation between "Junichiro Koizumi" and the "2004 G8 summit" in which Koizumi attends the 2004 G8 summit with leaders of other countries. On the web, images similar to these images outnumber images similar to the image depicted in Fig. 1. Therefore, the image depicted in Fig. 1 is not typical for representing the relation between "Junichiro Koizumi" and the "2004 G8 summit." Zhang et al. [14] proposed a evidence-based method for searching images representing a relation on the Web. For an image i indicating relation r, we first search a set I of images representing r on the Web using the evidence-based method. We then investigate the similarity between i and images in I to examine the visual typicality of i.

As another contribution of this paper, we propose a system for analyzing the appropriateness of images in Wikipedia[1].Given an image existing in a Wikipedia article, the system computes relatedness, consistency, typicality, and appropriateness for the image using the RCT model. The system also offers images retrieved from the Web that contain knowledge about the same relation as that indicated by the input image. Images retrieved from the Web could serve as references for users to add appropriate images or text to edit high-quality and image-rich Wikipedia articles. We discuss the system in detail in Section 2.

The rest of this paper is organized as follows. Section 3 introduces the method proposed by Zhang et al. [13,15,14] for analyzing, mining and searching knowledge about relations. Section 4 presents the RCT model for examining relatedness, consistency and typicality of images for Wikipedia articles through analyzing relations based on the methods introduced in Section 3. Section 5 reports experiments used for evaluating the RCT model described in Section 4. Section 6 reviews related work. Section 7 concludes this paper.

2   System for Analyzing Wikipedia Image Appropriateness

We now propose the system for analyzing the appropriateness of images for Wikipedia articles. The system also offers information retrieved from the Web, which could be helpful for users to edit high-quality Wikipedia articles. Fig. 4 portrays the system interface, in which we examine an image in the Japanese Wikipedia article "Shinzo Abe." The interface comprises two parts, denoted respectively as "A" and "B."

---

[1] The system is accessible at http://www.db.soc.i.kyoto-u.ac.jp/enishi/imageAnalysis.html

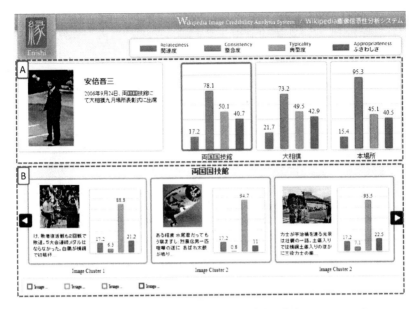

Fig. 4. Screen shot of the system for analyzing Wikipedia image appropriateness

The left side of part "A" displays the input image for analysis. The right side of part "A" presents the four values of relatedness, consistency, typicality, and appropriateness for the input image, with respect to each relation indicated by the input image. The four values are computed using the RCT model presented in Section 4. For example, the interface portrayed in Fig. 4 displays the four values indicated with bars for the input image, with respect to each of the three relations indicated by the image. The bars representing the four values for each relation are depicted inside a block.

By clicking the block drawing the bars for a relation in part "A", part "B" displays images that include knowledge about the relation. These images are retrieved from the Web using the method [14] proposed by Zhang et al. The system presents captions for the images, which are extracted to help users understand the images. Users can click on an image to link to the Web page that includes the image. The system also computes relatedness, consistency, typicality and appropriateness for each image. The images are classified into different clusters according to their visual similarity computed using the method proposed by Wang et al. [10, 7]. The system use the method proposed by Chen et al. [2] to cluster images.

The system would be useful in the following scenarios.

- Images having extremely low relatedness are meaningless for addition to Wikipedia. Users are able to use the system to check the relatedness of images for an article.
- If the consistency of an image i in article p with respect to relation r is low, then the image should be removed or some description of the image should be added to the article. The system offers images retrieved from the Web whose surrounding text includes knowledge about relation r. Then users can refer to the surrounding text of the images to add knowledge about relation r to article p.

– Assuming that there is no copyright problem, if the typicality of the input image is low, then users can use the system to search images having high typicality on the Web to replace the input image in Wikipedia.

3   Methods for Analyzing, Mining and Searching Knowledge about Relations

We propose methods for examining image appropriateness for Wikipedia articles through analyzing relations indicated by images, but first we introduce methods for analyzing, mining and searching knowledge about relations proposed by Zhang et al. [13, 15, 14].

3.1   Generalized Max-Flow Model for Measuring Relations

A *Wikipedia information network* is a directed graph whose vertices are articles of Wikipedia and whose edges are links between articles. Zhang et al. [13] model a relation between two objects in a Wikipedia information network using a generalized max-flow. The generalized max-flow problem [11] is identical to the classical max-flow problem except that every edge e has a gain $\gamma(e) > 0$; the value of a flow sent along edge e is multiplied by $\gamma(e)$. Let $f(e) \geq 0$ be the amount of flow f on edge e, and $\mu(e) \geq 0$ be the capacity of edge e. The capacity constraint $f(e) \leq \mu(e)$ must hold for every edge e. The goal of the problem is to send a flow emanating from the source into the destination to the greatest extent possible, subject to the capacity constraints. Let *generalized network* $G = (V, E, s, t, \mu, \gamma)$ be information network (V, E) with the source $s \in V$, the destination $t \in V$, the capacity function μ, and the gain function γ. Fig. 5 depicts an example of a generalized max-flow. It shows that 0.4 units and 0.2 units of the flow respectively arrive at "USA" along path (A) and path (B).

To measure the strength of a relation from object s to object t, Zhang et al. [13] use the value of a generalized maximum flow emanating from s as the source into t as the destination; a larger value signifies a stronger relation. We omit details of the model here because of space limitations. Zhang et al. [13] ascertained that the model can measure the strength of relations more correctly than previous methods [8, 6] can.

3.2   Generalized Flow Based Method for Mining Elucidatory Objects

Based on the generalized max-flow model, Zhang et al. proposed a method to mine disjoint paths that are important for a relation from object s to object t in Wikipedia [15].

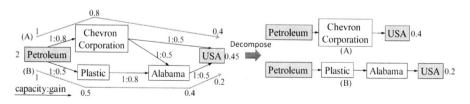

Fig. 5. Generalized max-flow and its decomposition

Zhang et al. first compute a generalized max flow f emanating from s into t on the Wikipedia information network. Flow f is then decomposed into flows on a set of paths. For example, the flow on the network depicted in Fig. 5 is decomposed into flows on two paths (A) and (B). The value of the decomposed flow on path (A) is 0.4; that on path (B) is 0.2. Finally, Zhang et al. output the top-k paths in decreasing order of the values of flows on paths to explain the relation between s and t. Zhang et al. define *elucidatory objects* of a relation as objects in the top-k paths, except the source and destination. Elucidatory objects of a relation r are objects constituting r; the elucidatory objects are able to explain r. Every elucidatory object o in a path p is assigned a weight $0 < w(o) < 1$, which equals the value of the decomposed flow on path p. A high weight signifies that the elucidatory object plays an important role in the relation.

3.3 Searching Images Explaining Relations on the Web

Zhang et al. [14] proposed an evidence-based method for searching sets of "image with surrounding text" (hereafter abbreviated as $IwST$) including knowledge about a relation between two objects s and t on the Web. Zhang et al. first searches images related to s and t using a keyword image search engine with query "$s\ t$." However, some $IwSTs$ include no knowledge about the relation between s and t. Zhang et al. [14] then infer that an $IwST$ includes knowledge about a relation, if the surrounding text of the $IwST$ includes many elucidatory objects of the relation. That is, elucidatory objects are evidence that is useful for judging whether a text includes knowledge about a relation.

We present the evidence-based method below.

Input: objects s and t, integer parameters m, and n. (1) Obtain a set O of elucidatory objects for the relation between s and t using the method discussed in Section 3.2. (2) Search the top-m images, say $m = 300$, using a keyword image search engine with query "$s\ t$." (3) Extract the surrounding text of each image. Let I be the set of the top-m $IwSTs$. (4) Remove $IwSTs$ whose surrounding text includes no s or t from I. (5) Compute a score $s(i)$ for every $i \in I$ to $s(i) = \sum_{o \in s,t,O'} w(o) \times log_e(e + f(o))$, where $O' \subseteq O$ is the set of elucidatory objects appearing in the surrounding text of i, and $f(o)$ is the appearance frequency of o in i. The weight $w(s)$ and $w(t)$ is set to the maximum weight of all objects in O. (6) **Output:** the top-n $IwSTs$ in I having high scores.

Zhang et al. [14] confirmed that the method is effective for searching $IwSTs$ including knowledge about relations.

4 RCT Model for Analyzing Wikipedia Image Appropriateness

As discussed in Section 1, for an image in article s whose caption includes links to articles $t_0, t_1, ..., t_m$, we assume that the image indicates the relation between s and each $t_j, 0 \leq j \leq m$. In this section, we present the RCT model for computing relatedness, consistency, typicality and appropriateness of an image for an article.

4.1 Relatedness

Relatedness represents the strength of the relation indicated by an image. We compute the strength $g(r)$ of relation r using the generalized max-flow model [13] introduced in

Section 3.1. The strength of relations between two objects in Wikipedia ranges from 0 to $\sqrt{max(d)}$, where $max(d)$ is the maximum of the number of links linking from or to an article in Wikipedia. Given an image i indicating a relation r, we normalize $s(r)$ to a value $0 \leq R(i, r) \leq 1$ as the relatedness of image i with respect to relation r, using the following equation.

$$R(i, r) = log \frac{g(r) + \alpha}{g(r)}; \quad if\ R(i, r) > 1, then\ R(i, r) = 1. \tag{1}$$

To normalize $R(i, r)$ to a value $0 \leq R(i, r) \leq 1$, we set α according to $max(d)$ of the Wikipedia dataset. For the dataset used in experiments discussed in Section 5, we set $\alpha = 5.8$.

4.2 Consistency

Consistency represents how many descriptions about the relation indicated by an image exist in the text of a Wikipedia article. We measure the consistency of the image with the text based on the method introduced in Section 3.3 by counting how many elucidatory objects of the relation appear in the text.

We present a method for examining consistency below.

Input: Image i indicating relation r in Wikipedia article p. (1) Obtain a set O of elucidatory objects for relation r. (2) Compute a score $s(i)$ using the following equation.

$$s(i, r) = \sum_{o \in O'} w(o)/d(o), \tag{2}$$

where $O' \subseteq O$ signifies the set of elucidatory objects appearing in the text of article p, and $d(o)$ denotes the distance from o to image i in article p. (3) **Output:** Normalize $s(i)$ to a value $0 \leq C(i, r) \leq 1$ as the consistency of image i with respect to relation r using the following equation.

$$C(i, r) = log \frac{s(i, r) + \beta}{s(i, r)}; \quad if\ C(i, r) > 1, then\ C(i, r) = 1. \tag{3}$$

Elucidatory objects appearing in text closer to image i in article p tend to relate to i more strongly. Therefore, we assign high scores to elucidatory objects closing to i in Equation 2. The distance in Equation 2 can be defined as the path length between the nodes including o and i, respectively, in the DOM tree of article p, or number of words or sentences between o and i in the text of article p. We set the distance as the be number of words between o and i in p, in the experiments discussed in Section 5.

To normalize $C(i, r)$ to a value $0 \leq C(i, r) \leq 1$, we set β according to the Wikipedia dataset. For the dataset used in Section 5, we set $\beta = 17.6$ after several experiments.

4.3 Typicality

An Image ti appearing in many web pages that contains description about a relation is typical for indicating the relation. Images which are visually similar to the image ti

are also typical. Inversely, images appearing in few web pages that contains description about a relation are untypical for indicating the relation. We observed that untypical images are inappropriate for indicating a relation or even unrelated to a relation. Given an image i and a set I of images which appears in web pages containing knowledge about a relaion, we measure the similarity among i and images in I to examine the typicality of i.

The RCT model used VisualRank [5] proposed by Jing et al. which computes the typicality of images. VisualRank employs PageRank on a network whose vertices representing images; every edge (u, v) is assigned a weight which is the visual similarity [10, 7] between images u and v. Then, a score vr is obtained for every image representing its typicality. A high score vr represents that the image is similar to many other images on the network. The intuition of VisualRank is that an image is typical if images similar to the image are also typical.

We present the method for examining typicality below.

Input: Image i indicating relation r. (1) Obtain a set I of images representing relation r using the evidence-based method introduced in Section 3.3. (2) Construct a network including image i and images in I, and compute score $vr(i)$ for i on the network. (3) **Output:** Normalize $vr(i)$ to a value $0 \leq T(i, r) \leq 1$ as the typicality of image i with respect to relation r using the following equation.

$$T(i,r) = \frac{vr(i)}{\max_{img \in i, I} vr(img)} \tag{4}$$

4.4 Appropriateness

If any one of the three values of relatedness, consistency and typicality of an image for an article is low, then the image is probably inappropriate for addition to the article. The RCT model computes appropriateness $A(i, r)$ for image i with respect to relation r, which are computed using the following equation.

$$A(i,r) = \sqrt[3]{R(i,r) \cdot C(i,r) \cdot T(i,r)} \tag{5}$$

5 Experiments

we conduct experiments to evaluate the RCT model on computing consistency and appropriateness by human subject. We do not evaluate the RCT model on computing relatedness because Zhang et al. [13] have confirmed that the generalized max-flow model used for computing relatedness can measure relations appropriately. Similarly, we do not evaluate the RCT model in terms of typicality because the effectiveness of VisualRank adopted for computing typicality has been ascertained by Jing et al. in [5].

5.1 Dataset

We first select 10 articles in different categories from a Japanese Wikipedia dataset (20090513 snapshot). The titles of the 10 articles are presented in Fig. 7. For each

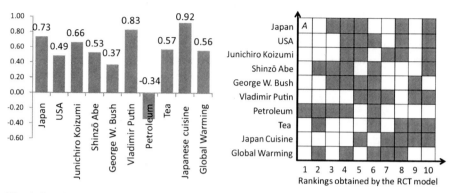

Fig. 6. Correlation coefficient for consistency   **Fig. 7.** Evaluation for appropriateness

article, we select 10 images appearing in the article for evaluation. If fewer than 10 images exist in an article, then we compensate the shortage with manually selected images, and create appropriate captions for the images. For example, Fig. 8 depicts 10 images selected for the article "USA."

In the RCT model, we assume that an image of article p shows relations between p and a link in the caption of the image. The system discussed in Section 2 analyzes an image with respect to all the possible relations between p and every link in the caption. For the experiments, we decided that humab subjects determine which relation is indicated by the image. For every image, we asked 10 participants to select one relation the image mainly indicates. Each participant independently selected one link from the links appearing in the caption of an image for article p. We then assume that the image indicates the relation between p and the link selected by most participants. For example, the captions of the images depicted in Fig. 8 are written in the column "Caption (Relation)." The links appearing in the captions are denoted by underlined letters, among which those denoted as large bold letters are links selected by the participants.

5.2   Evaluation of Consistency

We first evaluated whether the RCT model can compute the consistency of an image with an article appropriately. For each of the 10 selected articles, we asked 10 participants to judge whether the text of an article includes a description about each of the 10 relations indicated by images selected for the article. Every participant gives an integer score from 0–4 to each relation. A higher score represents that the text includes a better description of the relation. We then compute the average of the scores given by the 10 participants as the value of consistency obtained by human subjects.

For each of the 10 articles, we compute the Pearson's correlation coefficient between consistency obtained by human subjects and the RCT model. Fig. 6 depicts the coefficient for all 10 articles. Except for the coefficient for the article "Petroleum" which is inferior, the coefficients for 7 of the 10 articles are higher than 0.5. Especially, the coefficient for article "Japanese cuisine" is 0.92. The average of the coefficients for the 10 articles is 0.53. Therefore, we conclude that the RCT model can examine consistency of images for an article appropriately to a satisfactory degree.

5.3   Evaluation of Appropriateness

The RCT model computes the appropriateness of an image according to relatedness, consistency and typicality of the image, as discussed in Section 5.3. In this section, we evaluate the accuracy of the appropriateness computed using the RCT model.

We first tell participants the following question to force them to consider the appropriateness seriously: "if only 5 of the 10 images could remain in the article, which images do you select?" We then ask them to give an integer score 0–10 to 10 images according to the appropriateness for each of the 10 selected articles, where a higher score represents higher appropriateness. We then compute the ranking of the images using the average of the scores given by the 10 participants.

In Fig.7, we compare the appropriateness obtained by human subjects with those computed by the RCT model. Every cell depicted in Fig.7 represents an image. A row including 10 cells represents 10 images for an article. Numbers on the horizontal axis indicate the rankings of the images according to the appropriateness obtained using the RCT model for each article. A white cell signifies an image that is ranked among the top-5 according to the appropriateness obtained by human subjects; inversely, a black cell shows an image ranked lower than 5th. For example, a cell denoted by alphabet "A" represents the top-1 image for the article "Japan," as ranked by the RCT model. The cell is white, therefore, the image is ranked among top-5 by human subjects. The cells in the first left-hand column are all white, except that of the article "Petroleum". That is, the top-1 image for 9 of the 10 articles ranked by the RCT model are also ranked among the top-5 by human subjects. Most cells in the second and third left-hand columns are also white. However, 7 of the 10 cells in the first right-hand column are black. That is, images ranked lowest by the RCT model for 7 of the 10 articles are also ranked as low by human subjects. From the discussion presented above, we conclude that the RCT model can examine the appropriateness of images for an article to a satisfactory degree. Furthermore, we survey the appropriateness computed using the RCT model through a case study presented in the next section.

5.4   Case Study: Images for Article "USA"

We observe the 10 images for the article "USA," depicted in Fig. 8. The column "Human" presents the consistency and appropriateness obtained by human subjects. The column "RCT Model" presents the relatedness, consistency, typicality, and appropriateness computed using the RCT model. Participants assign a score of 0 as the consistency to the three images appearing in the 6th, 8th, and 10th rows because they could not find a description about the relations indicated by the three images in the article "USA." However, by considering the relatedness for the image appearing in the 6th row is strong, i.e., the strength of the relation between the "USA" and "the battle of Gettysburg" is strong, they assign score 6.80 as the appropriateness to the image. In the column "Ranking," the figures out of parentheses and the figures in parentheses respectively represent the rankings of the images according to the appropriateness assigned by human subjects and the ranking obtained using the RCT model. The top-3 images and the 10th image obtained by human subject and the RCT model are identical.

The four images ranked as 7th, 8th, and 10th by the RCT model have much lower typicality than other images have. Therefore, the four images are ranked lowest by

Ranking	Image	Caption (Relation)	Human		RCT Model			
			C[2]	A[4]	R[1]	C[2]	T[3]	A[4]
1 (2)		Barack Obama was inaugurated as the first black president in U.S. history.	2.56	8.80	0.59	0.35	0.74	0.53
2 (3)		View of the World Trade Center of the 9/11 terrorist attacks in New York City.	2.56	7.50	0.48	0.33	0.86	0.52
3 (1)		Dr. Martin Luther King giving his "I Have a Dream" speech.	1.56	7.40	0.51	0.33	1.00	0.55
4 (8)		A crowd gathers at Wall Street during the Great Depression.	2.89	7.20	0.35	0.20	0.12	0.21
5 (7)		Painting, Declaration of Independence. The painting can be found on the back of the U.S dollar $2 bill.	3.44	7.00	0.52	0.39	0.10	0.27
6 (4)		The battle of Gettysburg, which was part of the American Civil War.	0.00	6.80	0.39	0.27	0.96	0.47
7 (4)		US President Ronald Reagan and Soviet General Secretary Mikhail Gorbachov in 1985. In 1987, they signed the INF Treaty.	1.78	6.30	0.58	0.21	0.85	0.47
8 (8)		Yalta summit in 1945 with Winston Churchill, Franklin Roosevelt and Joseph Stalin.	0.00	4.80	0.36	0.23	0.11	0.21
9 (6)		An F-117 Nighthawk in flight, which was a icon of American Power.	0.22	4.60	0.17	0.15	1.00	0.29
10 (10)		The Mayflower, which set sail for America across the Atlantic Ocean.	0.00	4.50	0.20	0.30	0.12	0.19

[1] R: Relatedness; [2] C: Consistency; [3] T: Typicality; [4] A: Appropriateness

Fig. 8. Images for the article "USA"

the RCT model, although the relatedness and their consistency are not the lowest. For example, the 10th image as ranked by the RCT model indicates the relation between the Mayflower and the USA. We retrieved images associated with the relation on the Web using the method introduced in Section 3.3, four of which are depicted in Fig. 9. By reading the descriptions of the images, we know that the images strongly relate to the relation between the Mayflower and the USA. The 10th image is dissimilar from most of the images retrieved from the Web, including the four depicted in Fig. 9. Therefore, the 10th image receives low typicality. As introduced in Section 2, the system for analyzing the appropriateness of images displays these images retrieved from the Web. The system could be useful to find typical images indicating a relation.

Image	Description	Image	Description
	Plymouth Plantation is a living museum in Plymouth. Plymouth is best known for being the site of the colony established by the Pilgrims, passengers of the Mayflower.		The Mayflower Steps Arch on Plymouth's ancient Barbican. This is the approximate place where the Puritan Pilgrim Fathers set sail in The Mayflower for America in 1620.
	The Mayflower II is a replica of the 17th century ship Mayflower, celebrated for transporting the Pilgrims to the New World.		The Pilgrim Monument commemorates the history of the Mayflower Pilgrims, their arrival and stay in Provincetown Harbor, and the signing of the Mayflower Compact.

Fig. 9. Images associated with the relation between the MayFlower and the USA

6   Related Work

To the best of our knowledge, no method has been proposed in the literature for particularly analyzing the appropriateness of images in Wikipedia. However, many methods examine the quality or credibility of Wikipedia articles. Thomas Chesney examined the credibility of Wikipedia by asking participants to read an article and to assess its credibility [3]. Chesney reports that the accuracy of Wikipedia is high; however, 13 percent of the Wikipedia articles include mistakes. Wilkinson et al. [12] observe that high-quality Wikipedia articles are distinguished by a high number of edits, number of editors, and the intensity of cooperative behavior. Several other works measure the quality of Wikipedia articles by examining the reputations of Wikipedia editors [1, 4]. These works assume that editors gain a reputation when the editing that they perform on Wikipedia articles are preserved by subsequent editors. They lose that reputation when their edits are modified or removed in short order. Articles edited by editors having a high reputation are then regarded as high-quality articles.

Consideration of consistency between an image and its surrounding text has not been reported before. Regarding the typicality of images, several reported works search for typical images based on a query. An image I is a typical image for query Q if Q is an appropriate label for I. Tezuka et al. [9] search typical images for a query by analyzing visual features of images such as color features. In contrast to those methods, we consider the visual typicality of images representing a relation.

7   Conclusion

We proposed a RCT model for analyzing the appropriateness of images for Wikipedia articles. For an image in article s whose caption includes links to articles $t_0, t_1, ..., t_m$, we assume the image indicates the relation between s and each $t_i, 0 \leq i \leq m$. We then examine the appropriateness of the image in the three aspects, relatedness, consistency and typicality, through analyzing relations indicated by the image. Our experiments revealed that the RCT model can examine consistency of images for an article appropriately, and the model can compute appropriateness for images to a satisfactory degree.

As another contribution, we propose a system for analyzing Wikipedia image appropriateness using the RCT model. Given an image appearing in a Wikipedia article, the system computes appropriateness for the image. The system also search images on the Web that contain knowledge about the same relation as that indicated by the input image. Our system could be helpful for users to judge Image appropriateness and to add appropriate images or text to edit high-quality and image-rich Wikipedia articles.

Acknowledgment

This work was supported in part by the National Institute of Information and Communications Technology, Japan.

References

1. Adler, B.T., de Alfaro, L.: A content-driven reputation system for the wikipedia. In: Proc. of 16th WWW, pp. 261–270 (2007)
2. Chen, Y., Wang, J.Z., Krovetz, R.: Clue: cluster-based retrieval of images by unsupervised learning. IEEE Transactions on Image Processing 14(8), 1187–1201 (2005)
3. Chesney, T.: An empirical examination of wikipedia's credibility. First Monday 11(11) (2006)
4. Hu, M., Lim, E.P., Sun, A., Lauw, H.W., Vuong, B.Q.: Measuring article quality in wikipedia: models and evaluation. In: Proc. of 16th CIKM, pp. 243–252 (2007)
5. Jing, Y., Baluja, S.: Visualrank: Applying pagerank to large-scale image search. IEEE Transactions on Pattern Analysis and Machine Intelligence 30(11), 1877–1890 (2008)
6. Koren, Y., North, S.C., Volinsky, C.: Measuring and extracting proximity in networks. In: Proc. of 12th ACM SIGKDD Conference, pp. 245–255 (2006)
7. Li, J., Wang, J.Z., Wiederhold, G.: Irm: Integrated region matching for image retrieval. In: Proc. 8th of ACM Multimedia, pp. 147–156 (2000)
8. Nakayama, K., Hara, T., Nishio, S.: Wikipedia mining for an association web thesaurus construction. In: Benatallah, B., Casati, F., Georgakopoulos, D., Bartolini, C., Sadiq, W., Godart, C. (eds.) WISE 2007. LNCS, vol. 4831, pp. 322–334. Springer, Heidelberg (2007)
9. Tezuka, T., Maeda, A.: Typicality ranking of images using the aspect model. In: Bringas, P.G., Hameurlain, A., Quirchmayr, G. (eds.) DEXA 2010. LNCS, vol. 6262, pp. 248–257. Springer, Heidelberg (2010)
10. Wang, J.Z., Li, J., Wiederhold, G.: Simplicity: Semantics-sensitive integrated matching for picture libraries. IEEE Trans. on PAMI 23(9), 947–963 (2001)
11. Wayne, K.D.: Generalized Maximum Flow Algorithm. Ph.D. thesis. Cornell University, New York, U.S. (January 1999)
12. Wilkinson, D.M., Huberman, B.A.: Cooperation and quality in wikipedia. In: Proc. of WikiSym 2007, pp. 157–164 (2007)
13. Zhang, X., Asano, Y., Yoshikawa, M.: Analysis of implicit relations on wikipedia: Measuring strength through mining elucidatory objects. In: Kitagawa, H., Ishikawa, Y., Li, Q., Watanabe, C. (eds.) DASFAA 2010. LNCS, vol. 5981, pp. 460–475. Springer, Heidelberg (2010)
14. Zhang, X., Asano, Y., Yoshikawa, M.: Enishi: Searching knowledge about relations by complementarily utilizing wikipedia and the web. In: Chen, L., Triantafillou, P., Suel, T. (eds.) WISE 2010. LNCS, vol. 6488, pp. 480–495. Springer, Heidelberg (2010)
15. Zhang, X., Asano, Y., Yoshikawa, M.: Mining and explaining relationships in wikipedia. In: Bringas, P.G., Hameurlain, A., Quirchmayr, G. (eds.) DEXA 2010. LNCS, vol. 6262, pp. 1–16. Springer, Heidelberg (2010)

SDDB: A Self-Dependent and Data-Based Method for Constructing Bilingual Dictionary from the Web*

Jun Han, Lizhu Zhou, and Juan Liu

Department of Computer Science and Technology, Tsinghua University, Beijing
100084, China
{hanj04,tiantang326}@gmail.com, dcszlz@tsinghua.edu.cn

Abstract. As various data on the World Wide Web are becoming massively available, more and more traditional algorithm centric problems turn to find their solutions in a data centric way. In this paper, we present such a typical example - a Self-Dependent and Data-Based (SDDB) method for building bilingual dictionaries from the Web. Being different from many existing methods that focus on finding effective algorithms in sentence segmentation and word alignment through machine learning etc, SDDB strongly relies on the data of bilingual web pages from Chinese Web that are big enough to cover the terms for building dictionaries. The algorithms of SDDB are based on statistics of bilingual entries that are easy to collect from the parenthetical sentences from the Web. They are simply linear to the number of sentences and hence are scalable. In addition, rather than depending on pre-existing corpus to build bilingual dictionaries, which is commonly adopted in many existing methods, SDDB constructs the corpus from the Web by itself. This characterizes SDDB as an automatic method covering the complete process of building a bilingual dictionary from scratch. A Chinese-English dictionary with over 4 million Chinese-English entries and over 6 million English-Chinese entries built by SDDB shows a competitive performance to a popular commercial products on the Web.

Keywords: bilingual dictionaries, statistics, data-based.

1   Introduction

Bilingual dictionaries are widely used in people's daily life and many fields of computer science, such as machine translation[3], cross language information retrieval[15] and so on. Traditional dictionaries are built manually with human annotation and revision which is expensive, time-consuming and difficult to update. Recently many researchers have been mining bilingual dictionaries from web ([4],[8],[14]).

* This work was supported in part by National Natural Science Foundation of China under grant No. 60833003.

X. Du et al. (Eds.): APWeb 2011, LNCS 6612, pp. 213–224, 2011.
© Springer-Verlag Berlin Heidelberg 2011

The basic idea for building bilingual dictionary from the Web is first to find the bilingual text, such as the parenthetical bilingual sentences in Figure 1 and the well formed bilingual text in Figure 2, and then use a method to find the correct matching of terms or phrases between two languages in the text. For instance, from Figure 1 and Figure 2 the English-Chinese translation of (socialism, 社会主义)，（communism, 共产主义），（commissary in charge of studies，学习委员）etc. should be the right entries selected by the method. Clearly how efficiently and precisely the method can find the correct matching is a crucial issue.

To address above issue, many works have been presented. But most of them focus on designing better algorithms for sentence segmentation, word alignment, or on complex NLP (Natural Language Processing) models using machine learning techniques etc([6],[19],[16]). However, the efficiency of these methods are affected while data is too big, and many experiments demonstrated that the traditional improving is not as effective as expected ([11],[18]). A 2001 study([1]) showed that the worst algorithms performed better than the best algorithms if they were trained with a larger word database. Namely, for a given problem, the richness of data plus simple algorithms could alternatively solve the problem well or better than complex algorithms. This is particularly the case for Web study. As a matter of fact, in recent years many researchers have turned to works on how to use Web data ([2],[7],[9],[10],[12], [22]) to solve interesting problems.

Following this *Solving a Problem by Big Data* pattern, we present a **S**elf-**D**ependent and **D**ata **B**ased Method (SDDB) for Constructing Bilingual Dictionaries from the Web based on our following observations to the Chinese Web data - as billions of web pages and huge amount of parenthetical bilingual and well formed bilingual text on the Web are available, we can assume the following.

First, the Web has covered almost all the terms of traditional bilingual dictionaries. Thus, it is realistic to use the Web data alone to build a bilingual dictionary and using no additional recourse. Taking the terms inside parenthesis from the parenthetical bilingual sentences can form a quality corpus, which contains single words, phrases and even short sentences, for building dictionaries. On Chinese Web, this is particularly true for some popular languages such as English. In addition, from the Web we can also mind the most up-to-date OOV (Out of Vocabulary) terms not covered by traditional dictionaries.

Second, term translations can be extracted from parenthetical bilingual sentences to form the entries of dictionaries. For every term translation, its frequency i.e. the number of its occurrences in parenthetical bilingual sentences on the Web can be collected. In the case when one term, such English word tea, has many extracted Chinese translations such as (茶，喝茶，爱喝茶), we can simply select the right one by frequency ranking. That is, the more frequent the translation is used on the Web, the more likely it is right. We believe this is a rule when the data for counting the frequency is big enough to cover the terms of dictionaries.

From above two assumes, we can see that the quality of our dictionary is based on that the used web pages contains the right bilingual sentences and the bilingual sentences appear frequent, and we have no requirement for the quality of the web pages, which makes the procedure of SDDB more convenient.

Now there have been many researcher works in the area of extracting bilingual dictionaries from the Web. However, these works are quite different from SDDB in principle. First, they require the support of other language resources. Moreover, their focus is on the improvement of algorithms rather than on the use of big data. For example, [4] and [14] both used segmentation for Chinese and alignment model when they extract translation pairs. [4] used additional Chinese to English bilingual dictionary to determine translation units and used supervised learning to build models for transliterations and translations separately. [14] used an unsupervised learning method, whose complexity of computing link score to deal with multi-word alignment is $O(N^2M^2)$, where N is the number of sentence pairs, and M is the number of words each sentence contains. [8] extracted translation from well-formed bilingual texts. They used a linking algorithm which need additional dictionaries and SVM classifier to identify potential translation pairs(seeds) and then extracted bilingual data. To our knowledge, the complexity of SVM is at least $O(N^2)$, where N is the number of pairs. These methods mainly focus on the accuracy while overlook the efficiency.

In summary, SDDB is a novel method for building bilingual dictionaries based on big data from the Web. It can be characterized by (1) fully automatic. Without human intervention, it covers the whole process for building a bilingual dictionary, including the construction of corpus, extraction and frequency computing of term translations, and noise elimination. (2) Simple. The algorithm of SDDB is O(N) where N is the number of bilingual sentences. This linear property makes SDDB very scalable. (3) Language independent. Without using language concerned sentence segmentation and word alignment, the algorithm of SDDB needs little knowledge of languages. Thus, so long as the data is big, SDDB is easy to be adapted to other languages with different characters in addition to the English-Chinese dictionary example presented in this paper, because we need to distinguish the two languages using characters.

This paper is organized as follows. Section 1 is this introduction. Section 2 will give the full procedure of SDDB step by step. Section 3 will describe the algorithms of SDDB in detail. The testing results of the quality of an English-Chinese dictionary built through SDDB will be shown and compared with commercial dictionaries in Section 4. Section 4 also includes the result of scalability experiment of SDDB. Related work is listed in Section 5 followed by our conclusion.

2   Procedure of SDDB

First we will introduce some concepts which will be used in the following sections.

1. **parentheses sentence.** In figure 1, we traverse the text before "(" from right to left and stop until punctuation or the language change(such as from Chinese to English or vice versa) appears, the traversed text and the term in parentheses constitute parentheses sentence, such as "真正的社会主义(SOCIALISM)","国内现在是资本主义下和共产主义(communism)".

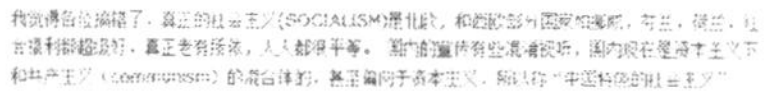

我就得名份编错了，真正的社会主义(SOCIALISM)理佳論，和国就都与国民如难和，有且、硬兰、
方硬判制徹設別，真正有所在，人人都更平等，我们通前有论義排技术，国约如在是過才主义下
和共产主义(communism)的岩岩保約的，甚至面向于资本主义。类似行"中国特仍到社会主义"

Fig. 1. Parentheses bilingual text

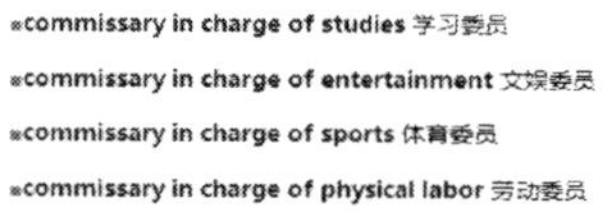

=commissary in charge of studies 学习委员

=commissary in charge of entertainment 文娱委员

=commissary in charge of sports 体育委员

=commissary in charge of physical labor 劳动委员

Fig. 2. Well-formed bilingual text

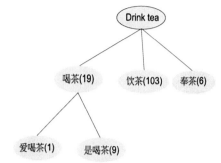

Fig. 3. Example I-tree

2. **sentence database.** all the parentheses sentences extracted from the web pages constitute the sentence database.
3. **C2E Sentence.** the parentheses sentence with Chinese terms in parentheses and English texts outside parentheses, such as "heavy Rain(暴雨)".
4. **E2C Sentence.** the parentheses sentence with Chinese texts outside parentheses and English terms in parentheses, such as "真正的社会主义(socialism)".
5. **Chinese-corpus.** the repository formed by the Chinese terms of all C2E sentences in the sentence database.
6. **prefix.** when traversing the text before parentheses from right to left starting from '(' in an E2C sentence one Chinese character by one Chinese character, for each character movement, we get a Chinese string C_s between the current position and '('. If we can find C_s in the Chinese-corpus, C_s is called a prefix of this E2C sentence.
7. **entry.** each prefix and English term of an E2C sentence form an entry.
8. **candidate entry.** the entry whose Chinese and English terms are correct translations of each other.
9. **entry frequency.** for an entry e, the entry frequency refer the number of sentences containing e in the sentence database.

As shown in Procedure SDDB, Line 2 extracts sentences and Line 3 constructs Chinese-corpus, Lines 4-13 count frequency of each entry. Line 14 groups all the entries by the English term, then in Lines 15-19, for each group of entries with a same English term ENG, the I-tree is constructed and pruned, which will be detailed in next section. Line 19 filter spam using etyma merging method. If the edit distance of two English translations of the same Chinese term is less than a specific threshold, the two translations will be merged. For example, "休斯顿" has translations like "housten" and "houston", and only "houston" is retained. Then the top-k translation of each Chinese term are selected as the final translations of this Chinese term. For example, there are such sentences like "别再说我爱着你(dance)", because the frequency is not the top-k, it will be filtered.

Our work used all the bilingual web pages indexed by **Sogou search engine**[1], and scanned the summary of these pages to extract about 1.2TB

---

[1] http://www.sogou.com

parentheses sentences. It should be noted that the size of web pages used in [4] is more than 300GB, which is much smaller than our work, because 1.2T parentheses sentences is only a small part of all web pages.

Procedure SDDB

Input: bilingual web pages BWP

1 **begin**

2   Extract all the parentheses sentences from BWP, and store these sentences in sentence database;

3   Build the Chinese-corps $CNCorp$ by extracting the Chinese terms from all C2E sentences in sentence database;

4   Let $CanSet$ be an item set, and the item of $CanSet$ has the format of $(ENG, CHN, freq)$. $CanSet = \Phi$;

5   **for** *each E2C sentence s in sentence database* **do**

6   **for** *each prefix p of s* **do**
                   /* s.ENG is the English term of s   */

7   **if** *(s.ENG, p, f) exists in CanSet* **then**

8   │ increase f by 1

9   **else**

10   │ create a new record $(s.ENG, p, 1)$ in $CanSet$

11   **end**

12   **end**

13   **end**

14   Group $CanSet$ by attribute ENG /* each group is formed by all
                   corresponding Chinese terms to one single English term   */

15   **for** *each group $g \in CanSet$* **do**

16   │ building an I-Tree R_{Itree} for g rooted at g.ENG;

17   │ $Pruning(R_{Itree})$

18   **end**

19   extracted all the candidate entries Ce_s from all pruned I-tree;

20   use etyma merging method to filter spam in Ce_s;

21 **end**

---

3   Pruning of I-Tree

3.1   Problem Definition

There may be multiple entries extracted from an E2C sentence while maybe only one entry is correct, such as "真正的社会主义(socialism)" and "社会主义(socialism)" ("社会主义(socialism)" is a candidate entry). And for some English terms, there are multiple different entries, for example, the entries with "drink tea" contain "爱喝茶" and "是喝茶". And we need select the right one.

Frequency is used to evaluate whether an entry is a candidate entry in our work because it is a fact that the more frequent a translation appears in the web, the more probable it is correct. So first we need to use an effective way to represent the entries. In SDDB method, an I-tree is constructed for each English term, and each I-tree node represent an entry.

Definition 1. *I-tree is used to represent all E2C sentences with a same English term. The I-tree has following features: (1)It is built for group of E2C sentences with a same English term ENG; (2)ENG is its root; (3)for the path $(C_1, C_2, \cdots, C_n)$ rooted at the direct child of ENG, every C_i is a candidate translation of ENG, and for $1 \le i \le n - 1$, C_i is a suffix of C_{i+1} and the frequency of C_{i+1} is included in the frequency of C_i; (4) there is no suffix relationship described in (3) between any two sibling nodes.*

For example, Table 1 shows the E2C sentences whose term in parentheses is "drink tea". Based on these E2C sentences, the I-tree as shown in Figure 3 is constructed.

It should be noted that: (1)"喝茶", "爱喝茶", "饮茶" and "奉茶" can be found in Chinese corpus. (2)C2E sentences are used to modify the frequency of I-tree node. Given a I-tree node N, suppose the term in N is t_c, and the term in root is t_e. If there is a C2E sentence s whose term outside parentheses is t_e and term in parentheses is t_c, we add the frequency of N by fp, where fp is the frequency of s. In this example, we can find C2E sentence "drink tea(饮茶)" whose frequency is 25, so the frequency of "饮茶" is not 78 but 103.

Through I-tree, the relationship between different entries with the same English term can be intuitively seen. Then we extract all candidate entries through pruning of I-tree.

3.2   Problem Solution

All entries are represented using different I-tree nodes. Now we need to considering how to prune these I-trees. First we introduce a principle as following about pruning of I-tree based on statistics.

Principle 1. *N is a child of root of I-tree, only one candidate entry can be extracted from the subtree of N.*

The above principle will cause that we miss some correct translations which will affect the recall of SDDB method. So we need evaluate how many translations we may miss. First if the subtree of N contains only one path, the English term is used as the translation of one node in this path on the web. Then we consider the case that subtree of N contains multiple paths. We collect about three million English phrases based on which three million I-trees are constructed. However, there are only seventy thousand I-trees contains such non-root node that the subtree of this node contains multiple paths, and the rate is only 2.3%. Then we randomly select one hundred such nodes. There are ten of these one hundred nodes are transliterations or abbreviations. For example, "malayalam" can be translated as "马拉亚拉姆语" or "马拉雅拉姆语". Finally only the most popular usage i.e. "马拉亚拉姆语" is retained, which will not affect the user experience. And there are other ten nodes in which we can find such node that there are multiple translations can be extracted from the subtree of this node. And in these ten nodes, the translations of six nodes are synonym and extracting only one translation will not affect user experience either. For example, "bus stop" can be translations as "公交车站" and "公交汽车站". The other four nodes contain translations with different meaning, for example "room service" can be translated as "客房服务" and "送餐服务" and some translations will be missed if we only retain one translation, however the rate of such term is only about 0.092%.

The entry frequency can be used to evaluate the probability that this entry is a correct translation. Because the existence of spam in web, a entry can be extracted as candidate entry only if the usage of this entry is dominated. And if there are no such dominated nodes, the common part will be selected as candidate entry. So we propose two principles to prune I-tree as Principle 2 and Principle 3.

Principle 2. *For an I-tree non-root node N, if the frequency of one child C is more than half of N's frequency, retain the node C and prune other children of N.*

Principle 3. *For an I-tree non-root node N, if there is no child whose frequency is more than half of N's frequency, retain the common part of these children (just N) and prune all these children.*

Each I-tree is pruned using the above two principles from the root to leaves of this I-tree. Then we can obtain the lemma as following.

Lemma 1. *Give an I-tree, N is one non-root node. After pruning the subtree of N according to principle 2 and principle 3, and there will be only one path in subtree of N.*

Algorithm 1. pruning of I-tree: pruning(N_n)

Input: I-tree node N_{in}
Output: candidate entry N_{out}

```
1  begin
2  |   S ← all children of N_in;
3  |   for each CC ∈ S do
4  |   |   if  freq(CC) > 0.5 * freq(N_in) then
5  |   |   |   return pruning(CC);
6  |   |   else
7  |   |   |   delete subtree of CC ;
8  |   |   end
9  |   end
10 |   return N_in;
11 end
```

After pruning I-tree according to principle 2 and principle 3, the leaf node of the retained path is selected as candidate entries. So for an I-tree R_{Itree}, n candidate bilingual entries will be extracted, where n is the number of children of R_{Itree}. The pruning algorithm is shown in Algorithm 1.

Here we use I-tree shown in Figure 3 as example to describe the pruning algorithm. In this I-tree, according to Algorithm 1, because there is no child of "喝茶" whose frequency is more than half of frequency of "喝茶", finally "喝茶", "奉茶" and "饮茶" are selected, and the candidate entries extracted from this I-tree are shown in Table 2.

Table 1. E-C-sentence example

No	sentence	frequency
1	喝茶（drink tea）	7
2	和喝茶（drink tea）	2
3	和饮茶（drink tea）	71
4	和热心地招呼我们奉茶（drink tea）	6
5	是喝茶（drink tea）	9
6	是一个爱喝茶（drink tea）	1
7	饮茶（drink tea）	6
8	我问你去不去饮茶（drink tea）	1

Table 2. Candidate sentence set

English	Chinese	frequency
drink tea	饮茶	103
drink tea	喝茶	19
drink tea	奉茶	6

4   Experimental Results

Through scanning twenty billion parentheses sentences, whose size is about 1.2TB, over four million Chinese to English translations and over six million English to Chinese translations are extracted finally.

Bilingual Translation have mostly been evaluated by manual examination of a small sample results. And there does not yet exist a common evaluation data set. In this paper, we first compare our work with **Youdao**[2] through manual annotation. Then we use wikipedia to evaluate our work, and the method is proposed by Lin[14]. In order to evaluate our translation quality on name entity, we collect 695 kinds of brands with both Chinese and English name as the standard set to evaluate our work. Last but not least, we detect the scalability of our work.

4.1   Comparison with Youdao

Youdao is a very popular web-based dictionary in China. In order to evaluate the quality of our mined dictionary, we crawled the query log of Youdao, and then get the translations in our mined dictionary and the web translations in

---

[2] http://dict.yodao.com, a popular commercial web dict in China.

Youdao. We crawled three groups of Chinese phrases which contain 73, 74 and 73 terms respectively and three groups of English phrases which contain 61, 70 and 61 terms respectively. Then we annotate these translations through blind test and evaluate the precision. The results are shown in Table 3 and Table 4.

Table 3. English to Chinese translation

Table 4. Chinese to English translation

		precision
English 1	SDDB	92.4%
	Youdao	93.1%
English 2	SDDB	95.6%
	Youdao	91.1%
English 3	SDDB	95.7%
	Youdao	91.0%
average of English	SDDB	94.6%
	Youdao	91.7%

		precision
Chinese 1	SDDB	97.1%
	Youdao	93.9%
Chinese 2	SDDB	90.4%
	Youdao	85.9%
Chinese 3	SDDB	96.0%
	Youdao	90.3%
average of Chinese	SDDB	94.5%
	Youdao	90.0%

From the experimental results we can see that the precision of SDDB increases at approximately 3 percent for English translate and 4.5 percent for Chinese translate. Because we do not know the number of all correct translations, we cannot calculate the recall of our system. However more users pay attention to the accuracy of dictionaries. The results indicate that the performance of SDDB outperforms the mature commercial applications of Youdao.

4.2   Evaluation with Wikipedia

Lin[14] used the translations in Wikipedia as evaluation on mining dictionary because the translations in Wikipedia contain far more terminologies and proper names than traditional bilingual dictionaries. We extract the titles of Chinese and English Wikipedia articles that are linked to each other. 78,284 such pairs are extracted by us and then we remove the pairs which are not translations or terms by rules consulted Lin's work [14]. After the data cleaning, 70,772 translation pairs are left as gold standard translations. For each Chinese and English word in the Wikipedia data, we look it up in our mined dictionary. The Coverage of the Wikipedia data is measured by the percentage of words for which one or more translations are found. We then see in theses words whether we can find a translation which is an Exact Match of the answer key in the Wikipedia data. We use the dict.cn[3] as comparison.

Table 5 and Table 6 show the Chinese-to-English and English-to-Chinese results. We can see that the coverage of SDDB has obvious improvement compared with Dict.cn and the growth are 18% and 10% respectively, which indicated that SDDB has very good coverage. Despite the translations of Dict.cn are manually drafted, SDDB method also has a considerable advantage on Exact Match, especially in English to Chinese results, which is about 11 percentage.

---

[3] http://dict.cn, a widely used manual drafted dictionary.

Table 5. Chinese to English Results

	Coverage	Exact Match
SDDB	54.1%	50.3%
Dict.cn	36.5%	54.6%

Table 6. English to Chinese Results

	Coverage	Exact Match
SDDB	52.8%	52.2%
Dict.cn	42.5%	40.9%

4.3 Evaluation with Brand Name

With the development of transnational enterprise, many brands have both Chinese and English names, such as "周大福(Chow Tai Fook)", "克莱斯勒(Chrysler)" etc. We can use brands as name entities to evaluate the translation quality of SDDB. We find 695 different brands with both Chinese and English names. For each brand name, we look it up in our mined dictionary, and we measure coverage by the percentage of brands for which one or more translations are found. Then we see in these brands whether our most frequent translation is an Exact Match of the brand name. Also we look up each brand name in Dict.cn as comparison. We compare the results to evaluate whether SDDB can cover more OOV terms. Table 7 and 8 show the Chinese to English translation and English to Chinese translation results.

Table 7. Chinese to English Results

	Coverage	Exact Match
SDDB	87.5%	81.7%
Dict.cn	43.9%	61.3%

Table 8. English to Chinese Results

	Coverage	Exact Match
SDDB	93.4%	73.1%
Dict.cn	65.0%	25%

We can see that the coverage and Exact Match of SDDB is far more than dict.cn because the brands in this experiment contain many up-to-date brands not included in traditional dictionaries. From the results we can conclude that SDDB can deal with the translation of name entities and OOV terms very well.

4.4 Scalability of SDDB

The data-driven methods are mainly statistical method and parallelization is a helpful way to increase processing speeds. In our experiment, we first calculate the running time when we use different computers to parallelize the experiment to evaluate the influence of the parallelization. Then we increase the data size gradually and calculate the time to evaluate the scalability while data increases. Fig 4 shows the time spent when we use several computers to run SDDB in parallel and Fig 5 shows the time spent for different size of data.

Here we only calculate the time spent from Line 3 to Line 20 in Procedure SDDB, which is the whole procedure after parentheses sentences are crawled. And data size is the size of all parentheses we used in the experiment. From Figure 4 we can see that through parallelization the time spent for SDDB has

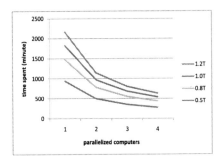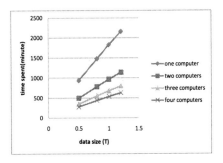

Fig. 4. time spent for different computers **Fig. 5.** time spent for different size of data

significant decrease. And Figure 5 indicates that the time spent increases almost linearly with increase of data size or computer numbers we use. The results demonstrate the better performance of data-driven method.

5   Related Work

As far as we know, there is no publication extracting bilingual data using linear algorithm and not using any additional resources Some methods only process particular data, such as person name[17], organization name[21] etc. Some methods use search engine to mine bilingual dictionaries([5], [13], [20]). They assumed the English term as input to search engine and extract translations from search results, which is difficult to build a large scale bilingual dictionaries.

[4] and [14] proposed two different methods to extract bilingual dictionaries from parentheses bilingual text. [8] extracted translations from well-formed bilingual text. And these methods all use word alignment to evaluate the accuracy of translation pairs. Different from previous work, we process both parentheses and well-formed bilingual texts. The complexity of our method is linear and we use no additional resources.The experimental results show that our method has high precision and coverage and very good scalability.

6   Conclusion

In this paper, we propose a novel method, SDDB, to extract translation dictionary from web data. Different from previous works, SDDB is a data-driven method, and never use semantic and additional resources. The method is linear and can be parallelized easily. Because no semantic is used during the construction of bilingual dictionary, the method can be transplanted to other languages conveniently. The experimental results indicate that data-driven method has very good scalability, but also can achieve better accuracy and coverage.

References

1. Banko, M., Brill, E.: Scaling to very very large corpora for natural language disambiguation. In: Proceedings of ACL. p. 33 (2001)
2. Baroni, M., Ueyama, M.: Building general-and special-purpose corpora by Web crawling. In: Proceedings of NIJL International Symposium. pp. 31–40 (2006)
3. Brown, P., Pietra, V., Pietra, S., Mercer, R.: The mathematics of statistical machine translation: Parameter estimation. Computational linguistics 19(2), 263–311 (1993)
4. Cao, G., Gao, J., Nie, J., Redmond, W.: A system to mine large-scale bilingual dictionaries from monolingual web. Proc. of MT Summit XI pp. 57–64 (2007)
5. Huang, F., Zhang, Y., Vogel, S.: Mining key phrase translations from web corpora. In: Proceedings of EMNLP. p. 490 (2005)
6. Huang, Z.: A fast clustering algorithm to cluster very large categorical data sets in data mining. In: Workshop on Research Issues on DMKD (1997)
7. Jansche, M., Sproat, R.: Named entity transcription with pair n-gram models. In: Proceedings of Named Entities Workshop. pp. 32–35 (2009)
8. Jiang, L., Yang, S., Zhou, M., Liu, X., Zhu, Q.: Mining bilingual data from the web with adaptively learnt patterns. In: Proceedings of ACL. pp. 870–878 (2009)
9. Jiang, L., Zhou, M., Chien, L., Niu, C.: Named entity translation with web mining and transliteration. In: Proc. of IJCAI. vol. 7, pp. 1629–1634 (2007)
10. Keller, F., Lapata, M.: Using the web to obtain frequencies for unseen bigrams. Computational linguistics 29(3), 459–484 (2003)
11. Lapata, M., Keller, F.: The Web as a baseline: Evaluating the performance of unsupervised Web-based models for a range of NLP tasks. In: Proc. of HLT-NAACL. pp. 121–128 (2004)
12. Lapata, M., Keller, F.: Web-based models for natural language processing. ACM TSLP 2(1), 3 (2005)
13. Li, H., Cao, Y., Li, C.: Using bilingual web data to mine and rank translations. IEEE Intelligent Systems 18(4), 54–59 (2003)
14. Lin, D., Zhao, S., Van Durme, B., Pasca, M.: Mining parenthetical translations from the web by word alignment. ACL08. pp994-1002 (2008)
15. Nie, J., Simard, M., Isabelle, P., Durand, R.: Cross-language information retrieval based on parallel texts and automatic mining of parallel texts from the Web. In: Proceedings of SIGIR. pp. 74–81 (1999)
16. Ramaswamy, S., Rastogi, R., Shim, K.: Efficient algorithms for mining outliers from large data sets. ACM SIGMOD 29(2), 438 (2000)
17. Sato, S.: Web-Based Transliteration of Person Names. In: Proceedings of WI-IAT. pp. 273–278 (2009)
18. Snow, R., O'Connor, B., Jurafsky, D., Ng, A.: Cheap and fast—but is it good?: evaluating non-expert annotations for natural language tasks. In: Proceedings of EMNLP. pp. 254–263 (2008)
19. Tsang, I., Kwok, J., Cheung, P.: Core vector machines: Fast SVM training on very large data sets. Journal of Machine Learning Research 6(1), 363 (2006)
20. Wu, J., Chang, J.: Learning to find English to Chinese transliterations on the web. In: Proc. of EMNLP-CoNLL. pp. 996–1004 (2007)
21. Yang, M., Liu, D., Zhao, T., Qi, H., Lin, K.: Web based translation of Chinese organization name. Journal of Electronics 26(2), 279–284 (2009)
22. Zhang, Y., Vines, P.: Using the web for automated translation extraction in cross-language information retrieval. In: Proceedings of SIGIR. pp. 162–169 (2004)

Measuring Similarity of Chinese Web Databases Based on Category Hierarchy*

Juan Liu, Ju Fan, and Lizhu Zhou

Tsinghua National Laboratory for Information Science and Technology,
Department of Computer Science and Technology, Tsinghua University,
Beijing 100084, China
{tiantang326, fanju1984}@gmail.com, dcszlz@tsinghua.edu.cn

Abstract. The amount of high-quality data in the Web databases has been increasing dramatically. To utilize such wealth of information, measuring the similarity between Web databases has been proposed for many applications, such as clustering and top-k recommendation. Most of the existing methods use the text information either in the interfaces of Web databases or in the Web pages where the interfaces are located, to represent the Web databases. These methods have the limitation that the text may contain a lot of *noisy* words, which are rarely discriminative and cannot capture the characteristics of the Web databases. To better measure the similarity between Web databases, we introduce a novel Web database similarity method. We employ the categories of the records in the Web databases, which can be automatically extracted from the Web sites where the Web databases are located, to represent the Web databases. The record categories are of high-quality and can capture the characteristics of the corresponding Web databases effectively. In order to better utilize the record categories, we measure the similarity between Web databases based on a unified category hierarchy, and propose an effective method to construct the category hierarchy from the record categories obtained from all the Web databases. We conducted experiments on real Chinese Web Databases to evaluate our method. The results show that our method is effective in clustering and top-k recommendation for Web Databases, compared with the baseline method, and can be used in real Web database related applications.

Keywords: Web database, similarity model, category hierarchy.

1 Introduction

As the amount of structured data in the Web databases (WDB for short) increases dramatically, it becomes more and more important to effectively organize the Web databases, and thus to utilize such wealth of information [1, 2, 4, 5, 12]. Recently, there has been an increasing interest to measure the similarity of Web

---

* This work was supported in part by National Natural Science Foundation of China under grant No. 60833003.

X. Du et al. (Eds.): APWeb 2011, LNCS 6612, pp. 225–236, 2011.

databases, and help users better explore the Web databases. A number of applications can benefit from an effective Web database similarity measure. The Web database clustering [3, 10, 9, 17] techniques can make use of the similarity measure to group the *similar* Web databases. Interface integration [19, 8, 6, 11, 18], the technique that integrates interfaces of multiple Web databases into a mediated interface, can use the similarity measure to identify similar Web databases to be integrated.

A key issue in addressing the similarity measure of Web databases is to find effective features to represent Web databases. Some interface integration techniques [10, 9, 17] focus on the interfaces of Web databases, i.e., HTML *forms*. They extract labels and elements from the interfaces and construct *interface schemas* as features of Web databases. A drawback of the above-mentioned features is that they can only be applied for forms whose contents are indicative of the database domain, and they cannot be applied for simple keyword-based interfaces. Thus, Barbosa et al. [3] proposed new types of features to represent a Web database. They make use of the text (i.e., the bag of words) in the interface and in the Web page (page for short) where the interface is located as features. In addition, they assume that multiple pages with a common backlink (i.e., a page that links to the pages) are similar with each other. However, the methods in [3] have the following limitations. Firstly, they only consider the text information, which is extracted from the home-page of the Web database, and thus they are limited to represent the underlying record content. For example, the sub-categories of "笔记本周边*(laptop accessories)*" is missed on the home-page of the WDB in Figure 1. Secondly, the text extracted from pages may contain a lot of *noisy* words, which are rarely discriminative and cannot capture the characteristics of the Web database. Take the Web database in Figure 1 as an example. Some words, e.g., "购物车*(shopping cart)*", "结账*(checking)*", "热卖商品*(hot sales)*", are useless to represent the corresponding Web database. Even worse, some words, e.g., "坚果*(nuts)*", may lead to errors in terms of the similarity measure. Thirdly, since some pages may link to Web databases with multiple domains (i.e., online directories), the link structure exploited in [3] may not reflect the similarity of Web databases.

To address the limitations of existing approaches, we introduce the *record categories* in a Web database as the feature of the database. Informally, the record categories of a Web database are the category words of the records in the database. A lot of Web databases, especially the e-commerce Web databases, provide such record categories, in order to facilitate users to browse the records in the databases. For example, Figure 1 provides the categories of the records in an e-commence Web databases. Users can use the category "双肩电脑包*(shoulders computer bags)*" to browse all the sub-categories belonging to computer bag as well as all records of shoulders computer bags in the databases. Employing record categories as Web-database features has the following advantages. Firstly, the record categories are provided by designers of the Web databases to allow users to better understand the data. Thus, the categories are carefully selected to capture the content of the Web databases as accurate as possible, and they are of high quality. For example, the record categories in the dashed rectangle of Figure

Fig. 1. Example of Category Terms

1 are very discriminative, and can better represent the Web database. Secondly, there exists hierarchy structure between these record categories, such as "电脑包*(computer bags)*" and "双肩电脑包*(shoulders computer bags)*", and we can improve the effectiveness of similarity measure by considering the relatedness of categories computed based on the category-hierarchy.

We study the research challenges that naturally arise for using record categories to measure the similarity of Web databases. The first challenge is an effective similarity model based on the record categories. The record categories of different WDBs may be very heterogenous in terms of both category names and hierarchy structures, even though the Web databases are very similar, because the designers of WDBs may have different criteria to categorize the underlying data. Therefore, it is very difficult to measure such heterogenous WDBs. To address this challenge, we introduce a novel similarity model, named *IOGM*. We propose to incorporate a unified category hierarchy, which can be derived from the record categories of WDBs, and we introduce a category hierarchy-based model to effectively measure the similarity between WDBs. The second challenge is the construction of the category hierarchy. Since we cannot derive the unified category hierarchy directly from the individual category hierarchies of WDBs, we introduce the *suffix* relationships of record categories, and develop an efficient algorithm to construct the unified category hierarchy based on the suffix relationships.

The contributions of our paper are summarized as follows.

 - We introduce the record categories, a novel feature, to well represent the Web databases.
 - We propose a category-hierarchy based similarity model to make full use of the record categories, and measure the similarity of WDBs effectively.
 - We propose a suffix-relationship based method to construct the category hierarchy.

– We conduct experiments in real data sets to compare our methods with the baseline Web-database similarity model.

The rest of the paper is organized as follows. Section 2 gives the related work. Section 3 introduces the category hierarchy based similarity model, and Section 4 shows the construction algorithm of the category hierarchy. We report the experiments in Section 5, and finally conclude in Section 6.

2   Related Work

Measuring the similarities of Web databases has been widely studied in the communities of the database and information retrieval. Some approaches of interface integration for Web databases consider the problem as the matching of different interface schemas[19, 8, 6, 11, 18]. These approaches extracted the interface schemas from the HTML-based interfaces of Web databases. Then, they exploit some schema matching techniques to match the attributes in different interfaces. However, these approaches can only be applied to the WDBs whose interface contents are indicative of the database domain, and they cannot be applied for simple keyword-based interfaces [3].

Some Web-database clustering approaches [3, 10, 9, 17] use the text in the interfaces to represent the corresponding WDBs, and propose to measure the similarity between these texts as the similarity between WDBs. The proposed approaches can avoid some drawbacks of interface integration. They, however, may be ineffective to measure the similarity between WDBs, because they bring many *noisy* words, which are rarely discriminative and cannot capture the characteristics of the Web database. Compared with the similarity models proposed in interface integration and Web-database clustering, we propose to use a novel feature, the record categories to represent the WDBs, and employ the category hierarchy to improve the performance of similarity measure. As shown in our experimental results, our similarity model can achieve better performance.

In addition, there are also several approaches employing the hierarchy structures. The method that is most related to our method is the OGM model in [7]. This model also employs a hierarchy structure to improve the similarity measure on top of sets of items. In addition, some approaches use the hierarchy structure to improve the performance of document similarity measure [15, 13] and document classification [15]. Compared with the above-mentioned approaches, we focus on employing the hierarchy structure to measure the similarity between two category sets obtained from the WDBs.

3   Category Hierarchy Based Similarity Model

We focus on the WDBs that contain a set of record categories (categories for short) provided by the designers of the WDBs. In particular, we take into account the *hierarchy* relationship of these categories. Take a WDB containing records in the bag domain as an example, as shown in Figure 1, we can obtain some record categories, e.g., *"包(bag)"*, *"电脑包(computer bag)"*, etc, by extracting the text corresponding to the categories from the Web pages. Obviously, the *"computer bag"* is a sub-categories of the *"bag"*, and thus they have a hierarchy relationship. The similarity measure of WDBs can benefit from the above-mentioned category hierarchy relationship, because we can use the hierarchy to measure the relatedness between each pair of categories. Consider another category *"单肩包(shoulder bag)"*. Although this category does not equal to the *"computer bag"*, both categories are all sub-categories of the *"bag"* according to the category hierarchy. Therefore, we can improve the effectiveness of similarity measure by considering the relatedness of categories computed based on the category hierarchy.

A naïve method incorporating the category hierarchy is to firstly represent each WDB as an individual category hierarchy, i.e., a tree of categories. Then, we can employ some similarity measures of trees to measure the similarity between the corresponding WDBs. However, this method cannot perform well due to the heterogenous structure of category hierarchies across WDBs. More specifically, the designers of WDBs usually use different criteria to summarize the categories of the underlying records, even though the records are very similar. For example, to categorize the records in the domain of *"bag"*, one designer may use the categories, *"women bags"* and *"men bags"*, and another one may use *"sport bags"* and *"computer bags"*. In addition, the hierarchy relationships of categories may be various across different WDBs. For example, in one WDB, the category *"women bags"* may contain the *"women computer bags"* and the *"women shoulder bags"*, but the category *"computer bags"* may contain *"women bags"* and *"shoulder bags"* in another WDB. Therefore, we cannot simply employ the hierarchy relationships provided by the WDBs.

To address the above-mentioned problem, we propose to firstly collect all the record categories without the hierarchy structure provided by the individual WDBs. Then, we develop some methods to construct a unified category hierarchy based on the collected categories. Using the constructed category hierarchy, we can improve the effectiveness of similarity measure of WDBs. In this section, we assume that we already have a unified category hierarchy, and concentrate on presenting the similarity model based on the category hierarchy. The details of category hierarchy constructions will be discussed in Section 4.

3.1 Similarity Model between WDBs

We consider the similarity between two WDBs as the similarity between the two sets of categories corresponding to the WDBs. Specifically, let $\Theta = \{C_1, C_2, \ldots, C_n\}$ denotes the set of categories corresponding to a WDB, where C_i is a record category in the WDB. We measure the similarity between two WDBs based on the similarity between the two sets of categories, i.e.,

$$sim(WDB_1, WDB_2) = sim(\Theta_1, \Theta_2). \tag{1}$$

Many similarity models on top of sets, e.g., Jaccard coefficient, Cosine similarity, etc., can be used to compute the similarity between two category sets,

$sim(\Theta_1, \Theta_2)$. However, these models only consider whether two categories are identical or not, and they neglect the interior relatedness between categories themselves. [7] proposes to employ a hierarchy tree to compute the similarity of two sets, and introduces a novel model, named Optimistic Genealogy Measure (OGM for short). We borrow the framework of OGM to measure the similarity of two category sets, Θ_1 and Θ_2, and propose the following similarity model, called Improved OGM (IOGM for short). The difference between OGM and IOGM is that they adopt different similarity function to measure the similarity of any two terms in Θ_1 and Θ_2. And their elements of the sets are also different: the elements in OGM are only the leaf nodes in the hierarchy tree, but the elements in IOGM are all the nodes in the hierarchy tree. The basic idea of our model is that we examine all categories in Θ_1. For each category $C_i \in \Theta_1$, we select the category C_j in Θ_2 with the maximum relatedness of C_i, and take the similarity between C_i and C_j as the similarity between C_i and Θ_2. For every category C_i in Θ_1, we compute the above-mentioned similarity of Θ_2, and then we sum the similarities together to obtain the overall similarity between Θ_1 and Θ_2.

Definition 1 (IOGM). *Consider two sets of categories, Θ_1 and Θ_2 corresponding to two WDBs. We measure the similarity between Θ_1 and Θ_2 as follows.*

$$sim(\Theta_1, \Theta_2) = \frac{\sum_{C_i \in \Theta_1} sim(C_i, \Theta_2)}{|\Theta_1|}, \tag{2}$$

and
$$sim(C_i, \Theta_2) = \max_{C_j \in \Theta_2} \sigma(C_i, C_j), \tag{3}$$

where $\sigma(C_i, C_j)$ is the relatedness of the two categories, C_i and C_j.

Notice that, $sim(\Theta_1, \Theta_2)$ is asymmetric. Next, we discuss the methods for estimating the relatedness of two categories based on the category hierarchy.

3.2   Estimating Category Relatedness Based on Category Hierarchy

Given a category hierarchy, we present a method to estimate the relatedness of two categories in this section. Formally, a category hierarchy is a tree-based structure of categories, where relationship between the parent and the child represents that the category corresponding to the child is a *sub-category* of the category corresponding to the parent. Figure 2 shows an example category hierarchy. We employ the category hierarchy to measure $\sigma(C_i, C_j)$ as follows.

Firstly, we introduce the least common ancestor LCA of two categories, C_i and C_j, in the category hierarchy. We define the LCA as the common ancestor of C_i and C_j with the maximal depth, where the depth is the number of nodes in the path from the root to the corresponding node in the category hierarchy. We use the category and the tree node exchangeably in the rest of the paper, if there is no ambiguity. For example, in Figure 2, $LCA(Y, H) = K$, because 1) K is the ancestor of both Y and H, and 2) K has the maximal depth in all ancestors of Y and H. We note that the depth of the LCA of the two categories can capture their similarity. In details, the deeper the depth of the LCA of C_i and C_j is, the more similar C_i and C_j are, because the two categories are sub-categories of a more specific super-category. Formally, we estimate the relatedness of C_i and C_j as follows.

$$\sigma(C_i, C_j) = \frac{2 \cdot depth(LCA(C_i, C_j))}{depth(C_i) + depth(C_j)}. \tag{4}$$

Notice that, when the two categories are same, we have $\sigma(C_i, C_i) = 1$ according to the formula.

In addition, we consider a special case that $LCA(C_i, C_j) = C_j$, which means that the first category is the *descendant* category of the second category according to the category hierarchy. In this case, we assume that the two categories are highly related to each other, because the super-category includes the descendant category in semantic (e.g., "包(bags)" includes "电脑包(computer bags)" in semantic). Therefore, we define that the similarity between two categories satisfying $LCA(X, Y) = Y$ is 1.

In summary, we define the similarity between C_i and C_j as follows.

$$\sigma(C_i, C_j) = \begin{cases} 1 & \text{if } LCA(C_i, C_j) = C_j, \\ \frac{2 \cdot depth(LCA(C_i, C_j))}{depth(C_i) + depth(C_j)} & \text{otherwise.} \end{cases} \tag{5}$$

4   Category Hierarchy Construction

In this section, we discuss the method for constructing the category hierarchy, which is used in the proposed similarity model in Section 3. The basic idea of our method is to utilize the suffix relationship between categories. We describe the suffix relationship in Section 4.1, and introduce an efficient construction method in Section 4.2.

4.1   Suffix Relationship

For two categories, we focus on determining whether they have a hierarchy relationship, that is, a category is the sub-category of the other category. To this end, we exploit the following grammatical observation in Chinese. If the longest common suffix of two categories is exactly the category with the shorter length, then the category with the shorter length is the sub-category of the one with the longer length. For example, in Figure 1, the longest suffix of "双肩电脑包(shoulders computer bag)" and "电脑包(computer bag)" is "电脑包(computer bag)", which is exactly the category with shorter length. Therefore, "双肩电脑包(shoulders computer bag)" is the sub-category of the "电脑包(computer bag)". Formally, we define the suffix relationship as follows.

Definition 2 (Suffix Relationship). *Suppose C_1 and C_2 are two record categories. If $C_2 = b_1 b_2 \cdots b_n C_1 (n \geq 1)$, where each b_i is a Chinese character, then we call that C_1 and C_2 are of suffix relationship, and C_1 is the suffix of C_2.*

Figure 3(a) shows an example of the Suffix Relationship. We note that for a given record category, there could be several suffixes. For example, the suffix of "双肩电脑包(shoulders computer bag)" may be "电脑包(computer bag)" or "包(bag)". In order to obtain the direct sub-categories of a category, we use the longest suffix of a category as its sub-category, which is defined as follows.

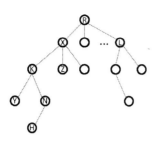

Fig. 2. Example for Similarity Function   **Fig. 3.** Example of Suffix Relationship and Category-Hierarchy

Definition 3 (Longest Suffix). *Suppose set $S(C_i)$ is the set of all the suffix of the category C_i, then the longest category in $S(C_i)$ is called the longest suffix of C_i.*

In particular, we name the record category which does not have a suffix, such as "包(bag)", as *isolated record category*. As we will see in 4.2, an isolated record category does not have any lower layer categories in the category hierarchy.

4.2   Category Hierarchy Construction

In this section, we present an algorithm that efficiently constructs the category-hierarchy based on the suffix relationship proposed in Section 4.1, as shown in Algorithm 1.

Input: $\Sigma = \bigcup_{i=1}^{n} \Theta_i$, the set of categories from all WDBs.
Output: CHT, the category-hierarchy
1 sort Σ by the length of C in ascending order;
2 create an virtual root for CHT;
3 **for** *each record category C in Σ* **do**
4   create a new tree node for C;
5   LS = find the longest suffix of C;
6   **if** *LS does not exist* **then**
7   add the C node to the children set of virtual root;
8   **else**
9   add the C node to the children set of LS node;
10   **end**
11 **end**
12 **return** the root of CHT;

Algorithm 1. CHT Construction Algorithm

The input of the algorithm is the set of categories from all WDBs, i.e., $\Sigma = \bigcup_{i=1}^{n} \Theta_i$. The output of the algorithm is a category-hierarchy, i.e., CHT. The algorithm firstly sorts the Σ according to the lengths of the categories in Σ in an ascending order (Line 1). This step guarantees that the father node of any category C has already been added to the category-hierarchy, CHT. The reason is that we visit the set Σ in ascending order of the length of categories. If the longest suffix length of a C, say LS, exists, then we have $length(LS) <$

$length(C)$, and so its longest suffix node has already been processed and added to the CHT. Otherwise LS does not exist, the father node in the CHT is the root node. Then the algorithm creates a virtual root node for the category-hierarchy in order to combine all the isolated record categories together (Line 2). Next, the algorithm processes every category C in Σ (from Line 3 to Line 13). In more details, it creates a new node for C in Line 4, and find the longest suffix for C in Line 6. Then, the algorithm determines whether C is an isolated category. If it is an isolated category, it adds tree node corresponding to C to the child set of the root. Otherwise, it adds tree node corresponding to C to the child set of the tree node corresponding to its longest suffix LS. Finally, the algorithm returns the root node of CHT. Figure 3(b) shows an example of the category-hierarchy constructed by Algorithm 1.

It is notable that we do not deal with the case that one word has more than one meanings(e.g., "苹果(apple)") in the construction of category hierarchy, which we may consider in the future work.

5   Experiments

In this section, we report our experiments to examine the effectiveness of our category hierarchy based model for measuring the similarity between two WDBs. To this end, we conducted two experiments based on our similarity model. Firstly, we applied our similarity model to find the top-k similar WDBs, given a query WDB, and examined whether the found WDBs were relevant to the given WDB. Secondly, we employed the similarity model for WDB clustering, which is an essential task for organizing the WDBs.

Table 1. Data Set

Domain	# of WDB	Domain	# of WDB	Domain	# of WDB	Domain	# of WDB
renting	11	medicine	11	digital products	34	ornament	41
cars	19	travel	15	home appliances	18	pets	11
office stuff	12	reading online	14	video online	13	shoes	10
toys	15	food	41	infant mom	27	gifts	32
clothes	19	health care	13	general merchandise	58	sports	21
flowers	22	books	26	cellphones	24	luxury	20
cosmetics	37	household	35	e-books	12	bags	12

5.1   Experiment Setup

Data Set. We examined the effectiveness of our category hierarchy based similarity model on top of a set of 623 Chinese WDBs, which we collected from the real Web using rule based method automatically. For each WDB, we manually annotated a domain label for it as a benchmark only for our clustering experiment. Note that the domain labels were only used for benchmark, and our proposed methods do not depend on these labels. Table 1 describes some details of the data set. From the table, we can see our experimental data contains 28 different domains, which covers most domains in Chinese WDBs.

Evaluation Metric. To evaluate the performance of top-k WDB recommendation based on our similarity model, we asked 15 volunteers to label whether the recommended WDBs were relevant to the query WDB (the volunteer labeled 1 for relevant WDB, and 0 for the irrelevant one). Then, we exploited an aggregation score, precision to evaluate the performance of top-k WDB recommendation, i.e.,

$$precision = \frac{\# \text{ of correct } WDBs}{k}. \tag{6}$$

We further averaged the precisions of all the 623 WDBs as the precision of top-k WDB recommendation.

To evaluate the quality of the clustering, we employed two standard measures, i.e., the entropy and the F-measure [3]. For each cluster, we compute the entropy by using the standard formula.

$$Entropy_j = -\sum_i p_{ij} log(p_{ij}), \tag{7}$$

where p_{ij} is the probability that a member of cluster j belongs to domain i. The entropy for all the clusters is the sum of the entropies of each cluster, weighted by the size of each cluster. Intuitively, the lower the entropy is, the better the clustering algorithm is. The F-measure provides a combined measure of precision and recall [16,3]. For a cluster X and a domain label D (see Table 1 for details), we computed F-measure, F as follows.

$$F = 2PR/(P+R),$$
$$P(X,D) = \frac{N_1}{N_2}, \text{ and } R(X,D) = \frac{N_1}{N_3},$$

where N_1 is the number of WDBs in cluster X which can be judged to the domain D, N_2 is the number of WDBs in cluster X, and N_3 is the number of WDBs in domain D. In general, the higher the F-measure value is, the better the clustering algorithm is.

Baseline Method. We implemented the similarity model employed in [3] (FC+PC for short), which is the most effective method for WDB clustering in the state-of-the-art. This similarity model extracted the text in the interfaces to represent the corresponding WDBs, and developed several methods to determine the weights of the text for measuring the similarity. We used FC+PC as the similarity model in our top-k experiment and clustering experiment, and compared the performances with our proposed similarity model.

5.2   Experimental Results

In this section, we report the experimental results of our category-hierarchy based similarity model.

Top-k WDB Recommendation. Figure 4(a) shows the performance of top-k WDB recommendation using our similarity model and the baseline model. From the figure, we can see that the precision of our similarity model is 20 percent higher than that of the FC+PC similarity model. The main reason that our method outperforms the baseline methods is that we exploit the record categories to represent the WDBs. Compared with the FC+PC model, our method can

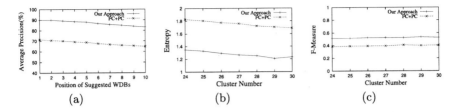

Fig. 4. Experiment Result:(a) The Precision of Top-k Recommendation; (b) The Entropy Result of Cluster; (c) The F-Measure Result of Cluster

make use of the text capturing the effective features of the WDBs. In contrast, the FC+PC method may bring many noisy terms, which would be harmful to the similarity measure. For example, there are many interfaces containing the text "搜索(search)", "折扣(discount)", etc. These terms are useless for measuring the similarity between WDBs. In addition, to make full use of the record categories, we take into account the category hierarchy, which can improve the performance of similarity measure, compared with the traditional similarity models on top of sets, e.g., Jaccard coefficient, Cosine similarity, which FC+PC used. The experimental results show that our category-hierarchy based similarity model are very effective, and the precision of top-1 WDB recommendation is 89.87%.

Clustering. Since our similarity value from IOGM is asymmetrical, we renormalize the similarity value by the following formula:

$$sim'(\Theta_1, \Theta_2) = sim'(\Theta_2, \Theta_1)$$
$$= \frac{sim(\Theta_1, \Theta_2) \cdot |\Theta_1| + sim(\Theta_2, \Theta_1) \cdot |\Theta_2|}{|\Theta_1| + |\Theta_2|}. \quad (8)$$

We ran the clustering algorithms based on our similarity model and FC+PC on top of our data set. We employed the Partitioning Around Medoids (PAM) as the clustering algorithm [14]. This algorithm starts by setting an initial set of medoids, and iteratively replaces one of the medoids by one of the non-medoids if it improves the total distance of the resulting clustering. We computed the distance between two WDBs as 1 minus their similarity value.

Figure 4(b) and (c) shows the entropy and F-measure values obtained by IOGM similarity model and by FC+PC model. We can see that both the entropy and F-measure obtained from our similarity model are much better than FC+PC. Our result is better because we consider more information about the content of the WDBs, and we also make use of the hierarchy structure of the record categories, which leads to a better similarity measure.

6   Conclusion

In this paper, we have introduced a novel model for measuring the similarity between Web databases. We proposed to employ the record categories, a novel feature provided by the designers of WDBs, to represent the corresponding WDBs. We proposed a model on top of the two sets of categories to measure the similarity of the two WDBs. In order to improve the performance of similarity measure, we proposed to exploit a category hierarchy to make full use of the hierarchy relationships of categories. In addition, we introduced an effective and efficient algorithm to construct a high-performance category hierarchy. We conducted experiments on real Chinese Web databases, and the experimental results have shown that our proposed similarity model has performance advantages, compared with the baseline method. Furthermore, as all steps in our method is automatically implemented, our method can be used in real applications easily.

References

1. Barbosa, L., Freire, J.: Siphoning hidden-web data through keyword-based interfaces. In: SBBD, pp. 309–321 (2004)
2. Barbosa, L., Freire, J.: Searching for hidden-web databases. In: WebDB, pp. 1–6 (2005)
3. Barbosa, L., Freire, J., da Silva, A.S.: Organizing hidden-web databases by clustering visible web documents. In: ICDE, pp. 326–335 (2007)
4. Chang, K.C.-C., He, B., Li, C., Patel, M., Zhang, Z.: Structured databases on the web: Observations and implications. SIGMOD Record 33(3), 61–70 (2004)
5. Chang, K.C.-C., He, B., Zhang, Z.: Toward large scale integration: Building a metaquerier over databases on the web. In: CIDR, pp. 44–55 (2005)
6. Dragut, E.C., Wu, W., Sistla, A.P., Yu, C.T., Meng, W.: Merging source query interfaces onweb databases. In: ICDE, p. 46 (2006)
7. Ganesan, P., Garcia-Molina, H., Widom, J.: Exploiting hierarchical domain structure to compute similarity. ACM Trans. Inf. Syst. 21(1), 64–93 (2003)
8. He, B., Chang, K.C.-C., Han, J.: Discovering complex matchings across web query interfaces: a correlation mining approach. In: KDD, pp. 148–157 (2004)
9. He, B., Tao, T., Chang, K.C.-C.: Clustering structured web sources: A schema-based, model-differentiation approach. In: Lindner, W., Fischer, F., Türker, C., Tzitzikas, Y., Vakali, A.I. (eds.) EDBT 2004. LNCS, vol. 3268, pp. 536–546. Springer, Heidelberg (2004)
10. He, B., Tao, T., Chang, K.C.-C.: Organizing structured web sources by query schemas: a clustering approach. In: CIKM, pp. 22–31 (2004)
11. He, H., Meng, W., Yu, C.T., Wu, Z.: Wise-integrator: An automatic integrator of web search interfaces for e-commerce. In: VLDB, pp. 357–368 (2003)
12. Hsieh, W.C., Madhavan, J., Pike, R.: Data management projects at google. In: SIGMOD Conference, pp. 725–726 (2006)
13. Ipeirotis, P.G., Gravano, L.: Classification-aware hidden-web text database selection. ACM Trans. Inf. Syst. 26(2) (2008)
14. Kaufman, L., Rousseeuw, P.: Finding groups in data, vol. 16. Wiley, New York (1990)
15. Lakkaraju, P., Gauch, S., Speretta, M.: Document similarity based on concept tree distance. In: Hypertext, pp. 127–132 (2008)
16. Larsen, B., Aone, C.: Fast and effective text mining using linear-time document clustering. In: KDD, pp. 16–22 (1999)
17. Peng, Q., Meng, W., He, H., Yu, C.T.: Wise-cluster: clustering e-commerce search engines automatically. In: WIDM, pp. 104–111 (2004)
18. Wu, W., Doan, A., Yu, C.T.: Merging interface schemas on the deep web via clustering aggregation. In: ICDM, pp. 801–804 (2005)
19. Wu, W., Yu, C.T., Doan, A., Meng, W.: An interactive clustering-based approach to integrating source query interfaces on the deep web. In: SIGMOD Conference, pp. 95–106 (2004)

Effective Term Weighting in ALT Text Prediction for Web Image Retrieval

Vundavalli Srinivasarao, Prasad Pingali, and Vasudeva Varma

Search and Information Extraction Lab
International Institute of Information Technology
Gachibowli, Hyderabad-32, India
srinivasarao@research.iiit.ac.in, {pvvpr,vv}@iiit.ac.in

Abstract. The number of images on the World Wide Web has been increasing tremendously. Providing search services for images on the web has been an active research area. Web images are often surrounded by different associated texts like ALT text, surrounding text, image filename, html page title etc. Many popular internet search engines make use of these associated texts while indexing images and give higher importance to the terms present in ALT text. But, a recent study has shown that around half of the images on the web have no ALT text. So, predicting the ALT text of an image in a web page would be of great use in web image retrieval. We treat the prediction of ALT text as the problem of automatic image annotation based on the associated texts. We propose a term weighting model that makes use of term co-occurrences in associated texts and predicts the ALT text of an image. Using our approach, we achieved a good improvement in performance over baseline.

1 Introduction

With the advent of digital devices like digital cameras, camera-enabled mobile phones, the number of images on the World Wide Web is growing rapidly. This led to an active research in the organization, indexing and retrieval of images on the World Wide Web. Providing search services for the web images has been difficult. Traditional image retrieval systems assign annotations to each image manually. Although it is a good methodology to retrieve images through text retrieval technologies, it is gradually becoming impossible to annotate images manually one by one due to the huge and rapid growing number of web images. Automatic Image Annotation has become an active research area since then. A common view is that semantics of web images are well correlated with their associated texts. Because of this, many popular search engines offer web image search based only on the associated texts.

ALT text is considered the most important of all associated texts. ALT attribute is used to describe the contents of an image file. It's important for several reasons: ALT attribute is designed to be an alternative text description for images. It represents the semantics of an image as it provides useful information to anyone using the browsers that cannot display images or image display disabled. It can help search engines to determine the best image to return for a user's query.

X. Du et al. (Eds.): APWeb 2011, LNCS 6612, pp. 237–244, 2011.

Many popular internet search engines like Google Image Search[1] make use of these associated texts while indexing the images and give higher importance to the terms present in ALT text. Google states the importance of ALT text in their official blog[2]: *"As the Googlebot does not see the images directly, we generally concentrate on the information provided in the "ALT" attribute."*

The ALT attribute has been used in numerous research studies of Web image retrieval. It is given the highest weight in [1]. In [2], the authors consider query terms that occur in ALT text and image names to be *'very relevant'* in ranking images in retrieval. Providing *'a text equivalent for every non-text element'* (for example, by means of the ALT attribute) is a checkpoint in the W3C's Web Content Accessibility Guidelines[3]. The authors of [3] also state the importance of using ALT text. However, a recent study [4] has shown that around half of the images on the web have no ALT text at all. The author collected 1579 images from Yahoo!'s random page service and 3888 images from the Google directory. 47.7% and 49.4% of images respectively had ALT text, of which 26.3% and 27.5% were null. It is clear from this study that most of the web images don't contain ALT text.

Therefore, predicting ALT text of an image in a web page would be of great use in web image retrieval. In this paper, we address the problem of predicting terms in ALT text by proposing a term weighting approach using term co-occurrences. We treat the prediction of terms in ALT text as the problem of automatic image annotation based on the image associated texts. This approach can be extended to any image dataset with associated texts.

The reminder of the paper is organized as follows. In Section 2, we describe our approach to term weighting using term co-occurrences. We describe the dataset, evaluate our system and prove the usefulness of it in Section 3. Section 4 gives an overview of related work. We conclude that our proposed approach can be useful in web image retrieval and give an account of our future directions in Section 5.

2   Term Weighting Model

In this section, we propose a term[4] weighting model to compute the term weights based on the term co-occurrences in image associated texts to predict the terms in ALT text. Terms in the texts are considered to be similar if they appear in similar contexts. Therefore, these similar words do not have to be synonyms or belong to the same lexical category. Before calculating term weights, associated texts need to be preprocessed. The first step is to remove common words(stopwords) using a list of stop words. The second step is to use stemming to normalize the representations of terms. In our work, we used Porter Stemmer [5] for stemming.

---

[1] http://images.google.com/

[2] http://googlewebmastercentral.blogspot.com/2007/12/using-alt-attributes-smartly.html

[3] Web content accessibility guidelines 1.0. Retrieved 26 August, 2005 from http://www.w3.org/TR/WAI-WEBCONTENT/

[4] We use *term* and *word* interchangeably in this paper.

According to the distributional hypothesis [6], semantically similar terms occur in same or similar contexts. We use this information in our term weighting model. A term is said to be important if it occurs in many associated texts and co-occurs with many other terms present in different associated texts.

For each image, we calculate term weights using the following equation.

$$w(t) = (\frac{\sum_i s(t, t_i)}{N})(Imp(t)) \qquad (1)$$

$s(t, t_i)$, the similarity between two terms t and t_i, is calculated using Jaccard Similarity as follows:

$$s(t, t_i) = \frac{|S_t \cap S_{t_i}|}{|S_t \cup S_{t_i}|} \qquad (2)$$

$S_t \cap S_{t_i}$ is the set of associated texts in which both t and t_i occur, $|S_t \cup S_{t_i}|$ can be calculated as $|S_t| + |S_{t_i}| - |S_t \cap S_{t_i}|$, and S_t is the set of associated texts which contain the term t. N is the total number of unique terms in all associated texts.

$Imp(t)$ is the importance of a term which is calculated as follows:

$$Imp(t) = \frac{\sum_i boost(a_i)}{|A|} \qquad (3)$$

$boost(a_i)$ is the boost of the associated text a_i which contains the term t and A is the set of associated texts which contain the term t. Based on the heuristic of importance(image caption, HTML title, image filename, anchor text, source page url, surrounding text in that order), we assign a boost to the extracted associated texts. Value of boost for each associated text is given based on the importance of the associated text as stated above. Once the weight for each term has been computed, the terms are ranked in descending order based on term weights and top k terms are selected as terms in ALT text.

3   Evaluation

In this section, we present the evaluation procedure of our approach. We briefly describe the data collection and preprocessing steps, present the evaluation procedure and finally results are discussed.

3.1   Data Collection and Preprocessing

A crawler is used to collect images from many websites. Images like banners, icons, navigation buttons etc, which are not so useful are not considered. The web documents are preprocessed to extract associated texts so that the images can be indexed and retrieved with these text features. The associated texts we considered are extracted from HTML page title, image filename, source page url, anchor text, image caption and surrounding text.

We used Guardian[5], Telegraph[6] and Reuters[7] as the source urls and collected a total of 100000 images: 11000 images from Reuters, 41000 images from Guardian and 48000 images from Telegraph which have ALT text. We selected these news websites because the ALT text provided in them is accurate and is very useful for evaluation. The pages in which the images are present, cover a wide range of topics including technology, sports, national and international news, etc. Stopwords are removed and stemming is used to filter the duplicate words from the extracted textual information.

3.2   Baseline Approaches

Most of the keyword extraction techniques use frequency based approaches like TF.IDF [7] or, simply term frequency. We implemented two approaches based on term frequency(TF) and document frequency(DF) respectively as the baseline approaches. TF is the number of occurrences of a term in all associated texts, where as DF is the number of associated texts in which the term appears. In general, if the term appears in many associated texts, it will have higher weight than the ones which appear in less number of associated texts. We compute TF and DF of each term present in associated texts of an image. Once tf and df for each term has been computed, the terms are ranked in descending order based on term weights and top k terms are selected as terms in ALT text.

IDF is used to enhance the significance of the terms in a given corpus which appear in less number of documents. In our approach, terms are used to derive the semantics of the embedded images in a single document. IDF is not used as a baseline approach in our analysis for the above reason.

3.3   Evaluation Procedure

In order to evaluate the effectiveness of our method, we compare the predicted terms produced by our approach against the terms extracted from ALT attribute of an image in the corresponding web page. In addition we also compare the performance of our method against the baseline approaches which are based on term frequency and document frequency respectively.

We present results using the top 5, 10, and 15 words. We adopt the recall, precision and F-measures to evaluate the performance in our experiments. If P_t is the set of terms predicted by our approach and A_t is the set of terms in ALT text, then in our task, we calculate precision, recall and F-measure as follows:

$$precision = \frac{Number\ of\ common\ terms\ between\ P_t\ and\ A_t}{Total\ number\ of\ terms\ in\ P_t} \tag{4}$$

$$recall = \frac{Number\ of\ common\ terms\ between\ P_t\ and\ A_t}{Total\ number\ of\ terms\ in\ A_t} \tag{5}$$

$$F - Measure = \frac{2 * precision * recall}{precision + recall} \tag{6}$$

---

[5] http://www.guardian.co.uk/
[6] http://www.telegraph.co.uk/
[7] http://www.reuters.com/

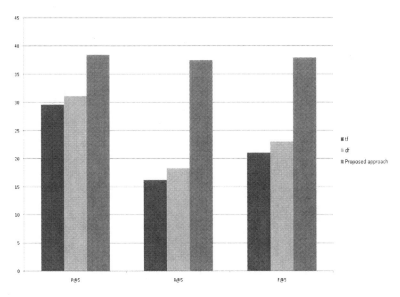

Fig. 1. Comparison of approaches for top 5 predicted terms

3.4   Analysis

As we can see from the results shown in Fig.1, Fig.2 and Fig.3, our approach gives good results at top 15 terms. From our experiments we found that it's just the order of importance of associated texts, not the values of boost of associated texts, that influences the performance of our approach. The predicted terms that are not present in the original ALT text may not completely be irrelevant. There could be cases where the predicted terms may represent the image, but they are not used in ALT text. Author of a web page may use some terms that comes to his mind when writing ALT text for an image and uses semantically related terms across other associated texts. In that case, the term that the author uses consistently across associated texts of an image will be predicted to be in original ALT text. For example, Consider an image of a cricketer. If the author of a web page uses the term 'batsman' in the ALT text and uses the term 'cricketer', which is semantically related to 'batsman' across other associated texts, the term 'cricketer' might be predicted by our approach. Manual evaluation has been done on 1000 images taking semantic relationship into account and found that the values of precision, recall and F-measure have been increased by 3.61%, 5.73% and 4.50% respectively. The results are given in Table 1.

Table 1. Results for the Proposed Approach

Semantic Relationship	Precision@15	Recall@15	F-Measure@15
No	32.48	69.17	44.21
Yes	36.09	74.90	48.71

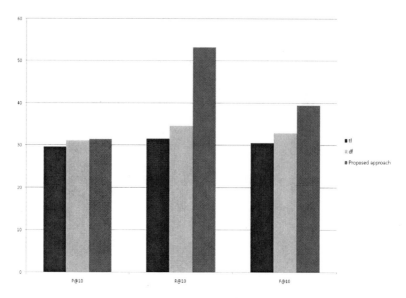

Fig. 2. Comparison of approaches for top 10 predicted terms

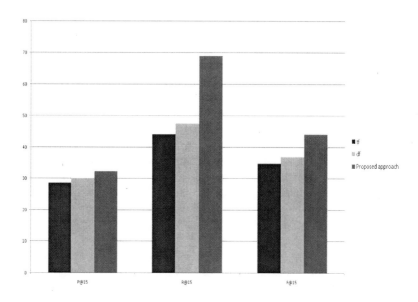

Fig. 3. Comparison of approaches for top 15 predicted terms

4   Related Work

Many web image search engines use keyword based search to retrieve images. Automatic Image Annotation is the key issue in keyword based image retrieval.

There has been plenty of work done in Automatic Image Annotation. Some of the early approaches [8, 9] to image annotation are closely related to image classification. Images are assigned a set of sample descriptions(predefined categories) such as people, landscape, indoor, outdoor, animal. Instead of focusing on the annotation task, they focus more on image processing and feature selection.

Co-occurrence Model [10], Translation Model [11], Latent Dirichlet Allocation Model [12], Cross-Media Relevance Model [13], Multiple Bernoulli Relevance Model [14], etc infer the correlations or joint probabilities between images and annotation keywords. Other works like linguistic indexing [15], and Multi-instanced learning [16] try to associate keywords(concepts) with images by learning classifiers. To develop accurate image annotation models, some manualy labeled data is required. Most of the approaches mentioned above have been developed and tested almost exclusively on the Corel[8] database. The latter contains 600 CD-ROMs, each containing about 100 images representing the same topic or concept, e.g., people, landscape, male. Each topic is associated with keywords and these are assumed to also describe the images under this topic.

[17] demonstrates some of the disadvantages of data-sets like Corel set for effective automatic image annotation. It is unlikely that models trained on Corel database will perform well on other image collections.

Web images differ from general images in that they are associated by plentiful of texts. As we mentioned in earlier sections, examples of these associated texts include image file name, ALT text, caption, surrounding text and page title, etc.

Various approaches [18,19] have been employed to improve the performance of web image annotation based on associated texts. Our work differs from previous work in that our approach is evaluated on a large number of images and works well for any image dataset with associated texts.

5   Conclusions and Future Work

This paper presented a term weighting approach that makes use of term co-occurrences in associated texts and predicts terms occurring in ALT text of an image. We compared the performance of our approach against two baseline approaches which use term frequency and document frequency respectively. Experiments on a large number of images showed that our model is able to achieve a good performance. One advantage of our approach is that it works well for any image dataset given the images have associated texts.

Having observed that semantic relatedness should be taken into account while evaluating, we would like to include this in the evaluation framework. We are working on making the dataset used in our experiments a standard one for web image retrieval. Our work is language independent. Our work can be extended to other domains like social bookmarking sites. The goal would then be to predict the tags of a url from the text content of the url.

---

[8] http://www.corel.com/

References

1. Cascia, M.L., Sethi, S., Sclaroff, S.: Combining textual and visual cues for content-based image retrieval on the world wide web. In: IEEE Workshop on Content-based Access of Image and Video Libraries, pp. 24–28 (1998)
2. Mukherjea, S., Hirata, K., Hara, Y.: Amore: A world wide web image retrieval engine. World Wide Web 2(3), 115–132 (1999)
3. Petrie, H., Harrison, C., Dev, S.: Describing images on the web: a survey of current practice and prospects for the future. In: Proceedings of Human Computer Interaction International, HCII 2005 (2005)
4. Craven, T.C.: Some features of alt texts associated with images in web pages. Information Research 11 (2006)
5. Porter, M.F.: An algorithm for suffix stripping. Program 14, 130–137 (1980)
6. Harris, Z.: Mathematical Structures of Language, vol. 21. Wiley, Chichester (1968)
7. Salton, G., Buckley, C.: Term-weighting approaches in automatic text retrieval. Information Processing & Management 24, 513–523 (1988)
8. Vailaya, A., Member, A., Figueiredo, M.A.T., Jain, A.K., Zhang, H.J., Member, S.: Image classification for content-based indexing. IEEE Transactions on Image Processing 10, 117–130 (2001)
9. Smeulders, A.W.M., Member, S., Worring, M., Santini, S., Gupta, A., Jain, R.: Content-based image retrieval at the end of the early years. IEEE Transactions on Pattern Analysis and Machine Intelligence 22, 1349–1380 (2000)
10. Hironobu, Y.M., Takahashi, H., Oka, R.: Image-to-word transformation based on dividing and vector quantizing images with words. In: Boltzmann Machines, Neural Networks, vol. 4 (1999)
11. Duygulu, P., Barnard, K., de Freitas, J.F.G., Forsyth, D.: Object recognition as machine translation: Learning a lexicon for a fixed image vocabulary. In: Heyden, A., Sparr, G., Nielsen, M., Johansen, P. (eds.) ECCV 2002. LNCS, vol. 2353, pp. 97–112. Springer, Heidelberg (2002)
12. Blei, D.M., Jordan, M.I.: Modeling annotated data. In: SIGIR 2003: Proceedings of the 26th Annual International ACM SIGIR Conference on Research and Development in Informaion Retrieval, pp. 127–134 (2003)
13. Jeon, J., Lavrenko, V., Manmatha, R., Callan, J., Cormack, G., Clarke, C., Hawking, D., Smeaton, A.: Automatic image annotation and retrieval using cross-media relevance models. SIGIR Forum, 119–126 (2003)
14. Feng, S., Manmatha, R., Lavrenko, V.: Multiple bernoulli relevance models for image and video annotation. In: CVPR 2004, vol. II, pp. 1002–1009 (2004)
15. Jia, L., Wang, Z.J.: Automatic linguistic indexing of pictures by a statistical modeling approach. IEEE Trans. Pattern Anal. Mach. Intell. 25, 1075–1088 (2003)
16. Yang, C., Dong, M.: Region-based image annotation using asymmetrical support vector machine-based multi-instance learning. In: Proceedings of the 2006 IEEE Computer Society Conference on Computer Vision and Pattern Recognition, vol. 2 (2006)
17. Tang, J., Lewis, P.: A study of quality issues for image auto-annotation with the corel data-set. IEEE Transactions on Circuits and Systems for Video Technology 1, 384–389 (2007)
18. Rui, X., Li, M., Li, Z., Ma, W.Y., Yu, N.: Bipartite graph reinforcement model for web image annotation. In: ACM Multimedia, pp. 585–594 (2007)
19. Shen, H.T., Ooi, B.C., Tan, K.-L.: Giving meaning to www images. In: ACM Multimedia, pp. 39–47 (2000)

Layout Object Model for Extracting the Schema of Web Query Interfaces

Tiezheng Nie, Derong Shen, Ge Yu, and Yue Kou

College of Information Science and Engineering, Northeastern University
110819 Shenyang, China
{nietiezheng,shenderong,yuge,kouyue}@ise.neu.edu.cn

Abstract. This paper presents a novel approach to extract the hierarchical schema of web interfaces by employing the visual features. In our approach, the Layout Object Model (LOM) is proposed to extract the schema of interface based on the geometric layout of interface elements. In the LOM, each field or label of interface is a layout object denoted with a rectangle in web browser. The schema of interface can be expressed by organizing these rectangles with a tree structure which is called as the LOM tree. So we extract the schema by constructing the LOM tree. The construction is start with generating the basic layout tree from the DOM tree. Then, we match labels for fields or groups of them by employing their layout relation and feature rules, and the LOM tree is constructed by adjusting the basic layout tree. Finally, we transforms the LOM tree of a web interface into a schema tree to extract the schema. The experimental results show that our approach can match labels and fields accurately, which is very useful for deep web applications.

Keywords: deep web, web interface, visual features, layout object model.

1   Introduction

In current applications, the deep web has become a very important resource for data analyzing, data integration and data mining since it contains high-quality information i.e. structured data records. However, data sources of the deep web provide their data by web interfaces that are presented with HTML and consist of labels and fields. To accurately access the deep web, we need extract the schema of web interfaces. This work is called understanding web interfaces in [8]. The schema of web interfaces has a tree structure, in which text tokens are regarded as semantic labels of input fields or their groups. The goal of extraction is accurately grouping labels and fields into a hierarchical presentation that express the schema of interface. For example, figure 1 shows a web interface of automobiles, in which the label "model" is used to express the semantic of its following input field. This work is very important for deep web data extraction [9, 10], deep web integration [11, 12], building unified query interfaces [13]. Some researches [1] have provides methods to discover web interfaces accurately. But they did not address the problem of understanding web interfaces. Existing methods [2,3] solve this problem with the structure of source code. We call

X. Du et al. (Eds.): APWeb 2011, LNCS 6612, pp. 245–257, 2011.

them as the code structure-based method (CSM). They parse the source code of interface into an element sequence or a DOM tree, and match label and field based on their distances in the structure of DOM tree. However, when the structure of web interface is complex, CSM will wrongly group labels and fields. For example, in figure 2, two labels are in the same branch of the DOM tree, and their corresponding fields are in another branch.

Fig. 1. A web interface example of automobile web source

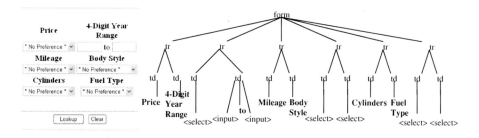

Fig. 2. An example of query interface with its DOM tree

Our observation shows that the designers of web interfaces express the schema with the layout of labels and fields rather than the code structure of page. It means the schema of an interface is hidden in its geometric layout structure which is more accurate than its code structure. Thus, in this paper, a Layout Object Model (LOM) is proposed to express labels and fields of interface with their display on a browser. In the LOM, each label and field is regard as a layout object that is denoted with a rectangular area on a browser. For example, the rectangle L1 in the middle figure 2 denotes the layout object of label "*Price*".

In this paper, the main contribution of this paper is the development of a method that utilizes the visual features of web interfaces constructing a LOM tree to express the schema. We call this method as the layout structure-based method (LSM). The LOM tree has similar features with the R-tree[17] in which each node denotes a rectangular area. The construction of the LOM tree is based on our insights on visual features of web interfaces. It is obvious that a LOM tree with appreciate structure can accurately express the schema of web interface, and this LOM tree can be easily transformed into a schema tree of web interface. The other contributions of this paper are as follows: 1 we define a set of rules based on the visual features for the construction of the LOM tree; 2 we propose algorithms for constructing the LOM tree and transforming it into the schema tree.

The rest of this paper is organized as follows. Section 2 shows related works. Section 3 presents the problem definition and models for understanding web interfaces. Section 4 describes the design rules and the algorithm to construct the LOM tree with layout information. In Section 5, we present our experimental results. Section 6 gives a conclusion of this paper.

2 Related Works

There have been some existing works to extract the query interface schema of deep web sources. We category them into two types: the code structured-based method and the layout structure-based method. The code structured-based method usually parses source codes of web interfaces into special structures. The approach presented in [4] regards the document of a query interface as an interface expression (IEXP). They match the prediction of the right label for a field on the interface expression. This approach is suitable for web interfaces with simple structure, but not the web interfaces with complex structure. The approach [14] uses the heuristic rule that considers the textual similarity between the name of a field and a label. This heuristic rule is not work when there is no semantic names or ids within elements of web pages. A recent work [5] improved this approach by considering the distance between labels and fields. It first formats the source codes of web interface into XML and builds a DOM tree structure. Then it evaluates the distance of between a label and a field on the DOM tree to match them. But the redundant html elements make the DOM tree more complex which will decrease the accuracy of the approach. Another approach [7] is based on applying data mining algorithms. It first selects a set of sample web interfaces, and extracts the distance, position relationship, element types of matched label and field. Then it trains a decision tree for extracting attributes from web interfaces. However, the design patterns of web interfaces will evolve with the technique development of web page. So the decision tree may be not useful for new web interfaces. [15] uses attributes to group sets of related fields, but its approach can't handle the nested attributes.

The layout structure-based method [6, 8] is mainly using the visual content features to predicate labels for fields of web interfaces. The LSM is more robust against HTML particularities and code evolution. The approach [6] introduces the notion of viewing query interfaces as visual languages, and hypothesis hidden syntax in query forms. It extracts heuristic rules by observing a set of interfaces. Its disadvantage is that too many special rules may have conflicts and produce error matching when web techniques evolved. The approach most similar to ours is [8] which builds a schema tree model to express web interface [2]. Given a query interface, they first construct a tree of fields by semantic order and a tree of labels based on semantic scopes of them. Then they develop an algorithm to derive an integrated tree structure that represents the schema tree of the interface based on integrating the tree of fields and the tree of text tokens. However, some interfaces have labels that are aligned to the center and not to the left. The semantic scopes of this kind of labels will be wrongly computed. Our model constructs the LOM tree with features of both layout and R-tree, which will avoid this problem.

3 Model Definition

3.1 The Schema Model of Web Interface

The schema of a web interface can be present with a hierarchical tree called schema tree [8] which expresses the semantic relationship between labels and fields. In the tree structure, leaves correspond to the fields and their own labels, and the order of sibling nodes corresponds to the display order of them. The internal nodes render a group of related fields or a group of related groups of fields. So a query attribute of interface composed with one or multiple fields is mapping to a sub tree in the schema tree. The schema tree presents the related labels of fields and the semantic context of labels. The semantic of labels with lower level are under the context of semantic of their parent labels. The hierarchical semantic structure of example in figure1 is shown in figure 3, in which the schema tree of web interface has 12 query attributes with four levels, and two "*From*" and "*to*" labels have different semantic, for they have context semantic in their parent nodes with label "*Year*" and "*Price Range*". In the model of schema tree, an internal node is a label which presents the semantic of labels and fields in its sub tree. The leaves of schema tree are fields with their own labels.

To model the hierarchical schema of web interface, we define a web interface WI as a set of attributes $A=\{A_1,...A_n\}$, in which A_i is a query attribute of WI. And t A_i is defined as $<GL, F, Gs>$, where L denotes the semantic label of A_i, $F=\{F_1,...,F_l\}$ denotes the fields directly contained by A_i, and $Gs=\{G_1,...,G_m\}$ denotes a set of groups which are the nested grouping of related fields. For each group G_i in Gs, to express the nested structure, it is also defined as $G_i = <GL', F', Gs'>$, where GL' denotes the semantic label of group. The field F_i is defined as $F_i = <L, f>$, where L is the own label of field f. Based on our definitions, the hierarchical semantic is expressed with a tree structure that we also call it as the schema tree in this paper.

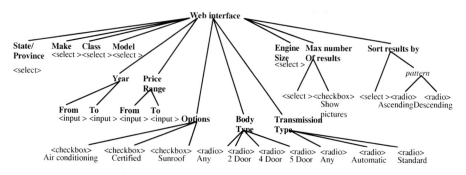

Fig. 3. The hierarchical semantic structure of example in figure1

3.2 The Layout Object Model and LOM Tree

In this paper, we propose the Layout Object Model for describing elements of web pages. Each of visual elements is regarded as a layout object which is denoted with a rectangular area in a browser. In the LOM, multiple layout objects can be grouped into a bigger layout object whose rectangular area is the minimum bounding rectangle of these layout objects. So, we can construct a layout tree by recursively grouping layout objects of a web page. Given a web interface, there are many layout trees that have different structures by grouping labels and fields with different order. It exist a layout tree whose structure is the same with the structure of its schema tree. We call this layout tree as the LOM tree in this paper. In the LOM tree, a leaf node is a basic layout object mapping to a label or field, and an inner node is a grouped layout object that is denoted by a MBR of its child nodes. The root of the LOM tree is the area that contains all elements of a web interface. The LOM tree can represent the hierarchical semantic relationship of labels and fields for a web interface. Therefore, we build the LOM tree to extract the schema of interfaces. Figure 4 shows the layout of the web interface in figure 2 and its LOM tree structure. *R1* denotes the area of interface and is the root node of the LOM tree. Rectangle *Ri* is mapping with six query attributes. *Li* and *Fi* denote the layout objects of labels and fields.

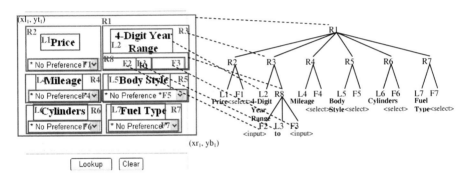

Fig. 4. The layout of web interface in figure 2 with its LOM tree

Then we formulize the LOM tree model of web interface as follows: an LOM tree of web interfaces is defined as *LT*, in which each node denotes a rectangular area in web page. The coordinate of a node in the LOM tree is defined as $Ri = <(xl_i, yu_i), (xr_i, yb_i)>$, where (xl_i, yu_i) is the coordinate of the left-upper corner of the rectangle, and (xr_i, yb_i) is the coordinate of the right-bottom corner of the rectangle. These two coordinates of points can present the boundary of a rectangle. xl_i and xr_i is the left bound and right bound of the rectangular area in horizontal, while yu_i and yb_i is the upper bound and bottom bound in vertical. For example, *R1* is the root node of the web interface in figure 4, and the boundary of its rectangle is presented by the points (xl_1, yu_1) and (xr_1, yb_1).

4 The Construction of the LOM Tree

In this section, we describe the methodology for the construction of the LOM tree. There are some issues to be addressed. First, the structure of the LOM tree must be matched with the structure of schema. The second problem is matching the semantic label for a given field.

4.1 The Design Rules for the LOM Tree

Based on the features of HTML and web query interfaces, we abstract them as a set of design rules that determine the construction of the LOM tree. Rules can be classified into basic rules and semantic rules. Basic rules restrict the structure of LOM tree by layout features. Semantic rules control the grouping of labels and fields by semantic features. Rule 1 to 3 belongs to the basic rules, and others are the semantic rules.

Rule 1: Labels and fields are organized top-down and left-to-right in the layout of web query interfaces.

Rule 2: Each two sibling nodes has no overlap area.

Rule 3: Let R is an internal node of a group, and the left-top point is the origin of the coordinate, the coordinate of R is $<(min(xl_i), min(yu_i)), (max(xr_i), max(yb_i))>$, where $<(xl_i, yu_i), (xr_i, yb_i)>$ is the coordinate of labels and fields in the group.

Rule 4: If label L_i and L_j has different text style on their text token, they are in the different semantic levels, and the MBRs of label L_i and L_j can not be grouped directly.

Rule 5: If label L_i has a larger or stronger font than label L_j, the semantic level of L_i is higher than that of L_j.

Rule 6: If field F_1 and F_2 are the members of the same group, text styles of their labels are also the same.

Rule 7: Fields with the same *name* value in HTML are in the same group.

Rule 8: If a label is used to denote the semantic of a group, its node is the sibling of the node of the group, and its MBR is on the left or above of the MBR of the group.

Rule 9: If a field has an own label, the orientation of label MBR is on the *left*, *right*, *above* or *below* of its MBR, and their nodes should be siblings in the SR-tree.

Rule 10: A label is used to denote either the semantics of a field or of a group of fields, but not both.

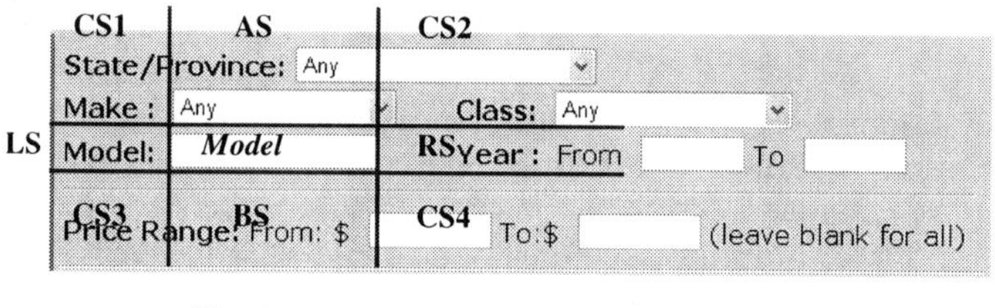

Fig. 5. Sectors around the MBR of a Field

The first rule is obvious for the predicted pattern of human browsing web pages. Rule 2 reflects the design pattern of web interface layout in two aspects: first, each pair of labels and fields has no overlap area; second, if a group of labels and fields denotes a query attribute, it has no overlap area with other groups of attributes. These patterns are used to avoid confusing user when user input values for his query. Rule 3 defines the formula to get the boundary of an internal node. To construct the accurate LOM tree, we must consider the semantic hierarchy for labels and fields. There are several heuristic rules (Rule 4 to 6) for employing the text style of label. The label of a group has a more distinct text style than the labels of its subgroups or fields. Rule 7 is used to group related fields based on the feature of HTML. The fields with the same name value are grouped as one parameter in the request of HTML. Rule 8 to 10 are defined to match related labels for fields and groups with the orientation of labels. Rule 8 presents the position relationship between a group and its label. We separate the area around the MBR of a field into 8 sectors, as shown in figure 5. The four sectors, denoted with *LS*(left sector), *RS*(right sector), *AS*(above sector) and BS(below sector), are potential to containing the label of the field. The corner sectors denoted with CS1 to CS4 are blind sectors, because labels entirely in those areas can not annotate the semantic of the center field in design pattern of web interfaces.

4.2 The Algorithm for the Construction

The most difficult challenge on constructing the LOM tree is to make its structure similar with the structure of semantic hierarchy of the web interface. In this paper, a basic layout tree is firstly constructed to satisfy the basic rules. Then, it constructs the LOM tree by adjusting the structure of the basic layout tree. In our approach, we generate the basic layout tree based on the code structure-based method.

4.2.1 Generating the Basic Layout Tree

The first step of our approach is generating the basic layout tree. With the code structure-based method, we first parse the source code of the web interface into a DOM tree, such as the left haft in figure 2. In the DOM tree, all vision elements of HTML are leaves, and an internal node contains one or multiple labels and fields. To transform the DOM tree into a layout tree, the internal nodes that annotate the text style of a label are combined with the node of label, and element nodes unrelated with the layout are eliminated from the DOM tree. After this operation, we only reserve nodes of labels and fields and internal nodes that contain labels and fields in their sub trees. Then, we mark each leave node with their coordinates of MBR which are calculated by Rule 3. Consequently, each node of the DOM tree is denoted with MBR, and the DOM tree can be regard as a layout tree. This layout tree satisfies the basic design rules we have defined. It is because the structure of the DOM tree of interface is designed to describe the layout of HTML elements in web browsers. Figure 6 shows an example of the transformation.

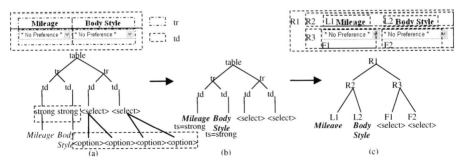

Fig. 6. The Transformation from the DOM tree to the Basic Layout Tree

4.2.2 Adjusting the Structure of the Basic Layout Tree to the LOM Tree

The basic layout tree is similar with the DOM tree of interface, so it can not be used to present the semantic hierarchy of interface schema. Therefore, we need adjust the structure of the basic layout tree by replace labels and fields. We develop an algorithm to derive the LOM tree. The detail of the algorithm is given in Algorithm 1.

In Algorithm 1, we first traverse the Basic layout tree with pre-order to build a field list *f-list* which determines the order of processing fields (Line 1). In most of interfaces, this order is the same with the layout order that is top-down and left-to-right. Then we group fields with the same *name* value based on Rule 7 (Line 2). To group a set of fields, we adjust their positions in the *f-list*: suppose f_i is the first field of group g that contains k fields, we move other fields of g orderly following f_i in *f-list*. Then f_{i+k+1} is the first of fields that are not in g and following f_i. The third step (Line 3) is to match candidate labels for each field of *f-list*. Candidate labels include twp types: candidate own labels L_{co} and candidate share labels L_{cs}. Given a field f_i, we discover its candidate own labels according to Rule 9. The own labels of f_i are in the four orientations that are *left*, *right*, *above* and *bellow*, and in each direction only one candidate label are selected based on the distance of two MBR.

Algorithm 1. Generating the LOM Tree

Input: the Basic Layout Tree *BT*, the DOM Tree *DT*
Output: the LOM Tree *LT*
1 Traverse *BT* with pre-order to get the field list *f-list*;
2 Group fields with the same *name* in *f-list*;
3 Match candidate labels for each field *f* of *f-list*;
4 **While** *f-list* != null **do**
5   Get a field f_i from *f-list* ;
6   **If** (f_i is in a group g_j) **Then**
7   matchGroup (g_j, *BT*);
8   doAdjustTree(g_j, *BT*);
9   remove g_j from *f-list*;
10   **else**
11   g_k= matchField (f_j, *BT*);
12   doAdjustTree(g_k, *BT*);
13   remove g_k from *f-list*;
14   **end if**
15 **End while**
16 *LT* = postAdjust(*BT*);

To develop the matching algorithm, we further define some matching rules *MRules* which include: (1) if a field *f* has an own label *l*, the MBR constructed with *l* and *f* must be not overlap or overlay with another field f'; (2) if a label *l* describes the semantic of a group *g*, the MBR constructed with *l* and *g* must be not overlap or overlay with another group g'; (3) if a label *l* describes the semantic of a field *f*, *l* is the closest label in their orientation, and *f* is the closest field in their orientation. If there is a field that has no candidate own labels, it indicates it is sharing a label with other fields.

For these fields, we match candidate share labels by searching closest labels in the four directions. The MBR of a field f and its candidate share label l must only contain or overlap other fields.

Then, the fourth step (Line 4 to 15) orderly accesses each field of f-list to match a label for it. Here, if the field f_i is in an identified group g, we match labels for all fields in g by procedure $matchGroup$. Procedure $matchGroup$ selects candidate own labels of fields in g with the same orientation and text style as their own labels according to Rule 6. However, there may be labels with different orientation satisfied the condition of own labels of a group. In this case, according to Rule 8, we select their right labels as their own labels, and assign the first left label as the label of group. Otherwise, if f_i is a single field in f-list, it may be an independent field or be sharing the same label with other fields. Procedure $matchField$ process f_i by checking whether it is sharing candidate labels with other fields. If f_i has only an exclusive candidate own label l, we regard f_i as an independent field and assign l as its own label. Otherwise, f_i is sharing a candidate label l with f_j. If l is the candidate share label of f_j and has the same orientation with f_i and f_j, we group them and assign l as their label. If l is the candidate own label of f_j, one of f_i and f_j has another candidate own label l' with the same orientation and style, we group it with l'. If both f_i and f_j has another candidate own label with the same style, we select the right label as their own label and assign the first left label as group label according to Rule 8. For fields that are assigned a group label, we generate a new group g_k with them in the algorithm.

Once labels are assigned to fields, procedure $doAdjustTree$ adjusts the structure of the basic layout tree to satisfy features of the LOM tree. The algorithm is given in Algorithm 2.

Algorithm 2. *doAdjustTree*

Input: group g, the Basic Layout Tree BT
Output: the Basic Layout Tree BT
1 **If** f_i of g has a own label l_i **Then**
2 **For each** f_i **In** g **do**
3   Move f_i as a sibling of l_i and insert an internal node for them;
4 **End for**
5 **Else**
6   Move all f_i of g as sibling leaves and create a parent internal node;
7   Adjust label l of g to be the sibling of parent node of all f_i;
8 **End If**
9 BT = adjustMBR(BT);

The adjustment of the example in figure 6 is shown in figure 7. Current group g contains $F1$ that has own label $L1$ (Figure7(a)). $F1$ is moved under $R2$ as a sibling of $L1$ and an internal node $F4$ is inserted (Figure7(b)). In this structure, lca of origin and target parent is $R1$ and MBR of $R2$ overlays MBR of $R3$ in top of Figure7(b). So, $R4$ is removed and $R1$ becomes parent of $R4$ and $L2$ (Figure7(c)). This structure of Basic R-tree is obviously satisfied Rule 2.

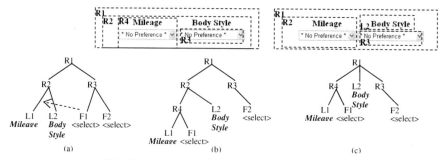

Fig. 7. An example of procedure *doAdjustTree*

When *f-list* is null, it means all fields have been grouped with labels and is replaced in an appropriate branch in *BT*. After that, we assign the current Basic layout tree as the LOM tree *LT*. However, the LOM tree may require additional adjustments (Line 16) to handle the semantic structure. One operation of adjustments is removing redundant nodes. Internal nodes that have the same MBRs with their exclusive child are removed from the LOM tree. Another operation is to group existing groups and assign label for the new group, which constructs the semantic hierarchy of interface in the LOM tree. In the LOM tree, a label that is not grouped with a field is regarded as the label of a group.

4.3   Transformation of the LOM Tree into the Schema Tree

The LOM tree presents the hierarchical semantic information of interface by its hierarchical structure on MBRs of nodes. In the LOM tree, all semantic labels is in leaves, which is not satisfied the definition of the schema tree. So we use an additional operation to transform the SR-tree into the schema tree of interface. In the transformation, a field f and its own label l in the LOM tree is combined as a field $F_i = <L_i, f_i>$ of the schema tree. For a given group label l in the LOM tree, we move it to replace its parent node and all sibling nodes of it are becoming its children. The group label l becomes an internal node and fields of group are in its sub tree. Therefore, a group g_i in the LOM tree is transformed into a group G_i or an attribute A_i of the schema tree by assigning the label of g_i as GL, fields of g_i as F, and subgroups of g_i as Gs. Finally, after we remove unmatched labels and unnecessary internal nodes from the SR-tree, the schema tree with the hierarchical semantic structure of interface is generated. Figure 3 shows the schema tree of the interface in figure 1.

5   Experiments

5.1   Dataset

The web interfaces used in our experiments is the dataset ICQ and WISE. ICQ contains query interfaces to the data sources on the Web over five domains: airfare, automobile, book, job and real estate. For each domain, 20 query interfaces are collected from invisibleweb.com and yahoo.com. Moreover, providers of ICQ have manually transformed interfaces into schema trees. WISE consists of 147 web interfaces from seven domains: books, electronics, games, movies, music, toys and watches. All three datasets are downloadable from their web sites. To get the positions of labels and fields in web pages, we use a java open source library named *cobra* [16] which is a part of the *lobo browser* project.

5.2   Evaluations

We evaluate our approach according to different metrics. The metrics are:

Labeling fields: This measure is used to evaluate the accuracy of matching labels for fields in our algorithm. We compute the ratio of the number of correctly labeled fields to the total number of fields.

Labeling groups: This measure is used to evaluate the accuracy of matching labels for groups in the interface. We compute the ratio of the number of correctly labeled groups to the total number of groups.

Structure similarity of schema tree: This measure evaluates the quality of the schema tree which is extracted by our algorithms. We compute this metrics based on the structural tree edit distance which is the minimum number of operations to convert a tree into another. The precision of schema tree is defined as $P_s = (N_l\text{-}D_s)/N_l$, where N_l is the number of node in extracted tree and D_s is the tree edit distance. The recall of schema tree is $R_s=(N_l\text{-}D_s)/N_s$, where N_s is the number of nodes in standard schema tree. Therefore, the F-measure is defined as $F_s=2P_sR_s/(P_s + R_s)$ which evaluates the accuracy of structure of the extracted tree.

Figure 8 shows the results of our experimental study on the dataset ICQ and WISE. We achieve high accuracy on both datasets in three metrics. The results of structure measure are worse. It is leaded by the illustrative text in interfaces which interfere with the structure of LOM tree. Moreover, the open source code used in our experiments has some layout mistakes when parsing some special web pages. So the accuracy of our algorithms can be improved by programming. Figure 9 shows the comparison between our layout method (denoted as LOM) and the code structure method (denoted as DOM). Our LOM method achieves higher accuracy in these interfaces, especially for interfaces with the top-down layout feature and hierarchical structure of schema.

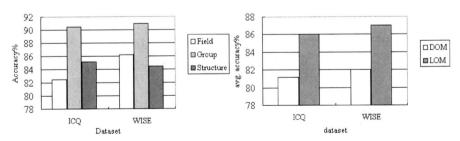

Fig. 8. Experimental results   **Fig. 9.** Comparison with CSM

6 Conclusion

In this paper, we present a novel approach for extracting the schema tree of web interface with the hierarchical semantic structure. Different from previous approaches, our approach regards the semantic layout of web interface as a layout tree structure. Based on the visual layout features of web interface, we design a set of rules for constructing the LOM tree to express semantic hierarchy. We regard each label and fields on interface as a MBR in web page layout, and construct the LOM tree from the DOM tree of interface according to rules. The LOM tree can be transformed into the schema tree of interface with high accuracy. We execute our experiments on a wide range of web interfaces with multiple domains. The experimental result shows that our approach extracts the schema of web interfaces in a higher accuracy. The future work is improving the accuracy of our algorithm by accurately generate the position of labels and fields.

Acknowledgments. This paper is supported by the National Natural Science Foundation of China under Grant No.60973021 and 61003060, the 863 High Technology Foundation of China under Grant No. 2009AA01Z131, the Fundamental Research Funds for the Central Universities under Grant No.N090104001.

References

1. Cope, J., Craswell, N., Hawking, D.: Automated discovery of query interfaces on the Web. In: The 14th Australasian Database Conference, Adelaide, pp. 181–189 (2003)
2. Wu, W., Yu, C., Doan, A., Meng, W.: An interactive clustering-based approach to integrating source query interfaces on the deep web. In: SIGMOD (2004)
3. Song, R., Liu, H., Wen, J.: Learning important models for Web page blocks based on layout and content analysis. SIGKDD Explorations 6(2), 14–23 (2004)
4. He, H., Meng, W., Yu, C.: Automatic Extraction of Web Search Interfaces for interface Schema Integration. In: Proc of the 13th International World Wide Web Conference, New York, pp. 414–415 (2004)
5. Hong, J., He, Z., Bell, D.A.: Extracting Web Query Interfaces Based on Form Structures and Semantic Similarity. In: Proc of the 2009 IEEE International Conference on Data Engineering, pp. 1259–1262. IEEE Computer Society, Washington, DC, USA (2009)
6. Zhang, Z., He, B., Chang, K.C.-c.: Understanding Web Query Interfaces: Best-Effort Parsing with Hidden Syntax. In: Proc. of the 2004 ACM SIGMOD International Conference on Management of Data, pp. 107–118. ACM, Paris (2004)
7. Nguyen, H., Nguyen, T., Freire, J.: Learning to Extract Form Labels. In: Proceedings of the VLDB Endowment, pp. 684–694. VLDB Endowment, Auckland (2008)
8. Dragut, E.C., Kabisch, T., Yu, C.: A Hierarchical Approach to Model Web Query Interfaces for Web Source Integration. In: Proc. of the VLDB Endowment, pp. 325–336. VLDB Endowment, Lyon (2009)
9. Wu, P., Wen, J.-R., Liu, H., Ma, W.-Y.: Query selection techniques for efficient crawling ofstructured web sources. In: ICDE (2006)
10. Barbosa, L., Freire, J.: An adaptive crawler for locating hidden web entry points. In: WWW (2007)

11. Wu, W., Yu, C., Doan, A., Meng, W.: An interactive clustering-based approach to integrating source query interfaces on the deep web. In: SIGMOD (2004)
12. He, B., Chang, K.C.-C., Han, J.: Discovering complex matchings across web query interfaces: a correlation mining approach. In: KDD (2004)
13. Dragut, E.C., Yu, C., Meng, W.: Meaningful labeling of integrated query interfaces. In: VLDB (2006)
14. Raghavan, S., Garcia-Molina, H.: Crawling the hidden web. In: VLDB (2001)
15. He, H., Meng, W., Yu, C., Wu, Z.: Constructing interface schemas for search interfaces of web databases. In: Ngu, A.H.H., Kitsuregawa, M., Neuhold, E.J., Chung, J.-Y., Sheng, Q.Z. (eds.) WISE 2005. LNCS, vol. 3806, pp. 29–42. Springer, Heidelberg (2005)
16. http://lobobrowser.org/cobra.jsp
17. Guttman, A.: R-trees: a dynamic index structure for spatial searching. In: Proceedings of the SIGMOD Conference, Boston, MA, pp. 47–57 (June 1984)

Tagging Image with Informative and Correlative Tags

Xiaoming Zhang[1], Heng Tao Shen[2], Zi Huang[2], and Zhoujun Li[1]

[1] School of computer, Beihang University, Beijing, 100191, China
yolixs@cse.buaa.edu.cn, lizj@buaa.edu.cn
[2] Information technology and electrical engineering, University of Queensland, Australia
{shenht,huang}@itee.uq.edu.au

Abstract. Automatic tagging can automatically label images with semantic tags to significantly facilitate multimedia search and organization. Existing tagging methods often use probabilistic or co-occurring tags, which may result in ambiguity and noise. In this paper, we propose a novel automatic tagging algorithm which tags a test image with an Informative and Correlative Tag (ICTag) set. The assigned ICTag set can provide a more precise description of the image by exploring both the information capability of individual tags and the tag-to-set correlation. Measures to effectively estimate the information capability of individual tags and the correlation between a tag and the candidate tag set are designed. To reduce the computational complexity, we also introduce a heuristic method to achieve efficient automatic tagging. The experiment results confirm the efficiency and effectiveness of our proposed algorithm.

Keywords: image tagging, tag information capability, tag correlation.

1 Introduction

Automatic image tagging has recently attracted much attention. However, it remains a challenging problem due to the unsatisfactory performance. Wu et al. [1] formulate image tagging as multi-label image annotation which is treated as a regression model with a regularized penalty. The approach in [6] performs a random walk over a tag similarity graph to refine the relevance scores. Wang et al. [3] select candidate tags and determine which tag is suitable using the visual content of the image based on random walk with restarts. In [5], for a given image, it's k nearest neighbours are retrieved by computing visual similarity of low-level features. With respect to their prior distribution among all images, tags that frequently appear among the nearest neighbours are considered relevant to the given image. However, those approaches often suggest the most common tags to users. In many cases, the most likely tag is also most general and least informative. The correlations between tags also aren't well exploited in these works.

In this paper, we propose a novel automatic tagging algorithm which assigns not the common tag set but the informative and correlative tag (ICTag) set. We propose a heuristic and iterative method to quickly discover the informative and correlative tag set. To effectively estimate the information capability of a tag, we utilize the change of posterior distributions of other tags after this tag is added to the candidate tag set.

X. Du et al. (Eds.): APWeb 2011, LNCS 6612, pp. 258–263, 2011.

We further use the relevance between the test image and the tag to filter out those tags that are "informative" but not relevant to the test image. To determine the correlation of a tag to the candidate tag set, we model the tag and the candidate tag set as two vectors in the semantic space which correlation can be measured by the cosine similarity.

2 Problem Formulation

In this paper, we use O to denote the test image, and D is used to represent an image in the training dataset. S_n is a tag set with n tags. The information which is given to O by S_n is define as $I(O, S_n)$, and the correlation of S_n is $C(O, S_n)$. We want to assign the most informative and correlative tag set to the test image:

$$\widehat{S}_n = \arg \max_{S_n \subset V} F(O, S_n)$$
$$= \arg \max_{S_n \subset V} ((1-\alpha) * I(O, S_n) + \alpha * C(O, S_n))$$

where V denotes the tag vocabulary of the training dataset, and α ($0<\alpha<1$) is a smoothing factor. We aim to maximize the objective function $F(O,S_n)$ which linearly combines the information capability and the correlation of the tag set.

Since the number of possible tag set S_n is exponential with respect to the size of the tag vocabulary. Thus we propose a heuristic and iterative algorithm to efficiently get an approximate solution. The information of S_n can be obtained from that of S_{n-1} added by the information which is given by the tag after it is added to S_{n-1}. Given a tag denoted as t, we have:

$$I(O, S_n) = I(O, S_{n-1}) + I(O, t)$$

where $S_n = S_{n-1} \cup \{t\}$ and $n>1$, and $I(O,t)$ denotes the information added to S_{n-1} by tag t, which is defined as the information capability of tag t.

Similarly, the correlation of S_n can be obtained from that of S_{n-1} added by the correlation of tag t to S_{n-1}, i.e.,

$$C(O, S_n) = C(O, S_{n-1}) + C(S_{n-1}, t)$$

where $C(S_{n-1},t)$ denotes the tag-to-set correlation from t to S_{n-1}.

Finally, the heuristic algorithm works as follows:

$$\hat{t} = \arg \max_{t \in V / S_{n-1}} F(O, t)$$
$$= \arg \max_{t \in V / S_{n-1}} ((1-\alpha) * I(O, t) + \alpha * C(S_{n-1}, t)) \tag{1}$$

The algorithm is initialized when $n=1$ with S_1 containing the most relevant tag to the test multimedia object without considering information capability and correlation since S_0 is nil. The total number of tags (i.e., the size of final tag set) assigned to the test multimedia object is often decided by users.

3 Tag Informative Capability

Based on the information theory, we assume that the less ambiguous a tag set is, the more information it has. Thus, we aim to tag a test image with a tag set which is least ambiguous. There are two scenarios where a tag set is ambiguous [2]. The first scenario is that the current tag set has more than one meaning. Examples of this ambiguity are word-sense ambiguity. For example, tag set {"apple", "photo"} can be a computer or fruit. Then we think that the tag set {"apple", "computer"} and {"apple", "fruit"} have are more informative. In this case, the tag "computer" or "fruit" adds more information to the tag set {"apple"} than other tags like "photo" because it can disambiguate the tag set. The second scenario is that the current tag set is not sufficiently specific. A good example of this ambiguous tag set is {"car"} since there are well-known examples of "TOYOTA" and "BMW" and others. Thus by adding the tag "TOYOTA" or "BMW", the tag set {"car", "BMW"} or {"car", "TOYOTA"} conveys more information than tag set {"car", "white"}.

The information capability of a tag indicates how much information it carries when it is added to the tag set and can be reflected from the ambiguity changes. Intuitively, a tag set is ambiguous if there are other tags such that adding one or more of them gives rise to very different co-occurring distributions over the rest tags. Then, the information capability of a tag can be estimated based on how differently the posterior distributions of other tags are changed after this tag is added to the tag set.

We compute the deviation between two posterior distributions using the KL divergence. The level of ambiguity for the tag set S_{n-1} can be measured by the KL divergence value to indicate that at what degree the tag set can be disambiguated by the added tag t. The greater the KL divergence value is, the greater the information capability tag t has. We aim to select a tag that gives the maximal information to S_{n-1}. Formally, the estimation of the information capability of tag t with respect to S_{n-1} is defined as:

$$I(O,t) \propto f(KL(P(V \mid S_{n-1} \cup \{t\}) \parallel P(V \mid S_{n-1})))$$
$$= f(KL(P(V \mid S_n) \parallel P(V \mid S_{n-1}))) \tag{2}$$

where $f(.)$ is a monotonically-increasing function that trades off the impact of the KL divergence with other factor. The calculation of KL(.) is defined as:

$$KL(P(V \mid S_n) \parallel P(V \mid S_{n-1})) = \sum_{t' \in V} P(t' \mid S_n) \log \frac{P(t' \mid S_n)}{P(t' \mid S_{n-1})} \tag{3}$$

In the following equations, we assume that the tags in S are conditional independent with each other, and use Bayes' rule to estimate $P(t/S)$:

$$P(t' \mid S) = \frac{P(S \mid t')P(t')}{P(S)} = \frac{P(t') \prod_{t_a \in S} P(t_a \mid t')}{\sum_{t_b \in V} P(t_b) \prod_{t_a \in S} P(t_a \mid t_b)}. \tag{4}$$

In the following, for any pair of tags t_a and t_b, we denote the number of images that contains t_a and t_b by $Z(t_a, t_b.)$. We have following formulation:

$$P(t'|t_a) = \frac{Z(t_a, t')}{\sum_{t_b \in V} Z(t_a, t_b)}, \quad P(t') = \frac{\sum_{t_a \in V} Z(t', t_a)}{\sum_{t_a \in V, t_b \in V} Z(t_a, t_b)}$$

In the above formula (2), we measure the information capability of a candidate tag based on the KL divergence of two posterior distributions only. However, it doesn't consider the relevance between the test image O and tag t. So, it is possible to find a tag with the maximal "information capability", but irrelevant to the test image. To solve this problem, we introduce the conditional probability $P(t/O)$ of the test image O to generate tag t. With this constraint, the tag which is unrelated to the test image O can be filtered out. Then, formula (2) is rewritten as:

$$I(O,t) \propto f((KL(P(V|S_n) \| P(V|S_{n-1}))) * P(t|O) \tag{5}$$

Assuming that $P(O)$ is uniformly distributed, we have

$$I(O,t) \propto f((KL(P(V|S_n) \| P(V|S_{n-1}))) * P(t,O) \tag{6}$$

$$P(t,O) \approx \frac{\sum_{i=1}^{k} L(t,D_i) Sim_{visual}(O,D_i)}{k}$$

where $Sim_{visual}(O,D_i)$ is the function to measure the visual similarity, $L(t, D_i)$ is an indicator function for tag $t \in V$, $L(t, D_i) = 1$ iff D_i was tagged with tag t. $L(t, D_i) = 0$ otherwise.

4   Tag-to-Set Correlation

In fact, tags do not appear only in pairs. To simplify our computation, we have assumed that conditional co-occurrences are independent in the estimation of tag information capability in Section 4. However, this method may induce deviation because it does not consider the correlation of tag set. The $C(S_{n-1},t)$ defined in Section 3 is proposed to measure the correlation between the candidate tag set S_{n-1} and tag t. This measure is used to select the tags that are correlative to the tag set S_{n-1} instead of individual tags. We take the entire candidate tag set as a single semantic concept S_c, and a concept combination process is use to combine all the tags in the candidate tag set to get the concept S_c. An important intuition in concept combination is that the important concept can dominate other concepts. Then the correlation is calculated as:

$$C(S_{n-1},t) = Sim_{text}(S_c,t)$$

To get the vector of combined concept S_c, we need to construct the vector of each tag first. The tag-to-object matrix which reflects the relationship between tags and images is described as $M_{|V| \times |D|}$, M_{ij} denotes whether the ith tag occurs in the tagging of the jth media object. Then, the semantic correlation between different tags:

$$U = MM^T. \tag{7}$$

$$U_{i,j} = \frac{U_{i,j}}{U_{i,i} + U_{j,j} - U_{i,j}}.$$

where $U_{i,j}$ denotes the co-occurrence frequency of tag t_i and tag t_j. Then each row of U is normalized. We use a heuristic concept combination process to construct the vector of the combination concept S_c. The concept combination is a form of vector addition whereby quality neighbour shared by both tags are emphasized. Given two tags:

$$\vec{t_i} = <v_{i1}, v_{i2}, ..., v_{i|T|}> = <w_{ti} * U_{ti,1}, w_{ti} * U_{ti,2}, ..., w_{ti} * U_{ti,|T|}>,$$

$$\vec{t_j} = <v_{j1}, v_{j2}, ..., v_{j2}> = <w_{tj} * U_{tj,1}, w_{tj} * U_{tj,2}, ..., w_{tj} * U_{tj,|T|}>.$$

$$W_t = \frac{|D(t)|}{\log(|T(t)|+1)} * P(t, O)$$

where $|D(t)|$ denotes the number of images that contain tag t in the K nearest neighbours of the test image O, $|T(t)|$ denotes the number of images that contain tag t in the training dataset. The resulting combined concept is denoted by $t_i \cup t_j$:

$$v_{ti \cup tj, k} = \begin{cases} \frac{\max(w_{ti}, w_{tj})}{w_{tj}} * (v_{ik} + v_{jk}) & v_{ik} \neq 0 \text{ and } v_{ik} \neq 0 \\ v_{ik} + v_{jk} & else \end{cases}$$

Then, the vector of the combined concept is normalized. Thus the combined concept S_c is a combination of $n-1$ "neighborhood" vectors, i.e., $(...((t_1 \cup t_2) \cup t_3) \cup ....) \cup t_{n-1})$. Therefore, the cosine of $\vec{U_t}$ and $\vec{S_c}$ is applied to measure the semantic similarity of tag set concept S_c and tag t:

$$Sim_{text}(S_c, t_i) = \frac{\vec{S_c} \bullet \vec{U_t}}{|\vec{S_c}| \times |\vec{U_t}|}$$

5 Experiments

In these experiments, we compare our approach with Neighbour Voting based image tagging (NVTag) [6] and Random Walk based image tagging (RWTag) in [8]. We use We split the dataset of image into two disjoint partitions: a training set with 150000 images and a test set with the remaining images. Traditional performance evaluation metrics including average precision and recall.

In Fig. 4 and Fig. 5, we compare the average precision and recall at top n result tags of different tagging algorithms. According to these figures, ICTag consistently outperforms both RWTag and NVTag. This is because that NVTag uses the frequency of a tag minus its prior frequency to restrain the tags, and then the common tags usually are assigned. This is the same to RWTag in which the tags has the most correlation to other tags are assigned to the test image. Thus both the algorithms always prefer the common tags. However, the ICTag prefers the informative tags when the number of assigned tags is increased. With the number of assigned tags increasing the improvement is more obviously.

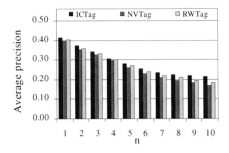 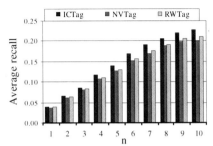

Fig. 4. Average precision at top n result tags   **Fig. 5.** Average recall at top n result tags

6 Conclusion

In this paper, we propose a novel automatic tagging algorithm which assigns the test multimedia object with the informative and correlative tags, based on the posterior probability computation using KL divergence and tag-to-set correlation analysis. The relevance of a tag to the test multimedia object is also considered in estimating the information capability. Measures to effectively estimate the information capability of individual tags and the correlation between a tag and the candidate tag set are introduced. Experiments on the existing datasets of NUS-WIDE web images show that our algorithm achieved superior tagging accuracy than existing algorithms.

References

1. Wu, F., Han, Y.H., Tian, Q., Zhuang, Y.T.: Multi-label Boosting for Image Annotation by Structural Grouping Sparsity. In: Proceedings of 18th ACM International conference on Multimedia, pp. 15–24 (2010)
2. Kilian, Q., Malcolm, S., Roelof, Z.: Resolving Tag Ambiguity. In: Proceeding of the 16th ACM International conference on Multimedia, pp. 111–120 (2008)
3. Wang, C., Jing, F., Zhang, L., Zhang, H.-J.: Image annotation refinement using random walk with restarts. In: Proceedings of 14th ACM International Conference on Multimedia, pp. 647–650 (2006)
4. Zhou, X., Wang, M., Zhang, Q., Zhang, J., Shi, B.: Automatic image annotation by an iterative approach: Incorporating keyword correlations and region matching. In: Proceedings of the 6th ACM International Conference on Image and Video Retrieval, pp. 25–32 (2007)
5. Li, X., Snoek, C.G., Worring, M.: Learning social tag relevance by neighbor voting. IEEE Transaction on Multimedia 11, 1310–1322 (2009)
6. Liu, D., Wang, M., Hua, X.S., Zhang, H.J.: Tag Ranking. In: Proceeding of the 18th ACM International Conference on World Wide Web, pp. 351–340 (2009)

Let Other Users Help You Find Answers: A Collaborative Question-Answering Method with Continuous Markov Chain Model

Xiaolong Wang and Ming Zhang

Institute of Network Computing
School of EECS, Peking University, Beijing, 100871, China
{dragonxlwang,mzhang_cs}@pku.edu.cn

Abstract. The answering communities, such as Yahoo! Answers, offer great intelligence to help people solve questions. Participants can express their judgements towards answers and the system also keeps a record for every user. Retrieving Question-Answer pairs (QA pairs) extracted from these forums can improve the quality of Question-Answering (QA) systems. In this paper, we propose a Collaborative Ranking (ColRank) algorithm employing the Continuous Markov Chain Model (CMCM) to combine the quality of QA pairs and relationships among them. Empirical results show that the innovative algorithm is effective and outperform the state of art Question-Answering baselines.

Keywords: Question Answering, Collaborative Ranking, Continuous Markov Model.

1 Introduction

Question Answering (QA) systems try to understand the human-proposed questions and come up with acceptable answers. A lot of interests are drawn to this topic and people expect to see further improvement. Approaches based on setting inference rules[1], applying ontologies[2,3] or introducing statistical models[4,5,6,7] were developed to address the problem. Nevertheless, most approaches suffer certain limitation resulting from huge human workload for annotating or expensive computational costs for supervised learning.

However, answering communities seem to be much better resources as compared with the pure text documents. People post and answer questions freely according to their interests and expertise. Yahoo! Answers, Baidu Zhidao (Chinese) and Naveer (Korean) are just some examples in this case. Most communities have human-involved grading systems to rate answers. Judgements from users facility us to assess the quality of answers. People interact by asking, answering, grading, etc. Users participation activities can be tracked through log records and users profiles delicately maintained by web service providers. In this paper, we utilize multiple collaborative information to identify QA pairs that are most useful to people.

X. Du et al. (Eds.): APWeb 2011, LNCS 6612, pp. 264–270, 2011.

Link-based analysis has been a popular approach to incorporate collaborative influence. *Page rank*[8] and *Hits*[9] are examples in case. To our best knowledge, the closest related link-based algorithm are *LexRank*[10] and *Biased LexRank*[11]. Most previous ranking models incorporate strength of relationships into evaluation, but fail to take the quality of nodes themselves into consideration. To meet this end, we propose the *Collaborative Rank Algorithm* (ColRank).

Within this context we make the following contributions:

- Our adapted *ColRank Algorithm* promotes the performance, as compared to the baselines *Biased LexRank* and *Collaborative LexRank*.
- We demonstrate that the collaborative information including user profile and user judgement towards answers are crucial to the performance of automatic QA systems.
- Our approach using Continuous Markov Chain (CMCM) Model successfully combine the QA pairs quality with the relationships among them to generate the rank of QA pairs.

The rest of this paper is organized as follows: Section 2 introduces some related work. The axiomatic formulation of Continuous Markov Chain Model we consider is detailed in Section 3. In section 4, we list the features used for the model and gives a description about the graph construction, and parameters configuration. We report our experimental setup and results in Section 5. Finally, Section 6 concludes with outlook on future research.

2   Model

2.1   Moving from Discreteness to Continuity

In *Biased LexRank*[10,11], is the first variant that aim to address the problem of Question Answering. It has the mathematical motivation of computing the stationary distribution in DMCM. In *Biased LexRank* (B-LR), the graph is constructed where each node represents a QA pair and edge is weighted by text similarity between two QA pairs. The order of stationary distribution for QA pairs is used as the rank to output. The basic formula is

$$\mathrm{BLR}(u|q) = d \cdot \frac{\mathrm{rel}(u|q)}{\sum\limits_{z \in C} \mathrm{rel}(z|q)} + (1-d) \cdot \sum\limits_{v \in adj[u]} \frac{\mathrm{sim}(v,u)}{\sum\limits_{z \in adj[v]} \mathrm{sim}(v,z)} \cdot \mathrm{BLR}(v|q) \quad (1)$$

where d is the damper factor determining the probability of restarting the random walk. For the link weight, they use cosine similarity computed through tf and idf scores to represent the strength of support between two QA pairs.[12]. Without using any non-textual clues, we believe that more collaborative features that are helpful for the problem we face are left out in. This idea finally inspires us to move from DMCM to CMCM.

2.2   Continuous Markov Chain Model

The CMCM cancels the limitation that jumping from one node to another can only occur at discrete time points. it enable the random walker to stay at a node for arbitrary time period. The transition probability matrix $P(t) = [p_{ij}(t)]_{n \times n}$ describes the transition property of CMCM. We further denote by Q the derivative of $P(t)$ with respect to t when t approaches to 0, which we call the **transition rate matrix**. It is proved to be a one-to-one correspondence between Q and $P(t)$ under previous assumptions. We decompose the CMCM as a process of successive staying at different nodes. Since the duration behavior are determined by $\{q_{ii}\}$, we still need to find ways to identify the transition behavior. We then introduce the Skeleton Process (a.k.a Embedded Markov Chain) to disclose the mechanism of transition in CMCM. Skeleton Process K corresponding to Q is defined as a Discrete Markov Chain Model (DMCM) where we assume any staying period is uniformly one time unit. Suppose we have already solved the stationary distribution $\widetilde{\pi}$ for the skeleton process K via approaches like Power Method[13]. we can prove that the stationary π distribution of the CMCM:

$$\pi = \frac{\widetilde{\pi} \cdot diag(\frac{1}{q_{11}}, \frac{1}{q_{22}}, \cdots, \frac{1}{q_{nn}})}{\| \widetilde{\pi} \cdot diag(\frac{1}{q_{11}}, \frac{1}{q_{22}}, \cdots, \frac{1}{q_{nn}}) \|} \tag{2}$$

3   Collaborative Rank

Previous work[14] noticed that there are numerous features that could be used in question answering and tried to select the most effective ones to fit in the model. The various features available usually have a great deal of redundancy. Generally speaking, we can group them into three categories and select as least of them as possible. Although here we take Yahoo! Answers as an example, the features selected are common in other answering communities as well.

– **Answer Quality:** Answering Community offers ways for users to vote for answers, which reflect their judgement of the answers quality. Common measurements include *thumb up, thumb down and stars*. We use a form of linear combination to compute the quality of answer A:

$$Q(A) = \lambda \cdot Star + (1 - \lambda) \cdot (Up - Down); \lambda \in [0, 1] \tag{3}$$

– **User Prestige:** To discriminate answers provided by an expert and an abuse answerer? we extract the *count of best answers* and the *total points* the answerer received from his/her profile. To evaluate the prestige of user U, it follows:

$$P(U) = \eta \cdot BestAnsCnt + (1 - \eta) \cdot TotalPt; \eta \in [0, 1] \tag{4}$$

– **Who-answered-who relationship:** *who-answered-who* (WAW) relationship has the potential to enhance our ability to suggest good answers.

Experiments showed that this kind of relationship consulting the question-answering history improves performance greatly. We now formalized this idea with the Equation (5):

$$R(A_{er}, Q_{er}) = \begin{cases} 1 \text{ the answerer } A_{er} \text{ once answered} \\ \quad \text{a question proposed by } Q_{er}, \\ 0 \text{ } otherwise; \end{cases} \quad (5)$$

In CMCM, we consider the graph $G = (V, E)$ where V represents the set of nodes and E denotes the edge set, as shown in Figure (1). It is a three layer graph. The first layer contains all the askers. The second layer denotes QA pairs and the third layer represents the answerers set.

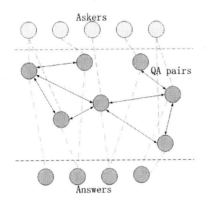

Fig. 1. Collaborative Rank

For the parameters k_{ij}, the transition attributes should incorporate the *who-answered-who* relationship and text similarity. The later part is added for purpose of smoothing because the *who-answered-who* relation score distributes sparsely. In our model, we have:

$$k_{ij} = \frac{\mu R(A_j, Q_i) + (1 - \mu)\text{sim}(\langle Q_i, A_i \rangle, \langle Q_j, A_j \rangle)}{\sum_k (\mu R(A_k, Q_i) + (1 - \mu)\text{sim}(\langle Q_i, A_i \rangle, \langle Q_k, A_k \rangle))}; \ \mu \in [0, 1] \quad (6)$$

The staying time assesses the quality of nodes. We proposed the estimation:

$$q_{ii} = -\frac{[\theta Q(A_i) + (1 - \theta)P(A_{er\,i})]^{-1}}{\sum_j [\theta Q(A_j) + (1 - \theta)P(A_{er\,j})]^{-1}}; \ \theta \in [0, 1] \quad (7)$$

4   Experiments

We collected QA pairs from Yahoo! Answers from Jun. 2010 to Sep. 2010. The dataset included 287,079 QA pairs in domain of "Computers & Internet", with

Table 1. Performance of B-LR, C-LR and ColRank When Ranking the Top 10 Answers

	B-LR	C-LR	ColRank
Precision	0.702	0.751	0.794
Recall	0.674	0.685	0.719
F_1	0.688	0.711	0.754
MRR	0.507	0.552	0.593
Improvement compared with B-LR:			
F_1	0%	3.3%	9.6%
MRR	0%	8.8%	16.9%
Improvement compared with C-LR:			
F_1	-	0%	6.1%
MRR	-	0%	7.4%

175,037 users considered. We perform our algorithm ColRank and the baselines. To measure the effectiveness, we compare the performance over two prevalent information retrieval metrics[15,16], which are F_1 ($F_1 = \frac{2PR}{P+R}$) and MRR (Mean Reciprocal Rank)

A natural competitor of our model is *Biased LexRank* (B-LR)[11]. We also derived it with a variant *Collaborative LexRank*, or C-LR for short, which values the WAW relationship as well.

We carried out evaluation on the B-LR, C-LR and our ColRank together and make comparisons. The results are listed in Table (1).

The performance of C-LR is higher compared to the baseline B-LR, gaining an increase 3.3% in F_1 and 8.8% in MRR. This demonstrates that the feature of WAW relationship is useful when ranking the answer candidates. In addition, adopting Continuous Markov Chain Model in ColRank offers further outstanding improvement. Compared with C-LR, an increase of 6.1% in F_1 and 7.4% in MMR is observed. The effectiveness of the ColRank is reasonable. Firstly, it explicitly integrate the quality of QA pairs as well as other features that measure the strength of links. In addition, our model makes it possible to consider multiple non-textual features at the same time. Through experiments, we see the non-textual features bringing about considerable benefit to the performance.

5   Conclusion and Future Work

In this paper, we propose a novel model based on Continuous Markov Chain Model for ranking in question answering. Unlike other approaches such as Biased LexRank, it can naturally incorporate the weight of nodes and link together. When ranking answer candidates, we adopt multiple non-textual features that possess potential to capture the user quality and user feedback. We further demonstrate the effectiveness of these features through experiments with various measurements. The experimental results also reveal that our proposed method outperforms Bisaed LexRank and Collaborative LexRank by about 16% and 7% in terms of MRR.

Our approach integrate textual and non-textual information in a unified framework. They are not limited to dataset from any specific answering community. An interesting future direction is to establish the "gold metrics" to address the problem of systematically feature selection. The goal is to extract features sufficiently generalized and effective that can be integrated to models like our ColRank.

Acknowledgement

This study is partially supported by the National High Technology Research and Development Program of China (863 Program No. 2009AA01Z143), the Specialized Research Fund for the Doctoral Program of Higher Education of China and Special Fund for Fast Sharing of Science Paper in Net Era by CSTD ("FSSP" Grant No. 20100001110203). We would also like to thank members of the Dlib group at Peking University and in particular Ziqi Wang and Yu He for their valuable suggestions and review of this paper.

References

1. Lin, D., Pantel, P.: Discovery of inference rules for question-answering. Natural Language Engineering 7(4), 734–749 (2001)
2. Hermjakob, U.: Parsing and question classification for question answering. In: Proceedings of the Workshop on Open-Domain Question Answering, vol. 12 (2001)
3. Lopez, V., Pasin, M., Motta, E.: AquaLog: An ontology-portable question answering system for the semantic web. In: Gómez-Pérez, A., Euzenat, J. (eds.) ESWC 2005. LNCS, vol. 3532, pp. 546–562. Springer, Heidelberg (2005)
4. Clarke, C.L.A., Cormack, G.V., Kemkes, G., Laszlo, M., Lynam, T.R., Terra, E.L., Tilker, P.L.: Statistical selection of exact answers. In: MultiText Experiments for TREC 2002 (2002)
5. Brill, E., Dumais, S., Banko, M.: An analysis of the askmsr question-answering system. In: EMNLP 2002: Proceedings of the ACL 2002 Conference on Empirical Methods in Natural Language Processing, vol. 10, pp. 291–298 (2002)
6. Paranjpe, D., Ramakrishnan, G., Srinivasan, S.: Passage scoring for question answering via bayesian inference on lexical relations. In: The Twelfth Text REtrieval Conference, pp. 305–310 (2003)
7. Ding, S., Cong, G., Lin, C.-Y., Zhu, X.: Using conditional random fields to extract contexts and answers of questions from online forums. In: Proceedings of the 46th Annual Meeting of the Association for Computational Linguistics, pp. 710–718 (2008)
8. Brin, S., Page, L.: The anatomy of a large-scale hypertextual web search engine. Computer Networks 30(1-7), 107–117 (1998)
9. Kleinberg, J.M.: Authoritative sources in a hyperlinked environment. Journal of the ACM 46(5), 604–632 (1999)
10. Erkan, G., Radev, D.R.: Lexpagerank: Prestige in multi-document text summarization. In: Proceedings of the 2004 Conference on Empirical Methods in Natural Language Processing, pp. 365–371 (2004)

11. Otterbacher, J., Erkan, G., Radev, D.R.: Biased lexrank: Passage retrieval using random walks with question-based priors. Information Processing and Management 45, 42–54 (2009)
12. Salton, G., Wong, A., Yang, C.S.: A vector space model for automatic indexing. Communications of the ACM 18, 613–620 (1975)
13. Ilse, I., Wills, R.M.: Analysis and computation of google's pagerank. In: 7th IMACS International Symposium on Iterative Methods in Scientific Computing
14. Jeon, J., Croft, W.B., Lee, J.H., Park, S.: A framework to predict the quality of answers with non-textual features. In: Proceedings of the 29th Annual International ACM SIGIR Conference on Research and Development in Information Retrieval, pp. 228–235 (2006)
15. Voorhees, E.: Overview of the trec 2001 question answering track. In Proceedings of the 10th Text Retrieval Conference (TREC 2010), pp. 157–165 (2001)
16. Li, X., Wang, Y.-Y., Acero, A.: Learning query intent from regularized click graphs. In: Proceedings of the 31st Annual International ACM SIGIR Conference on Research and Development in Information Retrieval, pp. 339–346 (2008)

Hybrid Index Structures for Temporal-Textual Web Search

Peiquan Jin, Hong Chen, Sheng Lin, Xujian Zhao, and Lihua Yue

School of Computer Science and Technology,
University of Science and Technology of China, 230027, Hefei, China
jpq@ustc.edu.cn

Abstract. Most Web pages contain temporal information. However, most of previous studies only consider the update time of Web pages rather than fully exploit different temporal features in Web. In this paper, we propose a novel approach to fusing different temporal features in Web pages to build an efficient index structure for temporal-textual Web search. Specially, we focus on *update time* and *content time*, and propose to use a hybrid index structure to organize textual keywords, update time, and content time. In particular, we study three mechanisms to implement a hybrid index structure for temporal-textual Web search: (1) first inverted file then MAP21-tree and B+-tree, (2) first inverted file then MAP21-tree, (3) expanded inverted file. We conduct experiments on a real dataset to evaluate the performance of those hybrid index structures. The experimental results show that the *first inverted file then MAP21-tree* index structure has the best query performance.

1 Introduction

Recently, time has been a focus in the area of Web information extraction and Web search [1]. However, most of the studies only consider the update time of Web pages rather than fully exploit different temporal features in Web. Traditional commercial search engines, such as Google, Bing, and Baidu, only support the crawled dates of Web pages, i.e., users can only query Web pages towards their creation dates in database. To our knowledge, there are few search systems considering the temporal information embedded in Web pages [2].

The current temporal text indexing is mainly towards the versioned document collections such as Web archives [3, 4]. There have been some indexing approaches on directly addressing the issue of temporal-textual indexing. Anick and Flynn [5] have pioneered this research to support versioning in a full-text index on bitmaps for terms in current versions, and delta change records to track incremental changes to the index backward over time. The disadvantage is the costly recreation of previous states. Recent work in [6] and its earlier proposals [14-17] concentrate on the problem of supporting text-containment queries and neglect the relevance scoring of results. Stack [7] reports practical experiences made when adapting the open source search engine *Nutch* to search Web archives. Weikum et al. address the temporal dimensions completely by extending the inverted files index to make it ready for temporal search and implement the time-travel text search in the FluxCapacitor prototype [8, 9].

X. Du et al. (Eds.): APWeb 2011, LNCS 6612, pp. 271–277, 2011.

In contrast, research in temporal databases has produced several index structures tailored for time-evolving databases. A comprehensive overview of the state-of-art is available in [10]. Unlike the inverted file index, their applicability to text search is not well understood.

The index structure of most related work about temporal-textual indexing is usually based on inverted file index [11]. However, the main difference between our work and previous researches is that we consider to index both update time and content time for Web pages, while previous temporal-textual indexes are focused on indexing update time, because they are designed for Web archive system or document versioning.

In this paper, we propose a novel approach to fusing different temporal features in Web pages, i.e., update time and content time, to build an efficient index structure for temporal-textual Web search. Our basic idea is to develop an efficient hybrid index structure to cope with temporal-textual queries, which makes an integration of traditional temporal index and textual index. The most famous textual index is the inverted file structure, so in this paper we use this structure as the basic textual index structure. For temporal index, we adopt the MAP21-Tree [12], which is an efficient temporal index structure in temporal database area. However, there are many choices when integrating inverted file with MAP21-Tree, and in some case we need to introduce B+-Tree as the index structure for update time. Hence, we aims at making a comparison study on those different integration mechanisms, and finally find the best hybrid index structure which has the best performance for temporal-textual queries.

2 Index Structures for Temporal-Textual Web Search

We aim at building hybrid index structures to integrate text keywords and temporal information of Web pages for temporal-textual Web search. We adopt the *inverted file* as the basic index structure for text keywords in Web pages, due to its widely use in document search. The temporal information contains update time and primary time, in which the update time is regarded as a time instant and the primary time is modeled as a time period. Primary time refers to the most appropriate content time associated with a Web page. The time granularity is set to day. As the update time is a time instant, we can use B+-tree to organize them or directly put them into inverted files. For the primary time, we adopt the MAP21-tree as the basic index structure. MAP21-tree is designed towards time period and has better performance than other temporal indexes such as R-tree [13].

Basically, there are five hybrid methods when considering B+-tree, inverted file, and MAP21-tree to index temporal information and text keywords together, which are:

(a) *inverted file, B+-tree and MAP-21 triple index.*
(b) *expanded inverted file.*
(c) *first inverted file then MAP21-tree and B+-tree.*
(d) *first inverted file then MAP21-tree.*
(e) *first MAP21-tree then inverted file.*

However, the (a) and (e) methods are not suitable in temporal-textual Web search. The method (a) has to build multiple index files, which will cost more storage spaces and result in poor search performance. The method (e) performs poorly when no time predicates are given in the query. So in this paper we focus on the remaining three methods. In additional, the method (b), namely expanded inverted file, can be considered as the representative of the previous work on building temporal index, because most existing work about temporal text indexing are focused on indexing update time and keywords together and typically use a expanded inverted file [8, 9]. Figure 1 to 2 illustrate the index structures corresponding to (b), (c), and (d), in which the symbol UT refers to update time, while PT refers to primary time.

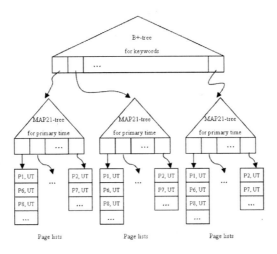

Fig. 1. Illustration of first inverted file then MAP21-tree

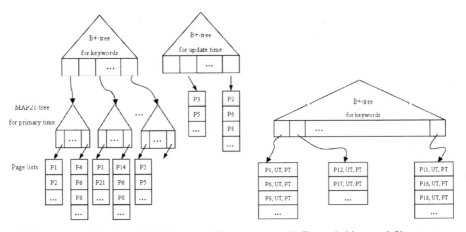

(a) First inverted file then MAP21-tree and B+-tree   (b) Expanded inverted file structure

Fig. 2. Illustration of the other two index structures

3 Experiments

In this section, we will implemented the three hybrid index structures and make comparison experiment to show the different performance of those index structures under different workloads. For simplicity, we use the following notation to represent the three hybrid index structures discussed in the previous section:

S_1: first inverted file then MAP21-tree and B+-tree.
S_2: first inverted file then MAP21-tree.
S_3: expanded inverted file.

We mainly evaluate the query performance of the three index structures. In order to get a comprehensive result, we use five types of queries in the experiment, which are:

Query 1: keywords only query
Query 2: keywords + update time instant
Query 3: keywords + primary time instant
Query 4: keywords + update time instant + primary time instant
Query 5: keywords + update time period + primary time period

Among those queries, Query 1 to 3 are partial queries, and Query 4 and 5 are complete temporal-textual queries.

In our experiments, we focus on the index size, query time, and rebuilt time of each index structure under given workloads and temporal-textual queries. Both query time and rebuilt time include disk I/O time and memory operation time.

3.1 Settings and Dataset

We choose the real dataset from the corpus provided by the SouGou lab (http://www.sogou.com/labs/) which records games, sports, IT, domestic and international news in May 2008 from some news sites.

In this experiment, we simply describe how to exact the update time and the primary time in one real Web page. In news Web pages, publish time is usually regarded as update time, and primary time is often appears in the first paragraph. The exaction of primary time is not the focus of this paper, so here we conduct a simple algorithm to extract primary time from Web pages. According to the simple algorithm, we always select primary time from the first paragraph of a Web page. If there is only time word in the first paragraph, we consider it as primary time. Otherwise, if there are two or more time words in the first paragraph, we choose the nearest time to the update time of the Web page as the primary time.

Keywords in Web page are exacted by a tool called ICTCLAS (http://ictclas.org/), which is the most efficient tool for the Chinese words segmentation. Each word is mapped a value in memory by ELFHASH function.

We use 250 thousand news Web pages as our real dataset and extract approximately 210 thousand different keywords. We run our experiment in a computer with an Intel Core 2.00 GHz CPU, 2 GB RAM, using Microsoft Window 7.

3.2 Comparison of Three Hybrid Index Structures

Firstly, we compare the index size (Mbytes) of three hybrid index structures. In our experiment, S1 has the smallest size, while S3 costs largest storage. The detailed size of S_1, S_2, and S_3 are 1346.29 Mbytes, 1425.14 Mbytes, and 1528.07 Mbytes, as shown in Table 1. Generally, we have the following result: $size(S_1) < size(S_2) < size(S_3)$, and our experimental result has also indicated this truth.

Secondly, we measure the rebuilt time to compare the creation time of the three structures, as the creation time of index is crucial in Web search. In this experiment, we first drop the original index structure, and then insert all extracted items, which are formed as <URL, keywords set, update time, primary time>, into the index structure, and compute the insertion time for each index structure. The rebuilt time of S_1, S_2, and S_3 needs 29235.10, 31310.14, and 14170.90 seconds respectively, as shown in Table 1. Since S_1 and S_2 have to construct MAP21-trees to maintain primary time, while S_3 only needs to create a single B+-tree, both S_1 and S_1 consume much more time than S_3.

Table 1. Index sizes and rebuilt time for three index structures

	Index Size (MBytes)	Rebuilt Time (s)
S_1	1346.29	29235.10
S_2	1425.14	31310.14
S_3	1528.07	14170.90

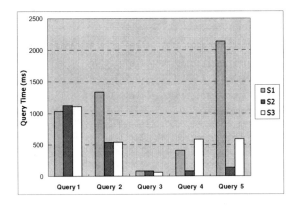

Fig. 3. Query time of the three index structures

Thirdly, we consider the query time of the three structures. Figure 3 shows the result of query time experiment. The query time includes three parts, i.e., looking-up on corresponding trees, reading page lists from disk, and merging page lists. The query time is calculated from query input to the output of the URLs results. Five types of queries are used in this experiment, which are denoted as Query 1 to Query 5, as described in the beginning of this section. We execute 300 queries and get the average run time as the experimental result of query time. Each query contains three keywords, each of which is a common keyword in the dataset, as indicated in the search log provided by the SouGou lab (http://www.sogou.com/labs/). Each query also contains one update time and one primary time, which are randomly generated but falls into a certain range of time.

According to our experiment, S_2 has the best query performance with respect to the complete temporal-textual queries, i.e., queries involving keywords, primary time, and update time. It only costs about 14% to 20% of time to complete such queries, compared with S_1 and S_3. Besides, S_2 has a comparable query performance when performing partial temporal-textual queries. Thus, in general S_2 has shown best performance in the measurement of query time.

4 Conclusions

In this work we have designed and implemented three hybrid index structures for temporal-textual Web search and studied the performance of these index structures. We conduct an experiment on a real dataset, and use five temporal-textual query types to evaluate the index size, query time, and rebuilt time of each hybrid index structure. The experimental results show that the index structure *"first inverted file then MAP21-tree"* has the relative better query performance among the three index structures, whereas it has a comparable index size and rebuilt time with the other two competitors. In conclusion, the index structure *"first inverted file then MAP21-tree"* may be an acceptable choice for temporal-textual Web search engines.

Acknowledgements

This work is supported by the National Science Foundation of China (no. 60776801 and 70803001), the Open Projects Program of National Laboratory of Pattern Recognition (20090029), the Key Laboratory of Advanced Information Science and Network Technology of Beijing (xdxx1005), and the USTC Youth Innovation Foundation.

References

1. Alonso, O., Gertz, M., Yates, R.B.: On the value of temporal information in information retrieval. In: Proc. of SIGIR 2007, pp. 35–41 (2007)
2. Deniz, E., Chris, F., Terence, J.P.: Chronica: a temporal Web search engine. In: Proc. Of ICWE 2006, pp. 119–120 (2006)
3. Herscovici, M., Lempel, R., Yogev, S.: Efficient Indexing of Versioned Document Sequences. In: Amati, G., Carpineto, C., Romano, G. (eds.) ECiR 2007. LNCS, vol. 4425, pp. 76–87. Springer, Heidelberg (2007)
4. Grandi, F.: Introducing an Annotated Bibliography on Temporal and Evolution Aspects in the World Wide Web. SIGMOD Record 33(2), 84–86 (2004)
5. Anick, P.G., Flynn, R.A.: Versioning a Full-Text Information Retrieval System. In: Proc. of SIGIR (1992)
6. Nørvåg, K., Nybø, A.O.N.: DyST: Dynamic and Scalable Temporal Text Indexing. In: Proc. of TIME (2006)
7. Stack, M.: Full Text Search of Web Archive Collections. In: Proc. of IWAW (2006)

8. Berberich, K., Bedathur, S.J., Neumann, T., Weikum, G.: FluxCapacitor: Efficient Time-Travel Text Search. In: Proc. Of VLDB, pp. 1414–1417 (2007)
9. Berberich, K., Bedathur, S.J., Neumann, T., Weikum, G.: A Time Machine for Text Search. In: Proc. Of SIGIR, pp. 519–526 (2007)
10. Salzberg, B., Tsotras, V.J.: Comparison of Access Methods for Time-Evolving Data. ACM Comput. Surv. 31(2), 158–221 (1999)
11. Zobel, J., Moffat, A.: Inverted Files for Text Search Engines. ACM Comput. Surv. 38(2), 6 (2006)
12. Nascimento, M., Dunham, M.: Indexing Valid Time Databases via B+-Trees. IEEE Transactions on Knowledge and Engineering 11(6), 929–947 (1999)
13. Beckmann, N., Kriegel, H.P., Schneider, R., Seeger, B.: The R-tree: An efficient and robust access method for points and rectangles. In: Proc. Of SIGMOD, pp. 322–331 (1990)
14. Nørvåg, K.: Space-Efficient Support for Temporal Text Indexing in a Document Archive Context. In: Koch, T., Sølvberg, I.T. (eds.) ECDL 2003. LNCS, vol. 2769, pp. 511–522. Springer, Heidelberg (2003)
15. Nørvåg, K.: Supporting temporal text-containment queries in temporal document databases. Data Knowl. Eng. 49(1), 105–125 (2004)
16. Nørvåg, K., Nybø, A.O.: Improving Space-Efficiency in Temporal Text-Indexing. In: Zhou, L.-z., Ooi, B.-C., Meng, X. (eds.) DASFAA 2005. LNCS, vol. 3453, pp. 791–802. Springer, Heidelberg (2005)
17. Nørvåg, K., Nybø, A.O.: Albert Overskeid Nybø. DyST: Dynamic and Scalable Temporal Text Indexing. In: Proc. of TIME, pp. 204–211 (2006)

Complex Event Processing over Unreliable RFID Data Streams*

Yanming Nie, Zhanhuai Li, and Qun Chen

School of Computer, Northwestern Polytechnical University,
Xi'an 710072, China
yanming@mail.nwpu.edu.cn, {lizhh,chenbenben}@nwpu.edu.cn
http://www.nwpu.edu.cn/jsj

Abstract. Existing RFID complex event processing (CEP) techniques always assume that raw RIFD data has been first cleansed to filter out all unreliable readings upfront. But this may cause delayed triggering of matched complex events. Furthermore, since the cleansed event streams need to be temporarily buffered for CEP evaluation, it may generate a large number of intermediate results. To address these issues, we propose an approach to perform CEP directly over unreliable RFID event streams by incorporating cleansing requirements into complex event specifications, and then employ a non-deterministic finite automata (NFA) framework to evaluate the transformed complex events. Experimental results show that our approach is effective and efficient.

Keywords: Unreliable RFID data streams, CEP, NFA.

1 Introduction

CEP can correlate individual RFID readings and transform them into semantic-rich complex events, and therefore plays a key role in monitoring applications. For example, shoplifting in a retail store can be detected by processing a complex event in such scenario: an item has been picked up at a shelf and then taken out of the store without being first checked out in a specific time window.

Raw RFID data is inherently incomplete, noisy, and need to be cleansed. We classify the inaccuracy of RFID data into two categories: **unreliability** and **uncertainty**. Here, unreliability refers to the erroneous readings that can be corrected by deterministic cleansing rules. In a retail store, for instance, if an item has been picked up from a shelf and checked out later on, any other shelf readings for this item in-between should be false positives. Instead uncertainty refers to the inconsistent RFID readings which cannot be determinately eliminated by cleansing rules due to their ambiguities. An instance of uncertainty occurs when the two different shelf readers detect an item simultaneously. So either of the two readings can be false positive, but available RFID data cannot arbitrate. In

---

* A preliminary version of this work appeared as a 4-page paper at Application of Research of Computers, Vol 26 No 7, 2009 [15].

X. Du et al. (Eds.): APWeb 2011, LNCS 6612, pp. 278–289, 2011.
© Springer-Verlag Berlin Heidelberg 2011

this paper, we focus only on RFID event stream containing unreliable readings, and leave the issues of CEP over uncertain RFID streams for future work.

The straightforward solution to CEP over unreliable RFID event stream is to filter out unreliable readings upfront, and then execute CEP. But this may cause CEP-enabled systems delaying to trigger corresponding response. Moreover, since cleansed events are needed to be temporarily recorded, it may generate large number of intermediate results. In this paper, we proposed an approach to incorporate cleansing requirements into CEP. Our approach is first to convert unreliability of RFID readings into its corresponding reliability constraints in complex event specifications, and then directly evaluate the transformed complex events over unreliable event streams. The main contributions are as following:

- We present a declarative **C**leansing **L**anguage for **U**nreliable RFID **E**vent **S**treams (**CLUES**), with which cleansing actions can be implicitly set without using a specific action clause.
- Based on CLUES, we propose an automated mechanism to enable evaluating complex event directly over unreliable RFID streams by implanting reliability check constraints into complex event specification.
- By extending the existing NFA implementation frameworks, we propose two approaches to evaluate complex events with reliability check constraints, i.e. a primitive approach and an advanced approach.

2   Related Work

Due to value-based constraints and sliding windows in RFID complex event specification, traditional evaluation frameworks [1][3] is no longer applicable. And their employed fixed data structures, such as tree [3], directed graph [1], finite automata [2], or Petri net [4], cannot adapt to necessary extensions required by RFID complex event specifications. To address these issues and to optimize CEP over huge-volume RFID data, Eugene et al [5] proposed a declarative specification language **SASE** and an NFA-based evaluation framework. Unfortunately, none of them can handle unreliable readings existed in RFID event streams.

RFID data is inherently incomplete, noisy, and need to be cleansed before being forwarded. **SMURF** [7] aims to capture the accurate time window of tag existence by viewing RFID stream as a statistical sample of the tags in physical world, and filters RFID data at the low level of edge device. In practice, however, RFID readings usually require to be analyzed and cleansed in a bigger context of business flow, so cleansing thus need to be executed probably within a large sliding window. A deferred approach was proposed in [8] for detecting RFID data anomalies by defining cleansing rules with **SQL-TS** and performing application-specific cleansing at query time. In [6], Wang et al provided with example rules for data filtering and cleansing. But these off-line and RDBMS-based solutions cannot be applied to RFID CEP. Other related work [11] [12] tried to perform interpretation and imputation over uncertain RFID data by fully exploiting the temporal and spatial relationships among the readings, but neither of them addressed the issues of RFID CEP.

There exist some work [14] [13] on event queries and evaluation over uncertain RFID data streams. A temporal model and some evaluation frameworks with corresponding optimizations were proposed in [13] to recognize patterns and perform CEP over the streams with imprecise timestamps. Event pattern detection and query evaluation techniques over correlated probabilistic streams was studied in [14]. As far as we know, there are no any related studies on CEP incorporate with data cleansing over unreliable RFID event streams yet.

3   Preliminaries

RFID data can be seen as a sequence of RFID tuples with the form of *(Oid, Rid, RTime)*, where *Oid* is the ID of an object, *Rid* is the ID of a reader, *Rtime* is the occurrence time of the reading. CEP is used to correlate individual RFID readings in event stream to form semantically meaningful complex events appropriate for end applications. Shoplifting monitoring in a retail store, for instance, requires to detect a sequence of occurrence or non-occurrence of events having same *Oid* within a specific time window. As an RFID complex event specification language, SASE [5] is declarative and has the overall structure as:

> *EVENT <event pattern>*
> *[WHERE <qualification>]*
> *[WITHIN <window>]*

The *EVENT* clause describe a sequence pattern, and its components are occurrence or non-occurrence of component events in a temporal order. The *WHERE* clause specifies constraints on those events. The *WITHIN* clause specifies the sliding window for the whole sequence of events. For example, the complex event corresponding to shoplifting in a retail store can be specified as Q_1:

> *EVENT SEQ(SHELF x, !(COUNTER y), EXIT z)*
> *WHERE x.Oid=y.Oid=z.Oid*
> *WITHIN a hour*

In Q_1, *SEQ* denotes sequence pattern. *SHELF*, *COUNTER* and *EXIT* are different event types. The sign ! denotes non-occurrence of an event (also called as a negation event).

The NFA approach of SASE [5] can flexibly implement attribute value comparisons between events. It creates an NFA for each query, uses instance stacks to record the events in different states, and partition the entries based on the attribute values to facilitate comparisons among events.

As for an RFID complex event, the value of *Oid* has a large domain, so the partitioning strategy is no longer applicable. Correlating the events with the same *Oid* thus requires to involve search operation on *Oid*. An example of NFA evaluation on a complex event with the same *Oid* is presented in Fig. 1. We denote an event type with a capital letter, a primitive event with a lowercase letter and a subscript (where the letter and subscript indicate event type and object ID, respectively). For instance, a_i is an event of type A and its *Oid* is i.

RFID readings usually contain false positives and false negatives. Directly applying the NFA approach to CEP over unreliable RFID event streams may produce incorrect results. We will detail our solutions in the next section.

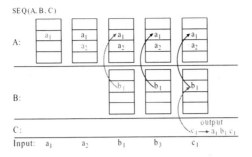

Fig. 1. An example of NFA evaluation

4   CLUES Cleansing Language

The unreliabilities of RFID event stream can be specified by cleansing rules. An effective specification language of cleansing rule is supposed to be declarative, sequence-based and easy-specifying. Unlike SQL-TS [8] which specifies cleansing rules in relational tables, CLUES is for RFID event streams. Its format is as:

IF <complex event specification>
THEN FORALL <Oid specification>
EXIST|NOT EXIST <event type>
WITHIN (BEFORE|AFTER) <window>

where the *IF* clause specifies the complex event. The *FORALL* clause specifies the objects may have unreliable readings. *<Oid specification>* can be a boolean combination of predicates with using comparison operators or containment operator (i.e. $\subset$). The keyword *EXIST* indicates that the reading may be false negative, while *NOT EXIST* false positive. The *WITHIN* clause specifies the temporal constraints for the specified complex event. The keywords *BEFORE* and *AFTER* are used to specify if the unreliable reading occurs before or after the complex event. The *<window>* specifies the size of valid time window between unreliable reading and the complex event. With the *BEFORE*, the time window refers to the occurrence time gap between unreliable reading and the first component event of the complex event. With the *AFTER*, the time window refers to the occurrence time gap between unreliable reading and the last component event. Also the *<window>* alone can be set as an absolute time interval.

The CLUES cleansing language is declarative, so we can construct event sequences and specify common RFID cleansing rules in a straightforward way. Note that there is no action clause defined in CLUES language, cleansing action has in fact been implicitly specified by the *EXIST* or *NOT EXIST* keywords. *EXIST* implies that the reading may be missing and thus need to be recovered if absented. In contrast, *NOT EXIST* implies that the reading may be a false positive, thus has to be dropped if presented. The *EXIST|NOT EXIST* clause specifies the inclusive or exclusive relationship between an unreliable reading and its triggering complex event. Various cleansing rules can thus be composed from the clauses of *EXIST|NOT EXIST*, *FORALL* and *WITHIN*, such as:

Example 1: Exclusive Rule. Consider an item is transported to location A (event type A). Suppose the transportation time is at least 1 minute and the items may be accidentally read by location B (event type B) nearby. We use the following cleansing rule R_e to remove all B readings before an A reading:

> IF a_1
> THEN FORALL $(Oid = a_1.Oid)$
> NOT EXIST B
> WITHIN BEFORE <1 minute>

According to R_e, any B event followed by an A event on the same object within one minute is supposed to be a false positive.

Example 2: Inclusive Rule. Suppose that an item detected at location A is always moved to location B within 2 minutes due to business flow. We use the following cleansing rule R_i to compensate some probably missed B readings:

> IF a_1
> THEN FORALL $(Oid = a_1.Oid)$
> EXIST B
> WITHIN AFTER <2 minutes>

According to R_i, any A event should be followed by a B event on the same object within two minutes.

Example 3: Cross Reading Rule. An item stays at location A, and may be accidentally read by location B. We use the following cleansing rule R_c to detect this type of cross readings:

> IF $(a_1.a_2$
>   WHERE $a_1.Oid = a_2.Oid$
>   WITHIN 3 minutes)
> THEN FORALL $(Oid = a_1.Oid)$
> NOT EXIST B
> WITHIN $[0 < a_2.Rtime - a_1.Rtime \leq 3\ minutes]$

According to R_c, if two A events on the same object occur within 3 minutes, any B event on that object in-between is supposed to be a false positive.

Example 4: Packaging Rule. Consider a pallet and its contained cases is being moved together along a certain business path. Suppose that pallet tags are always more readable, but due to material interfering and tag orientation, case tags may fail to be detected. We use the following cleansing rule R_p to recover missed A readings for the cases:

> IF (a_1 WHERE <$a_1.Oid$ indicates a pallet>)
> THEN FORALL $(Oid \subset a_1.Oid)$
> EXIST A
> WITHIN $[a_1.Rtime - 5\ seconds, a_1.Rtime + 5\ seconds]$

Here $Oid \subset a_1.Oid$ denotes that the object (case) with unreliable reading is contained in a_1 (pallet). According to R_p, as a pallet reading occurs, the readings for the contained cases should be detected before or after it within 5 seconds.

5 Complex Event Evaluation

With CLUES cleansing rules, CEP over unreliable RFID event streams can be solved through imposing reliability constraints on component events and then evaluating the transformed complex event.

5.1 Complex Event Transformation

Suppose that the complex event specified in the *IF* clause is denoted by P_c. For false positive, we can use $[!P_c]$ to represent its reliability constrain (! denotes non-occurrence of P_c), and the reliability of a reading depends on non-occurrence of P_c. Similarly, for false negative, we can use $[|P_c]$ (| denotes an option), a missed reading should be recovered by verifying occurrence of P_c. Semantically, both predicates $[!P_c]$ and $[|P_c]$ impose additional constraints on events.

Suppose that the event type of false positive readings specified in a CLUES cleansing rule is A. Its reliability is formally specified as:

$$A[!(P_c \ WHERE \ <Oid \ qualification> \ and \ <Rtime \ qualification>)]$$

where P_c is the complex event specified in the *IF* clause of cleansing rule. *<Oid qualification>*, derived from *<Oid specification>* in the *FORALL* clause, specifies additional qualifications on the *Oid* of component events in P_c. *<Rtime qualification>*, derived from the *WITHIN* clause, specifies additional constraints on the *Rtime* of component events in P_c.

Generally, P_c consists of multiple component events, and may has its own attribute value comparisons and time window specifications. The reliability constraint shares the same event sequence pattern as P_c, but its *WHERE* clause has to integrate the Oid and Rtime qualifications specified in cleansing rule with those of P_c using ∧. Consider the cleansing rule R_c in Section 4, a reading of type B should be imposed with a reliability constraint as:

$$B[!(A_1.A_2 \ WHERE \ (A_1.Oid = A_2.Oid) \ and$$
$$(B.Oid = A_1.Oid) \ and$$
$$(A_1.Rtime \leq B.Rtime \leq A_2.Rtime)$$
$$WITHIN \ 3 \ minutes)]$$

Note that the constraints $(B.Oid = A_1.Oid)$ and $(A_1.Rtime \leq B.Rtime \leq A_2.Rtime)$ are derived from the cleansing rule.

The unreliability constraint of false negative readings of type A is as:

$$A[|(P_c \ WHERE \ <Oid \ qualification> \ and \ <Rtime \ qualification>)]$$

Since a false negative reading is actually absented from event stream, *<Oid qualification>* and *<Rtime qualification>* do not impose additional qualifications on P_c. Instead, they delimit the probable *Oid* and *Rtime* values of the missed readings. Consider the cleansing rule R_i presented in Section 4, B events should be imposed with a reliability constraint as:

$$B[|(A \ WHERE \ (A.Oid = B.Oid) \ and$$
$$(0 \leq B.Rtime - A.Rtime \leq 2 \ minutes))]$$

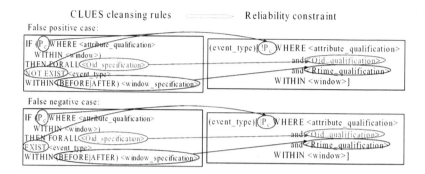

Fig. 2. Transforming CLUES rules into reliability constraints

Fig. 2 shows how to transform CLUES rules into reliability constraints. In false positive case, incorporating reliability constraint into complex event specification is straightforward as the event with constraints will be substantiated during evaluation. In false negative case, reliability constraints is to detect a missing event and specify the value of *Oid* and *Rtime*. The *Oid* of missing event may be $=$ or $\subset$ the *Oid* of P_c. But the occurrence time of a false negative reading depends on the occurrence time of its corresponding complex event specified in reliability constraint, as well as business flow. In this paper, we assume that a false negative may occur within the inferred time interval.

5.2   NFA-Based Evaluation Approaches

We use NFA-based evaluation framework for RFID complex event with reliability constraints. **The primitive approach** is to evaluate the complex event first, and then validate component events case-by-case using reliability constraints.

For false positive, the reliability constraint of a B reading is denoted as $B[!P_c]$. To check reliability constraint, the primitive approach has to record the instances of P_c. While validating a B event, it checks if it has a corresponding P_c instance by searching on *Oid* and *Rtime*. Note that if the *WITHIN* clause of cleansing rule has the keyword *BEFORE* or *AFTER*, the reliability constraint can be transformed into $B.!P_c$ or $!P_c.B$, respectively. For false negative, the reliability constraint is denoted as $B[|P_c]$. Since B event may be missing, the primitive approach has to detect all complex event instances with or without B event first. Then for each partially matched complex events, checks if there is a P_c instance which implies a corresponding B event occurred before.

The primitive approach produces all matching complex event instances without reliability constraints and record the instances of P_c specified in constraints. Next, we will present **an advanced approach** which interleaves reliability validation with CEP and evaluates reliability constraints eagerly, thus effectively reduce intermediate result size during evaluation.

For false positive $B[!P_c]$, the advanced approach detects the instances of P_c and uses them to filter out unreliable B events. It evaluates CEP and P_c

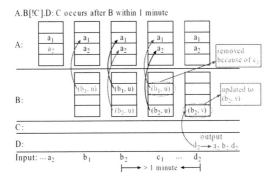

Fig. 3. A NFA running on complex event with false positive reliability check

simultaneously. Consider the case that B event is specified to occur before or within the time window of P_c in reliability constraint. The corresponding B events are considered as unreliable and removed from NFA as a P_c instance is found. A B event in NFA is considered as reliable only after it is out of the time window of $B[!P_c]$. So the P_c instances do not need to be recorded. For the case that B event is specified to occur after the time window of P_c, as a P_c instance is detected, its corresponding B event is unavailable, so the P_c instances have to be temporarily stored. The B event will be checked immediately against the existing P_c instances once available. A P_c instance can be dropped only after it is out of the sliding window of $B[!P_c]$. Each event in NFA is marked as *valid* or *uncertain*, and will be changed from *uncertain* to *valid* if it is out of the sliding window of $B[!P_c]$. It will immediately be removed once marked as *invalid*.

An evaluation example of $B[!P_c]$ is shown in Fig. 3. It supposes that the occurrence time gap between d_2 and b_2 is > 1 minute. (b_2, u) and (b_2, v) denotes b_2 to be *uncertain* and *valid*, respectively. Once d_2 is encountered, b_2 has fallen out of the sliding window of $B[!C]$, is thus considered as *valid*.

The advanced approach processes false negative $B[\|P_c]$ in a similar way. Whenever a P_c instance is detected, its corresponding B event, whose probable occurrence time is specified as a time interval, should be inserted into NFA if not exists. In case that B event is specified to occur before or within the time window of P_c, it will be clear whether the corresponding B event is missing once the P_c is detected. For the case that the B event is to occur after the time window of P_c, even though the corresponding B event of P_c does not exist in NFA by the time P_c is detected, it may occur later. The supposed B event inferred from P_c should be replaced with this real one as it occurs and be inserted into NFA.

Suppose that $B[\|P_c]$ is followed by D, and B event is specified to occur before or within the time window of P_c. When a D event occurs and its corresponding B event not existed in NFA, an imaginary B event will be generated and inserted. The imaginary B event is initially marked as *uncertain*, and updated to *valid* as its corresponding P_c is detected. An example of evaluating $B[\|P_c]$ is shown in Fig. 4, where the corresponding b_2 event of c_2 is missed in the event stream.

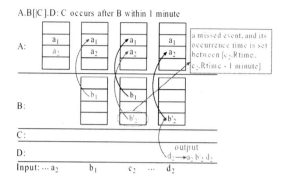

Fig. 4. A NFA running on complex event with false negative reliability check

6   Experimental Study

We conduct a preliminary experimental study to validate our evaluation approaches on two metrics: memory usage and processing time. Used memeory is measured by the total number of buffered event instances. We first generated synthetic RFID data by emulating tags moving along a path of $A{\to}B{\to}D$. The complex event to be evaluated is $A.B.D$, and its components have same Oid. The B events are supposed to be unreliable, and can be determined by reliability constraints involving C events. We consider both false positive reliability constraint (i.e. $B[!C]$) and false negative one (i.e. $B[\|C]$), where C event is specified to be after B event within a time window. All the tests is running on a Windows machine with Intel Core 2 Dual 2GHz CPU and 2GB RAM.

6.1   Advanced Approach vs. Primitive Approach

This test is to compare the performance between the advanced approach and the primitive one. Among the generated data, there are 50 thousand events can match $A.B.D$ with varying sliding window from 50 to 500, 10 thousand inaccurate $A.B.D$ whose B readings are followed by C readings with varying time window of $B[!C]$ from 20 to 200 for false positive case, 10 thousand $A.D$ whose B events in-between are falsely missed but can be recovered if with the corresponding C events for false negative case, and some other noisy readings.

Fig. 5 shows the results for $A.B[!C].D$, where *50/20* on the X-axis means that the sliding window of $A.B.D$ is 50 while the time window of $B[!C]$ is 20. The advanced approach is better on memory usage with the margins larger than 30% as sliding window > 300, as shown in Fig. 5(a). That's because the primitive approach has to record C readings and unreliable B readings, and the total number of these readings will grow accordingly as the sliding window increases. Even though the two approaches process C readings differently, they execute roughly the same number of event correlations and thus take nearly the same CPU time, as shown in Fig. 5(b).

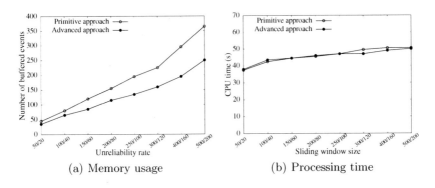

(a) Memory usage   (b) Processing time

Fig. 5. Advanced vs primitive on $A.B[!C].D$

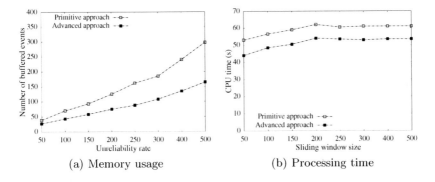

(a) Memory usage   (b) Processing time

Fig. 6. Advanced vs primitive on $A.B[|C].D$

Fig. 6 shows the results for $A.B[|C].D$. The primitive approach records all C events and complex events of $A.D$, while the advanced one only buffers the C events with preceding A events and D events preceded by either B or C events. So the advanced approach uses less memory, as shown in Fig. 6(a). The primitive approach produces many inaccurate complex events of $A.D$ which need to be checked against C events. While the advanced one processes C events eagerly and does not need such checking. We can see that the advanced approach obviously outperforms the primitive one on CPU time, as shown in Fig. 6(b).

6.2 Advanced Approach vs. Traditional NFA Approach

In this test, we study the performance differences between the advanced approach and the traditional NFA approach without reliability constraints. The test RFID data includes 100 thousand accurate complex events of $A.B.D$ and other complex events involving false B events. Unreliability rate for an event stream is defined to be the ratio of the number of complex events with false B events to the total number of matching complex events. We varies unreliability rate from 5% to 50%. Instead, the traditional approach assumes that each B event is reliable.

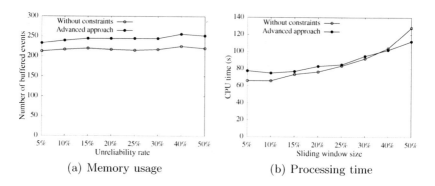

Fig. 7. Unreliable vs reliable in false positive case

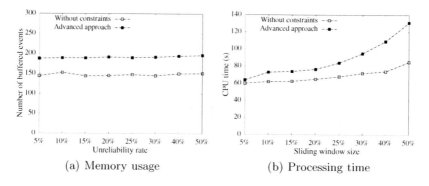

Fig. 8. Unreliable vs reliable in false negative case

Fig. 7 shows the results for the false positive case. The advanced approach does not records C events, but need to cache the constructed instances of $A.B.D$ until the B events is out of the time window of $B[!C]$. The advanced approach performs worse by around 15%, as shown in Fig. 7(a). Note that the used memory of the advanced approach does not grow with the increasing unreliability rates. That is because even though the total number of false positive B events becomes larger as unreliability rates increase, they are distributed in a larger event stream. Therefore higher unreliability rates do not necessarily mean more memory usage given a fixed sliding window. The traditional approach does not process C events, but constructs more complex events of $A.B.D$. So the differences of CPU time between the two are negligible, as shown in Fig. 7(b).

Fig. 8 shows the results for the false negative case. As the advanced approach records the C events with corresponding A events, it performs worse on memory usage, as shown in Fig. 8(a). The extra is proportional to the number of complex events of $A.C$ within the sliding window of $A.B.D$. Because the advanced approach processes more C events, it performs worse on CPU time, as shown in Fig. 8(b). As the number of C events increase with the increasing unreliability rates, the degradation is roughly linear to the increase of unreliability rates.

7 Conclusion

In this paper, we study the issues of CEP over unreliable RFID event streams. First, we present CLUES, an RFID cleansing language, to define unreliability of RFID readings, and then transform reading unreliability into reliability constraint in complex event specification. Thus CEP can be directly performed over unreliable RFID event streams. Finally, we extend the current NFA evaluation framework to facilitate RFID CEP with reliability constraints. The preliminary experimental results show that our approach is effective and efficient.

There still exist such possible extensions for the CLUSE as: 1) The *IF* clause can be extended to accommodate boolean combination of multiple complex events, and 2) The *EXIST|NOT EXIST* clause can be enhanced to allow specifying one or even multiple complex events. Also we plan to investigate the issue of CEP over RFID event streams with both unreliable and uncertain readings.

Acknowledgments. This work is supported by the NSFC under Grant No. 60720106001 and No. 60803043, and the National National High Technology Development 863 Program of China under Grant No. 2009AA1Z134.

References

1. Hinze, A.: Efficient filtering of composite events. In: BNCOD, pp. 207–225 (2003)
2. Gehani, N.H., Jagadish, H.V., Shmueli, O.: Composite event specification in active databases: Model and implementation. In: VLDB, pp. 327–338 (1992)
3. Chakravarthy, S., Krishnaprasad, V., Anwar, E., Kim, S.: Composite events for active databases: Semantics, contexts and detection. In: VLDB, pp. 606–617 (1994)
4. Gatziu, S., Dittrich, K.R.: Events in an active object-oriented database system. In: RIDS, pp. 23–39 (1993)
5. Wu, E., Diao, Y., Rizvi, S.: High-performance complex event processing over streams. In: SIGMOD, pp. 407–418 (2006)
6. Wang, F., Liu, P.: Temporal management of RFID data. In: VLDB, pp. 1128–1139 (2005)
7. Jeffery, S.R., Garofalakis, M.N., Franklin, M.J.: Adaptive cleaning for RFID data streams. In: VLDB, pp. 163–174 (2006)
8. Rao, J., Doraiswamy, S., Thakkar, H., et al.: A deferred cleansing method for RFID data analytics. In: VLDB, pp. 175–186 (2006)
9. Chen, Q., Li, Z., Liu, H.: Optimizing Complex Event Processing over RFID Data Streams. In: ICDE, pp. 1442–1444 (2008)
10. Agrawal, J., Diao, Y., Gyllstrom, D., Immerman, N.: Efficient pattern matching over event streams. In: SIGMOD, pp. 147–160 (2008)
11. Cocci, R., Tran, T., et al.: Efficient Data Interpretation and Compression over RFID Streams. In: ICDE, pp. 1445–1447 (2008)
12. Gu, Y., Yu, G., et al.: Efficient RFID data imputation by analyzing the correlations of monitored objects. In: Zhou, X., Yokota, H., Deng, K., Liu, Q., et al. (eds.) DASFAA 2009. LNCS, vol. 5463, pp. 186–200. Springer, Heidelberg (2009)
13. Zhang, H., Diao, Y., Immerman, N.: Recognizing Patterns in Streams with Imprecise Timestamps. In: VLDB (2010)
14. Re, C., Letchner, J., et al.: Event queries on correlated probabilistic streams. In: SIGMOD, pp. 715–728 (2008)
15. Chen, Y., Li, Z., Chen, Q.: Complex event processing over unreliable RFID data streams. Application Research of Computers 26(7), 2538–2539, 2542 (2009)

A Graph Model Based Simulation Tool for Generating RFID Streaming Data*

Haipeng Zhang[1], Joonho Kwon[2], and Bonghee Hong[1]

[1] Department of Computer Engineering
[2] Institute of Logistics Information Technology
Pusan National University, Jangjun-dong Kumjung-gu,
Busan, 609-735, Republic of Korea
{jsjzhp,jhkwon,bhhong}@pusan.ac.kr

Abstract. Since raw RFID streaming data is continuous, voluminous, incomplete, temporal and spatial, RFID applications cannot directly make use of it. As a key component of RFID systems, an RFID middleware needs to clean, filter and transform the raw RFID streaming data. In order to know whether these functions work well or not, we need to test the middleware using raw RFID streaming data generated from RFID environments. A general RFID environment contains readers which are deployed at different locations and tagged objects which are transited between locations. However, it is a complex and costly process to construct such an environment to collect RFID streaming data. In this paper, we present a graph model to simulate RFID environments. In the graph model, nodes correspond to RFID readers and edges determine tags' movements between nodes, virtual tags are automatically inputted and transited in the graph. We define several parameters for configuring the graph and generating raw RFID data. Raw RFID streaming data is automatically generated from the graph model when tags arrive in the nodes. Based on the graph model, we implement a simulation tool to generate raw RFID streaming data. The experimental results show that the generated RFID data streams can be used for testing the functions which are provided by RFID middleware.

Keywords: RFID middleware, graph model, environment simulation.

1 Introduction

RFID (Radio Frequency IDentification) is a promising and leading technology that provides fast, reliable and automatic identifying, location tracking and monitoring physical objects without line of sight. With its inherent advantages, RFID technology is being adopted and deployed in a wide area of applications, such as logistics and supply chain management, yard management, access control and healthcare.

---

* This work was supported by the grant of the Korean Ministry of Education, Science and Technology (The Regional Core Research Program/Institute of Logistics Information Technology).

X. Du et al. (Eds.): APWeb 2011, LNCS 6612, pp. 290–300, 2011.

In an RFID environment, RFID readers are deployed at different locations and communicate with RFID middleware via wire or wireless interfaces, and products with identified tags are transited between locations. Readers can generate a huge volume of streaming data when tags move in the interrogation range of them. Like other streaming data, RFID data is continuous and enormous, meanwhile it also has the following several distinct features [1, 2].

- *Simple and incomplete data*: Data generated from RFID readers is a stream of the form (*tag_id, reader_id, timestamp, event_type*), where *tag_id* refers to EPCs which uniquely identify the tagged products, and *reader_id* identifies a reader which reads the tag, *Timestamp* is the reading occurred time, and *event_type* refers to a tag enters or exits the reading range. Also RFID data is incomplete since the reading rates are below 100% in actual deployments. When lots of tags move through a reader, a few tags may be missed which causes the data is incomplete.
- *Temporal and spatial data*: RFID data is dynamically generated when tags move in the reading range. The simple form of RFID data can perverse the state change information of objects, such as the reading location, reading time, and location changing history along the time. All these information are temporal and spatial.

The raw RFID data is impossible to be used for user applications due to above special data features. We need to clean, filter and transform the data before sending it to user applications. According to EPCglobal standard [3], an RFID middleware provides these important functions for processing the raw RFID data. Therefore, it is very important to know whether an RFID middleware provides these functions or not before deployed, and whether it can process the RFID streaming data if provided. We need to test its functions using raw RFID streaming data. However, building real RFID environments to collect raw RFID data is a high cost way, since we need to deploy readers at different locations and moving tagged items between readers. Although a testing tool [4] can generate raw RFID data, it cannot guarantee the data contains special features which are very important for testing the functions provided by RFID middleware.

In this paper, we propose a directed graph model to simulate RFID environments. Using the graph model, we do not need to construct the real RFID environment to create RFID test data. In the graph model, nodes correspond to RFID readers and edges determine tags's movements between nodes. Virtual tags are continuously inputted in the graph and moved in the graph according to the directed edges. Nodes automatically generate RFID data when virtual tags arrive in. We define several parameters for configuring and controlling the graph, so that test data generated from our graph model can possess the real special features. Based on the graph model, we implement a simulation tool to generate raw RFID streaming data. We also perform several experiments to evaluate our generated RFID streaming data by using a few defined metrics. Our main contributions can be summarized as follows:

- Design a graph model to simulate the real RFID environments.
- Define several parameters to configure and control the graph model for generating RFID streaming data
- Design and implement a simulation tool, as well as define a few metrics to evaluate the generated RFID data.

The remainder of this paper is organized as follows. Section 2 introduces the preliminaries and related studies. In Section 3, we present our graph model and several defined parameters for configuring it. Section 4 introduces our implementation of simulation tool and experiments for evaluating generated RFID data. We conclude our work in Section 6.

2   Related Studies

In this section, we briefly introduce an RFID middleware. Then we discuss several related studies on RFID middleware test and virtualization of RFID environments.

2.1   RFID Middleware

As shown in Figure 1, an RFID system typically consists four components: RFID tags, readers, middleware and user applications. Each tag stores an electronic product codes (EPCs) [5] in the memory to unique identify itself. Readers are capable of reading the information stored in RFID tags placed in their vicinity and sending the streaming data to middleware through wired or wireless interfaces. Since the raw RFID streaming data is enormous and incomplete and contains implicit information, an RFID middleware cleans it, filters it and transforms it in usable formats. Then, user applications can use the data after processed in RFID middleware.

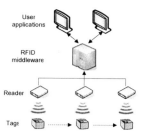

Fig. 1. RFID system

After processed by RFID middleware, RFID data can tell the information about what happened, where, when, and why. For example, a reader deployed at entrance of a room generates raw data form: "Tag 100 is read by reader A at 4:00pm". RFID middleware can transform this data into high-level event like "A LG laptop enters room A at 4:00pm" which is more useful to user applications. So middleware is very important in an RFID system. In order to know a middleware is capable for processing raw RFID streaming data, we need to test it.

2.2   RFID Middleware Test

A few methods related to RFID middleware test have been proposed. Oh et al. [6] analyzed quality factors for evaluating an RFID middleware and selected important quality factors such as functionality, efficiency, reliability, usability in constructing middleware through a quantitative method. However, they did not mention how evaluate these factors using test data. How to evaluate these factors using test data is beyond the scope of this paper. Lee et al. [4] define several parameters for the performance of RFID middleware and implement a software tool to test the RFID middleware. The performance testing tool cannot be used for other factors of RFID middleware mentioned in [6], and it randomly generates test data without considering special features of RFID streaming data. Park et al [7] design an architecture of RFID middleware evaluation toolkit based on virtual reader emulator that provides RFID test data and collects middleware status for evaluation. It cannot guarantee whether the generated data is suitable for RFID middleware test or not. Rifidi Tag Streamer [8] is a load testing tool that allows user to create virtual readers and tags to generate RFID data streams for an RFID system. It also does not consider the special features of RFID data. Collecting real RFID streaming data from test centers [9] spends time and human recourses since we need to deploy readers and move lots of products with tags. To address these problems, we can simulate and build a virtual RFID environment for creating RFID streaming data.

2.3 Virtualization of RFID Environment

As constructing real environments for generating RFID streaming data is a high cost process, our previous work [10] propose to use virtual environment for generating RFID data without using any real devices. A virtual environment contains three virtual components: virtual readers, virtual tags and virtual controller. Virtual readers and tags play the same roles of real devices. Virtual controller controls the movement of virtual tags. However, it is not convenient to represent and construct different virtual environments since the virtual controller manages virtual tags in a centralized way.

This problem motivates us to propose a graph model to simulate the real RFID environment. An RFID environment generally contains both static elements such as readers and tag moving paths, and dynamic elements such as the changing status of tags. Both static and dynamic elements of RFID environment should be well simulated in the virtual environment. We use a graph model to simulate the static elements of the RFID environment. In the graph model, nodes represent RFID readers and edges represent tag's moving path among nodes. We define several parameters to simulate the changing status of tags and configure the graph model. A virtual tag generator continuously produces tags and sends to the graph. In this way, we can easily construct an RFID environment by creating a graph and setting parameters (Fig. 2).

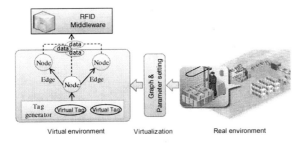

Fig. 2. Construction of virtual environment

3   Graph Model

In this section, we present our graph model for simulating RFID environment and defined parameters for configuring the graph.

3.1   Graph Representation

A graph model not only facilitates the construction of virtual RFID environment, but also provides a convenient way to configure itself and generate RFID data.

A graph model is a simple directed graph $G = (N, E)$ consists of two finite sets, where $N = \{n_1, n_2, ... , n_k\}$ and $E = \{e_1, e_2, ... , e_l\}$ such that $e_i = (n_i, n_j)$, $n_i, n_j \in N$, $i \neq j$. The elements of N are called *nodes* or *vertices*. The elements of E are called *edges*. In the graph model, an edge $e = (n_i, n_j)$ means virtual tags arrived at node n_i can be moved to node n_j, n_i is called tail and n_j is called head. The nodes which firstly receive virtual tags are called source nodes. After receiving tags, source nodes will generate tag data to RFID middleware and distribute tags to their adjacent nodes. The other nodes will perform in the same way of source nodes after tags arriving.

Figure 3 shows an example of the graph model to simulate an RFID environment. In a warehouse, there are two storage rooms with RFID readers deployed at both entry and exit doors. Products are read when they enter in the warehouse and can be observed again when they move in and out of the rooms. The products are eventually observed at shipping door when they go out of the warehouse. Using our graph model, we can easily simulate this RFID environment (Figure 3).

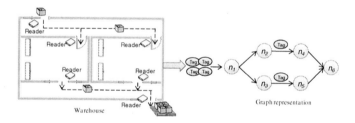

Fig. 3. An example of graph representation

Various RFID environments can be simulated by using the graph model. A graph cannot fully simulate an RFID environment since the graph only represents the static elements. RFID streaming data generated from the graph model lacks of state changing information of objects, such as reading locations, moving paths and durations at locations. We have to simulate the dynamic elements of RFID environment in the graph model. To achieve this, we define several parameters for the graph model.

3.2   Parameters for Configuring Graph Model

3.2.1   Requirements of Parameters

From previous explanation, we know that raw RFID streaming data possesses some special characteristics. That is why it needs to be processed and interpreted in the middleware. So data generated from virtual environment also should possess the same characteristics. In our virtual environment, it mainly contains two parts: static and dynamic elements which are represented as graph model and the changing status of tags, respectively. We need to define several parameters for configuring and controlling the virtual environment. The parameters are mainly related to both static and dynamic elements of virtual environment so that RFID data with special features can be generated from the virtual environment.

3.2.2 Parameter for Node

In the graph model, nodes represent the real readers. They play the same role of real readers. Nodes can read the data stored in virtual tag when tag arriving, and generate streaming data to RFID middleware. RFID readings are inherently incomplete since reading rates are below 100% in real deployments. Missed reading is frequently occurred in the real environments. To simulate the missed reading, we define a parameter for each node.

The missed reading ratio R_i of the node n_i is a probability that determines tags cannot be observed when they move into the node, such that $0 \leq R_i \leq 1$. In the real RFID environments, reader may have different reading rates since they cannot have the same strength of RF signal or they are deployed in various circumstances. Using the missed reading ratio, we can simulate different readers.

3.2.3 Parameters for Edge

RFID streaming data usually carries the status changing information, such as products' locations, moving path and their durations at locations. All these information are correlated temporally and spatially. The generated data should contain this information. In the graph model, edges represent the virtual paths which can transport tags from one node to the next node. All the status changing information is related to the edge. So we define parameters for configuring virtual path to simulate tag movement and tag status. For an edge $e=(n_i, n_j)$ in the graph, we define two parameters which are moving time $T_{i,j}$ and distribution weight $W_{i,j}$.

The distribution weight $W_{i,j}$ is a distribution probability assigned to edge $e = (n_i, n_j)$, such that $0 \leq W_{i,j} \leq 1$. It means how many percentage of tags from node n_i can be distributed to node n_j. In the real environments, tags from one reading location can be distributed to other different places. Corresponding to the graph model, a node n can have several adjacent nodes. It means the node n can distribute tags to its adjacent nodes. In order to simulate how to distribute tags, we define the distribution weight for each edge. For distribution weight, there is a fact defined in Eq. (1) that we should to know

$$\sum_{j=1}^{k} W_{i,j} = 1 \tag{1}$$

Where $e=(n_i, n_j) \in E$ for $j =1, 2, ... ,k$. If a node n has several adjacent nodes linked with edges, the Eq. (1) guarantees that the sum of distribute weight of edges having the same tail node n is equal to 1.

The moving time $T_{i,j}$ defines the time a tag spends on the edge $e = (n_i, n_j)$ moving from node n_i to node n_j. It is a range value that has minimum moving time and maximum moving time. In the real environment, there is a distance between two readers and tags can be transported by different ways, such as human, conveyor system, fork-lift or vehicle. So the moving time between two readers may be fixed or variable according to transportation ways. If an edge has fixed moving time, we can define the minimum moving time equal to maximum moving time. All tags moved on this edge have same moving time. Otherwise, tags moved on the edge may have variable moving time which is randomly generated between minimum moving time and maximum moving time.

4   Implementation and Experiments

In this section, we shall first explain the detailed implementation of our simulation tool. Then, we shall describe experiments for evaluating generated RFID data.

4.1   Design and Implementation

Based on the graph model, we design and implement a simulation tool to generate RFID data streams for testing RFID middleware. Using the simulation tool, we can easily construct a virtual RFID environment and configure it via setting different parameters which are defined in previous section. RFID streaming data is automatically generated from the simulation tool and sent to middleware system.

Our simulation test tool mainly contains three components: user graphic interface, virtual tag generator and graph manager. Through the graphic interface, user can construct various RFID environments and configure them via setting different parameters. Virtual tag generator continuously generates virtual tags with EPC codes and sends to graph controller. It consists of four sub-modules: EPC generator that generates unique EPC codes, EPC storage that stores the generated EPC codes, tag generator that generates virtual tags, and tag dispatcher that dispatches virtual tags to graph model. Graph manager is the core component of our simulation tool and responsible for generating and managing the graph. It simulates the virtual RFID environment, controls the movements of virtual tags among different nodes and generates RFID streaming data. It consists of two sub-modules: graph generator that generates the graph according to user's setting, graph controller that controls the generated graph. Based on the system design, we have implemented a simulation test tool in java.

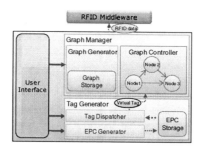

Fig. 4. Black diagram of simulation tool

4.2 Experiments

In order to verify the generated data can be used for testing the functionalities of RFID middleware, we define a few measurement metrics to evaluating the data and perform several experiments.

4.2.1 Measurement Metrics

Our simulation tool can continuously generate RFID data according to the constructed graph and inputted parameters. The generated data is supposed to use for testing the functions provided by RFID middleware. RFID middleware provides several functions for handling raw RFID streaming data since RFID data possesses different characteristics with other general streaming data. To know the data can be satisfied with the test objectives, we define measurement metrics which are mainly related to RFID data features to evaluate test data.

RFID streaming data is incomplete since reading rates are below 100%. A few tags may be missed when they move through the reading range. We use the missed reading ratio to simulate the missed readings in our graph model. In order to know whether the generated RFID data is incomplete or not, we need to calculate the generated missed reading ratio. If the calculated missed reading ratio is same with user pre-set value, we can say the node is well simulated and can generate RFID data. We define the absolute deviation of missed reading ratio D_{Ri} to measure a single node i. The is D_{Ri} defined in Eq. (2), where M_i is the number of missed tags from node i, N_i is the number of tags inputted in the node i, R_i is user pre-set missed reading ratio. D_{Ri} only evaluates the data generated from a single node. To measure the whole data, we need to evaluate all the nodes in the graph. We use the standard deviation of missed reading ratio S_R to measure the whole data generated from the graph mode. S_R is defined in Eq. (3), where n is the number of node in the graph model, D_{Ri} is defined in Eq. (2). For convenience, the notations we have introduced and others we will introduce later in this section are summarized in Table 1.

$$D_{R_i} = \left| \left(\frac{M_i}{N_i} \right) - R_i \right| \tag{2}$$

$$S_R = \sqrt{\sum_{i=1}^{n} D_{R_i}^2 \Big/ n} \tag{3}$$

Table 1. Summary of notations

Notation	Meaning
R_i	missed reading ratio of node i
D_{Ri}	absolute deviation of R_i
M_i	number of missed tags in node i
N_i	number of tags inputted in node i
S_R	standard deviation of R_i for all nodes
$W_{i,j}$	Distribution weight of edge $e_{i,j}$
$D_{Wi,j}$	ablolute deviation of $W_{i,j}$
$O_{i,j}$	number of tags distributed on edge $e_{i,j}$
$P_{i,j}$	number of expected tags distributed on edge $e_{i,j}$
S_W	standard deviation of $W_{i,j}$ for all edges

RFID data contains the state changing information of tags. We use the distribution weight of edges for controlling the location transition and traveling path of RFID tag. So the generated RFID data can contain the state changing information. In order to know whether tags are transported and distributed on edges or not, we use absolute deviation of distribution weight $D_{Wi,j}$ to measure a single edge $e_{i,j}$. The $D_{Wi,j}$ is defined in Eq. (4), where $O_{i,j}$ is the number of tags transported on the edge $e_{i,j}$, $P_{i,j}$ is the number of tags are expected to transport on the edge $e_{i,j}$, $W_{i,j}$ is user pre-set distribution weight of the edge $e_{i,j}$. $D_{Wi,j}$ only evaluates a single edge. To measure the whole data, we need to evaluate all the edges in the graph model. We use the standard deviation of distribution weights S_W to evaluate all the edges in the graph model. S_W is defined in Eq. (5), where m the number of edges in the graph model, E is the edge set, $D_{Wi,j}$ is defined in Eq. (4).

$$D_{W_{i,j}} = \left| \frac{O_{i,j}}{P_{i,j}} - W_{i,j} \right| \tag{4}$$

$$S_W = \sqrt{\sum_{\forall e_{i,j} \in E} D_{W_{i,j}}^2 \Big/ m} \tag{5}$$

The two standard deviations are used for measuring the difference between generated RFID data from our simulation tool and expected RFID data. The smaller values we can have, the better data we can get.

4.2.2 Experimental Results

By using these data measurement metrics defined above, we perform several experiments to evaluate the generated RFID data. All the experiments are conducted on a standard desktop computer with an Intel Core2 2.4GHz, 3GB of main memory and Windows 7 operation system. To perform the experiments, we create a graph with six nodes and six edges (Fig. 5), and set the parameters with different values. Firstly, we fix parameter values and generate different number of tags to evaluate. And then, we generate fixed number of tags with different parameter values.

In the first experiment, we fix the parameter settings and generate different number of tags. The initial setting of parameters setting is given in Table 2 and Table 3. We generate four data sets which have 100, 1000, 5000, 10000 tag data. Using Eq. (2) and

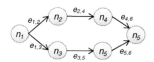

Fig. 5. Graph for testing

Table 2. Initial setting of nodes

Node	n_1	n_2	n_3	n_4	n_5	n_6
R_i	0.1	0.1	0	0.1	0.1	0.1

Eq. (4), we can calculate D_R for each node and D_W for each edge. Then using Eq. (3) and (5), we can get the standard deviation of all nodes and edges. Table 4 and Figure 6 show the calculated results of each data set. Form the Eq. (2) and (4) we can know that the small changing of M_i or $T_{i,j}$ will bring big results when the total number of tag is very small. So in the test results, the two standard deviations are higher when the number of tag is 100. The more tags we generate, the better test data we can get.

Table 3. Initial setting of edges

Edge	$e_{1,2}$	$e_{1,3}$	$e_{2,4}$	$e_{3,5}$	$e_{4,6}$	$e_{5,6}$
$W_{i,j}$	0.2	0.8	1	1	1	1

Table 4. Results of S_R and S_W with variable tag counts

# Tag	S_R	S_W
100	3.3%	0.57%
1000	1.1%	0.57%
5000	0.23%	0.23%
10000	0.23%	0.14%

Table 5. Results of S_R and S_W with variable parameters

# Tag	S_R	S_W
10000	0.34%	0.2%
10000	0.19%	0.29%
10000	0.47%	0.33%
10000	0.77%	0.27%

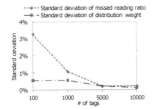 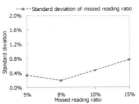 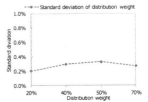

Fig. 6. Vary number of tag   **Fig. 7.** Vary missed reading ratio   **Fig. 8.** Vary distribution weight

For the second experiment, we give different parameter values and generate the same number of tags. Firstly, we vary the missed reading ratio which are set to 5%, 8%, 10% and 15%, fix the distribution weight using Table 3 and generate 10000 tags for each data set. The test results are shown in Table 5 and Fig. 7. Although the generated result is increasing as the missed reading ratio increasing, the standard deviation is still very small that less than 1%. Secondly, we vary the distribution weight of edge $e_{1.2}$ which is set to 20%, 40%, 50% and 70% and fix the missed reading ratio using Table 2, and generate 10000 tags for each data set. The distribution weight of edge $e_{1.3}$ is also variable according to Eq. (1). The test results are shown in Table 5 and Fig. 8. In Fig. 8, we see that the standard deviation of distribution weight varies very small as distribution weight changing. From all the test results, we can see that our simulation tool can provide RFID data with special data features.

5   Conclusion

Testing the functionalities of an RFID middleware is labor-intensive and expensive as we need to construct various RFID environments for generating RFID data streams. To facilitate data generation, we propose a graph model to simulate the real RFID environments. To make generated data have same characteristics with real RFID streaming data, we define several parameters for configuring and controlling the graph model. And then we implement a simulation test tool based on the graph model. To verify the correctness of generated RFID data we define several measurement metrics and perform several experiments. Through all the experiment results we find that the generated RFID data has very small difference with the expected data. Overall, using our simulation test tool, we can construct various test environments for generating RFID streaming data which can be used in RFID middleware test.

References

1. Wang, F.S., Liu, P.Y.: Temporal Management of RFID Data. In: Proceeding of the VLDB 2005, pp. 1128–1139 (2005)
2. Derakhshan, R., Orlowska, M., Li, X.: RFID Data Management: Challenges and Opportunities. In: IEEE International Conference on RFID, pp. 175–182 (2007)
3. EPCglobal standards, http://www.gs1.org/gsmp/kc/epcglobal
4. Lee, J.Y., Kim, N.S.: Performance Test Tool for RFID Middleware: Parameters, Design, Implementation, and Features. In: The International Conference on Advanced Communication Technology, ICACT 2006, vol. 1, pp. 149–152 (2006)
5. EPC Tag Data Standard, http://www.gs1.org/sites/default/files/docs/tds/tds_1_5-standard-20100818.pdf
6. Oh, G.O., Kim, D.Y., Kim, S.I., et al.: A Quality Evaluation Technique of RFID Middleware in Ubiquitous Computing. In: 2006 International Conference on Hybrid Information Technology (ICHIT 2006), vol. 2, pp. 730–735 (2006)
7. Park, C.K., Ryu, W.S., Hong, B.H.: RFID Middleware Evaluation Toolkit Based on a Virtual Reader Emulator. In: 1th International Conference on Emerging Databases, pp. 154–157 (2009)
8. Rifidi Tag Streamer, http://wiki.rifidi.org/index.php/Tag_Streamer_User%27s_Guide_1.1
9. RFID Test Centers, http://www.epcglobalinc.org/test_centers/accr_test_centers/
10. Zhang, H.P., Ryu, W.S., Hong, B.H.: A Test Data Generation Tool for Testing RFID Middleware. In: International Conference on Computers and Industrial Engineering (CIE), Awaji, Japan (2010)

Assisting the Design of XML Schema: Diagnosing Nondeterministic Content Models[*]

Haiming Chen and Ping Lu

State Key Laboratory of Computer Science
Institute of Software, Chinese Academy of Sciences
Beijing 100190, China
{chm,luping}@ios.ac.cn

Abstract. One difficulty in the design of XML Schema is the restriction that the content models should be deterministic, i. e., the unique particle attribution (UPA) constraint, which means that the content models are deterministic regular expressions. This determinism is defined semantically without known syntactic definition for it, thus making it difficult for users to design. Presently however, no work can provide diagnostic information if content models are nondeterministic, although this will be of great help for designers to understand and modify nondeterministic ones. In the paper we investigate algorithms that check if a regular expression is deterministic and provide diagnostic information if the expression is not deterministic. With the information provided by the algorithms, designers will be clearer about why an expression is not deterministic. Thus it contributes to reducing the difficulty of designing XML Schema.

Keywords: XML Schema, deterministic content models, diagnostic information.

1 Introduction

Extensible Markup Language (XML) has been popular for the Web and other applications. Usually in applications XML data are provided with schemas that the XML data must conform to. These schemas are important for solving problems and improving efficiency in many tasks of XML processing, for example, in query processing, data integration, typechecking, and so on. Among the many schema languages for XML, XML Schema is recommended by W3C and has been the most commonly used one. It is not easy, however, to design a correct XML Schema: investigation reveals that many XML Schema Definitions in practice have errors [3,7]. One difficulty in designing XML Schema is the restriction that the content models should be deterministic, i. e., the unique particle attribution (UPA) constraint, which means that the content models are deterministic regular expressions. In another XML schema language recommended by W3C, Document Type Definition (DTD), deterministic content models are also used.

---

[*] Work supported by the National Natural Science Foundation of China under Grants 61070038, 60573013, and the ISCAS Grand Project under number YOCX285056.

X. Du et al. (Eds.): APWeb 2011, LNCS 6612, pp. 301–312, 2011.

A regular expression is deterministic (or one-unambiguous) if, informally, a symbol in the input word should be matched uniquely to a position in the regular expression without looking ahead in the word. This determinism, however, is defined semantically without known syntactic definition for it, thus making it difficult for users to design.

Brüggemann-Klein [4] showed that deterministic regular expressions are characterized by deterministic Glushkov automata, and whether an expression is deterministic is decidable. In [5] an algorithm is provided to decide whether a regular language, given by an arbitrary regular expression, is deterministic, i. e., can be represented by a deterministic regular expression. An algorithm is further given there to construct equivalent deterministic regular expressions for nondeterministic expressions when they define deterministic languages. The size of the constructed deterministic regular expressions, however, can be exponentially larger than the original regular expressions. In [1] an algorithm is proposed to construct approximate deterministic regular expressions for regular languages that are not deterministic. Bex et al. [2] further provide improved algorithms for constructing deterministic regular expressions and constructing approximations.

All existing work, however, cannot provide diagnostic information for nondeterministic expressions. Consider if a design tool can locate the error positions and tell the type of error making the expression nondeterministic, just like what compilers or other program analysis tools do for programs, then it will be greatly helpful for designers to understand and modify nondeterministic expressions. Note here *error* is used to denote what make an expression nondeterministic.

In this paper we tackle the above issue. Our aim is to provide as much diagnostic information for errors as possible when expressions are nondeterministic. The idea is, if we can check expressions at the syntactic level, then it is easier to locate errors. Following [5], we designed a conservative algorithm, which will accept deterministic expressions, but may also reject some deterministic expressions. We improved the conservative algorithm by borrowing semantic processing and obtained an exact checking algorithm, which will accept all deterministic expressions, and reject only nondeterministic expressions. But it will not provide as precise diagnostic information for some nondeterministic expressions as for the other nondeterministic expressions. We further presented a sufficient and necessary condition for deterministic expressions, which leads to another exact checking algorithm for deterministic expressions. With the information provided by the algorithms, designers will be clearer about why an expression is not deterministic. Thus the difficulty of designing deterministic expressions, and, of designing XML Schema at large, is lowered.

We also implemented the algorithm in [5] which constructs an equivalent deterministic expression if an expression is not deterministic but defines a deterministic language, as an alternative way to obtain deterministic expressions.

We conducted several preliminary experiments, and the experimental results are presented. The main contributions of the paper are as follows.

(1) We propose the notion of diagnosing deterministic regular expressions. While similar notion has been familiar in other areas of software, this notion is missing for deterministic regular expressions due to the semantic nature.

(2) We prove several properties of deterministic regular expressions, which are the base of the algorithms.

(3) We present several algorithms for checking deterministic regular expressions and providing diagnostic information. The algorithms presented in the paper check if regular expressions are deterministic at the syntactic level of the expressions. The first algorithm uses stronger but syntactical conditions and is conservative. The second algorithm is exact, but may obtain imprecise diagnostic information for some regular expressions. The third algorithm is also exact, and may obtain diagnostic information for all nondeterministic regular expressions. The work of the paper can be considered as a first step towards syntactic solutions of detecting deterministic regular expressions.

The above algorithms can be used in schema design tools in which designers are able to find and fix bugs iteratively.

There is another issue that is connected with the present issue. That is, since deterministic regular expressions denote a proper subclass of regular languages, if a nondeterministic expression does not define a deterministic language, then the expression cannot have any equivalent deterministic expression. So when an expression is nondeterministic it is useful to tell the designer in the mean time whether the expression denotes a deterministic language.

Section 2 introduces definitions. The algorithms are presented in Section 3, with a discussion of diagnostic information and illustration of examples. Experiments are presented in Section 4. Section 5 contains a conclusion.

2   Preliminaries

Let Σ be an alphabet of symbols. The set of all finite words over Σ is denoted by Σ^*. The empty word is denoted by ε. A regular expression over Σ is $\emptyset, \varepsilon$ or $a \in \Sigma$, or is the union $E_1 + E_2$, the concatenation $E_1 E_2$, or the star E_1^* for regular expressions E_1 and E_2. For a regular expression E, the language specified by E is denoted by $L(E)$. Define $EPT(E) = true$ if $\varepsilon \in L(E)$ and $false$ otherwise. The size of E is denoted by the number of symbol occurrences in E, or the alphabetic width of E. The symbols that occur in E, which is the smallest alphabet of E, is denoted by Σ_E.

We require an expression to be star-reduced: any subexpression of the form $(E^*)^*$ is reduced to E^* in the expression.

For a regular expression we can mark symbols with subscripts so that in the marked expression each marked symbol occurs only once. For example $(a_1 + b_2)^* a_3 b_4 (a_5 + b_6)$ is a marking of the expression $(a + b)^* ab(a + b)$. The marking of an expression E is denoted by $\overline{E}$. The same notation will also be used for dropping of subscripts from the marked symbols: $\overline{\overline{E}} = E$. The subscribed symbols are called $positions$ of the expression. We extend the notation for words and automata in the obvious way. It will be clear from the context whether $\bar{\ }$ adds or drops subscripts.

Definition 1. *An expression E is deterministic if and only if, for all words $uxv, uyw \in L(\overline{E})$ where $|x| = |y| = 1$, if $x \neq y$ then $\overline{x} \neq \overline{y}$. A regular language is deterministic if it is denoted by some deterministic expression.*

For an expression E over Σ, we define the following functions:

$$first(E) = \{a \mid aw \in L(E), a \in \Sigma, w \in \Sigma^*\}$$
$$last(E) = \{a \mid wa \in L(E), w \in \Sigma^*, a \in \Sigma\}$$
$$follow(E, a) = \{b \mid uabv \in L(E), u, v \in \Sigma^*, b \in \Sigma\}, \text{for } a \in \Sigma$$

One can easily write equivalent inductive definitions of the above functions on E, which is omitted here.

Define $followlast(E) = \{b \mid vbw \in L(E), v \in L(E), v \neq \varepsilon, b \in \Sigma, w \in \Sigma^*\}$. An expression E is in *star normal form* (SNF) [4] if, for each starred subexpression H^* of E, $followlast(\overline{H}) \cap first(\overline{H}) = \emptyset$ and $\varepsilon \notin L(H)$.

3 Determining and Diagnosing Nondeterministic Expressions

3.1 Algorithms

The Glushkov automaton was introduced independently by Glushkov [6] and McNaughton and Yamada [8]. It is known that deterministic regular expressions can be characterized by Glushkov automata.

Lemma 1. ([5]) *A regular expression is deterministic if and only if its Glushkov automaton is deterministic.*

Lemma 1 has led to an algorithm to check if an expression is deterministic [4]. We call this algorithm the semantic checking algorithm in the paper.

If diagnostic information about why a regular expression is not deterministic is required, we need a syntactic characterization, or at least some syntactic properties, of deterministic expressions. Such a characterization, however, is not known presently. We started from a property of deterministic expressions in star normal form in [5] by modifying it to more general expressions. To do this, Lemma 2 is required.

Lemma 2. *For a regular expression E, if $followlast(E) \cap first(E) = \emptyset$ then $followlast(\overline{E}) \cap first(\overline{E}) = \emptyset$.*

The following is a modified version of the afore mentioned property proved in [5][1].

---

[1] The original property in [5] requires the expression to be in star normal form. It was mentioned in the proof that this condition can be removed with slight change of the property.

Lemma 3. *Let E be a regular expression.*
If $E = \emptyset, \varepsilon$, or $a \in \Sigma$, then E is deterministic.
If $E = E_1 + E_2$, then E is deterministic iff E_1 and E_2 are deterministic and $first(E_1) \cap first(E_2) = \emptyset$.
If $E = E_1 E_2$, then if $L(E) = \emptyset$, then E is deterministic, otherwise E is deterministic iff (1) E_1 and E_2 are deterministic, $\varepsilon \in L(E_1)$, $first(E_1) \cap first(E_2) = \emptyset$, and $followlast(E_1) \cap first(E_2) = \emptyset$, or (2) E_1 and E_2 are deterministic, $\varepsilon \notin L(E_1)$, and $followlast(E_1) \cap first(E_2) = \emptyset$.
If $E = E_1^*$, then E is deterministic and $followlast(\overline{E_1}) \cap first(\overline{E_1}) = \emptyset$ iff E_1 is deterministic and $followlast(E_1) \cap first(E_1) = \emptyset$.

The last case of Lemma 3 can be proved from a modification of the proof in [5] in addition with Lemma 2. The other cases are proved in [5].

This property, however, is not a sufficient and necessary condition of deterministic expressions, since in the last case the expression E is accompanied with an additional condition. Actually, in this case there are examples in which either E is deterministic and $followlast(E_1) \cap first(E_1) \neq \emptyset$, or E_1 is deterministic but E is not deterministic. In other words, the condition that E_1 is deterministic and $followlast(E_1) \cap first(E_1) = \emptyset$ is too strong to ensure E_1^* to be deterministic.

Proposition 1. *For a regular expression $E = E_1^*$,*
(1) E can be deterministic when $followlast(E_1) \cap first(E_1) \neq \emptyset$.
(2) If E is deterministic then E_1 is deterministic, but not vice versa.

On the other hand, up to date there is no known simpler condition for the last case of Lemma 3.

In order to check if an expression is deterministic and locate error position when the expression is not deterministic, one way is to directly use the property of Lemma 3, thus resulting in a conservative algorithm, i. e., if it accepts an expression, then the expression must be deterministic, but it may also reject some deterministic expressions.

To obtain an exact checking algorithm we make some compromise and use the following strategies, based on the above properties. When $e = e_1^*$ we first check if e_1 is deterministic using Lemma 3. If e_1 is not deterministic, then e is not either by Proposition 1. Furthermore, if the erroneous part in e_1 is not a starred subexpression, then precise diagnostic information can be obtained. If e_1 is deterministic, and $followlast(e_1) \cap first(e_1) = \emptyset$, then e is deterministic by Lemma 3. Otherwise, we encounter the only uncertain case: e_1 is deterministic and $followlast(e_1) \cap first(e_1) \neq \emptyset$, and shift to semantic level and use the semantic checking algorithm [4] to check if e is deterministic. If e is deterministic, then the algorithm proceeds smoothly without any impact of the semantic checking. If e is not deterministic, then we can only say that e is nondeterministic to the users, without any further diagnostic information. The resulting algorithm is exact: it will accept all deterministic expressions, and reject only nondeterministic expressions. Notice that the semantic checking algorithm runs in linear time [4], so will not much lower down the efficiency of the whole algorithm. The cost is, it will not provide as precise diagnostic information for some nondeterministic expressions as for the other nondeterministic expressions; When the erroneous subexpression is located by the semantic checking algorithm, diagnostic information will be less precise than in other situations. For example, when checking the expression $(c(ca + a)^*)^*$, the algorithm can only show that $(c(ca + a)^*)^*$ is not deterministic, and the subexpressions of it are deterministic.

Further, for any expression $E = E_1^*$, we give a sufficient and necessary condition of E being deterministic as follows.

Proposition 2. *For* $E = E_1^*$, E *is deterministic iff* E_1 *is deterministic and* $\forall y_1 \in followlast(\overline{E_1}), \forall y_2 \in first(\overline{E_1})$, *if* $\overline{y_1} = \overline{y_2}$ *then* $y_1 = y_2$.

By Proposition 2 and using Lemma 3 we get a sufficient and necessary condition for deterministic expressions. This gives another algorithm which provide diagnostic information for all nondeterministic expressions.

The above proposition actually requires that if $followlast(E_1)$ and $first(E_1)$ have common elements, the intersection can only be in the same positions of E_1. To ease the computation and obtain more diagnostic information, we present the following inductive computation of the condition.

Definition 2. *The function* $\mathcal{P}(E)$ *which returns true or false is defined as*

$\mathcal{P}(\varepsilon) = \mathcal{P}(\emptyset) = \mathcal{P}(a) = true \quad a \in \Sigma$
$\mathcal{P}(E_1 + E_2) = \mathcal{P}(E_1) \wedge \mathcal{P}(E_2) \wedge (followlast(E_2) \cap first(E_1) = \emptyset)$
$\qquad \wedge (followlast(E_1) \cap first(E_2) = \emptyset)$
$\mathcal{P}(E_1 E_2) = (\neg(EPT(E_1)) \wedge \neg(EPT(E_2)) \wedge (followlast(E_2) \cap first(E_1) = \emptyset)) \vee$
$\qquad (EPT(E_1) \wedge \neg(EPT(E_2)) \wedge \mathcal{P}(E_2) \wedge (followlast(E_2) \cap first(E_1) = \emptyset)) \vee$
$\qquad (\neg(EPT(E_1)) \wedge EPT(E_2) \wedge \mathcal{P}(E_1) \wedge (followlast(E_2) \cap first(E_1) = \emptyset)) \vee$
$\qquad (EPT(E_1) \wedge EPT(E_2) \wedge \mathcal{P}(E_1) \wedge \mathcal{P}(E_2) \wedge (followlast(E_2) \cap first(E_1) = \emptyset))$
$\mathcal{P}(E_1^*) = \mathcal{P}(E_1)$

Proposition 3. *Let an expression* E *be deterministic. The following two statements are equivalent.*
(1) $\forall y_1 \in followlast(\overline{E}), \forall y_2 \in first(\overline{E})$, *if* $\overline{y_1} = \overline{y_2}$ *then* $y_1 = y_2$.
(2) $\mathcal{P}(E) = true$.

In the following we give the three algorithms. The conservative one is *deterministic_c*, the second one is *deterministic*, and the third one is *deterministicpl*.

All of the algorithms take as input a regular expression, and output a Boolean value indicating if the expression is deterministic as well as diagnostic information if the expression is not deterministic. In the algorithm *deterministc_c*, *lambda(e)* is just the function $EPT(e)$ which returns true if $\varepsilon \in L(e)$ and false otherwise. *print_err* is not a real function here, it just indicates some statements in the implementation that print current error information. For example, in line 5 *print_err* should print that $first(e_1) \cap first(e_2)$ is not empty, and in line 9 *print_err* should print that $followlast(e_1) \cap first(e_2)$ is not empty. It is not difficult to indicate the positions of e_1 and e_2 by the parse tree of the whole expression. The difference between *deterministic_c* and the other algorithms only starts from line 16. In *deterministic*, *isdtre* is the semantic checking algorithm [4] to check if a regular expression is deterministic. It is used in

---

Algorithm 1. *deterministic_c(e)*: Boolean

Input: a regular expression e

Output: true if e is deterministic or false and diagnostic information otherwise

1. **if** $e = \emptyset, \varepsilon$, or a for $a \in \Sigma$ **then return** true
2. **if** $e = e_1 + e_2$ **then**
3.   **if** *deterministic_c*(e_1) and *deterministic_c*(e_2) **then**
4.   **if** $first(e_1) \cap first(e_2) = \emptyset$ **then return** true **else**
5.   \{*print_err;* **return** false\}
6.   **else return** false
7. **if** $e = e_1 e_2$ **then**
8.   **if** *deterministic_c*(e_1) and *deterministic_c*(e_2) **then**
9.   **if** $followlast(e_1) \cap first(e_2) \neq \emptyset$ **then** \{*print_err;* **return** false\} **else**
10.   **if** $lambda(e_1)$ **then**
11.   **if** $first(e_1) \cap first(e_2) \neq \emptyset$ **then** \{*print_err;* **return** false\} **else**
12.   **return** true
13.   **else return** true
14.   **else return** false
15. **if** $e = e_1^*$ **then**
16.   **if** *deterministic_c*(e_1) **then**
17.   **if** $followlast(e_1) \cap first(e_1) = \emptyset$ **then return** true **else**
18.   \{*print_err;* **return** false\}
19.   **else return** false

---

the case of $e = e_1^*$. In *deterministicpl*, $P(e)$ calculates $\mathcal{P}(e)$ and print diagnostic information if e is nondeterministic. When checking the previous expression $(c(ca + a)^*)^*$, the algorithm *deterministicpl* will provide the following information: $followlast((ca + a)^*) \cap first(c) \neq \emptyset$.

Theorem 1. *If deterministic_c(e) returns true, then e is deterministic.*

Proof. It follows directly from Lemma 3.   □

Theorem 2. *deterministic(e) returns true if and only if e is deterministic.*

Proof. It follows from Lemma 3, Lemma 1, and Proposition 1.   □

Theorem 3. *deterministicpl(e) returns true if and only if e is deterministic.*

Proof. It follows from Lemma 3, Proposition 2, and Proposition 3.   □

To implement the algorithms we first calculate the $first, followlast$ sets and EPT functions for each subexpressions, then run the algorithms presented above. Suppose the regular expression is E. The first calculation can be done on the syntax tree of E [4], which can be computed in $O(m^2)$ time where m is the size of E. In the algorithms, emptiness test of $first(E_1) \cap first(E_2)$ or $followlast(E_1) \cap first(E_2)$ for subexpressions E_1, E_2 can be completed in $O(2m)$ time with an auxiliary array indexed by every letters in the alphabet of E. The algorithms may conduct the test at every inner node on a traversal of the syntax tree of E, which totally takes $O(m^2)$ time. So the time complexity of the algorithms is $O(m^2)$. Notice here we do not take into account the time for printing diagnostic information.

Algorithm 2. *deterministic(e)*: Boolean

Input: a regular expression e

Output: true if e is deterministic or false and diagnostic information otherwise

    (1 – 14 are the same as *deterministic_c*)

15. if $e = e_1^*$ then

16.   if **not** *deterministic(e_1)* **then return** false

17.   if $followlast(e_1) \cap first(e_1) = \emptyset$ **then return** true **else**

18.   if *isdtre(e)* **then return** true **else**

19.   {*print_err*; **return** false}

Algorithm 3. *deterministicpl(e)*: Boolean

Input: a regular expression e

Output: true if e is deterministic or false and diagnostic information otherwise

    (1 – 14 are the same as *deterministic_c*)

15. if $e = e_1^*$ then

16.   if **not** *deterministicpl(e_1)* **then return** false

17.   if $P(e_1)$ **then return** true **else**

18.   **return** false

3.2 Reporting Errors

Three kinds of error information can be reported by the above algorithms:

- Error location. Using the parse tree of an expression, the subexpressions that cause an error can be located precisely.

- Types of errors. There are roughly the following types of errors:

(1) $first$-$first$ error, indicating $first(e_1) \cap first(e_2) \neq \emptyset$. It can further be classified into $first$-$first$-+ and $first$-$first$-., corresponding to a (sub)expression $e = e_1 + e_2$ and $e = e_1e_2$ respectively.

(2) $followlast$-$first$ error, indicating $followlast(e_1) \cap first(e_2) \neq \emptyset$. Similarly it is also classified into $followlast$-$first$-+, $followlast$-$first$-., and $followlast$-$first$-*.

(3) A starred (sub)expression is not deterministic, indicating the semantic checking error in *deterministic*.

(4) $followlast$-$first$-nd error, indicating an error of $followlast(e_1) \cap first(e_2) \neq \emptyset$ in $\mathcal{P}(e)$, corresponding to a violation of the condition is Proposition 2.

- Other diagnostic information. For example, for a type (1) error, the two $first$ sets can be provided. For a type (2) error, besides the $followlast$ and $first$ sets, the symbols in the $last$ set that cause the overlap of the $followlast$ and $first$ sets, and symbols in the $follow$ set of the previous symbols, can be provided.

In addition to the above information, other information like parse trees of expressions, the Glushkov automata, and the matching positions of a word against an expression can also be displayed, thus providing debugging facilities of expressions.

Of course, in an implementation of a tool, the above information can be implemented such that the users can select which information to display.

3.3   Examples

Example 1. Suppose one want to write a schema for papers with no more than two authors. The content model of the papers can be defined as
    `Title, Author?, Author, Date, Abstract, Text, References` [2]
which equals to the following regular expression
    `Title, (Author+empty), Author, Date, Abstract, Text, References`
    By using the above algorithms, the following information is displayed[3]:

```
error: the expression is not deterministic.
error found in: "Title, (Author+empty), Author"
hints: the sets of followlast((Title, (Author+empty))) and
       first(Author) have common elem
       followlast((Title, (Author+empty)))={Author(2)}
          trace: Title in last((Title, (Author+empty)))
                 follow((Title, (Author+empty)), Title)={Author}
       first(Author)={Author(3)}
```

Then the content model can be rewritten into the following:
    `Title, Author, Author?, Date, Abstract, Text, References`
which is deterministic.

Of course for some nondeterministic content models their equivalent deterministic ones are very difficult to find, as in the following example.

Example 2. $(a + b)^*a$ defines any string of a or b, including the empty word, followed by one a. The above algorithms will show that this expression is not deterministic, and this is because $followlast((a + b)^*) \cap first(a) \neq \emptyset$. For the expression it is difficult to write an equivalent deterministic regular expression.

However, using the diagnostic information, the designer can change the design to circumvent the error: $(a + b)^*c$, and c is defined as a in another rule.

So the diagnostic information can
- help the designer to locate errors and rewrite nondeterministic expressions into correct ones, and
- help the designer to understand the reasons of errors, or change design to circumvent the errors.

4   Experiments

We have implemented the algorithms and performed some experiments. The algorithms were tested with randomly generated regular expressions in different

---

[2] The comma (,) denotes concatenation.
[3] Concatenation and union are assumed to be left associative as usual in expressions.

Table 1. Numbers of deterministic regular expressions

size	10	20	30	40	50	60, 70, ...
deterministic_c	208	48	4	0	0	0
others	356	163	43	12	3	0

Table 2. Errors found by the algorithms

size	deterministic_c		deterministicpl			deterministic		
	first-first	fola-first	first-first	fola-first	ff-nd	first-first	fola-first	star-exp
10	107	37	135	5	4	135	5	4
20	184	153	305	12	20	305	12	20
30	244	213	422	9	26	422	9	26
40	263	225	448	16	24	448	16	24
50	266	231	450	23	24	450	23	24
60	274	226	454	23	23	454	23	23
70	271	229	451	18	31	451	18	31
80	267	233	463	15	22	463	15	22
90	259	241	460	20	20	460	20	20
100	300	200	465	21	14	465	21	14
110	285	215	472	15	13	472	15	13
120	280	220	466	20	14	466	20	14
130	282	218	469	12	19	469	12	19
140	290	210	469	10	21	469	10	21
150	299	201	461	22	17	461	22	17
160	296	204	465	17	18	465	17	18

sizes. The sizes of regular expressions range from 10 to 160 every time increased by 10 in the experiments, with 500 expressions in each size. The size of the alphabet was set to 40. The algorithms were implemented in C++. The experiments were run on Intel core 2 Duo 2.8GHz, 4GB RAM.

Table 1 shows the numbers of deterministic regular expressions determined by each algorithm, in which the first line denotes the sizes of expressions, the remaining lines indicate the numbers of deterministic regular expressions identified by different algorithms. 'deterministic_c' denotes the algorithm *deterministic_c*, 'others' denotes the semantic checking algorithm, the algorithm *deterministic*, and *deterministicpl*. We can observe the number of deterministic expressions decreases when the size of regular expressions grows. Intuitively, this is because the possibility of the occurrences of a same symbol increases in one expression when the size of expressions increases, which increases the possibility of nondeterminism. Actually, when size is greater than 50, all expressions are nondeterministic. The numbers of deterministic regular expressions identified by the algorithm *deterministic_c* are less than the numbers identified by the other algorithms, which coincides with that *deterministic_c* is conservative. The numbers of deterministic expressions determined by the algorithms *deterministic* and *deterministicpl* are identical with the numbers determined by the semantic checking algorithm, reflecting that *deterministic* and *deterministicpl* exactly detect all deterministic expressions.

The errors in the tested regular expressions found by the algorithms are shown in Table 2. The first column in the table shows the size of regular expressions. The other columns include the numbers of different types of errors caught by *deterministic_c*, *deterministic* and *deterministicpl* in the experiment, in which `first-first`, `fola-first`, `star-exp`, and `ff-nd` correspond to the types (1), (2), (3), and (4) of errors presented in Section 3.2, respectively. It shows that the most common type of errors in the experiment is the *first-first* errors. Also the numbers of errors for each of `first-first` and `fola-first` of *deterministicpl* and *deterministic* are identical, and the numbers of errors for `ff-nd` and `star-exp` are identical too. This is because the two algorithms both exactly check whether an expression is deterministic, and differ only in the processing of stared subexpressions. When a starred subexpression has an error, each of the algorithms will detect one error.

Figure 1 shows the average time for detecting one nondeterministic regular expressions by the algorithms. Each value is obtained by the time to check the total amount of nondeterministic regular expressions in each size divided by the number of nondeterministic regular expressions in that size. The time used to print diagnostic information in the programs is not included. The algorithms spend averagely less than 6 milliseconds for a regular expression of size 160, thus are efficient in practice. It is not strange that the semantic checking algorithm runs faster, since the algorithms presented in this paper will do more than the semantic algorithm. On the other hand, the implementation of the algorithms presented in the paper still have much room for improvement. In the experiment *deterministicpl* runs almost as faster as *deterministic_c* and *deterministic*. Thus we can use *deterministicpl* for diagnostic tasks.

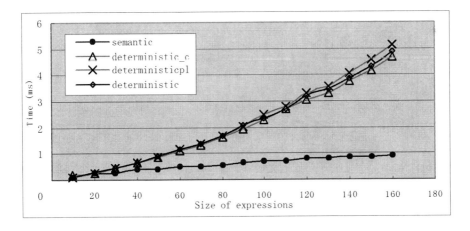

Fig. 1. Average running time of the algorithms

5 Conclusion

Due to its semantic definition, a deterministic regular expression is hard to design and understand, and semantic checking techniques can only answer yes or no. The paper presented several algorithms as an attempt to diagnose nondeterministic regular expressions, making it possible to analyze and give hints to errors thus reducing the difficulty of designing deterministic content models. This would be convenient for designers to utilize their knowledge and intuition. In the future, it would be useful to find more syntactic conditions for deterministic regular expressions, in the hope of more detailed revealing of errors. A presently unclear question is, can we use other more intuitive conditions to replace the emptiness condition of $followlast\text{-}first$ or $first\text{-}first$ intersection? The approaches to display diagnostic information effectively also constitute a significant aspect. The diagnostic information offered by the algorithms may also be used to generate counter examples of nondeterministic content models, which is also helpful for designers, but is not discussed in the paper. It is possible to integrate the above techniques in tools to provide analyzing and debugging facilities for content models.

References

1. Ahonen, H.: Disambiguation of SGML content models. In: Nicholas, C., Wood, D. (eds.) PODDP 1996 and PODP 1996. LNCS, vol. 1293, pp. 27–37. Springer, Heidelberg (1997)
2. Bex, G.J., Gelade, W., Martens, W., Neven, F.: Simplifying XML schema: effortless handling of nondeterministic regular expressions. In: SIGMOD 2009, pp. 731–744 (2009)
3. Bex, G.J., Neven, F., Bussche, J.V.: DTDs versus XML schema: a practical study. In: WebDB 2004, pp. 79–84 (2004)
4. Brüggemann-Klein, A.: Regular expressions into finite automata. Theoretical Computer Science 120, 197–213 (1993)
5. Brüggemann-Klein, A., Wood, D.: One-unambiguous regular languages. Information and Computation 142(2), 182–206 (1998)
6. Glushkov, V.M.: The abstract theory of automata, Russian Math. Surveys 16, 1–53 (1961)
7. Martens, W., Neven, F., Schwentick, T., Bex, G.J.: Expressiveness and complexity of XML Schema. ACM Transactions on Database Systems 31(3), 770–813 (2006)
8. McNaughton, R., Yamada, H.: Regular expressions and state graphs for automata. IRE Trans. on Electronic Computers 9(1), 39–47 (1960)

Improving Matching Process in Social Network Using Implicit and Explicit User Information

Slah Alsaleh, Richi Nayak, Yue Xu, and Lin Chen

Computer Science Discipline, Queensland University of Technology
Brisbane, Australia
{s.alsaleh,r.nayak,xu.yue}@qut.edu.au,
133.chen@student.qut.edu.au

Abstract. Personalised social matching systems can be seen as recommender systems that recommend people to others in the social networks. However, with the rapid growth of users in social networks and the information that a social matching system requires about the users, recommender system techniques have become insufficiently adept at matching users in social networks. This paper presents a hybrid social matching system that takes advantage of both collaborative and content-based concepts of recommendation. The clustering technique is used to reduce the number of users that the matching system needs to consider and to overcome other problems from which social matching systems suffer, such as cold start problem due to the absence of implicit information about a new user. The proposed system has been evaluated on a dataset obtained from an online dating website. Empirical analysis shows that accuracy of the matching process is increased, using both user information (explicit data) and user behavior (implicit data).

Keywords: social matching system, recommender system, clustering users in social network.

1 Introduction and Related Work

A social matching system can be seen as a recommender system that recommends people to other people instead of recommending products to people. Recommending people is more complicated and sensitive compared with recommending products (Tobias, 2007). The reason behind this is the special nature of the relationship between the users in social matching systems like online dating networks. Analysis of an underlying online dating network shows that it is not always true when two users share the same attributes (such as values that describe themselves including age, personality, and smoking habit), then they can be recommended to each other. As a result, the standard recommendation techniques may not be appropriate to match people in online dating networks. Another issue is the increasing numbers of social network members along with their information, which builds up the computational complexity. In Facebook, for example, there are more than 500 million active users, with 50% of the active users logging on in any given day (www.facebook.com, 2010). Another example is RSVP which considered as the largest dating network in

X. Du et al. (Eds.): APWeb 2011, LNCS 6612, pp. 313–320, 2011.
© Springer-Verlag Berlin Heidelberg 2011

Australia, with more than 2 million members and an average of 1,000 new members every day (http://www.rsvp.com.au/, 2010).

In such social networks, a lot of personal information about users is required. This information can be divided into explicit and implicit information. The explicit information is collected by asking the user to answer a series of questions which represent his/her characteristics and preferences. Many social networks also record user's behaviour on the network (such as sending messages, watching profiles). A user's online activities are referred to as his/her implicit information. Three learning approaches are widely used in social matching systems: rules-based approach, content-based approach and collaborative approach (Brusilovsky, Kobsa, & Nejdl, 2007).

Even though social matching systems are increasingly used and attract more attention from both academic and industrial researchers, many aspects still need to be explored and developed. A major issue is the accuracy of matching users in social networks. A two-way matching process is needed to deal with the compatibility between users whom have been recommended. The majority of existing social matching systems consider the similarity between user x and user y and produce the recommendation without checking that user y is also compatible with user x. Terveen and McDonald (2005) have also raised some general questions, namely: what is the information that should be used to achieve high quality matching? How does the system make a good match? Is it possible to evaluate the matching process and then use the evaluation outcome as a feedback? The authors argue that data mining techniques can be used to improve recommendation in social networks (Terveen & McDonald, 2005).

Clustering is one of the data mining techniques that can be used to improve the matching process in social networks (Eirinaki & Vazirgiannis, 2003). It reduces the data size and cuts down the complexity that the matching process has. Moreover, a clustering technique assists the recommendation process in social networks by overcoming some existing problems such as cold start and sparsity. For example, cold start problem occurs when a new user joins a social networks and the system has not gather enough information about the user to be matched. Assigning this user to an existing cluster which already has been matched with the appropriate opposite gender cluster allows him\her to receive recommendation instantly. For these reasons, clustering is used in this paper to group together users who have similar characteristics.

This paper proposes a social matching system that considers both explicit data (information that users provide about themselves) and implicit data (user activities on the social networks) to improve the matching process in social networks. It utilizes the dating type of social matching system in which opposite gender users are recommended to each other. Explicit data are used to cluster male and female users into homogonous groups. Then the male clusters are matched with the female clusters using implicit data and users are recommended to each other. The proposed system is evaluated on a dataset obtained from an online dating website. Empirical analysis demonstrated that the proposed system improves the accuracy of the recommendations from 13.9% to 50.32%, reduces the matching complexity by 93.95% and overcomes cold-start problem. The matching process in a majority of social networks is based on explicit data that users provide to describe themselves (Boyd & Ellison, 2008). However, the proposed system proved that using implicit data that represents the users' activities in the matching process increases the recommendation accuracy by 36.42%.

2 The Proposed Method

A.   Preliminaries

There are many types of social networks according to their purpose of use. An online dating system is a type of social network that aims to introduce people to their potential partners. Online dating systems usually keep two different types of data about their users; implicit data and explicit data. The implicit data are collected by recording the users' activities while they are interacting with the system. These data include profiles that a user has seen and messages that they have sent. The explicit data are usually obtained through an online questionnaire; the users are asked to answer a number of questions which represent their personality traits and interests. The explicit data can be divided into two subsets, data about the user and data about the preferred partner.

Let U be the users in a social network, $U = \{u_1, u_2, \dots u_n\}$, where u_i is either a male or a female user. Each user has a set of attributes $attr(u_i) = \{o_1, o_2, \dots o_x, p_1, p_2, \dots p_g\}$ where o_i is an attribute that specifically describes the user (such as age, height, education,... etc) and p_i is an attribute that describes the preferred partner. Users are allowed to have one value for each o_i attribute; however, they may have a null, single or multiple values for each p_i attribute.

A user u_i also performs some activities on the network such as viewing another user's profile $V(u_1, u_2)$ where u_1 viewed the profile of u_2. This activity shows that u_1 may be interested in contacting u_2. Other activities usually include communicating with short pre-defined messages $(kiss)^1$ and long free-text messages $(email)^2$. Sending a kiss $K(u_1, u_2, k_t, k_r)$ is another activity that confirms further interest where u_1 sends a kiss k_t that contains a pre-defined message to u_2. Users can select one of a variety of k_t that represents the users' feelings. u_2 can reply to the kiss sent by u_1 by using one of several pre-defined messages k_r which vary between a positive and a negative reply. k_r may be null indicating the target user has not responded to the sender's request. Furthermore, u_1 is allowed to send emails $E(u_1, u_2)$ to another user u_2.

In this paper, both the user profiles and user activities are employed to improve the recommendations in the proposed system. User profiles are used to cluster similar users in groups according to their profiles' similarities. Clustering users into groups overcomes two major problems in the matching process, which are cold start problem and the data sparsity, and results in increasing the matching accuracy and efficiency, as the findings show. Once users are clustered, data obtained from the users' activities (implicit data) are used to match the users' clusters with each other. The clustering and matching phases are discussed in the following in more detail.

B.   Clustering users in social networks

In order to cluster users in social networks the data need to be pre-processed. The pre-processing includes data integration and transformation. Once the data have been integrated and transformed, they are then divided into two groups: male users denoted as $U_m = \{u_1, u_2, \dots u_m\}$, and female users denoted as $U_f = \{u_1, u_2, \dots u_f\}$. This prepares the data to be clustered in the next phase. The male and female users are clustered

---

[1] In this paper, we call the pre-defined message as kiss.
[2] In this paper, the long free-text messages are called as email.

separately in order to match the male clusters with the female clusters using the clusters centroid, and then recommend users to each other. The male users are clustered based on their own attributes $\{o_1, o_2, ... o_x\}$, while the female users are clustered based on their preferred partners' attributes $\{p_1, p_2, ... p_g\}$. This is done to ensure high accuracy matching. Using both their own and preferred attributes in clustering the users makes it very difficult to find a male cluster that is similar to a female cluster. For example, while two users are similar in their own attributes, they may be very different in what they are looking for in their partners.

C. Matching clusters and recommendation

After clustering male and female users, two processes are performed: matching clusters and recommending clusters' members to each other. Let C_m be the clustering solution grouping the male users containing c clusters, $C_m = \{C_{m1}, C_{m2}, ... C_{mc}\}$. Let a_{ml} be the centroid vector of cluster C_{ml} represented as $\{a_{ml1}, a_{ml2}, ... a_{mlx}\}$. Let C_f be the clustering solution grouping the female users containing d clusters, $C_f = \{C_{f1}, C_{f2}, ... C_{fd}\}$. Let a_{fl} be the centroid vector of cluster C_{fl} represented as $\{a_{fl1}, a_{fl2}, ... a_{flg}\}$.

This process of matching all clusters is based on the communications between clusters' members utilizing the implicit data. As mentioned previously, social networks contain some implicit data that can be used to improve the recommendation. Users interact with each other by sending kisses $K(u_1, u_2, k_t, k_r)$, emails $E(u_1, u_2)$ and by viewing each other's profiles $V(u_1, u_2)$. These implicit data can be used to match the users' clusters. In the proposed system, the successful kiss interaction $K(u_1, u_2, k_t, k_r)$ between a male user's cluster and the rest of female users' clusters are assessed to determine the pair of clusters that has more interactions between them; these are then recommended to each other. A kiss is defined as a successful kiss when it received a positive reply k_r which means that the opposite party is also interested in this relationship. The communication score $CommScore$ can be presented as follows.

$$CommScore(C_{m1}, C_{f1}) = \sum_{\substack{0 \le i \le \#C_{m1} \\ 0 \le j \le \#C_{f1}}} K(u_i, u_j, k_t, k_r)$$

Where

$$K(u_i, u_j, k_t, k_r) = \begin{cases} 1, & k_r \text{ is positive} \\ 0, & Otherwise \end{cases}$$

When the matching process is completed, the recommendation phase takes place. In this phase, recommendations are presented to the users in ranking order of users' compatibility scores. The compatibility scores are calculated between the members of the pair of matched (or compatible) clusters according to members' profile and preference similarity. The user profile vector of the male user $u_m \in C_{mc}$ is compared with the preference vector of all female users $u_f \in C_{fd}$. The compatibility score $UserSim$ can be presented as follows.

$$UserSim(u_m, u_f) = \sum_{i=1}^{x} sim(o_{mi}, p_{fi})$$

$$Where\ sim(o_{mi}, p_{fi}) = \begin{cases} 1, & o_{mi} = p_{fi} \\ 0, & Otherwise \end{cases}$$

Figure 1 explains the proposed matching algorithm

Input: Male and female dataset;
Output: List of ranked recommendation for each user;
Begin
1 Cluster male users based on their own attributes $\{o_1, o_2 ,... o_x\}$ into c number of groups
2 Cluster female user based on their preferences $\{p_1, p_2 ,... p_g\}$ into d number of groups
3 For each male cluster C_{mi}
    Find a female cluster C_{fi} that has highest successful communication with C_{mi}

4 For each user $u_m \in C_{mi}$
 Calculate the compatibility with all user in C_{fi} where C_{fi} is *best matched cluster with C_{mi}*
5 Present top n recommendations to the users in ranking order of users' compatibility scores
UserSim.

Fig. 1. High level definition of the proposed method

3 Experiments and Discussion

The proposed system has been tested on a dataset obtained from a real online dating website that contains more than two million users. A subset of data were used showing all the active users in a period of three months, as the data analysis shows that an average user is active for three months and at least initiates one communication channel. Therefore, the proposed system targets the users who are active within three months.

A. Evaluation Measures

Success rate and recall are used to evaluate the accuracy of the proposed social matching recommendation system. They evaluate the accuracy of recommending users in recommendation systems. The success rate was used to measure the probability of recommendation being successful as indicated by the (positive) kiss returned by the recommended partner. Recall was also used to measure the probability of recommending the right partners. Both are mathematically presented below.

$$\text{Success Rate} = \frac{\left|\{Successfull\_Kisses\} \cap \{All\_Kisses\}\right|}{\left|\{All\_kisses\}\right|}$$

$$\text{Recall} = \frac{\left|\{Kissed\_partners\} \cap \{Recommended\_Partners\}\right|}{\left|\{Kissed\_partners\}\right|}$$

B. Experiment Design

Experiments were conducted on a machine with 3GB of RAM and a 3.00GHz Intel Cor2Due processor. Experiments were conducted to evaluate the accuracy of recommendation in social network using the proposed method. The results are compared by the baseline results when the proposed method is not used. The baseline success rate that users achieve in their communications without using the proposed system is 13.9%. The proposed method uses the implicit information to match the clusters. The proposed method is also compared with a variation when the clusters are matched with the explicit information only. The profile and preference similarity are used to compare the clusters' centroids for both male and female clustering solutions and matching was done based on their explicit profile information rather than implicit information. In this case, the matching score $MatchScore$ is used as follows.

$$MatchScore(C_{m1}, C_{f1}) = \sum_{i=1}^{x} sim\,(a_{m1i}, a_{f1i})$$

$$Where\ sim\,(a_{m1i}, a_{f1i}) = \begin{cases} 1, & a_{m1i} = a_{f1i} \\ 0, & Otherwise \end{cases}$$

C. Results and Discussion

The first phase of the experiment was clustering male and female users individually into 100 clusters. We selected 100 clusters as this is the most appropriate value where the majority of clusters are homogeneous and have a similar average number of members per group. However, we had several clusters that contain higher number of users which indicates that there are some popular attributes that many users share. Oracle Data Miner (ODM) was used to cluster both male and female users. K-means clustering algorithm was used to cluster both datasets by using the Euclidean distance function. K-means was chosen because of its simplicity and speed which allows it to run on large datasets, as in social networks. The output of the clustering phase is summarized in Table 1.

Table 1. Clustering output

	Male users	Female users
No. of clusters	100	100
Average no. of users per cluster	324	89
Maximum no. of users per cluster	1069	1283
Minimum no. of user per cluster	38	2

The second phase of the experiment was matching male clusters with the female clusters and then recommending users to each other. In this phase, clusters were matched using successful communication. In this stage, we calculated the success rate when we matched the clusters based on the communication as well as when we matched them based on their similarity. The success rate increased from 0.31 when clusters were matched based on similarity to 0.78 when clusters were matched based on communication. This demonstrate that matching clusters based on the communications is more efficient as the success rate is more than double when compared to matching clusters based on their similarity. The main reason for the low success rate (when considering the similarity) is the sparsity that social network data have. As a result, the proposed system utilizes the communications to match clusters with each other.

Once the male clusters are matched with the female clusters, the recommendation task takes place. The top n female recommendation will be presented to each member of the male clusters. The ranking is based on the similarity between male and female users, as explained in Section 2. As shown in Table 2, the proposed system achieves a high success rate compared to the baseline system, especially with the top 1 to top 10

Table 2. Success rate and recall Vs. baseline

	Top 1	Top 5	Top 10	Top 20	Top 50	All
Success rate	50.32%	38.17%	28.14%	21.52%	17.04%	15.52%
Recall	2.45%	3.06%	3.89%	5.14%	5.79%	7.18%
Baseline success rate	13.9%					

recommendations. In terms of recall, our system gained the best recall when recommending all users with a percentage of 7.18%. The recall then decreases to reach 2.45% with the top 1 recommendation.

With the improvement in success rate, the proposed system also reduces the computational complexity of matching users in social networks. The baseline system has to compare every male user with all the female users. However, the proposed system limits the comparisons to be with an assigned cluster. For the dataset used in this experiment, instead of conducting 1443519630 comparing processes, we were able to reduce the number to be 92466720 which reduces the proposed system computational complexity by 93.95%.

4   Conclusion and Future Work

Many social matching systems are based on recommender system techniques. The difference is that the social matching systems recommend people to each other rather than recommend items to people. Recommending people to each other is much more complicated and challenging. Therefore, current social matching systems suffer from two issues, the computational complexity and the matching accuracy.

In this paper, a new hybrid social matching system was presented that uses implicit and explicit data to improve the recommendation process. The clustering technique was also used to reduce the computational complexity in the proposed system. Experimental results showed that the proposed system provides satisfactory and high-quality recommendations with reasonable computational complexity.

Matching users in social networks based on their implicit data is a promising research area with many aspects that need to be explored. Furthermore, data mining techniques including clustering and association rules can be employed to develop ways to match users in social networks.

References

Harper, F.M., Sen, S., Frankowski, D.: Supporting social recommendations with activity-balanced clustering. Paper presented at the The 2007 ACM conference on Recommender systems (2007)

Oinas-Kukkonen, H., Lyytinen, K., Yoo, Y.: Social Networks and Information Systems: Ongoing and Future Research Streams. Journal of the Association for Information Systems 11(2), 3 (2010)

Terveen, L., McDonald, D.W.: Social matching: A framework and research agenda. ACM Transactions on Computer-Human Interaction (TOCHI) 12(3), 401–434 (2005)

Boyd, D.M., Ellison, N.B.: Social network sites: Definition, history, and scholarship. Journal of Computer-Mediated Communication 13(1), 210–230 (2008)

Eirinaki, M., Vazirgiannis, M.: Web mining for web personalization. ACM Transactions on Internet Technology 3(1), 1–27 (2003)

Grcar, M.: User Profiling: Web Usage Mining. In: Proceedings of the 7th International Multi-conference Information Society IS (2004)

Lathia, N., Hailes, S., Capra, L.: Evaluating collaborative filtering over time. Paper presented at the Workshop on The Future of IR Evaluation (SIGIR), Boston (2009)

Mislove, A., Marcon, M., Gummadi, K.P., Druschel, P., Bhattacharjee, B.: Measurement and analysis of online social networks. Paper presented at the The 7th ACM SIGCOMM conference on Internet measurement (2007)

Mobasher, B.: Data mining for web personalization. In: Brusilovsky, P., Kobsa, A., Nejdl, W. (eds.) Adaptive Web 2007. LNCS, vol. 4321, p. 90. Springer, Heidelberg (2007)

Morgan, E.M., Richards, T.C., VanNess, E.M.: Comparing narratives of personal and preferred partner characteristics in online dating advertisements. Computers in Human Behavior 26(5), 883–888 (2010)

Dynamic Network Motifs: Evolutionary Patterns of Substructures in Complex Networks

Yutaka Kabutoya, Kyosuke Nishida, and Ko Fujimura

NTT Cyber Solutions Laboratories, NTT Corporation,
1-1 Hikari-no-oka, Yokosuka-shi, Kanagawa, 239-0847 Japan

Abstract. We propose an entirely new approach to understanding complex networks; called "dynamic network motifs (DNMs)." We define DNMs as statistically significant local evolutionary patterns of a network. We find such DNMs in the networks of two web services, Yahoo Answers and Flickr, and discuss the social dynamics of these services as indicated by their DNMs.

Keywords: link analysis, subgraph mining.

1 Introduction

Many studies in the fields of engineering and science have tackled complex network analysis. In recent years, improvements in computational power have advanced the field of network mining, i.e. identifying all subgraphs in a given network [6,7].

Network motif (NM) analysis is one of the subgraph mining methods proposed by Milo et al. [4,5]. Motifs are small (usually from three to seven nodes in size) connected subgraphs that appear in the given network at higher frequency than in the equivalent (in terms of the number of nodes and edges) random networks. NM analysis is attracting a lot of attention because it enables a fuller understanding of complex networks in terms of their local structure.

A recent extension to data mining attempted to cover the evolution of subgraphs [2]. In [2], since the statistical significance of the evolution of subgraphs is not still defined, no discussion was given on what evolution of subgraphs is important for a given network.

To achieve this target, we propose a method, called *dynamic network motif (DNM)* analysis, that can analyze the evolution of subgraphs statistically. We define DNMs as statistically significant local evolutionary patterns of a network.

In this paper, we apply our method to networks based on datasets of two actual web services, Yahoo Answers[1] and Flickr[2]. The DNMs found in the network from Yahoo Answers are distinct from the DNMs found in the network from Flickr: This is due to the difference in communication properties, which influences the evolution of the network. We consider that the networks of Yahoo Answers and Flickr are evolving under the pressures of expertise and politeness, respectively.

---

[1] http://answers.yahoo.com
[2] http://www.flickr.com

X. Du et al. (Eds.): APWeb 2011, LNCS 6612, pp. 321–326, 2011.

2   Conventional Method: Network Motifs

Before introducing our method, we review network motifs (NMs), which are the basis of our method. NMs are defined to be statistically significant patterns of local structures in real networks. They are being studied by many researchers, including Milo et al. [5].

Milo et al. started with network $G = \langle V, E \rangle$ where the interaction between nodes is represented by directed edges. The types of n-node subgraphs, which are defined to be subgraphs that have n nodes, are connected, and are not isomorphic to one another, were introduced. For example, there are 13 types of 3-node subgraphs as shown in Figure 1, and 199 types of 4-node subgraphs. The network is scanned for all possible n-node subgraphs (in many studies, $3 \leq n \leq 7$), and the number of occurrences of each type is recorded. To focus on those that are likely to be important, the real network is compared to K equivalent randomized networks. NMs are defined to be the type that recur in the real network with significantly higher frequency than in the equivalent networks.

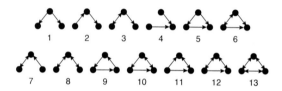

Fig. 1. All 13 types of triads

Furthermore, to compare the local structures of networks from different fields, Milo et al. [4] introduced the concept of the significance profile (SP) of a network. Here, m_i and $\hat{m}_{ik}$ denote the number of occurrences of type i in the real network and in the kth equivalent randomized network, respectively. First, the Z-score of the occurrences of type i is computed as:

$$z_i = \frac{m_i - \mu_i}{\sigma_i}, \tag{1}$$

where $\mu_i = \frac{1}{K}\sum_{k=1}^{K} \hat{m}_{ik}$ and $\sigma_i^2 = \frac{1}{K}\sum_{k=1}^{K}(\hat{m}_{ik} - \mu_i)^2$. The SPs are computed by normalizing the vector of Z-score $\boldsymbol{z} = \{z_i\}_i$ to length 1.

3   Proposed Method: Dynamic Network Motifs

We propose a method to analyze evolution of n node subgraphs (this paper mainly addresses the case of $n = 3$). We define dynamic network motifs (DNMs) as statistically significant evolutionary patterns of local structures in given networks.

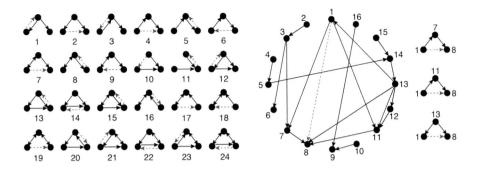

Fig. 2. All 24 evolutionary types of triads

Fig. 3. Edge $\langle 1, 8 \rangle$ yields three triad-evolution

Suppose we have network $G = \langle V, E \rangle$ where each edge $e = \langle u, v \rangle \in E$ represents an interaction from u to v that took place at a particular time, $t(e)$. Though potentially there are multiple interactions from u to v, we record only the oldest one. For two times $t_0 < t_1$, let $G_0 = \langle V, E_0 \rangle$ and $G_1 = \langle V, E_1 \rangle$ denote the subgraph of G consisting of all edges with a time-stamp before t_0 and t_1, respectively.

In this paper, we regard each single interaction that took place between t_0 and t_1, i.e. $e \in E_1 - E_0$, as evolution of network G_0. The evolutionary type of a n node subgraph is determined by the type of the subgraph in G_0 and the evolution $e \in E_1 - E_0$. The number of evolutionary types increases with subgraph size. For example, three node subgraphs offer 24 possible evolutionary types and four node subgraphs offer more than 200 evolutionary types. Let J denote the number of evolutionary types. Figure 2 shows all 24 evolutionary types of three node subgraphs, where black and red dashed lines indicate existing edges in E_0 and new edges in $E_1 - E_0$, respectively.

We scan G_0 and $E_1 - E_0$ for all possible evolution of n node subgraphs, and record the number of occurrence of each evolutionary type. In general, a single new edge is related to multiple subgraphs. In the case of Figure 3, for example, pair $\langle 1, 8 \rangle$ yields three triad-evolution. Focusing on the three triads of $\langle 1, 7, 8 \rangle$, $\langle 1, 11, 8 \rangle$, and $\langle 1, 13, 8 \rangle$, the pairs are related to evolutionary types 4, 4, and 2, respectively.

The number of occurrences of evolutionary type j in $E_1 - E_0$ of G_0 is given by:

$$d_j = \sum_{\langle u, v \rangle \in E_1 - E_0} x_{uvj}, \qquad (2)$$

where x_{uvj} represents the number of occurrences of evolutionary type j related to new edge $\langle u, v \rangle$ (e.g., $x_{uvj} = 2$ where $u = 1$, $v = 8$, and $j = 4$ in the case of Figure 3). Similar in NM analysis, to compute the statistical significance of evolutionary type j, we compare $E_1 - E_0$ to the equivalent randomized sets of edges $\hat{E}_k$ in terms of the number of occurrences of each evolutionary type in

Table 1. Number of nodes, two times t_0 and t_1, and number of edges of social networks

	$|V|$	t_0	$|E_0|$	t_1	$|E_1|$
Yahoo Answers	5,835	2010/07/21	18,796	2010/07/22	37,320
Flickr	7,776	2010/06/15	82,868	2010/06/18	173,344

G_0. The randomized sets of edges are generated in a way similar to randomized networks in NM analysis.

The statistical significance of evolutionary type j is given by:

$$\zeta_j = \frac{d_j - \nu_j}{\tau_j}, \tag{3}$$

where $\hat{d}_{jk}$ represents the number of occurrences of evolutionary type j for $\hat{E}_k$, $\nu_j = \frac{1}{K}\sum_{k=1}^{K}\hat{d}_{jk}$, and $\tau_j^2 = \frac{1}{K}\sum_{k=1}^{K}(\hat{d}_{jk} - \nu_j)^2$. The SPs of evolutionary types are computed by normalizing the vector of Z scores $\boldsymbol{\zeta} = \{\zeta_j\}_{j=1}^{J}$ as in NM analysis.

4   Experiments

4.1   Datasets

We exploited two social media sources, Yahoo Answers and Flickr, to gather the experimental data. Our Yahoo Answers data set is based on a snapshot crawled in the period of July 20th, 2010 to July 22nd, 2010, and our Flickr data set is based on a snapshot crawled using Flickr API[3] in the period of June 12th, 2010 to June 18th, 2010.

We extract a dense part from each graph as experimental data using parameter κ (set to three in each case). Dense is defined as occurring when all nodes have at least κ edges in each graph by Liben-Nowell et al. [3]. Table 1 lists details of social networks.

4.2   NM and DNM Analyses

In this paper, we focus on three node subgraphs. The analysis about subgraphs that have more than three nodes is future work.

NM analysis allows us to clarify the properties of interactions among nodes in terms of local structures in networks. Figure 4 displays network motif profiles of the Yahoo Answers network and the Flickr network.

From this figure, we can see that both media have strongly defined cliques (see triad 13 in the figure) compared to random networks. This triad, which represents the case that three people collaborate with each other, is a typical motif of social networks [4]. As well, we can see that only Yahoo Answers has

---

[3] http://www.flickr.com/services/api

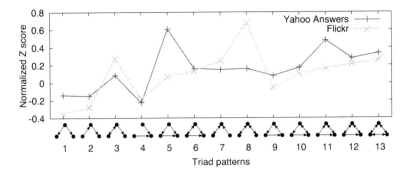

Fig. 4. Network motif profiles for social networks

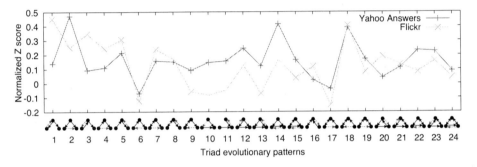

Fig. 5. Dynamic network motif profiles for social networks

a high number of feed forward loops (see triad 5 in the figure). This result matches that shown by Adamic et al. [1]. This motif indicates the expertise of Yahoo Answers users; people with high levels of expertise are willing to help people of all levels, whereas people of lower expertise help only those with even less expertise. On the other hand, only the Flickr network has the motif of two mutual dyads (see triad 8 in the figure). This motif indicates existence of core users who communicate mutually with the two other people in the triad.

DNM analysis enables us to discover the hidden properties of interaction among nodes in terms of the evolutionary patterns of substructures in networks. Figure 5 shows the dynamic network motif profiles of the Yahoo Answers network and the Flickr network.

First, we focus on common DNMs for both networks. From this figure, we can see that triad evolutionary type 18 is a common DNM for both networks, which is an intuitive result. This DNM is created by a user who wants to be invited into a community sending messages to one member after another. Note that neither triad evolutionary type 23 nor type 24 are DNM for both networks; therefore, interestingly, such users cannot necessarily become a member of the community.

Furthermore, we focus on the DNMs unique to each network. We can see that triad evolutionary type 2 is a DNM for the Yahoo Answers network whereas type 4 is not. This indicates that people with high levels of expertise help others faster than people of lower expertise. As well, we can see that type 1 is a DNM for the Flickr network whereas neither type 3 nor type 11 is. This DNM indicates that core users tend to send comments to others before they receive comments. This implies that core users are well-mannered in that they reply to others' comments.

5   Conclusion

We proposed an entirely novel notion to analyze complex networks more deeply, we call it *dynamic network motifs (DNMs)*. We defined DNMs as statistically significant local evolutionary patterns of a network. We applied our method to real networks from two web services, Yahoo Answers and Flickr, found DNMs in the networks, and analyzed the interaction properties of the two web services based on the DNMs.

In future work, we will conduct further experiments. First, we hope to confirm the efficiency of DNMs in the case of more than three node subgraphs. Second, we intend to use datasets gathered from social network services (SNS) such as Facebook[4].

References

1. Adamic, L., Zhang, J., Bakshy, E., Ackerman, M.: Knowledge sharing and Yahoo Answers: everyone knows something. In: Proceedings of the 17th International Conference on World Wide Web, pp. 665–674 (2008)
2. Inokuchi, A., Washio, T.: A fast method to mine frequent subsequences from graph sequence data. In: Proceedings of the 8th IEEE International Conference on Data Mining, pp. 303–312 (2008)
3. Liben-Nowell, D., Kleinberg, J.: The link-prediction problem for social networks. Journal-American Society for Information Science and Technology 58(7), 1019–1031 (2007)
4. Milo, R., Itzkovitz, S., Kashtan, N., Levitt, R., Shen-Orr, S., Ayzenshtat, I., Sheffer, M., Alon, U.: Superfamilies of evolved and designed networks. Science 303(5663), 1538–1542 (2004)
5. Milo, R., Shen-Orr, S., Itzkovitz, S., Kashtan, N., Chklovskii, D., Alon, U.: Network motifs: simple building blocks of complex networks. Science 298(5594), 824–827 (2002)
6. Nijssen, S., Kok, J.: A quickstart in frequent structure mining can make a difference. In: Proceedings of the 10th ACM SIGKDD International Conference on Knowledge Discovery and Data Mining, pp. 647–652 (2004)
7. Yan, X., Han, J.: gSpan: Graph-based substructure pattern mining. In: Proceedings of the 2002 IEEE International Conference on Data Mining, pp. 721–724 (2002)

---

[4] http://www.facebook.com

Social Network Analysis on KAD
and Its Application

Xiangtao Liu[1,2], Yang Li[1], Zhezhong Li[1], and Xueqi Cheng[1]

[1] Institute of Computing Technology,
Chinese Academy of Sciences, 100190 Beijing, China
[2] Department of Computer Science and Technology,
Huaqiao University, 362021 Quanzhou, China
xtliupaper@hotmail.com, {liyang,lizhezhong}@software.ict.ac.cn,
cxq@ict.ac.cn

Abstract. In recent years, peer-to-peer (P2P) file sharing applications (e.g., eMule) have dominated the Internet. eMule deploys its distributed network (i.e., KAD) based on Kademlia, a robust distributed hash table (DHT) protocol, to facilitate the delivery of content. In this paper, we conduct a series of analyses on the social network of KAD which is formed based on the information of how peers share files in KAD. Our analyses reveal many interesting characteristics of KAD. We further discuss the uses and misuses of these characteristics. Especially, we propose a novel approach to improve the routing performance of KAD based on the characteristic of its community structure.

Keywords: Peer-to-peer, KAD, social network analysis, application.

1   Introduction

In today's Internet, peer-to-peer (P2P) file sharing applications become more and more popular [1]. According to the 2008/2009 Internet traffic report of Ipoque [2], 43% ~ 70% of the Internet traffics are from P2P applications and services, where eMule constitutes the majority (up to 47%). To facilitate the delivery of content, eMule deploys its distributed network (i.e., KAD) based on Kademlia [3], a robust distributed hash table (DHT) protocol. Along with the popularity of eMule, KAD is also being widely used.

Social network analysis has been applied in many research fields including P2P networks. Khambatti et al. [4] was the first to apply social network analysis on P2P networks; they proposed efficient methods to discover the formation of self-configuring communities in P2P networks. Afterwards, many community formation and discovery methods (e.g., [5]) in P2P network were put forward.

Moreover, the results of social network analysis can be utilized to improve the performance of traditional applications. For instance, Liu et al. [6] proposed that web content could be delivered much faster through building interest-based communities. Tian et al. [7] proposed that in BitTorrent network, the lifetime of a torrent could be greatly prolonged, when taking priority to share files among

X. Du et al. (Eds.): APWeb 2011, LNCS 6612, pp. 327–332, 2011.

peers with similar-number chunks. Yu et al. [8] made social network analysis on the graph formed by sybil peers and honest peers which are connected by trust relationship. Then they proposed a decentralized approach, SybilGuard, to limit the corruptive influence of Sybil attacks (i.e., the forging of multiple identities).

Despite the wide use of KAD, existing works mainly focused on the analyses of its common characteristics, such as the peer churn (i.e., the phenomenon that peers joining or leaving a network frequently). In this study, we try to identify characteristics of the social network of KAD, which is formed based on the information of how peers share files in KAD (see details in Section 2). We further discuss how these characteristics could be used and misused. Especially, we propose a new routing table maintenance algorithm (i.e., RTMA-2, see details in Section 4) to improve the routing performance of KAD based on the characteristic of its community structure.

2   Methodology

We use Rainbow, a P2P crawler for Kademlia-based DHT networks, to collect data of KAD. Rainbow had run on a server from May 29, 2009 to June 9, 2009. We obtained a data set of how $39,941$ peers were sharing $7,172,189$ files. Note that we did not download the content of files, but just collected the meta-information of files including filename and file size. Then we formed a graph of KAD, through connecting peers with edges if they share at least one file, and giving the weight on this edge as the number of shared files. In this way, we obtained a weighted non-directed graph named "fullgraph", which has $39,941$ peers and $30,151,958$ edges. Next we set a threshold value 1 to filter those unstable relationships whose weight is 1, and to simplify the analysis of "fullgraph". Finally, we formed the largest connected subgraph (LCS, i.e., the social network of KAD) whose weights are over 1, with the peer number $n = 36,757$ and the edge number $m = 6,523,043$. The rest analyses of this paper will mainly focus on LCS.

We conduct a series of analyses on the characteristics of LCS, which include the degree distribution, the clustering coefficient, the mean shortest distance and the community structure. Specifically, given a graph $\mathbb{G} = (\mathbb{V}, \mathbb{E})$, where $\mathbb{V}$ is the vertex set which satisfies $\mathbb{V} = \{v_i | i = 1, 2, \ldots, n\}$ and $\mathbb{E}$ is the edge set which satisfies $\mathbb{E} = \{e_i | i = 1, 2, \ldots, m\}$, the degree of a vertex v_i is the number of edges incident on v_i; the frequency of degree k, denoted by p_k, is the fraction of vertices in the graph that have degree k; the degree distribution refers to the p_k vs. degree relation; the clustering (or transitivity) coefficient (i.e., C) measures the extent of inter-connectivity between the neighbors of a vertex; the mean shortest distance satisfies $l = \frac{2 \sum_{i>j} d_{ij}}{n(n+1)}$, where d_{ij} is the shortest distance from vertex i to j; on the characteristic of community structure, we focus on how peers are distributed in communities and how the identified characteristic of community structure could be applied to improve the routing performance of KAD.

3   Experimental Results

We find that the degree distribution (Fig. 1) fits a power law distribution $p_k \sim k^{-\alpha}$ with $\alpha = 0.9938$. Furthermore, we observe that peers with high degree tend to own more files. For example, the top 10 highest degree peers averagely own 10,761 files, which is much (33.7 times) larger than 319 files which all peers averagely own. Thus we infer that KAD is vulnerable to targeted attacks, which are launched targeting at the peers playing important roles (e.g., to remove the high-degree peers from KAD), and is robust against random attacks, which are launched without target.

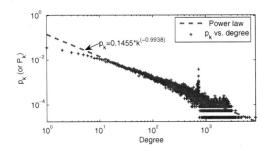

Fig. 1. The degree distribution (log-log scale)

Moreover, we reveal that for LCS: (1) the clustering coefficient $C = 0.49$, which is a high value and indicates that LCS is a network with high transitivity; (2) the mean shortest distance $l = 2.80$. According to Watts et al. [9], a network with a small mean shortest distance and a large clustering coefficient is a small world network, thus the small l and high C of LCS suggest that LCS is a small world network. Consequently, this will make the sharing of files between peers in KAD very efficient, yet this may also be exploited by malicious attackers to spread viruses more quickly through KAD, according to the general properties of the small world network.

In [5], the quality of community structure is measured by the modularity, a scalar value between -1 and 1 which measures the density of links inside communities as compared to links between communities [10]. Usually, modularity takes values 0.3 to 0.7 to mean a distinguishable community structure. In our experiment, we produced 31 communities with a high modularity value of 0.5928, through running the fast community discovery algorithm [5] on LCS. We further observe that the top 3 largest communities account for 91% of peers, and that there are 20 communities each of which contains less than 10 peers. This indicates that the distribution of the sizes of communities fits power law to some extent.

Moreover, we observe that peers in a community are mainly distributed in only a few countries (Fig. 2). For example, 86% of peers in community No.24 are distributed in France. Thus, we infer that peers tend to share files with the peers

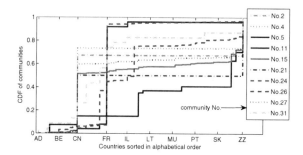

Fig. 2. The country distribution of peers in top 10 largest communities (ZZ denotes reserved IP addresses)

in the same country. We consider this is due to that peers in the same country are more closely-linked and have more similar tastes than peers in different countries. As we know, the allocations of IP address space are globally managed by the Internet Assigned Numbers Authority (IANA). It usually allocates neighboring IP address blocks to Internet service providers or other entities in each country. Thus, we deduce that IP addresses[1] in the same country should be near. Therefore, we conclude that peers tend to share files with IP-address-near peers.

4 Application

Based on the above conclusion that peers tending to share files with IP-address-near peers, we may improve the routing performance of KAD, trying to make more IP-address-near peers as neighbors through modifying the routing table maintenance algorithm of KAD.

We illustrates the routing table (i.e., the structure which contains the neighboring peers of the local peer) of KAD in Fig. 3. Note that the routing table is an unbalanced binary tree, where the root denotes the upmost node in level 0, and the zone denotes an arbitrary node in the routing table. If a zone is a leaf zone, it has a k-bucket which contains at most k contacts (k is set as 10 in eMule); if not, it has two subnodes and has no k-bucket. For example, one can see that the non-leaf zone x in Fig. 3 has two subtrees named subtree[0] and subtree[1].

We consider the original routing table maintenance algorithm (i.e., RTMA-1) as a black-box algorithm. To improve the routing performance of KAD, we modify RTMA-1 into the new routing table maintenance algorithm (i.e., RTMA-2, see Alg. 1), through adding two operations to RTMA-1. Specifically, the first operation is to compute the IP-address distances of contacts (i.e., the candidate neighbor peers which are returned from UDP messages), according to the formula: IP-address distance = |IP address of contact − IP address of local peer|; the

---

[1] IP address "x.y.z.u" is denoted by a network-order unsigned integer "$256^3 \times x + 256^2 \times y + 256 \times z + u$".

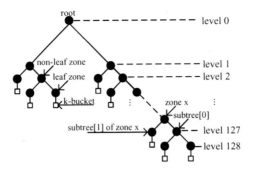

Fig. 3. The routing table of KAD

second operation is to rank those 2 to 20 contacts returned from UDP message, according to their IP-address distances from the local peer. After these operations, those IP-address-near contacts will be processed first. Since the routing table of KAD can only contain at most 6,360 contacts [11], the chance of RTMA-2 for those IP-address-near contacts being inserted into the routing table will be higher than that of RTMA-1. Consequently, RTMA-2 will have advantages over RTMA-1 in the aspects of routing speed and the speed of locating sources of files.

Algorithm 1. RTMA-2

Input: 2 to 20 contacts returned from UDP message
Output: new routing table

1: compute the IP-address distances of contacts
2: rank contacts according to their IP-address distance from the local peer
3: RTMA-1

---

5   Conclusion

In this paper, we conducted a series of social network analyses on the KAD (i.e., the DHT network deployed by eMule) on many aspects: the degree distribution, the clustering coefficient, the mean distance and the community structure. We revealed many interesting characteristics of the social network of KAD (i.e., LCS), such as:

- the degree distribution of LCS fits a power law;
- LCS demonstrates high transitivity and is a small world network;
- the peers in a community of LCS are mainly distributed in only a few countries.

We also discussed the uses and misuses of these characteristics. Especially, we have shown that the characteristic of the community structure of LCS could be applied to improve the routing performance of KAD.

Acknowledgment

This work is supported in part by China NSF under Grants No. 60803085 and No. 60873245.

References

1. Karagiannis, T., Broido, A., Faloutsos, M., Claffy, K.: Transport layer identification of P2P traffic. In: Proceedings of the Internet Measurement Conference (IMC), Taormina, Sicily, Italy, pp. 121–134 (2004)
2. Ipoque, http://www.ipoque.com/resources/internet-studies
3. Maymounkov, P., Mazières, D.: Kademlia: A peer-to-peer information system based on the XOR metric. In: Druschel, P., Kaashoek, M.F., Rowstron, A. (eds.) IPTPS 2002. LNCS, vol. 2429, pp. 53–65. Springer, Heidelberg (2002)
4. Khambatti, M., Ryu, K.D., Dasgupta, P.: Peer-to-peer communities: formation and discovery. In: International Conference on Parallel and Distributed Computing Systems, Cambridge, USA, pp. 161–166 (2002)
5. Blondel, V.D., Guillaume, J.L., Lambiotte, R., Lefebvre, E.: Fast unfolding of community in large networks. Journal of Statistical Mechanics: Theory and Experiment 2008(10), 440–442 (2008)
6. Liu, K., Bhaduri, K., Das, K., Nguyen, P., Kargupta, H.: Client-side web mining for community formation in peer-to-peer environments. In: The Twelfth ACM SIGKDD International Conference on Knowledge Discovery and Data Mining, Philadelphia, USA, pp. 11–20 (2006)
7. Tian, Y., Wu, D., Wing Ng, K.: Modeling, analysis and improvement for Bittorrent-like file-sharing networks. In: IEEE International Conference on Computer Communications (INFOCOM), Barcelona, Catalunya, Spain, pp. 1–11 (2006)
8. Yu, H., Kaminsky, M., Gibbons, P.B., Flaxman, A.: SybilGuard: defending against Sybil attacks via social network. IEEE/ACM Trans. on Networking 16, 576–589 (2008)
9. Watts, J.D., Strogatz, H.S.: Collective dynamics of 'small-world' networks. Nature 393(6684), 440–442 (1998)
10. Newman, M.E.J.: Modularity and community structure in networks. Proceedings of the National Academy of Sciences 103(23), 8577–8582 (2006)
11. Brunner, R.: A performance evaluation of the Kad-protocol. Master's thesis, University Mannheim, Germany (2006)

Modeling and Querying Probabilistic RDFS Data Sets with Correlated Triples

Chi-Cheong Szeto[1], Edward Hung[1], and Yu Deng[2]

[1] Department of Computing, The Hong Kong Polytechnic University, Hong Kong
{csccszeto,csehung}@comp.polyu.edu.hk
[2] IBM T.J. Watson Research Center, P.O. Box 704, Yorktown Heights,
NY 10598 USA
dengy@us.ibm.com

Abstract. Resource Description Framework (RDF) and its extension RDF Schema (RDFS) are data models to represent information on the Web. They use RDF triples to make statements. Because of lack of knowledge, some triples are known to be true with a certain degree of belief. Existing approaches either assign each triple a probability and assume that triples are statistically independent of each other, or only model statistical relationships over possible objects of a triple. In this paper, we introduce probabilistic RDFS (pRDFS) to model statistical relationships among correlated triples by specifying the joint probability distributions over them. Syntax and semantics of pRDFS are given. Since there may exist some truth value assignments for triples that violate the RDFS semantics, an algorithm to check the consistency is provided. Finally, we show how to find answers to queries in SPARQL. The probabilities of the answers are approximated using a Monte-Carlo algorithm.

1 Introduction

Resource Description Framework (RDF) [10] is a World Wide Web Consortium (W3C) Recommendation and is a data model to represent information on the Web. It uses Uniform Resource Identifier (URI) references[1] to identify things and RDF triples of the form (s, p, o) to make statements, where s, p, o are the subject, property and object respectively. (Tom, rdf:type, Professor) is an example of RDF triples that would appear in the university domain, which we will use throughout the paper to illustrate some concepts. RDF Schema (RDFS) is an extension of RDF. It provides a vocabulary to describe application-specific classes, properties, class and property hierarchies, and which classes and properties are used together. For example, (Professor, rdf:type, rdfs:Class) means that Professor is a class. We often divide RDF triples into two sets. One contains triples the subjects of which are either classes or properties. We call them

---

[1] QNames are used in this paper as abbreviations for URI references. QName prefixes are omitted except rdf: and rdfs:, which are assigned to the namespace URIs
http://www.w3.org/1999/02/22-rdf-syntax-ns# and
http://www.w3.org/2000/01/rdf-schema# respectively.

X. Du et al. (Eds.): APWeb 2011, LNCS 6612, pp. 333–344, 2011.
© Springer-Verlag Berlin Heidelberg 2011

schema triples. The other contains triples the subjects of which are instances. We call them instance triples. Schema triples are more permanent and definitional in nature while the instance triples are more dynamic.

Because of lack of knowledge, some triples are known to be true with a certain degree of belief. Huang et al.'s probabilistic RDF database [5] encodes probabilistic knowledge by assigning each triple a probability and assuming that triples are statistically independent of each other. However, it cannot encode statistical relationships among correlated triples. It also does not comply with the RDFS semantics. Probabilistic RDF (pRDF) [12] encodes probabilistic knowledge by specifying the probability distribution over possible objects of a triple. Any triangular norm can be used to compute the probability of a conjunction of triples. pRDF restricts the probability distributions to those over triples with the same subject and property.

In this paper, we introduce probabilistic RDFS (pRDFS), which has greater flexibility in encoding probabilistic knowledge. It assumes schema triples are always true and models statistical relationships among correlated instance triples by specifying the joint probability distributions over them. The probability distributions cannot be specified arbitrarily since there may exist some truth value assignments for triples that violate the RDFS semantics. We provide an algorithm to check the consistency. We also describe how to compute answers with probabilities to queries in SPARQL [11], which is a W3C Recommendation and is a query language for RDF.

The remainder of this paper is organized as follows. Sections 2 and 3 define the syntax and semantics of pRDFS respectively. Section 4 discusses how to check the consistency of a pRDFS theory. Section 5 describes query evaluation based on a pRDFS theory. Section 6 evaluates experimentally the execution time performance of consistency checking and query answering on top of a pRDFS theory. Section 7 reviews the related work. Finally, Section 8 concludes this paper.

2   pRDFS Syntax

Let $\mathbb{V}$ be a set of vocabulary, which consists of a set of URI references $\mathbb{U}$ and a set of literals $\mathbb{L}$. $\mathbb{C}$, $\mathbb{P}$ and $\mathbb{I}$ are the sets of classes, properties and individuals respectively. They are subsets of $\mathbb{U}$ and we assume that they are mutually disjoint. A pRDFS schema H is a set of RDF triples, whose subjects are classes or properties. A pRDFS instance (R, θ) consists of a set of RDF triples R, whose subjects are individuals and a mapping θ describing the uncertainty of R. Each triple in R can be interpreted as a boolean variable and takes a value from {true, false}, abbreviated as $\{T, F\}$. Let Partition$(R) = \{R'_1, \ldots, R'_n\}$ be a partition of R such that any two different elements of the partition are statistically independent. Let τ be a truth value assignment function for the triples in R. It is from R to $\{T, F\}$. $\tau|_{R'_i}$ is the restriction of τ to R'_i. Let $P_{R'_i}$ be the joint probability distribution function of R'_i that maps $\tau|_{R'_i}$ to a probability value. Because of the independence assumption, the joint probability distribution function of R,

$P_R(\tau)$ can be written as $P_{R_1'}(\tau|_{R_1'}) \times \cdots \times P_{R_n'}(\tau|_{R_n'})$. We now define the mapping θ, which is from R to $\{P_{R_1'}, \ldots, P_{R_n'}\}$ and maps t to P_{R_i} where $t \in R_i$. A pRDFS theory is a triple (H, R, θ), where H and (R, θ) are the pRDFS schema and instance respectively.

Example 1. Let $h_1 =$ (Student, rdf:type, rdfs:Class), $h_2 =$ (Professor, rdf:type, rdfs:Class), $h_3 =$ (Course, rdf:type, rdfs:Class), $h_4 =$ (takesCourse, rdf:type, rdf:Property), $h_5 =$ (teacherOf, rdf:type, rdf:Property), $r_1 =$ (semanticWeb, rdf:type, Course), $r_2 =$ (Tom, rdf:type, Professor), $r_3 =$ (May, rdf:type, Professor), $r_4 =$ (Tom, teacherOf, semanticWeb), $r_5 =$ (May, teacherOf, semanticWeb), $r_6 =$ (John, rdf:type, Student), $r_7 =$ (Mary, rdf:type, Student), $r_8 =$ (John, takesCourse, semanticWeb) and $r_9 =$ (Mary, takesCourse, semanticWeb). $H = \{h_1, \ldots, h_5\}$ is a pRDFS schema. $R = \{r_1, \ldots, r_9\}$ is a pRDFS instance. Professors Tom and May both specialize in semantic web. Their chance to be the teacher of semantic web is described by $P_{\{r_4, r_5\}} = \{$ ($\{(r_4,\text{T}), (r_5,\text{T})\}$, 0), ($\{(r_4,\text{T}), (r_5,\text{F})\}$, 0.5), ($\{(r_4,\text{F}), (r_5,\text{T})\}$, 0.5), ($\{(r_4,\text{F}), (r_5,\text{F})\}$, 0) $\}$. Students John and Mary have the same interest and are good friends. Their chance to take semantic web together is described by $P_{\{r_8, r_9\}} = \{$ ($\{(r_8,\text{T}), (r_9,\text{T})\}$, 0.4), ($\{(r_8,\text{T}), (r_9,\text{F})\}$, 0.1), ($\{(r_8,\text{F}), (r_9,\text{T})\}$, 0.1), ($\{(r_8,\text{F}), (r_9,\text{F})\}$, 0.4) $\}$. We know r_1, r_2, r_3, r_6, r_7 are true. Therefore, $P_{\{r_i\}} = \{$ ($\{(r_i,\text{T})\}$, 1), ($\{(r_i,\text{F})\}$, 0) $\}$, $i = 1, 2, 3, 6, 7$. Partition(R) = $\{$ $\{r_1\}, \{r_2\}, \{r_3\}, \{r_4, r_5\}, \{r_6\}, \{r_7\}, \{r_8, r_9\}$ $\}$. (H, R, θ) is a pRDFS theory, where $\theta = \{$ $(r_1, P_{\{r_1\}})$, $(r_2, P_{\{r_2\}})$, $(r_3, P_{\{r_3\}})$, $(r_4, P_{\{r_4, r_5\}})$, $(r_5, P_{\{r_4, r_5\}})$, $(r_6, P_{\{r_6\}})$, $(r_7, P_{\{r_7\}})$, $(r_8, P_{\{r_8, r_9\}})$, $(r_9, P_{\{r_8, r_9\}})$ $\}$. The probability that Tom is the only teacher of semantic web and both John and Mary do not take semantic web, $P_R($ $\{$ (r_1,T), (r_2,T), (r_3,T), (r_4,T), (r_5,F), (r_6,T), (r_7,T), (r_9,F), (r_{10},F) $\}$) can be computed as $P_{\{r_1\}}(\{(r_1,\text{T})\}) \times P_{\{r_2\}}(\{(r_2,\text{T})\}) \times P_{\{r_3\}}(\{(r_3,\text{T})\}) \times P_{\{r_4, r_5\}}(\{(r_4,\text{T}), (r_5,\text{F})\}) \times P_{\{r_6\}}(\{(r_6,\text{T})\}) \times P_{\{r_7\}}(\{(r_7,\text{T})\}) \times P_{\{r_8, r_9\}}(\{(r_8,\text{F}), (r_9,\text{F})\}) = 0.5 \times 0.4 = 0.2$.

3   pRDFS Semantics

A world W is a set of triples that follows the semantics of RDFS. Specifically, it satisfies the following criteria.

1. RDFS axiomatic triples $\subseteq W$.
2. If $(x, \text{rdfs:domain}, y)$, $(u, x, v) \in W$, then $(u, \text{rdf:type}, y) \in W$.
3. If $(x, \text{rdfs:range}, y)$, $(u, x, v) \in W$, then $(v, \text{rdf:type}, y) \in W$.
4. If $(x, \text{rdf:type}, \text{rdf:Property}) \in W$, then $(x, \text{rdfs:subPropertyOf}, x) \in W$.
5. If $(x, \text{rdfs:subPropertyOf}, y)$, $(y, \text{rdfs:subPropertyOf}, z) \in W$, then $(x, \text{rdfs:subPropertyOf}, z) \in W$.
6. If $(x, \text{rdfs:subPropertyOf}, y)$, then $(x, \text{rdf:type}, \text{rdf:Property})$, $(y, \text{rdf:type}, \text{rdf:Property}) \in W$ and $\forall a \ \forall b \ ((a, x, b) \in W \implies (a, y, b) \in W)$.
7. If $(x, \text{rdfs:subClassOf}, y)$, then $(x, \text{rdf:type}, \text{rdfs:Class})$, $(y, \text{rdf:type}, \text{rdfs:Class}) \in W$ and $\forall a \ ((a, \text{rdf:type}, x) \in W \implies (a, \text{rdf:type}, y) \in W)$.
8. If $(x, \text{rdf:type}, \text{rdfs:Class}) \in W$, then $(x, \text{rdfs:subClassOf}, x) \in W$.

9. If $(x,$ rdfs:subClassOf, $y)$, $(y,$ rdfs:subClassOf, $z) \in W$, then $(x,$ rdfs:subClassOf, $z) \in W$.

In this paper, we ignore the handling of blank nodes, containers and datatypes. Ω is the set of all possible worlds. A pRDFS interpretation is a mapping $I : \Omega \to [0,1]$, such that $\sum_{W \in \Omega} I(W) = 1$. I satisfies a pRDFS theory (H, R, θ) iff $P_R(\tau) = \sum_{W \in \Omega \mid A \subseteq W \wedge H \subseteq W \wedge \forall t \in R(\tau(t)=T \implies t \in W \wedge \tau(t)=F \implies t \notin W)} I(W)$ for any truth value assignment τ, where A is the set of RDFS axiomatic triples. A pRDFS theory is consistent if it has a satisfying interpretation. A theory (H, R_1, θ_1) entails another theory (H, R_2, θ_2) iff every satisfying interpretation of (H, R_1, θ_1) is a satisfying interpretation of (H, R_2, θ_2).

Example 2 (Consistent theory). Consider the pRDFS theory in Example 1. Let worlds $W_i = A \cup H \cup \{r_1, r_2, r_3, r_6, r_7\} \cup B_i$ for $i = 1, \ldots, 8$, where A is the set of RDFS axiomatic triples, $B_1 = \{r_4, r_8, r_9\}$, $B_2 = \{r_4, r_8\}$, $B_3 = \{r_4, r_9\}$, $B_4 = \{r_4\}$, $B_5 = \{r_5, r_8, r_9\}$, $B_6 = \{r_5, r_8\}$, $B_7 = \{r_5, r_9\}$ and $B_8 = \{r_5\}$. The theory has a satisfying interpretation I, which maps W_1, W_4, W_5, W_8 to 0.2, W_2, W_3, W_6, W_7 to 0.05, and all other worlds to 0. Therefore, it is consistent.

Example 3 (Inconsistent theory). Consider another pRDFS theory (H, R, θ), where $H = \{h_1, \ldots, h_5\}$, $R = \{r_1, \ldots, r_4\}$, $h_1 = $ (Professor, rdf:type, rdfs:Class), $h_2 = $ (Department, rdf:type, rdfs:Class), $h_3 = $ (worksFor, rdf:type, rdf:Property), $h_4 = $ (headOf, rdf:type, rdf:Property), $h_5 = $ (headOf, rdfs:subPropertyOf, worksFor), $r_1 = $ (Tom, rdf:type, Professor), $r_2 = $ (department of computing, rdf:type, Department), $r_3 = $ (Tom, headOf, department of computing), $r_4 = $ (Tom, worksFor, department of computing), $\theta = \{$ $(r_1, P_{\{r_1\}})$, $(r_2, P_{\{r_2\}})$, $(r_3, P_{\{r_3\}})$, $(r_4, P_{\{r_4\}})$ $\}$, $P_{\{r_i\}} = \{$ $(\{(r_i,T)\}, 1)$, $(\{(r_i,F)\}, 0)$ $\}$ for $i = 1$ and 2, $P_{\{r_3\}} = \{$ $(\{(r_3,T)\}, 0.7)$, $(\{(r_3,F)\}, 0.3)$ $\}$, and $P_{\{r_4\}} = \{$ $(\{(r_4,T)\}, 0.8)$, $(\{(r_4,F)\}, 0.2)$ $\}$. Consider the truth value assignment $\tau = \{$ (r_1,T), (r_2,T), (r_3,T), (r_4,F) $\}$. $P_R(\tau) = P_{\{r_3\}}(\{(r_3,T)\}) \times P_{\{r_4\}}(\{(r_4,F)\}) = 0.7 \times 0.2 = 0.14$. However, there is no world that corresponds to this truth value assignment. Suppose such a world W exists. $h_5, r_3 \in W$ and $r_4 \notin W$. According to the RDFS semantics, if $h_5, r_3 \in W$, then $r_4 \in W$. This is a contradiction. The theory considered in this example has no satisfying interpretation and so is inconsistent.

4   pRDFS Consistency

This section describes an algorithm used to check the consistency of a pRDFS theory (H, R, θ). Inconsistency arises when a theory assigns a non-zero probability value to a truth value assignment which assigns the false value to a triple t and assigns the true value to a set of other triples which, together with the schema, derives t.

Algorithm 1 shows an algorithm to check the consistency. The basic idea is to find all sets of triples each of which derives t, where t is in R and can take the false value. For each set of triples P, the algorithm checks whether the probability that $\forall p \in P(p = T) \wedge t = F$ is zero or not. If the probability of one set of triples is non-zero, the theory is inconsistent. Otherwise, the theory is consistent.

Algorithm 1. Check whether a given pRDFS theory (H, R, θ) is consistent (return true) or not (return false).

1:   $H \leftarrow H \cup$ transitive closures of rdfs:subClassOf and rdfs:subPropertyOf.
2:   **for** each triple $t = (s, p, o) \in R$ that can take the false value, where $p = $ rdf:type **do**
3:      /* rule rdfs9 */
4:      $B \leftarrow \{(s, p, o') \in R \mid (o', \text{rdfs:subClassOf}, o) \in H \wedge o \neq o'\}$.
5:      **if** $\exists b \in B(P_R(\{(b, T), (t, F)\}) > 0$, **then** return false.
6:      /* combination of rules rdfs9 and rdfs2 */
7:      $O \leftarrow \{o' \mid (s, p, o') \notin R \wedge (o', \text{rdfs:subClassOf}, o) \in H \wedge o \neq o'\}$.
8:      $C \leftarrow \{(s, p', o'') \in R \mid o' \in O \wedge (p', \text{rdfs:domain}, o') \in H\}$.
9:      **if** $\exists c \in C(P_R(\{(c, T), (t, F)\}) > 0$, **then** return false.
10:     /* combination of rules rdfs9, rdfs2 and rdfs7 */
11:     $P \leftarrow \{p' \mid o' \in O \wedge \nexists o''((s, p', o'') \in R) \wedge (p', \text{rdfs:domain}, o') \in H\}$.
12:     $D \leftarrow \{(s, p'', o') \in R \mid p' \in P \wedge (p'', \text{rdfs:subPropertyOf}, p') \in H \wedge p' \neq p''\}$.
13:     **if** $\exists d \in D(P_R(\{(d, T), (t, F)\}) > 0$, **then** return false.
14:     /* combination of rules rdfs9 and rdfs3 */
15:     $C \leftarrow \{(s', p', s) \in R \mid o' \in O \wedge (p', \text{rdfs:range}, o') \in H\}$.
16:     **if** $\exists c \in C(P_R(\{(c, T), (t, F)\}) > 0$, **then** return false.
17:     /* combination of rules rdfs9, rdfs3 and rdfs7 */
18:     $P \leftarrow \{p' \mid o' \in O \wedge \nexists s'((s', p', s) \in R) \wedge (p', \text{rdfs:range}, o') \in H\}$.
19:     $D \leftarrow \{(s', p'', s) \in R \mid p' \in P \wedge (p'', \text{rdfs:subPropertyOf}, p') \in H \wedge p' \neq p''\}$.
20:     **if** $\exists d \in D(P_R(\{(d, T), (t, F)\}) > 0$, **then** return false.
21:     /* rule rdfs2 */
22:     $B \leftarrow \{(s, p', o') \in R \mid (p', \text{rdfs:domain}, o) \in H\}$.
23:     **if** $\exists b \in B(P_R(\{(b, T), (t, F)\}) > 0$, **then** return false.
24:     /* combination of rules rdfs2 and rdfs7 */
25:     $P \leftarrow \{p' \mid \nexists o'((s, p', o') \in R) \wedge (p', \text{rdfs:domain}, o) \in H\}$.
26:     $C \leftarrow \{(s, p'', o') \in R \mid p' \in P \wedge (p'', \text{rdfs:subPropertyOf}, p') \in H \wedge p' \neq p''\}$.
27:     **if** $\exists c \in C(P_R(\{(c, T), (t, F)\}) > 0$, **then** return false.
28:     /* rule rdfs3 */
29:     $B \leftarrow \{(s', p', s) \in R \mid (p', \text{rdfs:range}, o) \in H\}$.
30:     **if** $\exists b \in B(P_R(\{(b, T), (t, F)\}) > 0$, **then** return false.
31:     /* combination of rules rdfs3 and rdfs7 */
32:     $P \leftarrow \{p' \mid o' \in O \wedge \nexists s'((s', p', s) \in R) \wedge (p', \text{rdfs:range}, o) \in H\}$.
33:     $C \leftarrow \{(s', p'', s) \in R \mid p' \in P \wedge (p'', \text{rdfs:subPropertyOf}, p') \in H \wedge p' \neq p''\}$.
34:     **if** $\exists c \in C(P_R(\{(c, T), (t, F)\}) > 0$, **then** return false.
35: **end for**
36: **for** each triple $t = (s, p, o) \in R$ that can take the false value, where $p \neq $ rdf:type **do**
37:     /* rule rdfs7 */
38:     $B \leftarrow \{(s, p', o) \in R \mid (p', \text{rdfs:subPropertyOf}, p) \in H \wedge p \neq p'\}$.
39:     **if** $\exists b \in B(P_R(\{(b, T), (t, F)\}) > 0$, **then** return false.
40: **end for**
41: return true.

The RDF specification [10] provides a set of deduction rules with respect to the RDFS semantics. Specifically, Algorithm 1 considers the following rules to find all sets of triples each of which derives t.

1. $\dfrac{(a,\ \text{rdfs:domain},\ x)\ (u,\ a,\ y)}{(u,\ \text{rdf:type},\ x)}$ rdfs2

2. $\dfrac{(a,\ \text{rdfs:range},\ x)\ (u,\ a,\ v)}{(v,\ \text{rdf:type},\ x)}$ rdfs3

3. $\dfrac{(u,\ \text{rdfs:subPropertyOf},\ v)\ (v,\ \text{rdfs:subPropertyOf},\ x)}{(u,\ \text{rdfs:subPropertyOf},\ x)}$ rdfs5

4. $\dfrac{(a,\ \text{rdfs:subPropertyOf},\ b)\ (u,\ a,\ y)}{(u,\ b,\ y)}$ rdfs7

5. $\dfrac{(u,\ \text{rdfs:subClassOf},\ x)\ (v,\ \text{rdf:type},\ u)}{(v,\ \text{rdf:type},\ x)}$ rdfs9

6. $\dfrac{(u,\ \text{rdfs:subClassOf},\ v)\ (v,\ \text{rdfs:subClassOf},\ x)}{(u,\ \text{rdfs:subClassOf},\ x)}$ rdfs11

It ignores rules that handle the blank nodes, containers and datatypes. It first uses rules rdfs5 and rdfs11 to find transitive closures of rdfs:subClassOf and rdfs:subPropertyOf in Line 1. Then, it divides the triples $t \in R$ that can take the false value into two groups to process. Lines 2-35 handle triples with property rdf:type while Lines 36-40 handle triples the properties of which are not rdf:type. In both cases, backward chaining approach is used to find all sets of triples each of which derives t. The former uses combinations of rules rdfs2, rdfs3, rdfs7 and rdfs9 while the latter uses rule rdfs7.

Example 4 (Cause of inconsistency). Consider the theory in Example 3. Algorithm 1 examines triples r_3, $r_4 \in R$, which take the false value with the probabilities of 0.3 and 0.2 respectively. There is no $P \subseteq (H \cup R)$ that derives r_3. $\{h_3, r_3\}$ derives r_4 by the deduction rule rdfs7. Since the schema triple h_3 is assumed to be always true, we only need to check the probability that $r_3 = \text{T}$ and $r_4 = \text{F}$. The probability is $0.14 > 0$, so $\{h_3,\ r_3,\ r_4\}$ is the cause of inconsistency of the theory.

5   pRDFS Query Evaluation

This section describes the evaluation of a SPARQL query on a pRDFS theory. We first review the evaluation on an RDF data set D. Let V be the set of query variables. A triple pattern is a member of the set $(\mathbb{U} \cup V) \times (\mathbb{U} \cup V) \times (\mathbb{U} \cup \mathbb{I} \cup V)$. Again, we ignore the handling of blank nodes. A simple graph pattern G is a set of triple patterns. Let $\text{var}(G) \subseteq V$ be the set of variables in G. A solution μ is a function from V to $(\mathbb{U} \cup \mathbb{I})$. $G(\mu)$ denotes a set of triples obtained by replacing every variable v in G with $\mu(v)$. In this paper, we restrict our discussion to a SPARQL query of the form: "select V_s where G", where $V_s \subseteq \text{var}(G)$ is a set of selected variables and G is a simple graph pattern. The result of the query is a set of solutions $\{\mu|_{V_s} \mid G(\mu) \subseteq D\}$, where $\mu|_{V_s}$ is the restriction of μ to V_s.

A pRDFS theory can express the negation of a triple $\neg t$ by specifying that $t = \text{F}$ whereas an RDFS document cannot. To ask questions about the negated triples in a pRDFS theory, we introduce a way to express negated triples in the SPARQL query. For a triple $t = (s, p, o)$, the negation of t is expressed by adding a special symbol $\neg$ before the property, that is, $\neg t = (s, \neg p, o)$. The result of

Algorithm 2. Find the result of a SPARQL query "select V_s from G" on a pRDFS theory (H, R, θ). Parameters ϵ and δ are used to approximate probability.

1: $D \leftarrow H \cup R \cup \{\neg t \mid t \in R \wedge P_R(\{(t, F)\}) > 0\}$.
2: $M \leftarrow \{\mu \mid G(\mu) \subseteq D \wedge \nexists t(t, \neg t \in G(\mu))\}$.
3: $S \leftarrow \{\mu|_{V_s} \mid \mu \in M\}$.
4: $\mathbb{R} \leftarrow \emptyset$.
5: **for** each $s \in S$ **do**
6:      $M_s \leftarrow \{\mu \in M \mid \mu|_{V_s} = s\}$ ($= \{\mu_1, \ldots, \mu_m\}$).
7:      $P(\bigwedge_{t \in G(\mu_i)} t) \leftarrow P_R(\{(t, T) \mid t \in G(\mu_i) \wedge t \in R\} \cup \{(t, F) \mid \neg t \in G(\mu_i) \wedge t \in R\})$,
        for $i \in \{1, \ldots, m\}$.
8:      **if** $m = 1$ **then**
9:          $P(s) \leftarrow P(\bigwedge_{t \in G(\mu_1)} t)$.
10:     **else**
11:         $F_s \leftarrow \bigvee_{\mu \in M_s} \bigwedge_{t \in G(\mu)} t$ ($= \{\bigwedge_{t \in G(\mu_1)} t \vee \cdots \vee \bigwedge_{t \in G(\mu_m)} t\}$).
12:         $C \leftarrow 0$.
13:         $N \leftarrow 4m\ln(2/\delta)/\epsilon^2$.
14:         **repeat**
15:             choose i randomly from $\{1, \ldots, m\}$ such that i is chosen with probability
                $P(\bigwedge_{t \in G(\mu_i)} t) / \sum_{\mu \in M_s} P(\bigwedge_{t \in G(\mu)} t)$.
16:             choose a truth value assignment τ randomly from the set of truth value
                assignments that satisfies $\bigwedge_{t \in G(\mu_i)} t$.
17:             **if** $\forall j < i(\tau$ does not satisfy $\bigwedge_{t \in G(\mu_j)} t)$ **then**
18:                 $C \leftarrow C + 1$.
19:             **end if**
20:         **until** $\lceil N \rceil$ times
21:         $P(s) \leftarrow C \sum_{\mu \in M_s} P(\bigwedge_{t \in G(\mu)} t) / N$.
22:     **end if**
23:     $\mathbb{R} \leftarrow \mathbb{R} \cup \{(s, P(s)\}$.
24: **end for**
25: **return** $\mathbb{R}$.

a query on a pRDFS theory is a set of pairs $S = \{(\mu|_{V_s}, P(\mu|_{V_s})) \mid G(\mu) \subseteq D \wedge \nexists t(t, \neg t \in G(\mu))\}$, where $P(\mu|_{V_s})$ is the probability of the solution $\mu|_{V_s}$ and $D = H \cup R \cup \{\neg t \mid t \in R \wedge P_R(\{(t, F)\}) > 0\}$.

Algorithm 2 shows an algorithm to find the result. It first constructs a data set D, which consists of schema triples H, instance triples R and negated triples. Then, it searches for intermediate solutions μ such that $G(\mu)$ is a subset of D and stores them in M. Moreover, it eliminates solutions μ such that $\exists t(t, \neg t \in G(\mu))$ because the probabilities of such solutions are zero. The final solutions S include the restriction of $\mu \in M$ to selected variables V_s.

The algorithm then finds the probability of each solution $s \in S$, denoted by $P(s)$. M_s in Line 6 is a subset of M. The restriction of $\mu \in M_s$ to V_s is the same as s. Let $M_s = \{\mu_1, \ldots, \mu_m\}$. If $m = 1$, the exact probability $P(s)$ is computed as in Line 7, where P_R is specified by θ. If $m > 1$, the calculation of $P(s)$ is formulated as a disjunctive normal form (DNF) probability problem. $F_s = \bigvee_{\mu \in M_s} \bigwedge_{t \in G(\mu)} t$ is the DNF formula for solution s. It is with respect to boolean variables in $\bigcup_{\mu \in M_s} G(\mu)$. The probability distribution on the set of all truth value assignments for the boolean variables is defined by θ, together with the assumption that all $t \in H$ are true. A truth value assignment satisfies F_s if F_s is true under the assignment. The DNF probability problem is to find the probability $P(F_s)$ that a truth value assignment randomly chosen satisfies F_s. $P(s)$ equals $P(F_s)$. The computation of exact probability of this problem is very hard. Therefore, Algorithm 2 in Lines 11-21 uses the coverage algorithm [6], which is a polynomial time Monte-Carlo algorithm with respect to m. The coverage algorithm is an ϵ, δ approximation algorithm, which computes an estimate $\tilde{P}(s)$ such that the probability $P((1 - \epsilon)P(s) \leq \tilde{P}(s) \leq (1 + \epsilon)P(s)) \geq 1 - \delta$. ϵ specifies the relative closeness of $\tilde{P}(s)$ to $P(s)$.

Example 5 (Negated triple). This example demonstrates a use of negated triples in queries. The query "select $?x$ where { (John, takesCourse, $?x$), (Mary, ¬takesCourse, $?x$) }" asks what courses that John takes but Mary does not. The result is { (semanticWeb, 0.1) }.

Example 6 (Query evaluation). Suppose the query "select $?x$ where { ($?x$, teacherOf, $?z$), ($?y$, takesCourse, $?z$) } " is made on the pRDFS theory (H, R, θ) in Example 1. The inputs to Algorithm 2 are the query, the theory, $\epsilon = 0.01$ and $\delta = 0.99$. D in Line 1 is constructed as $\{h_1, \ldots, h_5, r_1, \ldots, r_9, \neg r_4, \neg r_5, \neg r_8, \neg r_9\}$. The intermediate solutions M are { $\mu_1, \ldots, \mu_4$ }, where $\mu_1 = \{$ ($?x$, Tom), ($?y$, John), ($?z$, semanticWeb) }, $\mu_2 = \{$ ($?x$, Tom), ($?y$, Mary), ($?z$, semanticWeb) }, $\mu_3 = \{$ ($?x$, May), ($?y$, John), ($?z$, semanticWeb) } and $\mu_4 = \{$ ($?x$, May), ($?y$, Mary), ($?z$, semanticWeb) }. The selected variable V_s is { $?x$ }, so the final solutions S are { { ($?x$, Tom) }, { ($?x$, May) } }. For solution $s = \{$ ($?x$, Tom) }, M_s in Line 6 is { μ_1, μ_2 } and the DNF formula F_s is $(r_4 \wedge r_8) \vee (r_4 \wedge r_9)$. F_s has two conjunctive clauses, that is, $m = 2$. The probabilities of these two clauses are $P(r_4 \wedge r_8) = P_R(\{ (r_4,T), (r_8,T) \}) = 0.25$ and $P(r_4 \wedge r_9) = P_R(\{ (r_4,T), (r_9,T) \}) = 0.25$. The coverage algorithm in Lines 11-21 is used to approximate $P(s)$. $N = 56255.8$. After the loop in Lines 14-20 is repeated $\lceil N \rceil$ times, C equals 33855 and $\tilde{P}(s)$ is computed as 0.3009 (The exact value is 0.3). We have ({ ($?x$, Tom) }, 0.3009) as a member of the result $\mathbb{R}$. Similarly, the approximate probability of solution $s = \{$ ($?x$, May) } is computed as 0.2980. Hence, $\mathbb{R} = \{$ ({ ($?x$, Tom) }, 0.3009), ({ ($?x$, May) }, 0.2980) }.

6   Experimental Study

This section studies the execution time performance of checking the consistency of a pRDFS theory and answering queries on a pRDFS theory. The data set that we use is the Lehigh University Benchmark (LUBM) [4], which has a schema[2] for the university domain, and an instance data generator. The generator can output data sets of different sizes, and a university is a minimum unit of data generation. Note that the instance triples do not contain any blank nodes.

---

[2] http://www.lehigh.edu/~zhp2/2004/0401/univ-bench.owl

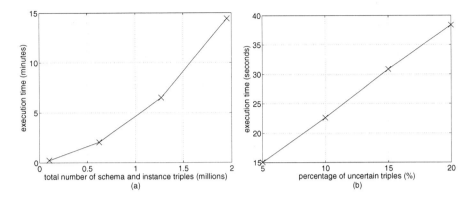

Fig. 1. Execution time of checking the consistency of a pRDFS theory as (a) the total number of schema and instance triples increases, and (b) the percentage of uncertain triples increases

To generate the probabilistic knowledge, we randomly select four triples each time from the instance data and assume these four triples are statistically correlated. We call them uncertain triples. We enumerate all combinations of truth value assignments for them. The probabilities of the combinations are randomly generated and then normalized such that their sum equals 1. The probabilistic knowledge is encoded in RDF triples using a specialized vocabulary[3].

The experiments were carried out on a computer with an Intel Pentium processor E5200 and 8 GB memory. The software code was written in Java and used Jena [1], which is an RDF toolkit including an RDF/XML parser, reasoners and a SPARQL query engine. Each experiment was run for 5 times and the average execution time was taken.

6.1   Consistency

This section examines how the execution time of checking the consistency of a pRDFS theory (Algorithm 1) scales with the data set size (the total number of schema and instance triples) and the number of uncertainties (the percentage of uncertain triples). Algorithm 1 is modified so that it does not stop when an inconsistency is found. It finds all causes of inconsistencies.

Two experiments were performed. In the first one, we varied the number of universities in the instance data from 1 to 15, and kept the percentage of uncertain triples at 5% of the total number of schema and instance triples. The result is shown in Fig. 1(a), which indicates that the execution time scales exponentially with the data set size. In the second experiment, we varied the percentage of uncertain triples from 5% to 20% and kept the number of universities in the instance data at 1. The result is shown in Fig. 1(b), which indicates that the execution time scales linearly with the number of uncertainties.

---

[3] http://www.comp.polyu.edu.hk/~csccszeto/prdfs/prdfs.owl

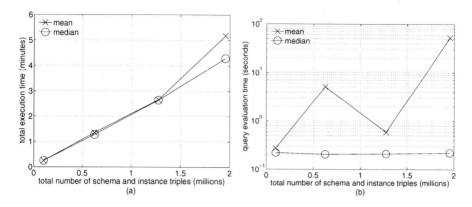

Fig. 2. (a) Total and (b) query evaluation time as the total number of schema and instance triples increases

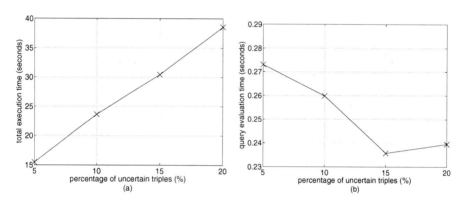

Fig. 3. (a) Total and (b) query evaluation time as the percentage of uncertain triples increases

6.2   Query Evaluation

This section examines how the execution time of query answering on a pRDFS theory (Algorithm 2) scales with the data set size and the number of uncertainties. Parameters ϵ and δ of Algorithm 2 are set to 0.1 and 0.9 respectively.

20 queries are randomly generated for each data set. We randomly select 3 connected instance triples to form the graph pattern of a query. The triples are connected in the way that at least one individual of each triple is the same as an individual of another triple. Then, we randomly select two elements from the triples, create a variable for each of them, and replace all occurrences of each element with its corresponding variable. For the two variables, one of them is the selected variable while the other is the non-selected one.

Two experiments were performed. In the first one, we varied the number of universities from 1 to 15, and kept the percentage of uncertain triples at 5%.

Fig. 2 shows both the total and query evaluation time. The total time was obtained by measuring all steps from loading the data set into the memory to query evaluation. Moreover, Fig. 2 shows both the mean and median of the time to answer 20 different queries because a few queries require much longer time than others to answer. Most of the query evaluation time of these few queries was spent on the computation of the probability of a very long DNF formula. The total time scales linearly with the data set size, but the query evaluation time does not depend on it. In the second experiment, we varied the percentage of uncertain triples from 5% to 20% and kept the number of universities at 1. Fig. 3 shows that the total time scales linearly with the number of uncertainties but the query evaluation time does not depend on it.

7   Related Work

There are attempts to model probabilistic knowledge in RDF/OWL [5,12]. Using Bayesian networks [9] is another approach. A vocabulary is provided in [3] to represent probabilistic knowledge in RDF, which can be mapped to a Bayesian network to do inference. PR-OWL [2] provides a vocabulary to represent probabilistic knowledge in OWL based on Multi-Entity Bayesian Networks (MEBN). MEBN is a first-order Bayesian logic that combines first-order logic with Bayesian networks.

OWL [8] has richer schema capabilities than RDFS. OWL DL is a sublanguage of OWL. It has the maximum expressiveness without losing decidability of key inference problems. It is based on the expressive description logic $\mathcal{SHOIN}(\mathbf{D})$. A lot of work has been done to represent probabilistic knowledge in description logics. We choose to recall the expressive probabilistic description logic P-$\mathcal{SHOIN}(\mathbf{D})$ [7] because it is the most expressive in terms of the description logic and probabilistic knowledge. P-$\mathcal{SHOIN}(\mathbf{D})$ uses the conditional constraint $(A|B)[l, u]$ to express probabilistic knowledge at both schema and instance levels, where A and B are class expressions that are free of probabilistic individuals. At the schema level, $(A|B)[l, u]$ is defeasible and means that the probability of A given B lies between l and u. At the instance level, $(A|B)[l, u]$ is associated with a probabilistic individual o. It is strict and means that the probability that o belongs to A given o belongs to B lies between l and u. Note that each probabilistic individual is associated with its own set of conditional constraints. Hence, P-$\mathcal{SHOIN}(\mathbf{D})$ cannot model statistical relationships among correlated statements involving different probabilistic individuals.

8   Conclusion

In this paper, we introduce pRDFS to model statistical relationships among correlated instance triples in RDFS data sets. We define its syntax and semantics. A consistency checking algorithm is provided to identify truth value assignments that violate the RDFS semantics. Experimental study shows that the execution time scales linearly with the number of uncertainties but exponentially with the data set size. Moreover, a query evaluation algorithm is provided to find answers to queries in SPARQL. Experimental study shows that the query evaluation time does not depend on the data set size and the number of uncertainties. However, a few queries require much longer time than others to compute the probabilities of answers.

Acknowledgements. The work described in this paper was partially supported by grants from the Research Grants Council of the Hong Kong Special Administrative Region, China (PolyU 5174/07E, PolyU 5181/06E).

References

1. Carroll, J.J., Dickinson, I., Dollin, C., Reynolds, D., Seaborne, A., Wilkinson, K.: Jena: implementing the semantic web recommendations. In: Feldman, S.I., Uretsky, M., Najork, M., Wills, C.E. (eds.) WWW (Alternate Track Papers & Posters), pp. 74–83. ACM, New York (2004)
2. Costa, P.C.G., Laskey, K.B.: PR-OWL: A framework for probabilistic ontologies. In: Bennett, B., Fellbaum, C. (eds.) FOIS. Frontiers in Artificial Intelligence and Applications, vol. 150, pp. 237–249. IOS Press, Amsterdam (2006)
3. Fukushige, Y.: Representing probabilistic relations in RDF. In: da Costa, P.C.G., Laskey, K.B., Laskey, K.J., Pool, M. (eds.) ISWC-URSW, pp. 106–107 (2005)
4. Guo, Y., Pan, Z., Heflin, J.: An evaluation of knowledge base systems for large OWL datasets. In: McIlraith, S.A., Plexousakis, D., van Harmelen, F. (eds.) ISWC 2004. LNCS, vol. 3298, pp. 274–288. Springer, Heidelberg (2004)
5. Huang, H., Liu, C.: Query evaluation on probabilistic RDF databases. In: Vossen, G., Long, D.D.E., Yu, J.X. (eds.) WISE 2009. LNCS, vol. 5802, pp. 307–320. Springer, Heidelberg (2009)
6. Karp, R.M., Luby, M., Madras, N.: Monte-carlo approximation algorithms for enumeration problems. Journal of Algorithms 10(3), 429–448 (1989)
7. Lukasiewicz, T.: Expressive probabilistic description logics. Artificial Intelligence 172(6-7), 852–883 (2008)
8. Web ontology language (OWL), http://www.w3.org/2004/OWL/
9. Pearl, J.: Probabilistic Reasoning in Intelligent Systems: Networks of Plausible Inference. Morgan Kaufmann Publishers Inc., San Francisco (1988)
10. Resource description framework (RDF), http://www.w3.org/RDF/
11. SPARQL query language for RDF, http://www.w3.org/TR/rdf-sparql-query/
12. Udrea, O., Subrahmanian, V.S., Majkic, Z.: Probabilistic RDF. In: IRI, pp. 172–177. IEEE Systems, Man, and Cybernetics Society (2006)

A Tableau Algorithm for Paraconsistent and Nonmonotonic Reasoning in Description Logic-Based System

Xiaowang Zhang[1], Zuoquan Lin[1], and Kewen Wang[2]

[1] School of Mathematical Sciences, Peking University, China
[2] School of Information and Communication Technology, Griffith University, Australia

Abstract. This paper proposes a paraconsistent and nonmonotonic extension of description logic by planting a nonmonotonic mechanism called *minimal inconsistency* in paradoxical description logics, which is a paraconsistent version of description logics. A precedence relation between two paradoxical models of knowledge bases is firstly introduced to obtain minimally paradoxical models by filtering those models which contain more inconsistencies than others. A new entailment relationship between a KB and an axiom characterized by minimal paradoxical models is applied to characterize the semantics of a paraconsistent and nonmonotonic description logic. An important advantage of our adaptation is simultaneously overtaking proverbial shortcomings of existing two kinds extensions of description logics: the weak inference power of paraconsistent description logics and the incapacity of nonmonotonic description logics in handling inconsistencies. Moreover, our paraconsistent and nonmonotonic extension not only preserves the syntax of description logic but also maintains the decidability of basic reasoning problems in description logics. Finally, we develop a sound and complete tableau algorithm for instance checking with the minimally paradoxical semantics.

1 Introduction

Description logics (DLs) [1] are a family of formal knowledge representation languages which build on classical logic and are the logic formalism for Frame-based systems and Semantic Networks. E.g. DLs are the logical foundation of the Web Ontology Language (OWL) in the Semantic Web [2] which is conceived as a future generation of the World Wide Web (WWW). As is well known, ontologies or knowledge bases (KBs) in an open, constantly changing and collaborative environment might be not prefect for a variety of reasons, such as modeling errors, migration from other formalisms, merging ontologies, ontology evolution and epistemic limitation etc [3,4,5,6,7]. That is, it is unrealistic to expect that real ontologies are always logically consistent and complete. However, DLs, like classical logics, are not good enough to represent some non-classical features of real ontologies [4] such as paraconsistent reasoning and nonmonotonic reasoning.

In order to capture these non-classical features of ontologies or KBs, several extensions of DLs have been proposed. They can be roughly classified into two two categories. The first (called *paraconsistent approach*) is extending paraconsistent semantics into DLs to tolerate inconsistencies occurring in ontologies, e.g., based on

X. Du et al. (Eds.): APWeb 2011, LNCS 6612, pp. 345–356, 2011.

Belnap's four-valued logic [8,9], Besnard and Hunter's quasi-classical logic [10], Elvang-Gøransson and Hunter's argumentative logic [11] and Priest's paradoxical logic [12]. While these paraconsistent semantics can handle inconsistencies in DLs in a way, they share the same shortcoming: their reasoning ability is too weak to infer useful information in some cases. For instance, the resolution rules do not work in four-valued DLs [8,9] and paradoxical DLs [12]; the application of proof rules are limited in a specific order in quasi-classical DLs [10]; and the proof systems are localized in sub-ontologies in argumentative DLs [11]. Moreover, the reasoning in most of existing paraconsistent DLs is monotonic and thus they are not sufficient to express evolving ontologies coming from a realistic world.

Considering a well known example about *tweety*: let $\mathcal{K}$ be a KB whose TBox is $\{Bird \sqsubseteq Fly, Bird \sqsubseteq Wing\}$ and ABox is $\{Brid(tweety), \neg Fly(tweety)\}$. In short, $\mathcal{K}$ tells us that all birds can fly, all birds have wings and tweety is a bird and cannot fly. It is easy to see that $\mathcal{K}$ is inconsistent. In our view, it might be reasonable that *tweety* has wings since the fact that *tweety* cannot fly doesn't mean that *tweety* hasn't wings, e.g., penguin has wings but it cannot fly. However, $Wing(tweety)$ could not be drawn from $\mathcal{K}$ in four-valued DLs [9] or in paradoxical DLs [12]. $Wing(tweety)$ is unknown in argumentative DLs [11]. Though $Wing(tweety)$ might be inferred in quasi-classical DLs [10], both $Bird(tweety)$ and $Fly(tweety)$ are taken as "contradiction" (both *true* and *flase*). In this sense, the quasi-classical inference might bring over contradictions. In addition, assume that we get a new information $\neg Bird(tweety)$ about *tweety*. That is, we have known that tweety is not bird. Intuitively, we would not conclude that either *tweety* can fly or *tweety* has wings. A new KB could be is obtained by adding $\neg Bird(tweety)$ in $\mathcal{K}$. However, conclusions from the new KB are the same as $\mathcal{K}$ in quasi-classical DLs. In other words, reasoning based on quasi-classical semantics cannot capture nonmonotonic feature of a true world.

The second (called *nonmonotonic approach*) is extending DLs with nonmonotonic features, e.g., based on Reiter's default logic [13], based on epistemic operators [14,15] and based on McCarthy's circumscription [16]. They provide some versions of nonmonotonic DLs. However, they are still unable to handle some inconsistent KBs because they are based on classical models. In other words, the capability of inconsistency handling is limited. For instance, in the *tweety* example, the original KB tells us that all birds can fly. When we find that penguin are birds which are unable to fly, we will usually amend the concept of bird by treating penguin as an exception (of birds) in those nonmonotonic DLs. It is impossible that we could enumerate all exceptions from incomplete KBs. Therefore, nonmonotonic mechanisms, in our view, might not be competent for deal with inconsistencies.

As argued above, paraconsistent DLs and nonmonotonic DLs have their advantages and disadvantages. It would be interesting to investigate paraconsistent nonmonotonic DLs by combining both paraconsistent and nonmonotonic approaches. While such ideas are not new in knowledge representation, it is rarely investigated how to define a paraconsistent nonmonotonic semantics for DLs. The challenge of combining paraconsistent DLs with nonmonotonic DLs is to preserve the features of classical DLs while some non-classical features such as nonmonotonicity and paraconsistency in DLs are incorpoted. Ideally, such a semantics should satisfy the following properties: (1) It is based on the original syntax of DLs (e.g. no new modal operators are introduced); (2) It is paraconsistent, i.e., every KB has at least one model (no matter it is consistent or inconsistent); (3) It is nonmonotonic; (4) It is still decidable. We note, however, the current approaches to incorporating nonmonotonic mechanism in DL reasoning hardly preserve the standard syntax of DLs, e.g., adding model operators in [14,15], open default rules in [13] and circumscription patten in [16]. In this sense, some proposals of introducing nonmonotonic reasoning in paraconsistent semantics have been considered in propositional logic, e.g., extending default reasoning with four-valued semantics in [17] and with quasi-classical semantics in [18]. However, it is not straightforward to adapt these proposals to DLs. Therefore, defining a paraconsistent and non-monotonic semantics for DLs that is of sufficient expressivity and computationally tractable is a non-trivial task.

In this paper, we propose a nonmonotonic paraconsistent semantics, called *minimally paradoxical semantics*, for DLs. The major idea is to introduce a concept of minimal inconsistency presented in the logic of paradox [19] in DLs so that some undesirable models of DL KBs that are major source of inconsistency are removed. Specifically, given a DL knowledge base, we first define a precedence relationship between paradoxical models of the ontology by introducing a partial order w.r.t. a principle of minimal inconsistency, i.e., the ontology satisfying inconsistent information as little as possible. Based on this partial order, we then select all minimally models from all paradoxical models as candidate models to serve for characterizing our new inference. We can show that our inference is paraconsistent and nonmonotonic. Thus nonmonotonic feature is extended in paraconsistent reasoning of DLs without modifying the syntax of DLs. Furthermore, we develop a decidable, sound and complete tableau algorithm for ABoxes to implement answering queries. In this paper, though we mainly consider DL $\mathcal{ALC}$ since it is a most basic member of the DL family, we also argue that our technique for $\mathcal{ALC}$ can be generalized in the other expressive DL members. The paraconsistent logic adopted in the paper is the paradoxical DL $\mathcal{ALC}$ [12] because it is closer to classical $\mathcal{ALC}$ compared to other multi-valued semantics, e.g., Belnap's four-valued semantics.

2   Description Logic $\mathcal{ALC}$ and Paradoxical Semantics

Description logics (DLs) is a well-known family of knowledge representation formalisms. In this section, we briefly recall the basic notion of logics $\mathcal{ALC}$ and the paradoxical semantics of $\mathcal{ALC}$. For more comprehensive background reasoning, we refer the reader to Chapter 2 of the DL Handbook [1] and paradoxical DLs [12].

Description Logic $\mathcal{ALC}$. Let $\mathcal{L}$ be the language of $\mathcal{ALC}$. Let N_C and N_R be pairwise disjoint and countably infinite sets of *concept names* and *role names* respectively. Let N_I be an infinite set of *individual names*. We use the letters A and B for concept names, the letter R for role names, and the letters C and D for concepts. $\top$ and $\bot$ denote the *universal concept* and the *bottom concept* respectively. Complex $\mathcal{ALC}$ concepts C, D are constructed as follows:

$$C, D \rightarrow \top \mid \bot \mid A \mid \neg C \mid C \sqcap D \mid C \sqcup D \mid \exists R.C \mid \forall R.C$$

An interpretation $\mathcal{I}_c = (\Delta^{\mathcal{I}_c}, \cdot^{\mathcal{I}_c})$ consisting of a non-empty domain $\Delta^{\mathcal{I}_c}$ and a mapping $\cdot^{\mathcal{I}_c}$ which maps every concept to a subset of $\Delta^{\mathcal{I}_c}$ and every role to a subset of $\Delta^{\mathcal{I}_c} \times \Delta^{\mathcal{I}_c}$, for all concepts C, D and a role R, satisfies conditions as follows: (1) top concept: $\top^{\mathcal{I}_c} = \Delta^{\mathcal{I}_c}$; (2) bottom concept: $\bot^{\mathcal{I}_c} = \emptyset^{\mathcal{I}_c}$; (3) negation: $(\neg C)^{\mathcal{I}_c} = \Delta^{\mathcal{I}_c} \setminus C^{\mathcal{I}_c}$; (4) conjunction: $(C_1 \sqcap C_2)^{\mathcal{I}_c} = C_1^{\mathcal{I}_c} \cap C_2^{\mathcal{I}_c}$; (5) disjunction: $(C_1 \sqcup C_2)^{\mathcal{I}_c} = C_1^{\mathcal{I}_c} \cup C_2^{\mathcal{I}_c}$; (6) existential restriction: $(\exists R.C)^{\mathcal{I}_c} = \{x \mid \exists y, (x,y) \in R^{\mathcal{I}_c} \text{ and } y \in C^{\mathcal{I}_c}\}$; (7) value restriction: $(\forall R.C)^{\mathcal{I}_c} = \{x \mid \forall y, (x,y) \in R^{\mathcal{I}_c} \text{ implies } y \in C^{\mathcal{I}_c}\}$

An $\mathcal{ALC}$ knowledge base (KB, for short) is a finite set of axioms formed by concepts, roles and individuals. A concept assertion is an axiom of the form $C(a)$ that assigns membership of an individual a to a concept C. A role assertion is an axiom of the form $R(a,b)$ that assigns a directed relation between two individuals a, b by the role R. An ABox contains a finite set of concept assertions and role assertions. A concept inclusion is an axiom of the form $C_1 \sqsubseteq C_2$ that states the subsumption of the concept C_1 by the concept C_2. A TBox contains a finite set of concept inclusions. An KB contains an ABox and a TBox. An interpretation $\mathcal{I}_c$ satisfies a concept assertion $C(a)$ if $a^{\mathcal{I}_c} \in C^{\mathcal{I}_c}$, a role assertion $R(a,b)$ if $(a^{\mathcal{I}_c}, b^{\mathcal{I}_c}) \in R^{\mathcal{I}_c}$, a concept inclusion $C_1 \sqsubseteq C_2$ if $C_1^{\mathcal{I}_c} \subseteq C_2^{\mathcal{I}_c}$. An interpretation that satisfies all axioms of a KB is called a *model* of the KB. Given a KB $\mathcal{K}$, we use $Mod(\mathcal{K})$ to denote a set of all models of $\mathcal{K}$. A KB $\mathcal{K}$ is *inconsistent* iff $Mod(\mathcal{K}) = \emptyset$. We say a KB $\mathcal{K}$ entails an axiom ϕ iff $Mod(\mathcal{K}) \subseteq Mod(\{\phi\})$, denoted $\mathcal{K} \models \phi$.

Paradoxical Semantics for $\mathcal{ALC}$. Compared with classical two-valued ($\{t, f\}$) semantics, the paradoxical semantics is three-valued ($\{t, f, \ddot{\top}\}$, $\ddot{\top}$ expressing both *true* and *false*) semantics where each concept C is interpreted as a pair $\langle +C, -C \rangle$ of (not necessarily disjoint) subsets of a domain $\Delta^{\mathcal{I}}$ and the union of them covering whole domain, i.e., $+C \cup -C = \Delta^{\mathcal{I}}$. We denote $\text{proj}^+(C^{\mathcal{I}}) = +C$ and $\text{proj}^-(C^{\mathcal{I}}) = -C$.

Intuitively, $+C$ is the set of elements which are known to belong to the extension of C, while $-C$ is the set of elements which are known to be not contained in the extension of C. $+C$ and $-C$ are not necessarily disjoint but mutual complemental w.r.t. the domain. In this case, we do not consider that *incomplete information* since it is not valuable for users but a statement for insufficient information.

Formally, a *paradoxical interpretation* is a pair $\mathcal{I} = (\Delta^{\mathcal{I}}, \cdot^{\mathcal{I}})$ with $\Delta^{\mathcal{I}}$ as domain, where $\cdot^{\mathcal{I}}$ is a function assigning elements of $\Delta^{\mathcal{I}}$ to individuals, subsets of $\Delta^{\mathcal{I}} \times \Delta^{\mathcal{I}}$) to concepts and subsets of $(\Delta^{\mathcal{I}} \times \Delta^{\mathcal{I}})^2$ to roles, so that $\cdot^{\mathcal{I}}$ satisfies conditions as follows: (1) $(\top)^{\mathcal{I}} = \langle \Delta^{\mathcal{I}}, \emptyset \rangle$; (2) $(\bot)^{\mathcal{I}} = \langle \emptyset, \Delta^{\mathcal{I}} \rangle$; (3) $(\neg C)^{\mathcal{I}} = \langle \text{proj}^-(C^{\mathcal{I}}), \text{proj}^+(C^{\mathcal{I}}) \rangle$; (4) $(C_1 \sqcap C_2)^{\mathcal{I}} = \langle \text{proj}^+(C_1^{\mathcal{I}}) \cap \text{proj}^+(C_2^{\mathcal{I}}), \text{proj}^-(C_1^{\mathcal{I}}) \cup \text{proj}^-(C_2^{\mathcal{I}}) \rangle$; (5) $(C_1 \sqcup C_2)^{\mathcal{I}} = \langle \text{proj}^+(C_1^{\mathcal{I}}) \cup \text{proj}^+(C_2^{\mathcal{I}}), \text{proj}^-(C_1^{\mathcal{I}}) \cap \text{proj}^-(C_2^{\mathcal{I}}) \rangle$; (6) $(\exists R.C)^{\mathcal{I}} = \langle \{x \mid \exists y, (x,y) \in \text{proj}^+(R^{\mathcal{I}}) \text{ and } y \in \text{proj}^+(C^{\mathcal{I}})\}, \{x \mid \forall y, (x,y) \in \text{proj}^+(R^{\mathcal{I}}) \text{ implies } y \in \text{proj}^-(C^{\mathcal{I}})\} \rangle$; (7) $(\forall R.C)^{\mathcal{I}} = \langle \{x \mid \forall y, (x,y) \in \text{proj}^+(R^{\mathcal{I}}) \text{ implies } y \in \text{proj}^+(C^{\mathcal{I}})\}, \{x \mid \exists y, (x,y) \in \text{proj}^+(R^{\mathcal{I}}) \text{ and } y \in \text{proj}^-(C^{\mathcal{I}})\} \rangle$

The correspondence between truth values in $\{t, f, \ddot{\top}\}$ and concept extensions can be easily observed: for an individual $a \in \Delta^{\mathcal{I}}$ and a concept name A, we have that

- $A^{\mathcal{I}}(a) = t$, iff $a^{\mathcal{I}} \in \text{proj}^+(A^{\mathcal{I}})$ and $a^{\mathcal{I}} \notin \text{proj}^-(A^{\mathcal{I}})$;
- $A^{\mathcal{I}}(a) = f$, iff $a^{\mathcal{I}} \notin \text{proj}^+(A^{\mathcal{I}})$ and $a^{\mathcal{I}} \in \text{proj}^-(A^{\mathcal{I}})$;
- $C^{\mathcal{I}}(a) = \ddot{\top}$, iff $a^{\mathcal{I}} \in \text{proj}^+(A^{\mathcal{I}})$ and $a^{\mathcal{I}} \in \text{proj}^-(A^{\mathcal{I}})$.

For instance, let $\Delta = \{a, b\}$ be a domain and A, B two concept names. Assume that $\mathcal{I}$ is a paradoxical interpretation on Δ such that $A^{\mathcal{I}} = \langle \{a, b\}, \{b\} \rangle$ and $B^{\mathcal{I}} = \langle \{a\}, \{a, b\} \rangle$. Then $A^{\mathcal{I}}(a) = t$, $A^{\mathcal{I}}(b) = B^{\mathcal{I}}(a) = \ddot{\top}$ and $B^{\mathcal{I}}(b) = f$.

A paradoxical interpretation $\mathcal{I}$ satisfies a concept assertion $C(a)$ if $a^{\mathcal{I}} \in \text{proj}^+(C^{\mathcal{I}})$, a role assertion $R(a, b)$ if $(a^{\mathcal{I}}, b^{\mathcal{I}}) \in \text{proj}^+(R^{\mathcal{I}})$, a concept inclusion $C_1 \sqsubseteq C_2$ iff $\Delta^{\mathcal{I}} \backslash \text{proj}^-(C_1^{\mathcal{I}}) \subseteq \text{proj}^+(C_2^{\mathcal{I}})$. A paradoxical interpretation that satisfies all axioms of a KB is called a *paradoxical model* of the KB. Given a KB $\mathcal{K}$, we use $Mod^P(\mathcal{K})$ to denote a set of all paradoxical models of $\mathcal{K}$. For instance, in the above example, $\mathcal{I}$ is a paradoxical model of an ABox $\{A(a), A(b), B(a)\}$ while it is not a paradoxical model of $\{A(a), A(b), B(a), B(b)\}$. We also find that $\mathcal{I}$ is also a paradoxical model of a TBox $\{A \sqsubseteq B\}$. A KB $\mathcal{K}$ *paradoxically entails* an axiom ϕ iff $Mod^P(\mathcal{K}) \subseteq Mod^P(\{\phi\})$, denoted $\mathcal{K} \models_{LP} \phi$.

We say a paradoxical interpretation $\mathcal{I}$ is *trivial* if $A^{\mathcal{I}} = \langle \Delta^{\mathcal{I}}, \Delta^{\mathcal{I}} \rangle$ for any concept name A. Note that contradictions which have form of $A \sqcap \neg A(a)$ for some concept name A and some individual a in $\mathcal{ALC}$ are only satisfied by trivial paradoxical interpretations. In general, trivial paradoxical interpretations/models cannot provide any information about querying since all queries are answered as "$\ddot{\top}$". For simplifying discussion, we mainly consider KBs without contradictions.

3 Minimally Paradoxical Semantics for $\mathcal{ALC}$

Paradoxical description logic is proposed in [12] as a paraconsistent version of description logics. In this section, we present a *minimally paradoxical semantics* for $\mathcal{ALC}$ by introducing a nonmonotonic reasoning mechanism *minimally inconsistency* (see [20]) to extend $\mathcal{ALC}$ with both paraconsistent and nonmonotonic features. The basic idea of our proposal is to introduce a preference relation on paradoxical models and to filter those models that cause more inconsistencies. Indeed, our proposal is able to maintain more consistent information in reasoning than previous approaches.

In a paradoxical DL, every paradoxical model can be represented as a set of concept assertions and role assertions similar to classical models. Since the constructor of the negation of a role $\neg R$ is absent in $\mathcal{ALC}$ (see [9]), for simplicity to convey our idea, we mainly consider concept assertions in defining the preference relation in this paper.

Next, we introduce a partial order $\prec$ between two paradoxical interpretations to characterize their difference on concept assertions.

Definition 1. *Let $\mathcal{I}$ and $\mathcal{I}'$ be two paradoxical interpretations in $\mathcal{ALC}$. We say $\mathcal{I}$ is more consistent than $\mathcal{I}'$, denoted $\mathcal{I} \prec \mathcal{I}'$ iff*

- *$\mathcal{I}$ and $\mathcal{I}'$ have the same domain Δ;*
- *if $A^{\mathcal{I}}(a) = \ddot{\top}$ then $A^{\mathcal{I}'}(a) = \ddot{\top}$ for any concept name $A \in N_C$ and any individual $a \in N_I$;*
- *there exists a concept name $A \in N_c$ and an individual $a \in \Delta$ such that $A^{\mathcal{I}'}(a) = \ddot{\top}$ but $A^{\mathcal{I}}(a) \neq \ddot{\top}$.*

Intuitively, the first condition states that if $\mathcal{I}_1$ and $\mathcal{I}_2$ do not share a common domain, then they are not comparable; the second condition ensures that if $\mathcal{I}_1 \prec \mathcal{I}_2$ then $\mathcal{I}_2$

contains no less inconsistencies than $\mathcal{I}_1$ does; and the third condition shows that if $\mathcal{I}_1 \prec \mathcal{I}_2$ then $\mathcal{I}_1$ contains less inconsistencies than $\mathcal{I}_2$.

For instance, let $\Delta = \{a, b\}$ be a domain and A a concept name. Let $\mathcal{I}_1$ and $\mathcal{I}_2$ be two paradoxical interpretations such that $A^{\mathcal{I}_1} = \langle\{a\}, \{b\}\rangle$ and $A^{\mathcal{I}_2} = \langle\{a, b\}, \{b\}\rangle$. Then we can easily see $\mathcal{I}_1 \prec \mathcal{I}_2$. If $\mathcal{I}_3$ is a paradoxical interpretation such that $A^{\mathcal{I}_3} = \langle\{b\}, \{a\}\rangle$. Then $\mathcal{I}_1 \not\prec \mathcal{I}_3$ and $\mathcal{I}_3 \not\prec \mathcal{I}_1$. That is, $\mathcal{I}_1$ and $\mathcal{I}_3$ are incomparable. So, in general $\prec$ is a partial order.

Note that $\prec$ is anti-reflexive, anti-symmetric and transitive. In other words, (1) $\mathcal{I} \not\prec \mathcal{I}$ for any paradoxical interpretation $\mathcal{I}$; (2) if $\mathcal{I} \prec \mathcal{I}'$ then $\mathcal{I}' \not\prec \mathcal{I}$; and (3) if $\mathcal{I} \prec \mathcal{I}'$ and $\mathcal{I}' \prec \mathcal{I}''$ then $\mathcal{I} \prec \mathcal{I}''$.

We denote $\mathcal{I} \preceq \mathcal{I}'$ as either $\mathcal{I} \prec \mathcal{I}'$ or $\mathcal{I} = \mathcal{I}'$. So $\preceq$ is also a partial order.

Definition 2. *Let $\mathcal{K}$ be a KB in $\mathcal{ALC}$. A paradoxical model $\mathcal{I}$ of $\mathcal{K}$ is minimal if there exists no other paradoxical model $\mathcal{I}'$ of $\mathcal{K}$ such that $\mathcal{I}' \prec \mathcal{I}$. We use $Mod_{min}^P(\mathcal{K})$ to denote the set of all minimal paradoxical models of $\mathcal{K}$.*

Intuitively, minimal paradoxical models are paradoxical models which contain minimal inconsistency. Since a non-empty KB always has a paradoxical model, it also has a minimal paradoxical model. That is, no matter whether a non-empty KB $\mathcal{K}$ is consistent or not, we have $Mod_{min}^P(\mathcal{K}) \neq \emptyset$ while $Mod(\mathcal{K}) = \emptyset$ for any inconsistent KB $\mathcal{K}$.

Example 1. *Let $\mathcal{K} = (\{A \sqsubseteq B\}, \{A(b), \forall R.\neg B(a), R(a, b)\})$ be a KB and $\Delta = \{a, b\}$ be a domain. We assume $\mathcal{I}$ is a paradoxical interpretation such that $A^{\mathcal{I}} = \langle\{b\}, \emptyset\rangle$, $B^{\mathcal{I}} = \langle\{b\}, \{b\}\rangle$, and $R^{\mathcal{I}} = \langle\{(a, b)\}, \emptyset\rangle$. It can be verified that $\mathcal{I}$ is the only minimally paradoxical model of $\mathcal{K}$. Thus $Mod_{min}^P(\mathcal{K}) = \{\mathcal{I}\}$.*

For a consistent KB, we can show that the set of its classical models corresponds to the set of its minimally paradoxical models. To this end, we first show how to transform each classical interpretation $\mathcal{I}_c$ into a paradoxical interpretation $\mathcal{I}$ as follows: for a concept name $A \in N_C$, a role name $R \in N_R$ and an individual name $a \in N_I$, define

$$\begin{cases} a^{\mathcal{I}} = a^{\mathcal{I}_c}, & \text{for any individual } a \in N_I; \\ a^{\mathcal{I}} \in +A, & \text{if } a^{\mathcal{I}_c} \in A^{\mathcal{I}_c}; \\ a^{\mathcal{I}} \in -A, & \text{if } a^{\mathcal{I}_c} \notin A^{\mathcal{I}_c}; \\ (a^{\mathcal{I}}, b^{\mathcal{I}}) \in +R \text{ and } -R = \emptyset, \text{ if } (a^{\mathcal{I}_c}, b^{\mathcal{I}_c}) \in R^{\mathcal{I}_c}. \end{cases} \quad (1)$$

For instance, let $\Delta = \{a, b\}$ and $\mathcal{I}_c$ be a classical interpretation such that $A^{\mathcal{I}_c} = \{a\}$ and $B^{\mathcal{I}_c} = \{a, b\}$. We transform $\mathcal{I}_c$ into a paradoxical interpretation $\mathcal{I}$ where $A^{\mathcal{I}} = \langle\{a\}, \{b\}\rangle$ and $B^{\mathcal{I}} = \langle\{a, b\}, \emptyset\rangle$.

If $\mathcal{K}$ is consistent, we use $Mod^T(\mathcal{K})$ to denote the collection of all paradoxical models of $\mathcal{K}$ transformed as above. It is interesting that each classical model corresponds to exactly one minimally paradoxical model for a consistent KB.

Proposition 1. *For any consistent $\mathcal{ALC}$ KB $\mathcal{K}$, $Mod_{min}^P(\mathcal{K}) = Mod^T(\mathcal{K})$.*

To see the validity of this result, we note that every paradoxical model in $Mod^T(\mathcal{K})$ is minimal because paradoxical models of a consistent KB are incomparable.

Based on the notion of minimally paradoxical models, we can define the following entailment relation.

Definition 3. *Let $\mathcal{K}$ be a KB and ϕ an axiom in $\mathcal{ALC}$. We say $\mathcal{K}$ minimally paradoxically entails ϕ iff $Mod^P_{min}(\mathcal{K}) \subseteq Mod^P_{min}(\{\phi\})$, denoted by $\mathcal{K} \models^m_{LP} \phi$.*

Intuitively, the minimally paradoxical entailment ($\models^m_{LP}$) characterizes an inference relation from a KB to an axiom by their minimally paradoxical models. Because the inference focuses on those paradoxical models in which inconsistency is minimized, $\models^m_{LP}$ can give consideration to both the classical entailment $\models$ and the paradoxical entailment $\models_{LP}$. When a KB is consistent, $\models^m_{LP}$ is equivalent to $\models$ since no model does contain an inconsistency. When a KB is inconsistent, $\models^m_{LP}$ inherits $\models_{LP}$ since every minimally paradoxical model contains at least an inconsistency. In this sense, $\models^m_{LP}$ is more reasonable than $\models$ and $\models_{LP}$. For instance, in the example about *tweety* (presented in Section 1). We have $\mathcal{K} \models^m_{LP} Wing(tweety)$ and $\mathcal{K} \cup \{\neg Bird(tweety)\} \not\models^m_{LP} Wing(tweety)$.

Note that the minimally paradoxical entailment is determined by restricting paradoxical entailment to the subclass of minimal paradoxical models. Thus the entailment relation is nonmonotonic.

For instance, given an ABox $\mathcal{A} = \{A(a), \neg A \sqcup B(a)\}$ where A, B are concept names and $\Delta = \{a\}$ a domain. Let $\mathcal{I}$ be a paradoxical interpretation s.t. $A^{\mathcal{I}} = \langle\{a\}, \emptyset\rangle$ and $B^{\mathcal{I}} = \langle\{a\}, \emptyset\rangle$. It easily check $\mathcal{I}$ is only one minimally paradoxical model of $\mathcal{A}$. Then $\{A(a), \neg A \sqcup B(a)\} \models^m_{LP} B(a)$. However, if we assume that $\mathcal{A}' = \mathcal{A} \cup \{\neg A(a)\}$ then there exist three minimally paradoxical models of $\mathcal{A}'$ $\mathcal{I}_i(i = 1, 2)$ where $A^{\mathcal{I}_i} = \langle\{a\}, \{a\}\rangle$ and $B^{\mathcal{I}_1} = \langle\{a\}, \emptyset\rangle$ and $B^{\mathcal{I}_2} = \langle\emptyset, \{a\}\rangle$. Thus $\mathcal{A}' \not\models^m_{LP} B(a)$.

In addition, $\models^m_{LP}$ is paraconsistent because each consistent KB has at least one minimally paradoxical model.

Theorem 1. $\models^m_{LP}$ *is paraconsistent and nonmonotonic.*

The next result shows that, if a KB is consistent, then the entailment $\models^m_{LP}$ coincides with the classical entailment.

Proposition 2. *Let $\mathcal{K}$ be a consistent KB and ϕ an axiom in $\mathcal{ALC}$. Then*

$$\mathcal{K} \models^m_{LP} \phi \; iff \; \mathcal{K} \models \phi.$$

This proposition directly follows from Proposition 1. Note that for an inconsistent KB, $\models^m_{LP}$ differs from $\models$. Thus, our new entailment relation naturally extends the classical entailment to all KBs in $\mathcal{ALC}$ while the classical reasoning in consistent KBs is still preserved.

Under classical entailment, anything can be inferred from an inconsistent KB. Thus, it is straightforward to see the following corollary.

Corollary 1. *Let $\mathcal{K}$ be a KB and ϕ an axiom in $\mathcal{ALC}$. If $\mathcal{K} \models^m_{LP} \phi$ then $\mathcal{K} \models \phi$.*

However, the converse of Corollary 1 is not true in general when $\mathcal{K}$ is inconsistent. For instance, we have $\{A(a), \neg A(a), \neg A \sqcup B(a), \neg B(a)\} \models B(a)$
while $\{A(a), \neg A(a), \neg A \sqcup B(a), \neg B(a)\} \not\models^m_{LP} B(a)$.

The resolution rule is important for automated reasoning in DLs. It is well-know that the inference rules *modus ponens, modus tollens* and *disjunctive syllogism* special cases of the *resolution rule*.

The resolution rule is not valid for $\models_{LP}^m$ in the following sense, while it is invalid in paradoxical DLs. In other words, the inference power of paradoxical DLs is strengthened by the concept of minimally paradoxical models.

Proposition 3. *Let C, D, E be concepts and a an individual in $\mathcal{ALC}$.*

$$\{C \sqcup D(a), \neg C \sqcup E(a)\} \models_{LP}^m D \sqcup E(a).$$

However, the resolution rule is not valid in general under minimally paradoxical semantics. For instance, $\{A(a), \neg A(a), A \sqcup B(a)\} \not\models_{LP}^m B(a)$.

We remark that minimally paradoxical semantics does not only preserve the reasoning ability of classical semantics for consistent knowledge but also tolerate inconsistencies (possibly) occurring KBs.

The results shown in this section demonstrate that the entailment relation $\models_{LP}^m$ is much better than the paradoxical entailment defined in [12].

4   Minimal Signed Tableau Algorithm

The minimally paradoxical semantics introduced in last section is based on minimal paradoxical models. A naive algorithm for reasoning under the new paraconsistent semantics could be developed by finding all minimal models from (possibly infinite) paradoxical models of a KB. However, such an algorithm would be very inefficient if it is not impossible. Instead, in this section, we develop a tableau algorithm for the new semantics. Tableau algorithms are widely used for checking satisfiability in DLs. Especially, *signed tableau algorithm*) has been developed in [12] for paradoxical $\mathcal{ALC}$. Our new tableau algorithm for minimally paradoxical semantics is obtained by embedding the minimality condition into the signed tableau algorithm. The challenge of doing so is how to find redundant clashes (i.e., a clash is caused by an inconsistency) and remove them from the signed tableau.

We first briefly recall the signed tableau algorithm for instance checking in ABoxes (the details can be found in [12]). The signed tableau algorithm is based on the notion of signed concepts. Note that roles are not signed because they represent edges connecting two nodes in tableaux. A sign concept is either $\mathbf{T}C$ or $\mathbf{F}C$ where the concept C is in NNF (i.e., negation ($\neg$) only occurs in front of concept names). Each signed concept can be transformed into its NNF by applying De Morgan's law, distributive law, the law of double negation and the following rewriting rules:

$$\mathbf{T}(C \sqcap D) = \mathbf{T}C \sqcap \mathbf{T}D, \ \mathbf{T}(C \sqcup D) = \mathbf{T}C \sqcup \mathbf{T}D, \ \mathbf{T}\forall R.C = \forall R.\mathbf{T}C, \ \mathbf{T}\exists R.C = \exists R.\mathbf{T}C$$
$$\mathbf{F}(C \sqcap D) = \mathbf{F}C \sqcup \mathbf{F}D, \ \mathbf{F}(C \sqcup D) = \mathbf{F}C \sqcap \mathbf{F}D, \ \mathbf{F}\forall R.C = \exists R.\mathbf{F}C, \ \mathbf{F}\exists R.C = \forall R.\mathbf{F}C$$

We use $\mathbf{T}\mathcal{A}$ to denote a signed ABox whose concept names are marked with $\mathbf{T}$, i.e., $\mathbf{T}\mathcal{A} = \{\mathbf{T}C \mid C \in \mathcal{A}\}$. A *signed tableau* is a forest whose trees are actually composed of nodes $\mathcal{L}(x)$ containing signed concepts and edges $\mathcal{L}(x, y)$ containing role names. Given an ABox $\mathcal{A}$ and an axiom $C(a)$, the signed tableau algorithm starts with $\mathcal{F}_\mathcal{A}$ as the initial forest of $\mathbf{T}\mathcal{A} \cup \{\mathbf{F}C(a)\}$. The algorithm then applies the signed expansion rules, which are reformulated in Table 1. The algorithm terminates if it encounters a clash: $\{\mathbf{T}A, \mathbf{F}A\} \subseteq \mathcal{L}(x)$ or $\{\mathbf{F}A, \mathbf{F}\neg A\} \subseteq \mathcal{L}(x)$ where A is a concept name. Finally,

Table 1. Expansion Rules in Signed Tableau

$\sqcap_{\mathbf{T}}$-rule	If: $\mathbf{T}(C_1 \sqcap C_2) \in \mathcal{L}$, but not both $\mathbf{T}C_1 \in \mathcal{L}(x)$ and $\mathbf{T}C_2 \in \mathcal{L}(x)$.
	Then: $\mathcal{L}(x) := \mathcal{L}(x) \cup \{\mathbf{T}C_1, \mathbf{T}C_2\}$.
$\sqcup_{\mathbf{T}}$-rule	If: $\mathbf{T}(C_1 \sqcup C_2) \in \mathcal{L}(x)$, but neither $\mathbf{T}C_i \in \mathcal{L}(x)$ for $i = 1, 2$.
	Then: $\mathcal{L}(x) := \mathcal{L}(x) \cup \{\mathbf{T}C_i\}$ ($i = 1$ or 2).
$\exists_{\mathbf{T}}$-rule	If: $\mathbf{T}\exists R.C \in \mathcal{L}(x)$, but there is no node $\mathcal{L}(z)$ s.t. $\mathbf{T}C \in \mathcal{L}(z)$ and $R \in \mathcal{L}(x, z)$.
	Then: create a new node $\mathcal{L}(y) := \{\mathbf{T}C\}$ and $\mathcal{L}(x, y) := \{R\}$.
$\forall_{\mathbf{T}}$-rule	If: $\mathbf{T}\forall R.C \in \mathcal{L}(x)$ and $R \in \mathcal{L}(x, y)$, but $\mathbf{T}C \notin \mathcal{L}(y)$.
	Then: $\mathcal{L}(y) := \mathcal{L}(y) \cup \{\mathbf{T}C\}$.
$\sqcap_{\mathbf{F}}$-rule	If: $\mathbf{F}(C_1 \sqcap C_2) \in \mathcal{L}(x)$, but neither $\mathbf{F}C_i \in \mathcal{L}(x)$ for $i = 1, 2$.
	Then: $\mathcal{L}(x) := \mathcal{L}(x) \cup \{\mathbf{F}C_i\}$ ($i = 1$ or 2).
$\sqcup_{\mathbf{F}}$-rule	If: $\mathbf{F}(C_1 \sqcup C_2) \in \mathcal{L}$, but not both $\mathbf{F}C_1 \in \mathcal{L}(x)$ and $\mathbf{F}C_2 \in \mathcal{L}(x)$.
	Then: $\mathcal{L}(x) := \mathcal{L}(x) \cup \{\mathbf{F}C_1, \mathbf{F}C_2\}$.
$\exists_{\mathbf{F}}$-rule	If: $\mathbf{F}\exists R.C \in \mathcal{L}(x)$ and $R \in \mathcal{L}(x, y)$, but $\mathbf{F}C \notin \mathcal{L}(y)$.
	Then: $\mathcal{L}(y) := \mathcal{L}(y) \cup \{\mathbf{F}C\}$.
$\forall_{\mathbf{F}}$-rule	If: $\mathbf{F}\forall R.C \in \mathcal{L}(x)$, but there is no node $\mathcal{L}(z)$ s.t. $\mathbf{F}C \in \mathcal{L}(z)$ and $R \in \mathcal{L}(x, z)$.
	Then: create a new node $\mathcal{L}(y) := \{\mathbf{F}C\}$ and $\mathcal{L}(x, y) := \{R\}$.

we obtain a completion forest The problem whether $\mathcal{A}$ paradoxically entails $C(a)$ is decided by checking whether the completion forest is closed, i.e., checking whether every tree of the completion forest contains at least one clash. The algorithm preserves a so-called *forest model* property, i.e., the paradoxical model has the form of a set of (potentially infinite) trees, the root nodes of which can be arbitrarily interconnected.

In the following, we develop a preference relation on trees of the completion forest to eliminating the trees with redundant inconsistencies.

Let $\mathcal{F}$ be a completion forest and $\mathbf{t}$ a tree of $\mathcal{F}$. We denote $IC(\mathbf{t}) = \{\mathbf{T}A \mid \{\mathbf{T}A, \mathbf{T}\neg A\} \subseteq \mathcal{L}(x) \text{ for some node } \mathcal{L}(x) \in \mathbf{t}\}$. Intuitively, $IC(\mathbf{t})$ is the collection of contradictions in $\mathbf{t}$.

Definition 4. *Let $\mathbf{t}_1$ and $\mathbf{t}_2$ be two trees of a identical completion forest. We denote $\mathbf{t}_1 \prec_{IC} \mathbf{t}_2$ if $IC(\mathbf{t}_1) \subset IC(\mathbf{t}_2)$ and $\mathbf{t}_1 \preceq_{IC} \mathbf{t}_2$ if $IC(\mathbf{t}_1) \subseteq IC(\mathbf{t}_2)$. If $\mathbf{t}_1 \prec_{IC} \mathbf{t}_2$ then we say $\mathbf{t}_2$ is* redundant *w.r.t. $\mathbf{t}_1$.*

Intuitively, if $\mathbf{t}_2$ is redundant w.r.t. $\mathbf{t}_1$ then $\mathbf{t}_2$ contains more inconsistencies than $\mathbf{t}_1$ does.

A tree $\mathbf{t}$ of $\mathcal{F}$ is a *minimally redundant tree* of $\mathcal{F}$ if $\mathbf{t} \preceq_{IC} \mathbf{t}'$ for each tree $\mathbf{t}'$ in $\mathcal{F}$ and there is not any other tree $\mathbf{t}''$ in $\mathcal{F}$ s.t. $\mathbf{t}'' \prec_{IC} \mathbf{t}$. A minimally redundant tree is a tree we want to keep. A *minimal completion forest* of $\mathcal{F}$, denoted by $\mathcal{F}^m$, is composed of all minimal trees of $\mathcal{F}$. It can be easily verfied that $\mathcal{F}^m$ always exists. Given an ABox $\mathcal{A}$ and an axiom $C(a)$, the process of computing the minimal completion forest of $\mathbf{T}\mathcal{A} \cup \{\mathbf{F}C(a)\}$ is called the *minimal signed tableau algorithm*.

A paradoxical interpretation $\mathcal{I}$ *satisfies* a tree $\mathbf{t}$ iff for any node $\mathcal{L}(x)$ and any edge $\mathcal{L}(x, y)$ in $\mathbf{t}$, we have
(1) $x^{\mathcal{I}} \in \text{proj}^+(A^{\mathcal{I}})$ if $\mathbf{T}A \in \mathcal{L}(x)$ and $x^{\mathcal{I}} \in \text{proj}^-(A^{\mathcal{I}})$ if $\mathbf{T}\neg A \in \mathcal{L}(x)$;
(2) $x^{\mathcal{I}} \in \Delta^{\mathcal{I}} - \text{proj}^+(A^{\mathcal{I}})$ if $\mathbf{F}A \in \mathcal{L}(x)$ and $x^{\mathcal{I}} \in \Delta^{\mathcal{I}} - \text{proj}^-(A^{\mathcal{I}})$ if $\mathbf{F}\neg A \in \mathcal{L}(x)$;
(3) $(x^{\mathcal{I}}, y^{\mathcal{I}}) \in \text{proj}^+(R^{\mathcal{I}})$ if $R \in \mathcal{L}(x, y)$.

Based on the above definition, it follows that if $\mathcal{I}$ is a paradoxical interpretation satisfying t then $A^{\mathcal{I}}(a) = \ddot{\top}$ iff $\{\mathbf{T}A(a), \mathbf{T}\neg A(a)\} \subseteq \mathbf{t}$ for any concept name $A \in N_C$ and any individual $a \in N_I$. As a result, there exists a close relation between $\prec$ (defined over paradoxical models) and $\prec_{IC}$ (defined over trees).

Theorem 2. *Let* t *and* t′ *be two trees of a completion forest* $\mathcal{F}$. *If* $\mathcal{I}$ *and* $\mathcal{I}'$ *be two paradoxical interpretations satisfying* t *and* t′ *respectively, then* $\mathcal{I} \prec \mathcal{I}'$ *iff* t $\prec_{IC}$ t′.

Proof.(Sketch) Note that the ABox contains only concept assertions. We show only that the theorem is true for atomic concept A. It is straightforward to prove that the conclusion is also true for a complex concept by induction.

($\Leftarrow$) If t $\prec_{IC}$ t′, then $IC(\mathbf{t}) \subset IC(\mathbf{t}')$. We want to show that $\mathcal{I} \prec \mathcal{I}'$. Assume that $\{\mathbf{T}A(a), \mathbf{T}\neg A(a)\} \subseteq \mathbf{t}$ then $\{\mathbf{T}A(a), \mathbf{T}\neg A(a)\} \subseteq \mathbf{t}'$. Thus $A^{\mathcal{I}}(a) = \ddot{\top}$ implies $A^{\mathcal{I}'}(a) = \ddot{\top}$. There is a concept assertion $B(b)$, where B is a concept name and b an individual name, such that $\{\mathbf{T}B(b), \mathbf{T}B(b)\} \subseteq \mathbf{t}'$ but $\{\mathbf{T}B(b), \mathbf{T}B(b)\} \not\subseteq \mathbf{t}$. Thus $B^{\mathcal{I}'}(b) = \ddot{\top}$ implies $B^{\mathcal{I}}(a) \neq \ddot{\top}$. Therefore, $\mathcal{I} \prec \mathcal{I}'$ by Definition 1.

($\Rightarrow$) If $\mathcal{I} \prec \mathcal{I}'$, we need to show that t $\prec_{IC}$ t′, i.e., $IC(\mathbf{t}) \subset IC(\mathbf{t}')$. For any concept assertion $A(a)$, if $A^{\mathcal{I}}(a) = \ddot{\top}$ then $A^{\mathcal{I}'}(a) = \ddot{\top}$. Thus $\{\mathbf{T}A(a), \mathbf{T}\neg A(a)\} \subseteq \mathbf{t}$ then $\{\mathbf{T}A(a), \mathbf{T}\neg A(a)\} \subseteq \mathbf{t}'$. There is a concept assertion $B(b)$, where B is a concept name and b an individual name, such that $B^{\mathcal{I}'}(b) = \ddot{\top}$ but $B^{\mathcal{I}}(a) \neq \ddot{\top}$. Thus $\{\mathbf{T}B(b), \mathbf{T}B(b)\} \subseteq \mathbf{t}'$ but $\{\mathbf{T}B(b), \mathbf{T}B(b)\} \not\subseteq \mathbf{t}$. Therefore, t $\prec_{IC}$ t′ since $IC(\mathbf{t}) \subset IC(\mathbf{t}')$.

Now we are ready to define the concept of minimally closed completion forests. A completion forest $\mathcal{F}$ is *minimally closed* iff every tree of $\mathcal{F}^m$ is closed.

If the completion forest of $\mathbf{T}\mathcal{A} \cup \{\mathbf{F}C(a)\}$ by applying the minimal signed tableau algorithm is minimally closed, then we write $\mathcal{A} \vdash_{LP}^m C(a)$.

We show that our minimal signed tableau algorithm is sound and complete.

Theorem 3. *Let* $\mathcal{A}$ *be an ABox and* $C(a)$ *an axiom in* $\mathcal{ALC}$. *We have*

$$\mathcal{A} \vdash_{LP}^m C(a) \text{ iff } \mathcal{A} \models_{LP}^m C(a).$$

Proof.(Sketch) We consider only atomic concept A here. Let $\mathcal{F}$ be a completion forest for $\mathbf{T}\mathcal{A} \cup \{\mathbf{F}A(a)\}$ by applying the minimal signed tableau algorithm. We need to prove that $\mathcal{F}$ is minimally closed, i.e., every tree of the minimal forest $\mathcal{F}^m$ is closed, iff $\mathcal{A} \models_{LPm} A(a)$.

($\Rightarrow$) Assume that $\mathcal{F}$ is minimally closed. On the contrary, supposed that $\mathcal{A} \not\models_{LP}^m A(a)$, that is, there exists a minimally paradoxical model $\mathcal{I}$ of $\mathcal{A}$ and $\mathcal{I} \not\models_{LP}^m A(a)$, then $\mathcal{I} \not\models_{LP} A(a)$ since every minimal paradoxical model is always a paradoxical model. There exists a tree t which is satisfied by $\mathcal{I}$ in $\mathcal{F}$ and t is not closed by the proof of Theorem 7 which states that the signed tableau algorithm is sound and complete w.r.t. paradoxical semantics for $\mathcal{ALC}$ (see [12]). We assert that t is not redundant. Otherwise, there might be another tree t′ $\prec_{IC}$ t. We define a paradoxical interpretation $\mathcal{I}'$ satisfing t′. It is easy to see that $\mathcal{I}'$ and $\mathcal{I}$ have the same domain by induction on the structure of complete $\mathcal{F}$. By Theorem 2, we have $\mathcal{I}' \prec \mathcal{I}$, which contradicts the minimality of $\mathcal{I}$. Thus the completion forest $\mathcal{F}$ for $\mathbf{T}\mathcal{A} \cup \{\mathbf{F}A(a)\}$ contains at least a tree t that is neither closed nor redundant, which contadicts with the assumption that $\mathcal{F}$ is minimally closed. Thus $\mathcal{A} \models_{LP}^m A(a)$.

($\Leftarrow$) Let $\mathcal{A} \models_{LP}^{m} A(a)$. On the contrary, supposed that $\mathcal{F}$ is not minimally closed. Then there exists a tree $\mathbf{t}$ of complete $\mathcal{F}$ such that $\mathbf{t}$ is neither closed nor redundant. Since $\mathbf{t}$ is not closed, by the proof of Theorem 7 (see [12]), we can construct a paradoxical interpretation $\mathcal{I}$ such that $\mathcal{I}$ is a paradoxical model of $\mathcal{A}$. However, $\mathcal{I} \not\models_{LP} A(a)$, a contradiction. Supposed that $\mathcal{I}$ is not minimal, there exists a paradoxical model $\mathcal{I}'$ of $\mathcal{A}$ such that $\mathcal{I}' \prec \mathcal{I}$ in the same domain. Thus there exists a tree $\mathbf{t}'$ that is satisfied by $\mathcal{I}'$ with $\mathbf{t}' \prec_{IC} \mathbf{t}$ by Theorem 2, which contradicts to the assumption that $\mathbf{t}$ is not redundant. Then $\mathcal{I} \not\models_{LP} A(a)$, i.e., $\mathcal{I} \notin Mod^{P}(\{A(a)\})$. Because $\{A(a)\}$ is consistent and A is a concept name, $Mod^{P}(\{A(a)\}) = Mod_{min}^{P}(\{A(a)\})$. Thus $\mathcal{I}$ is a minimally paradoxical model of $\mathcal{A}$ but $\mathcal{I} \not\models_{LP}^{m} A(a)$. Therefore, $\mathcal{F}$ is minimally closed.

Example 2. *Let* $\mathcal{A} = \{C \sqcup D(a), \neg C \sqcup E(a)\}$ *be an ABox and* $D \sqcup E(a)$ *an axiom. There are four trees of the completion forest* $\mathcal{F}$ *of* $\{\mathbf{T}\mathcal{A} \cup \{\mathbf{F}(D \sqcup E)(a)\}$ *where*
$\mathbf{t}_1 = \{\mathbf{T}C(a), \mathbf{T}\neg C(a), \mathbf{F}D(a), \mathbf{F}E(a)\};$ $\mathbf{t}_2 = \{\mathbf{T}C(a), \mathbf{T}E(a), \mathbf{F}D(a), \mathbf{F}E(a)\};$
$\mathbf{t}_3 = \{\mathbf{T}D(a), \mathbf{T}\neg C(a), \mathbf{F}D(a), \mathbf{F}E(a)\};$ $\mathbf{t}_4 = \{\mathbf{T}D(a), \mathbf{T}E(a), \mathbf{F}D(a), \mathbf{F}E(a)\}.$
Since the minimal completion forest $\mathcal{F} = \{\mathbf{t}_2, \mathbf{t}_3, \mathbf{t}_4\}$ *are closed,* $\mathcal{F}$ *is minimally closed. Thus* $\{C \sqcup D(a), \neg C \sqcup E(a)\} \vdash_{LP}^{m} D \sqcup E(a)$. *Therefore,* $\{C \sqcup D(a), \neg C \sqcup E(a)\} \vdash_{LP}^{m} D \sqcup E(a)$ *by Theorem 3.*

5   Conclusion and Future Work

In this paper, we have presented a nonmonotonic and paraconsistent semantics, called minimally paradoxical semantics, for $\mathcal{ALC}$, which can be seen a naturally extension of the classical semantics. The suitability of our semantics is justified by several important properties. In particular, the new semantics overcomes some shortcomings of existing paraconsistent DLs and nonmonotonic DLs. Based on the signed tableau, we have developed a sound and complete algorithm, named minimal signed tableau, to implement paraconsistent and nonmonotonic reasoning with DL ABoxes. This is achieved by introducing a preference relation on trees of completion forests in signed tableau. This new approach can be used in developing new tableau algorithms for other nonmonotonic DLs. There are several issues for future work: First, we plan to implement the minimal signed tableau for DLs. Before this is done, we will first develop some heuristics for efficient implementation; Second, we will explore applications of the new paraconsistent semantics in ontology repair, revision and merging.

Acknowledgments

The authors appreciate the referees for their helpful and constructive comments. This work was supported by NSFC under 60973003, 60496322 and ARC under DP1093652.

References

1. Baader, F., Calvanese, D., McGuinness, D.L., Nardi, D., Patel-Schneider, P.F. (eds.): The Description Logic Handbook: Theory, Implementation, and Applications. Cambridge University Press, Cambridge (2003)

2. Berners-Lee, T., Hendler, J., Lassila, O.: The Semantic Web. Scientific American (2001)
3. Bertossi, L.E., Hunter, A., Schaub, T. (eds.): Inconsistency Tolerance. LNCS, vol. 3300. Springer, Heidelberg (2005)
4. Schaffert, S., Bry, F., Besnard, P., Decker, H., Decker, S., Enguix, C.F., Herzig, A.: Paraconsistent reasoning for the semantic web. In: Proc. of ISWC-URSW 2005, Ireland, pp. 104–105 (2005)
5. Huang, Z., van Harmelen, F., ter Teije, A.: Reasoning with inconsistent ontologies. In: Proc. of IJCAI 2005, UK, pp. 454–459. Professional Book Center (2005)
6. Haase, P., van Harmelen, F., Huang, Z., Stuckenschmidt, H., Sure, Y.: A framework for handling inconsistency in changing ontologies. In: Gil, Y., Motta, E., Benjamins, V.R., Musen, M.A. (eds.) ISWC 2005. LNCS, vol. 3729, pp. 353–367. Springer, Heidelberg (2005)
7. Qi, G., Du, J.: Model-based revision operators for terminologies in description logics. In: Proc. of IJCAI 2009, USA, pp. 891–897 (2009)
8. Schlobach, S., Cornet, R.: Non-standard reasoning services for the debugging of description logic terminologies. In: Proc. of IJCAI 2003, Mexico, pp. 355–362. Morgan Kaufmann, San Francisco (2003)
9. Ma, Y., Hitzler, P., Lin, Z.: Algorithms for paraconsistent reasoning with OWL. In: Franconi, E., Kifer, M., May, W. (eds.) ESWC 2007. LNCS, vol. 4519, pp. 399–413. Springer, Heidelberg (2007)
10. Zhang, X., Xiao, G., Lin, Z.: A tableau algorithm for handling inconsistency in OWL. In: Aroyo, L., Traverso, P., Ciravegna, F., Cimiano, P., Heath, T., Hyvönen, E., Mizoguchi, R., Oren, E., Sabou, M., Simperl, E. (eds.) ESWC 2009. LNCS, vol. 5554, pp. 399–413. Springer, Heidelberg (2009)
11. Zhang, X., Zhang, Z., Xu, D., Lin, Z.: Argumentation-based reasoning with inconsistent knowledge bases. In: Farzindar, A., Kešelj, V. (eds.) Canadian AI 2010. LNCS, vol. 6085, pp. 87–99. Springer, Heidelberg (2010)
12. Zhang, X., Lin, Z., Wang, K.: Towards a paradoxical description logic for the Semantic Web. In: Link, S., Prade, H. (eds.) FoIKS 2010. LNCS, vol. 5956, pp. 306–325. Springer, Heidelberg (2010)
13. Baader, F., Hollunder, B.: Embedding defaults into terminological knowledge representation formalisms. J. Autom. Reasoning 14(1), 149–180 (1995)
14. Donini, F.M., Lenzerini, M., Nardi, D., Nutt, W., Schaerf, A.: An epistemic operator for description logics. Artif. Intell. 100(1-2), 225–274 (1998)
15. Donini, F.M., Nardi, D., Rosati, R.: Description logics of minimal knowledge and negation as failure. ACM Trans. Comput. Log. 3(2), 177–225 (2002)
16. Bonatti, P.A., Lutz, C., Wolter, F.: Description logics with circumscription. In: Proc. of KR 2006, UK, pp. 400–410. AAAI Press, Menlo Park (2006)
17. Yue, A., Ma, Y., Lin, Z.: Four-valued semantics for default logic. In: Lamontagne, L., Marchand, M. (eds.) Canadian AI 2006. LNCS (LNAI), vol. 4013, pp. 195–205. Springer, Heidelberg (2006)
18. Lin, Z., Ma, Y., Lin, Z.: A fault-tolerant default logic. In: Fisher, M., van der Hoek, W., Konev, B., Lisitsa, A. (eds.) JELIA 2006. LNCS (LNAI), vol. 4160, pp. 253–265. Springer, Heidelberg (2006)
19. Priest, G.: Minimally inconsistent LP. Studia Logica: An Inter. J. for Symbolic Log. 50(2), 321–331 (1991)
20. Ginsberg, M.L. (ed.): Readings in nonmonotonic reasoning. Morgan Kaufmann Publishers Inc., San Francisco (1987)

Key Concepts Identification and Weighting in Search Engine Queries

Jiawang Liu and Peng Ren

Department of Computer Science
Shanghai Jiao Tong Uninversity
ljwsummer@gmail.com, renpeng@apex.sjtu.edu.cn

Abstract. It has been widely observed that queries of search engine are becoming longer and closer to natural language. Actually, current search engines do not perform well with natural language queries. Accurately discovering the key concepts of these queries can dramatically improve the effectiveness of search engines. It has been shown that queries seem to be composed in a way that how users summarize documents, which is so much similar to anchor texts. In this paper, we present a technique for automatic extraction of key concepts from queries with anchor texts analysis. Compared with using web counts of documents, we proposed a supervised machine learning model to classify the concepts of queries into 3 sets according to their importance and types. In the end of this paper, we also demonstrate that our method has remarkable improvement over the baseline.

Keywords: Machine learning, NLP, anchor analysis, key concepts.

1 Introduction

As the search queries are becoming closer to natural language, discovering key concepts from search queries and assigning reasonable weighting to them is an important way to understand user's search goals. Key concepts identification and weighting are key points in a serious of query analysis issues. Not only can it be directly used by search engine, but also it acts an important role in several research issues, such as query rewriting, query expansion, etc. It's the foundation of many query processing problems.

Current commercial search engines process billions of queries per day [1], most of them are free text. There are a lot of useless terms in those queries. For example, for the query "What's the weather of Chicago in Sunday", it is not friendly to search engine. If the search engine could identify the key concept of this query is "Chicago weather" and there's no need to retrieve documents use other terms. The relevance of results would be remarkably improved. There are often several concepts in a query. We use noun phrases extracted from the queries as concepts. Our goal is assigning 3 different levels weight to them according to their importance. There is no need to assign different weight to different key concepts of a query, for they are all the most important parts of the query. We will illustrate it in the following example.

X. Du et al. (Eds.): APWeb 2011, LNCS 6612, pp. 357–369, 2011.

Table 1. An example of TREC topic

$<num>$	Number: 336
$<title>$	Black Bear Attacks
$<desc>$	A relevant document would discuss the frequency of vicious black bear attacks worldwide and the possible causes for this savage behavior.

In the TREC topic of Table 1, we treat the $<desc>$ part as a raw query. We expect it can generate the following results after processing.

Table 2. Expected result

weight	concepts
3	black bear, attacks
2	possible causes
1	other concepts

Weight 3 means the most important concepts of the query. Weight 2 is assigned to the concepts which are important to the query, but not the necessary parts. We identify the concepts with Weight 2 in order to guarantee the recall of the retrieved documents. Meanwhile, we assign Weight 1 to the useless ones.

There are three primary contributions in this paper: (1)we propose a supervised machine learning technique for automatic extracting key concepts; (2)discovering key concepts from natural language queries with anchor analysis, since anchor and queries are organized by similar language; (3)not only discovering the most important concepts in queries, but also classifying them into 3 levels. It can guarantee the recall rate of documents.

The rest of this paper is organized as follows: In section 2 we discuss previous work on key concepts identification and weighting. We describe our proposed model's details in Section 3 and our experimental results are shown in Section 4. And finally, section 5 summarizes the conclusions and gives a look to future work.

2   Related Work

Most traditional information retrieval models, such as language modeling, treat query terms as independent and of uniform importance [2]. For example, inverse document frequency(idf) is thought of as a simple query term weighting model, however, it hardly contains much information. Meanwhile, it is not clear if idf is a proper standard to judge the importance of phrases and other generic concepts [3]. Furthermore, most of concepts detection work treat all concepts matches in queries equally, rather than weighting them [16,17,18,19,4,8,12]. As queries are becoming more complex, it is clearly much unreasonable for these queries.

Some previous work on key concept detection use linguistic and statistical methods to discover core terms in $< desc >$ queries and use name entity recognition to convert $< desc >$ queries into structured INQUERY queries [20]. While our proposed model towards all kinds of general search queries.

Key concept detection, focused on verbose queries, has been a subject in recent work. Bendersky et. al [6] proposed a supervised machine learning technique for key concept detection. In their model, nearly all features were statistical count in documents($tf, idf, etc.$). However, documents and search queries are organized by so much different language. Kumaran et al. [9] use learning to rank all subsets of the original query (sub-queries) based on their predicted quality for query reduction. Although similar to our work, the weighting techniques discussed in these papers are based on document counts, while our model is based on anchor analysis.

In sum, most of previous work on key concept detection via statistical methods (such as inverse document frequency) [10, 11, 13] or machine learning methods are based on direct web counts in documents [6]. However, queries and documents are organized by so much different language [14]. Even the different parts of a document are organized differently, such as title, body, anchor, etc. Both anchor text and queries seem to be composed in a way that how users summarize documents. In contrast to previous work, we do not treat all parts of documents as the same. We use a supervised machine learning technique for key concepts detection and use a diverse mix of features which extracted via anchor analysis.

3   Model

In this section we present our model for key concepts detection in natural language queries. First we describe the details of the supervised machine learning technique with anchor analysis. Then we detail the feature selection method that we used.

3.1   Anchor Analysis

Most of text in documents and queries are organized by different languages. Main text in documents try to detail problems or describe something, while queries often ask for information by summarized words. A query can often be regarded as a very brief summary of a document which may be the best fit of user's expectation. Anchor texts are so similar to queries on this point. So we try to extract some significant features via anchor analysis. We extract anchor texts and the relationship between anchor and documents from billions of web pages[1]. The following are features we considered in our model.

Concept Frequency($cf(c_i)$). Concept Frequency($cf(c_i)$) in all anchor texts. It is supposed that key concepts will tend to have a higher *concept frequency* in anchor text than non-key concepts.

---

[1] The web pages of Yahoo Search.

Inverse Host Frequency$(ihf(c_i))$. Concept ihf in the corpus. It inspired by the inverse document frequency. The inverse document frequency is a measure of the general importance of the term (obtained by dividing the total number of documents by the number of documents containing the term, and then taking the logarithm of that quotient). The *inverse host frequency* tend to filter out common concepts. In our paper, the form of $ihf(c_i)$ we used is

$$ihf(c_i) = \log_2 \frac{|H|}{|\{h : c_i \in h\}| + 1} \tag{1}$$

where $|H|$ is the total number of hosts in the corpus. $|\{h : c_i \in h\}|$ is the number of hosts which the concept c_i points to.

Concept Frequency-Inverse Anchor Frequency$(cfiaf(c_i))$. The concept frequency only considers the raw concept frequency in all anchor texts and treats different anchor-url pair equally even if some urls are very heavily linked. We define $(a_i, d_j) = 1$ when there are at least one link between anchor a_i and url d_j. Intuitionally, more general urls would be linked with more anchors. Using information theory, we define the entropy of a url d_j as

$$
\begin{aligned}
H(d_j) &= - \sum_{a_i \in A} p(a_i|d_j) \log_2 p(a_i|d_j) \\
&= - \sum_{a_i \in A} \frac{1}{\sum_{a_i \in A}(a_i, d_j)} \log_2 \frac{1}{\sum_{a_i \in A}(a_i, d_j)}
\end{aligned}
\tag{2}
$$

where we suppose $p(a_i|d_j) = \frac{1}{\sum_{a_i \in A}(a_i,d_j)}$, A means the set of anchor that linked to url d_j. So the maximum entropy of url d_j is

$$H(d_j)_{max} = \log_2 \sum_{a_i \in A}(a_i, d_j) \tag{3}$$

Then the *inverse anchor frequency* is defined as

$$iaf(d_j) = \log_2 |D| - H(d_j)_{max} = \log_2 \frac{|D|}{\sum_{a_i \in A}(a_i, d_j)} \tag{4}$$

where $|D|$ is the total number of urls in our corpus. $iaf(d_j)$ is inspired by Entropy-biased model [23]. The most important function of $iqf(d_j)$ is lessen the impact of some very popular urls and trying to balance the bias of them. Finally, $cfiaf(c_i)$ is defined as

$$cfiaf(c_i) = \sum_{d_j \in D} cf(c_i) iaf(d_j) \tag{5}$$

where $cf(c_i)$ is the number of urls which the concept c_i linked.

The number of concept's different destination host($hf(c_i)$). We define the source host of anchor is the host it appears. Destination host of anchor is the host it point to. The host frequency is defined as

$$hf(c_i) = \sum_{sh_j \in SH, dh_k \in DH} (sh_j, c_i, dh_k) \tag{6}$$

where sh_j is the source host of concept c_i, dh_k is the destination host of concept c_i. $(sh_j, c_i, dh_k) = 1$, if $sh_j \neq dh_k$.

3.2 Other Features

Inverse document frequency $idf(c_i)$. IDF is commonly used in information retrieval as a weighting function [13]. It is defined as

$$idf(c_i) = \log_2 \frac{|D|}{|\{d : c_i \in d\}| + 1} \tag{7}$$

where $|D|$ is the number of documents[2] and $|\{d : t_i \in d\}|$ is the number of documents where the concept c_i appears.

Residual IDF $ridf(c_i)$. Residual IDF [6] is the difference between IDF and the value predicted by a Poisson model [22].

$$ridf(c_i) = idf(c_i) - \log_2 \frac{1}{1 - e^{-\theta_i}} \tag{8}$$

where $(1 - e^{-\theta_i})$ is the probability of a document with at least one occurrence of concept c_i under a Poisson distribution. $\theta_i = \frac{tf(c_i)}{|D|}$, $|D|$ is the total number of documents. It is supposed that only the non-content concepts' distribution fits the Poisson distribution.

$\frac{qe(c_i)}{qp(c_i)}$. [6] $qp(c_i)$ is the number of a concept c_i was used as a part of some queries in the query log. $qe(c_i)$ is the number of a concept was used as a query in the query log. We extract these two features from billions of queries from the query log[3]. It is supposed that the ratio of the key concepts would be higher.

3.3 Feature Selection Method

Pairwise Comparison. Pairwise Comparison [21] measures the similarity of two ranking results. Let E is the set of all elements. A and B are two ranking results of E. The Pairwise Comparison of A and B is defined as

$$T = \{(x_1, x_2) | x_1, x_2 \in E\} \tag{9}$$

---

[2] In our corpus $|D| = 4.5$ billion.
[3] Yahoo Search 2009 search query log excerpt.

$$C^+ = \{(x_1, x_2) \in T | x_1 \prec_A x_2, x_1 \prec_B x_2\} \tag{10}$$

$$C^- = \{(x_1, x_2) \in T | x_2 \prec_A x_1, x_2 \prec_B x_1\} \tag{11}$$

$$Pairwise\_Comparison(A, B) = \frac{\| C^+ \cup C^- \|}{\| T \|} \times 100\% \tag{12}$$

where $\forall a_i, a_j \in A$, if $i > j$, then $a_i \prec_A a_j$.

Feature Selection with Pairwise Comparison. The Feature Selection method is based on greedy algorithm. It worked well in some similar machine learning issues [15]. It is defined as follows:

Algorithm 1. Feature Selection with Pairwise Comparison

C = {f1, f2, ..., fn}
S = ∅
while C ≠ ∅ **do**
    max = PC(∅) = Pairwise_Comparison(Bas, S)
    select = NULL
    for f **in** C **do**
        PC(f) = Pairwise_Comparison(Bas, S ∪ {f})
        if PC(f) > max **then**
            max = PC(f)
            select = f
        end if
    end for
    if select == NULL **then**
        break
    else
        C = C - {select}
        S = S ∪ {select}
    end if
end while
return S

Bas is a ranking result as the basis of the algorithm. For each feature f in C, we calculate the Pairwise Comparison between Bas and $S \bigcup \{f\}$. We let $Pairwise\_Comparison(A, \emptyset) = 0$, if $A \neq \emptyset$.

4   Experiments

In this section, we describe the details about our experiment results of our work. First, we provide a brief introduction of the corpora we used in our experiments.

Then we assess our model with 3 experiments. Section 4.2 assess the effectiveness of our supervised machine learning model with a non-supervised weighting approach. Section 4.3 analyze the utility of various features used in our model with Feature Selection Method outlined in Section in 3.3 and a feature contribution experiment. Section 4.4 compare the result of our model with another well known supervised machine learning model [6].

4.1 Experiments Preparation

The data source we used in our experiments are provided in Table 3. ROBUST04 is a newswire collection and W10g is a web collection. We also extract 5000 queries from search log[4] for assessing the effectiveness of our supervised machine learning model.

Table 3. TREC collections and Queries of Search Log

Name	Topic Numbers
ROBUST04	250 (301-450, 601-700)
W10g	100 (451-550)
QUERIES of Search Log	5000

We employ *Support Vector Machine* as our machine learning algorithm. The details of the our supervised machine learning approach are outlined in Section 3.

4.2 Concept Weighting Results

In this section, we assess our supervised machine learning model via comparing with a non-supervised weighting method. In the first step of this experiment, we examine how well our proposed model classify the concepts into 3 sets outlined in Section 1. Then we test whether our proposed model outperforms a simple non-supervised weighting approach where $idf(c_i)$ weighting are directly used to classify the concepts. All $< desc >$ queries in TREC collections and queries from search log are labeled by professional surfers.

In the first step of the experiment, we use our proposed model to classify the concepts into 3 sets according their importance.

We use a cross-validation approach to get the final test results. For the TREC collections, all $< desc >$ queries are divided into subsets of 50 queries. For *queries of search log*, each subset contains 1000 queries. Each subset will be regard as a test set in turn, while the rest of subsets serve as a training set. Experiments are run independently for each data collection. At last, average results on all test sets will be reported as our final results.

In the second step of the experiment, we use a simple non-supervised approach, $idf(c_i)$ directly, to classify the concepts. We assign each concepts to the corresponding result set according to the $idf(c_i)$ ranking results. Each subset of collection is tested and the average results are reported.

---

[4] Yahoo 2009 Search Log.

Table 4. Comparison Results

NAME	WEIGHT	Our Model		$idf(c_i)$	
		ACC	RECALL	ACC	RECALL
ROBUST04	3	82.20%	81.00%	54.10%	50.20%
	2	80.33%	79.40%	52.33%	54.20%
	1	93.40%	90.33%	71.42%	75.27%
W10g	3	84.67%	82.20%	65.58%	67.33%
	2	83.67%	81.40%	66.49%	68.58%
	1	94.47%	91.20%	74.47%	70.26%
QUERIES of Search Log	3	85.10%	83.50%	67.40%	62.49%
	2	84.70%	81.77%	65.33%	60.33%
	1	94.67%	92.31%	75.18%	69.84%

Table 4 is the comparison of accuracy and recall results for ROBUST05, W10g collections and queries of search log, when using svm algorithm with the features detailed in Section 3 and a single idf for concept weighting.

Table 4 shows that for test results of all collections our proposed model outperforms $idf(c_i)$ ranking. We note that most of misclassification cases are the queries contain multiple key concepts. It's the reason that the effectiveness on ROBUST04 collection is not so good as other two collections. For the topics of ROBUST04 collection tend to contain more verbose $< desc >$ queries, multiple key concepts are more common in it.

4.3   Feature Analysis

In this section, we analyze the utility of the various features we used in our supervised machine learning model. First, we use feature selection method detailed in Section 3.3 to analyze the features. Then, in order to assess the feature contribution, for each feature, we evaluate the influence to the effectiveness of the model with the rest of features.

Feature Selection. Pairwise Comparison and Greedy algorithm has been proven worked well in some similar machine learning problems [15]. We employ them as our feature selection method. The experiments described in previous section are repeated with different features combination. In each iteration, the best feature will be moved to selected features set. The details of the selection algorithm has been described in Section 3.3. Table 5 reports the feature selection experimental results. We can get several conclusions from the results of the feature selection.

First, The results in Table 5 indicates that all features have positive impact to overall accuracy. However, we note that not all features are positive to every collection. For example, $ridf(c_i)$ have a little negative impact for the ROBUST04 collection. Second, for each feature, it has different contribution to each collection. It depends on the types of collections and features. For example, the W10g collection prefer the feature $ridf(c_i)$, however, $ridf(c_i)$ has a little negative impact on ROBUST04 collection. Third, though some features in our model are

Table 5. Feature Selection with Pairwise Comparison

Feature	ROBUST04	W10g	Queries	Overall
$\{cf\}$	52.31%	55.67%	56.77%	54.92%
$\{ihf\}$	55.41%	58.40%	57.10%	56.97%
{cfiaf}	**63.70%**	**66.77%**	**68.44%**	**66.30%**
$\{hf\}$	51.77%	59.67%	57.20%	58.20%
$\{idf\}$	56.71%	60.10%	62.30%	59.70%
$\{ridf\}$	55.40%	52.77%	56.70%	54.96%
$\{\frac{qe}{qp}\}$	59.70%	53.71%	60.10%	57.84%
$\{cfiaf\} + \{cf\}$	69.23%	70.30%	72.13%	70.55%
{cfiaf} + {ihf}	**72.41%**	**75.60%**	**76.10%**	**74.70%**
$\{cfiaf\} + \{hf\}$	70.22%	74.11%	72.17%	72.17%
$\{cfiaf\} + \{idf\}$	71.01%	74.77%	77.10%	74.29%
$\{cfiaf\} + \{ridf\}$	69.15%	72.27%	74.11%	71.84%
$\{cfiaf\} + \{\frac{qe}{qp}\}$	71.67%	74.47%	75.60%	73.91%
$\{cfiaf, ihf\} + \{cf\}$	73.11%	77.21%	77.32%	75.88%
$\{cfiaf, ihf\} + \{hf\}$	74.44%	76.33%	77.12%	75.96%
{cfiaf, ihf} + {idf}	**76.70%**	**82.72%**	**78.64%**	**79.35%**
$\{cfiaf, ihf\} + \{ridf\}$	73.34%	78.37%	78.89%	76.87%
$\{cfiaf, ihf\} + \{\frac{qe}{qp}\}$	75.11%	77.47%	79.67%	77.42%
$\{cfiaf, ihf, idf\} + \{cf\}$	76.88%	83.17%	80.50%	80.18%
$\{cfiaf, ihf, idf\} + \{hf\}$	78.92%	83.47%	82.33%	81.57%
$\{cfiaf, ihf, idf\} + \{ridf\}$	76.40%	82.97%	79.10%	79.49%
{cfiaf, ihf, idf} + $\{\frac{qe}{qp}\}$	**79.92%**	**83.67%**	**83.10%**	**82.23%**
$\{cfiaf, ihf, idf, \frac{qe}{qp}\} + \{cf\}$	80.20%	83.84%	83.44%	82.49%
{cfiaf, ihf, idf, $\frac{qe}{qp}$} + {hf}	**81.20%**	**84.13%**	**84.67%**	**83.33%**
$\{cfiaf, ihf, idf, \frac{qe}{qp}\} + \{ridf\}$	79.87%	84.04%	83.00%	82.30%
{cfiaf, ihf, idf, $\frac{qe}{qp}$, hf} + {cf}	**81.82%**	**85.31%**	**86.21%**	**84.45%**
$\{cfiaf, ihf, idf, \frac{qe}{qp}, hf\} + \{ridf\}$	81.18%	85.45%	83.67%	83.43%
{cfiaf, ihf, idf, $\frac{qe}{qp}$, hf, cf} + {ridf}	**81.10%**	**86.77%**	**86.22%**	**84.70%**

biased towards different collections, our proposed model is basically not biased towards any specific query type. It more general than simple non-supervised approaches.

Feature Contribution. In order to assess the contribution and the impact of each feature, we repeat the experiments with different features set. In each iteration, a single feature is moved from the full feature set and the rest of features used to assess the effectiveness of the model. Table 6 reports the feature contribution experiments result. The positive value indicates the accuracy is increased after removing the feature, while negative value indicates the accuracy is decreased after removing the feature.

From the results, we can get some similar conclusions in Feature Selection experiments. First, features are biased toward different data collections. Same conclusion was also got in Feature Selection Experiments. Second, most of non-biased features do not perform very well, while the biased features are not

Table 6. Feature Analysis Experiments Results

	$cf(c_i)$	$ihf(c_i)$	$cfiaf(c_i)$	$hf(c_i)$	$idf(c_i)$	$ridf(c_i)$	$\frac{qe(c_i)}{qp(c_i)}$
ROBUST04	-1.71%	-2.40%	-3.71%	-1.80%	-3.37%	+3.07%	-1.15%
W10g	-3.60%	-4.07%	-8.94%	-3.71%	-8.67%	-6.91%	-4.45%
Queries of Search Log	-4.10%	-1.97%	-6.70%	+2.90%	-1.61%	-3.31%	-6.97%

general enough. It the restriction of non-supervised approach. However, supervised machine learning approach as both biased features and non-biased features the input and get a general model for key concepts detection.

4.4 Key Concepts Identification Results

In this section, we evaluate the effectiveness of our proposed model via comparing with another supervised machine learning approach [6]. Bendersky et. al [6] proposed a supervised machine learning approach for discovering key concepts in verbose queries. They assign weighting to each concept and select the concept with the highest weight as the key concept of the query. They employ the *AdaBoost.M1* meta-classifier with *C4.5 decision trees* as base learners [7]. They extract just one key concept from each verbose query, however, there are often not only one key concept in a natural language query. We try to extract all key concepts from queries. It's the difference between our work and Bendersky et. al's work [6]. Bendersky et. al's work [6] are similar to us and can significant improve retrieval effectiveness for verbose queries in their experiments. So we choose the method as the baseline.

In the experiments, we select top k concepts with baseline approach as the key concepts set. Our proposed model is designed to extract the key concepts set directly. We use the cross-validation approach outlined in Section 4.2 to get the experiment results.

Table 7 reports the *accuracy* and the *mean average precision*(MAP) results when either the baseline *AdaBoost.M1* or our model. For the key concepts set may contain more than one key concept, so we use MAP to assess the effectiveness. The results in Table 7 and Fig. 1 show that for all tested collections our proposed approach outperforms the baseline method. We note that most of misclassification cases are ambiguous concepts. For the queries that contain several key concepts, it is hard to judge whether an ambiguous concept belongs to the key concepts set.

Table 7. Key Concepts Identification Results

	Our Model		AdaBoost.M1	
NAME	ACC	MAP	ACC	MAP
ROBUST04	81.20%	84.00%	72.50%	83.8%
W10g	85.30%	86.67%	80.67%	86.80%
QUERIES of Search Log	87.40%	88.33%	82.20%	85.33%

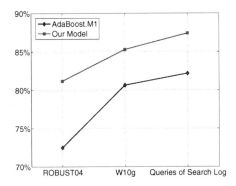

Fig. 1. Comparison of our model and AdaBoost.M1

5 Conclusions and Future Work

In this paper we proposed a supervised machine learning approach for detecting the key concepts from natural language queries. We use 2 standard TREC collections and 1 Query collection to assess the effectiveness of our model.

For modern search engines, one of the main issues in query analysis is detecting the key concepts from queries. It's the foundation of many query processing problems. The most difficult part of the problem is the results would be highly amplified in the retrieved documents. If the key concepts were detected wrong, it would lead a lack of focus on the main topics of the query in the retrieved documents.

We use features with anchor analysis and other common features as the inputs of our supervised machine learning algorithm. It is known that queries and documents are organized by different language, while anchor language is so much similar to query language. Both of them try to summarize documents. Moreover, the link relationship between anchors and urls are similar to the relationship between queries and clicked urls in search engine log. It has been proved that the anchor analysis approach performed well for key concepts identification and weighting problem in our experiments.

There are still some issues needed to be addressed as future work. First, we plan to extract more query-dependent features as the input of our proposed model. In the experiments, we observed that some concepts are always non-key concepts. Intuitively, concepts in different queries should be assign different weight. However, most of features of machine learning algorithm are global features. They do not perform well on these queries. We need to extract more significant query-dependent features as the input of machine learning algorithm. Second, besides key concepts set, we plan to pay more attention to second grade concepts in verbose queries, such as *"download"* in the former query. It is not the key concept, however, it indicates that the user want a download page, but not a introduction page. These second grade concepts are not the most important concepts for query, however, it also provide a lot of information indicate the users search goals.

Acknowledgments

We are grateful to Honglei Zeng's work on data preparation. Gui-Rong Xue also provided many very useful suggestions. We would also like to thank for Xin Fu and Prof. Bing Liu's encouragement.

References

1. Hu, J., et al.: Understanding User's Query Intent with Wikipedia. In: WWW 2009 (2009)
2. Bendersky, et al.: Learning Concept Importance Using a Weighted Dependence Model. In: WSDM 2010 (2010)
3. Pickens, J., Croft, W.B.: An exploratory analysis of phrases in text retrieval. In: Proc. of RIAO 2000 (1999)
4. Mishne, G., et al.: Boosting web retrieval through query operations. In: Losada, D.E., Fernández-Luna, J.M. (eds.) ECIR 2005. LNCS, vol. 3408, pp. 502–516. Springer, Heidelberg (2005)
5. Aurial, et al.: Support Vector Machines as a Technique for Solvency Analysis (2008)
6. Bendersky, M., et al.: Discovering Key Concepts in Verbose Queries. In: SIGIR 2008 (2008)
7. Freund, Y., Schapire, R.E.: Experiments with a new boosting algorithm. In: Machine Learning: Proceedings of the Thirteenth International Conference, pp. 148–156 (1996)
8. Peng, J., et al.: Incorporating term dependency in the dfr framework. In: SIGIR 2007 (2007)
9. Kumaran, et al.: Reducing Long Queries Using Query Quality Predictors. In: SIGIR 2009 (2009)
10. Hiemstra, D.: Term-specific smoothing for the language modeling approach to information retrieval: the importance of a query term. In: SIGIR 2002 (2002)
11. Mei, Q., Fang, H., Zhai, C.: A study of poisson query generation model for information retrieval. In: SIGIR 2007 (2007)
12. Tao, et al.: An exploration of proximity measures in information retrieval. In: SIGIR 2007 (2007)
13. Salton, G., Buckley, C.: Term-weighting approaches in automatic text retrieval. Inf. Process. Manage. 24(5), 513–523 (1988)
14. Huang, J., Gao, J., Miao, J., Li, X., Wang, K., Behr, F.: Exploring Web Scale Language Models for Search Query Processing. In: WWW 2010 (2010)
15. Ren, P., Yu, Y.: Web site traffic ranking estimation via SVM. In: Huang, D.-S., Zhang, X., Reyes García, C.A., Zhang, L. (eds.) ICIC 2010. LNCS, vol. 6216, pp. 487–494. Springer, Heidelberg (2010)
16. Bai, J., Chang, Y., et al.: Investigation of partial query proximity in web search. In: WWW 2008 (2008)

17. Bendersky, M., Croft, W.B., Smith, D.A.: Two-stage query segmentation for information retrieval. In: Proc. SIGIR 2009 (2009)
18. Cummins, R., O'Riordan, C.: Learning in a pairwise term-term proximity framework for information retrieval
19. Metzler, D., et al.: A Markov Random Field model for term dependencies. In: SIGIR 2005 (2005)
20. Allan, J., Callan, J., Bruce Croft, W., Ballesteros, L., Broglio, J., Xu, J., Shu, H.: INQUERY at TREC-5. pp. 119-132. NIST (1997)
21. Pairwise Comparison, http://en.wikipedia.org/wiki/Pairwise_comparison
22. Kenneth, et al.: Poisson mixtures. Natural Language Engineering 1(2), 163–190 (1995)
23. Deng, H., King, I., Lyu, M.R.: Entropy-biased Models for Query Representation on the Click Graph. In: SIGIR 2009 (2009)

SecGuard: Secure and Practical Integrity Protection Model for Operating Systems[*]

Ennan Zhai[1,2], Qingni Shen[1,3,4**], Yonggang Wang[3,4], Tao Yang[3,4], Liping Ding[2], and Sihan Qing[1,2]

[1] School of Software and Microelectronics, Peking University, China
[2] Institute of Software, Chinese Academy of Sciences, China
[3] MoE Key Lab of Network and Software Assurance, Peking University, China
[4] Network & Information Security Lab, Institute of Software, Peking University, China
ennan@nfs.iscas.ac.cn, qingnishen@ss.pku.edu.cn,
{wangyg,ytao}@infosec.pku.edu.cn, dlp@iscas.ac.cn,
qsihan@ss.pku.edu.cn

Abstract. Host compromise is a serious security problem for operating systems. Most previous solutions based on integrity protection models are difficult to use; on the other hand, usable integrity protection models can only provide limited protection. This paper presents SecGuard, a secure and practical integrity protection model. To ensure the security of systems, SecGuard provides provable guarantees for operating systems to defend against three categories of threats: network-based threat, IPC communication threat and contaminative file threat. To ensure practicability, SecGuard introduces several novel techniques. For example, SecGuard leverages the information of existing discretionary access control information to initialize integrity labels for subjects and objects in the system. We developed the prototype system of SecGuard based on Linux Security Modules framework (LSM), and evaluated the security and practicability of SecGuard.

1 Introduction

Background. As the increment of the Internet-scale, computer systems are faced with more threats. For example, the Internet worms can compromise and propagate hosts by compromising vulnerable computer systems. Compromised hosts may be organized to launch large-scale network attacks. Most existing efforts defend against such attacks by using the network-level techniques (e.g., firewalls and NIDS). However, the study in [1] claims that network-level solutions cannot resist such attacks fundamentally, because: 1) software on hosts are buggy, and 2) discretionary access control (DAC) mechanism is insufficient against network-based attacks. Therefore, the problem should be addressed by introducing mandatory access control (MAC) mechanism into operating systems. Existing MAC models (e.g., DTE [2,3], SELinux [4], Apparmor [5,6], and LIDS [7]) are very complex to configure and difficult to use. For example, there are many different categories of objects in SELinux; moreover, after configuring such MAC models, some

---

[*] The first three authors of this paper are alphabetically ordered according to first names.
[**] Corresponding Author.

X. Du et al. (Eds.): APWeb 2011, LNCS 6612, pp. 370–375, 2011.
© Springer-Verlag Berlin Heidelberg 2011

existing applications will not be used. On the other hand, there has also been some efforts on practical MAC models (e.g., LOMAC [8] and UMIP [1]). However, these solutions only provide heuristic approaches without strong guarantees (e.g., provable guarantees). Furthermore, these models are evaluated only against synthetic attacks and designed based on some strong assumptions. For example, UMIP model allows the remote system administration through secure shell daemon (sshd) to be completely trustworthy (this means the integrity level of that process can not drop). Nevertheless, attackers can actually always successfully exploit bugs in such daemon program, and then "overwhelm" the system. In summary, it is still an open question that how to design a secure and practical MAC model to protect the integrity of operating systems.

Our approach and contributions. This paper presents SecGuard, a secure and practical integrity protection model for operating systems. SecGuard aims to resist three categories of threats: network-based threat, IPC communication threat, and contaminative file threat[1]. SecGuard has the following contributions: 1) SecGuard secures operating systems from three categories of threats: network-based threat, IPC communication threat, and contaminative file threat; 2) SecGuard is a practical MAC model, and it is easier to be configured and used than the existing MAC models; 3) SecGuard provides provable guarantees; therefore, the security of the model can be ensured in theory; and 4) SecGuard has been developed as a prototype system in Linux, and we present some representative designs and evaluations.

Roadmap. The rest of this paper is organized as follows. Threat scenarios and assumptions are described in Section 2. Section 3 shows details of SecGuard. Our evaluations are given in Section 4. Finally, we conclude in Section 5.

2   Threat Scenarios and Assumptions

Threat Scenarios. SecGuard aims to defend against three categories of threats: 1) Network-based threat. Because the applications of system may contain some bugs, attackers are able to utilize the network to inject malicious code into our hosts. Even though the attackers will not launch the active attack, careless users still might also download the malicious code into their local hosts from insecure network; 2) IPC communication threat. When two processes communicate with each other, one process might read the IPC object owned by the other process. However, the IPC object might contain malicious codes which can destroy the integrity of systems; and 3) Contaminative file threat. The most common way to destroy the integrity of system is one particular process may read the system objects carrying malicious code, and thus the data owned by this process will be modified by the malicious code.

Assumptions. Three assumptions: 1) We assume that network server and client programs contain bugs and can be exploited by the attacker; 2) We assume that users may make careless mistakes in their operations, e.g., downloading a malicious file from the Internet and reading it; and, 3) We assume that the attacker cannot physically access the host. Based on the above assumptions, SecGuard aims to ensure *attackers can only*

---

[1] The existing study in [8] has pointed these three categories of attacks are the main threats in operating systems.

Table 1. Definitions of Symbols of Initialization Algorithm

dac_list	DAC information list of system
$other - bits(9 - bits)$	the 7th, 8th, and 9th bits of 9-bits of DAC authority
$group - bits(9 - bits)$	the 4th, 5th, and 6th bits of 9-bits for DAC authority
$user - bits(9 - bits)$	the 1st, 2nd, and 3rd bits of 9-bits for DAC authority
$object_{(i)}.s\_i$	the important integrity level of $object_{(i)}$
$object_{(i)}.s\_c$	the current integrity level of $object_{(i)}$
$user_{(j)}.s\_i$	the important integrity level of $user_{(j)}$
$user_{(j)}.s\_c$	the current integrity level of $user_{(j)}$
$object_{(i)}.owner.s\_i$	the important integrity level of the user who is the owner of $object_{(i)}$
$object_{(i)}.owner.s\_c$	the current integrity level of the user who is the owner of $object_{(i)}$

obtain limited privileges (not root information) and cannot compromise the operating system integrity under most attack behaviors.

3   Design of SecGuard

In this section, we discuss all the details of SecGuard. Because we have developed SecGuard as a prototype system for Linux by using the Linux Security Module (LSM) framework [9], the description of SecGuard is mainly based on our design for Linux. Actually, SecGuard can be easily adopted to other UNIX variants with minor changes.

Design of SecGuard's integrity labels. SecGuard assigns two integrity labels to subjects (e.g., processes) in the system: *important integrity label of subject* (denoted as $s\_i(s)$) and *current integrity label of subject* (denoted as $s\_c(s)$). Both of $s\_i(s)$ and $s\_c(s)$ have two levels (i.e., values): high or low. Meanwhile, SecGuard also assigns objects (e.g., files) with two integrity labels. They are *important integrity label of object* (denoted as $o\_i(o)$) and *current integrity label of object* (denoted as $o\_c(o)$) respectively. The same as $s\_i(s)$ and $s\_c(s)$, both of $o\_i(o)$ and $o\_c(o)$ also have two levels: high or low. Notice that we define *important integrity level* as the level of important integrity label, and define *current integrity level* as the level of current integrity label.

How to initialize the level of integrity labels for subjects? In SecGuard, only the important integrity levels of root processes (system-level processes) are high, and their current integrity levels are high in the startup. Normal processes' important integrity levels should be set to low, and their current integrity levels are also set to low. When a process (subject) is created, it will inherit both the important integrity level and current integrity level from its parent process. In sub-process's life cycle, its important integrity level can not be changed. On the other hand, the current integrity level of subject can be changed dynamically according to the security policy of SecGuard model.

How to initialize the level of integrity labels for objects? SecGuard introduces a novel *Initialization Algorithm* for initializing the level of integrity labels of objects in the system. The algorithm leverages existing DAC information of system to initialize the configuration of integrity level for objects. Notice that we only pay attention to the 9-bits mechanism for DAC, and the current DAC enhanced by ACL mechanism is not

---

Algorithm 1. Initialization Algorithm for Objects

---

1 **for** *each object$_{(i)}$ in dac_list* **do**
2   **if** *the other-bits(9-bits) of object$_{(i)}$ is writable* **then**
3   *object$_{(i)}$.s_i ← low;*
4   *object$_{(i)}$.s_c ← low;*
5   *continue;*
6   **if** *the group-bits(9-bits) of object$_{(i)}$ is writable* **then**
7   **for** *each user$_{(j)}$ in group of object$_{(i)}$* **do**
8   **if** *user$_{(j)}$.s_i = low && user$_{(j)}$.s_c = low* **then**
9   *object$_{(i)}$.s_i ← low;*
10   *object$_{(i)}$.s_c ← low;*
11   *break;*
12   *continue;*
13   **if** *the user-bits(9-bits) of object$_{(i)}$ is writable* **then**
14   *object$_{(i)}$.s_i ← object$_{(i)}$.owner.s_i;*
15   *object$_{(i)}$.s_c ← object$_{(i)}$.owner.s_c;*

---

our consideration, since the information provided by ACL mechanism is not used by Initialization Algorithm. To elaborate the Initialization Algorithm clearly, we present the meanings of symbols of the algorithm in Table 1. The Initialization Algorithm initializes the configuration of integrity level for objects in the system based on three key steps which are shown in Algorithm 1. More details of Initialization Algorithm are shown in [10].

SecGuard's security properties. In SecGuard, when a subject accesses an object or communicates with a subject, the accessed object or subject must be in available state; If a subject can read an object, the current integrity level of the subject must dominate (i.e., be higher than or equal to) the current integrity level of the object and the important integrity level of the subject must dominate the important integrity level of the object; If a subject can modify the content of an object, the current integrity level of the subject

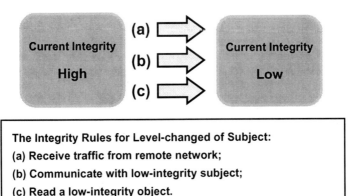

Fig. 1. SecGuard's Security Policies

must dominate the important integrity level of the modified object. Due to the limited space, we move formal description of SecGuard's security properties to [10].

SecGuard's security policies. The level-changed conditions of a subject's current integrity are shown in Fig. 1: 1) When a subject receives the traffic from network (e.g., downloads some files or scripts), the subject drops its current integrity level to low; 2) After a subject, whose current integrity level is high, communicates with a subject whose current integrity level is low, the former drops its current integrity level to low, and 3) When a subject, whose current integrity level is high, reads an object whose current integrity level is low, the subject drops its current integrity level to low. Due to the limited space, we move formal description of SecGuard's security policies to [10].

Provable guarantees. We provide provable guarantees to SecGuard. Due to the limited space, we move all the formal proofs to [10].

4 Evaluation

We have developed SecGuard as a prototype protection system for Linux based on the Linux Security Module (LSM) framework. Due to the space constraints, we only present some representative designs and evaluations. We move all the details of our experiments to [10].

Prototype. The basic implementation of SecGuard is as follows. Each process has two integrity labels, and when the one is created, it inherits the important integrity level from its parent process. SecGuard can not restrict the process whose current integrity level is high; however, a low current integrity process cannot perform any sensitive operation in the system. If a process can send a request, it must be authorized by both the DAC of the system and SecGuard. Due to the space constraints, we cannot present details for the implementation of SecGuard.

Evaluation of practicability of SecGuard. In order to evaluate the practicability of SecGuard, we built a server with *Fedora Core 8*, and enabled SecGuard as a security module loaded during the system booting. The existing programs of the system have not been affected after our security module loading. After that we installed some normal applications and the system can still provide services. SecGuard contributes several features of practicability on the operating system: the novel initialization algorithm, without complex configuration of integrity labels, and existing application programs and common practices for using can still be used under SecGuard.

Evaluation of security of SecGuard. Actually, our formal proof has provided strong guarantees to SecGuard (see [10] for details). Here, we use Linux Rootkit Family (LRK) to attack our system. The LRK is a well-known rootkit of user-mode and it can replace many system programs and introduce some new programs to build backdoors and to hide adversaries. LRK can be installed successfully and replaces the current SSH daemon in the system when SecGuard was closed. Then, we connect to the server as root with the predefined password. When SecGuard run, installation is failed and system returns a permitted error. Thus, our system remained security under SecGuard.

5   Conclusion

This paper presents SecGuard, a novel secure and practical integrity protection model for operating systems. Aiming to three threats in systems: network-based threat, IPC communication threat, and contaminative file threat, SecGuard provides a robust defense for operating systems, and leverages information in the existing discretionary access control mechanism to initialize integrity labels both for processes and files in systems. Furthermore, SecGuard provides provable guarantees to SecGuard. Finally, we describe the implementation of SecGuard for Linux and evaluations.

Acknowledgment

We thank the anonymous reviewers for helpful comments. This work is supported in part by the National Natural Science Foundation of China under Grant No. 60873238, No. 60970135 and No. 61073156, as well as Accessing-Verification-Protection oriented secure operating system prototype under Grant No. KGCX2-YW-125.

References

1. Li, N., Mao, Z., Chen, H.: Usable mandatory integrity protection for operating systems. In: IEEE Symposium on Security and Privacy, pp. 164–178 (2007)
2. Badger, L., Sterne, D.F., Sherman, D.L., Walker, K.M.: A domain and type enforcement unix prototype. Computing Systems 9(1), 47–83 (1996)
3. Badger, L., Sterne, D.F., Sherman, D.L., Walker, K.M., Haghighat, S.A.: Practical domain and type enforcement for UNIX. In: IEEE Symposium on Security and Privacy (1995)
4. NSA: Security enhanced linux, http://www.nsa.gov/selinux/
5. Apparmor application security for Linux, http://www.novell.com/linux/security/apparmor/
6. Cowan, C., Beattie, S., Kroah-Hartman, G., Pu, C., Wagle, P., Gligor, V.D.: Subdomain: Parsimonious server security. In: LISA, pp. 355–368 (2000)
7. LIDS: Linux intrusion detection system, http://www.lids.org/
8. Fraser, T.: Lomac: Low water-mark integrity protection for cots environments. In: IEEE Symposium on Security and Privacy, pp. 230–245 (2000)
9. Wright, C., Cowan, C., Smalley, S., Morris, J., Kroah-Hartman, G.: Linux security modules: General security support for the linux kernel. In: USENIX Security Symposium, pp. 17–31 (2002)
10. Zhai, E., Shen, Q., Wang, Y., Yang, T., Ding, L., Qing, S.: Secguard: Secure and practical integrity protection model for operating systems. Technical Report PKU-TR-08-710, Peking University School of Software and Microelectronics (March 2010), http://infosec.pku.edu.cn/~zhaien/TRSecGuard.pdf

A Traceable Certificateless Threshold Proxy Signature Scheme from Bilinear Pairings

Tao Yang[1,2], Hu Xiong[1,2], Jianbin Hu[1,2,*], Yonggang Wang[1,2], Yong Deng[1,2], Biao Xiao[3], and Zhong Chen[1,2]

[1] MoE Key Lab of Network and Software Assurance, Peking University, China
[2] Network & Information Security Lab, Institute of Software, Peking University, China
[3] School of Computer and Information Technology, Northern Jiaotong University, China
{ytao,xionghu,dengyong}@pku.edu.cn,
{hjbin,wangyg,chen}@infosec.pku.edu.cn, xiaobiao66@sina.com

Abstract. Using our (t, n) threshold proxy signature scheme, the original signer can delegate the power of signing messages to a designated proxy group of n members. Any t or more proxy signers of the group can cooperatively issue a proxy signature on behalf of the original signer, but $t - 1$ or less proxy signers cannot. Recently, in order to eliminate the use of certificates in certified public key cryptography and the key-escrow problem in identity-based cryptography, the notion of certificateless public key cryptography was introduced. In this paper, we present a traceable certificateless threshold proxy signature scheme based on bilinear pairings. For the privacy protection, all proxy signers remain anonymous but can be traceable by KGC through a tag setting. We show the scheme satisfies the security requirements in the random oracle model. To the best of our knowledge, our scheme is the first traceable certificateless threshold proxy signature scheme from bilinear pairings.

Keywords: certificateless cryptography; proxy signature; threshold.

1 Introduction

The concept of proxy signature was first introduced by Mambo et al. in 1996 [8]. Since Mambo *el al.*'s scheme, many proxy signature schemes have been proposed[2,3,7]. Based on the ideas of secret sharing and threshold cryptosystems, Zhang and Kim et al. independently constructed the first threshold proxy signatures in [11] and [5], respectively. ID-based cryptography which was introduced in 1984 by Shamir [9]: the public key of each user is easily computable from a string corresponding to this user's identity, (such as an email address), while the private key associated with that ID is computed and issued secretly to the user by a trusted third party called private key generator (PKG). The first work on ID-based threshold proxy signature was proposed by Xu *et al.* [10].

---

* Corresponding Author.

X. Du et al. (Eds.): APWeb 2011, LNCS 6612, pp. 376–381, 2011.
© Springer-Verlag Berlin Heidelberg 2011

To simplify the certificate management in traditional public key infrastructure and overcome the drawback of key escrow in ID-PKC simultaneously, Al-Riyami and Paterson[1] proposed a paradigm called certificateless public key cryptography(CL-PKC) in 2003. The concept was introduced to suppress the inherent key-escrow property of identity-based public key cryptosystems (ID-PKC)without losing their most attractive advantage which is the absence of digital certificates and their important management overhead. Like ID-PKC, certificateless cryptography does not use public key certificate[1,12], it also needs a third party called Key Generation Center(KGC) to help a user to generate his private key. In their original paper[1], Al-Riyami and Paterson presented a CLS scheme. Certificateless cryptography have some advantages over traditional PKC and ID-PKC in some aspects. In this paper, we introduce the notion of threshold proxy signature into certificateless cryptography and propose a concrete certificateless threshold proxy signature scheme with traceability.

2   Construction of Our Scheme

We employ some ideas of the certificateless signature scheme in [12], and the ID-based threshold proxy signature scheme in [6]. The proposed scheme involves five roles: the Key Generation Center (KGC), an original signer($\mathcal{OS}$) with identity A, a set of proxy signers $\mathcal{PS} = \{B_1, B_2, \cdots, B_n\}$ with n members (where ID_{B_i} is the identity of the ith ($1 \leqslant i \leqslant n$) member), a set of real signers $\mathcal{RS} = \{C_1, C_2, \cdots, C_t\}$ with t members (where ID_{C_j} is the identity of the jth ($1 \leqslant j \leqslant t$) member), and a verifier. Our scheme consists of the following 8 phases.

MasterKeyGen: Given a security parameter $k \in \mathbb{Z}$, the algorithm works as follows:

1. Run the parameter generator on input k to generate a prime q, two groups $\mathbb{G}_1$, $\mathbb{G}_2$ of prime order q, two different generator P and Q in $\mathbb{G}_1$ and an admissible pairing $\hat{e} : \mathbb{G}_1 \times \mathbb{G}_1 \rightarrow \mathbb{G}_2$.
2. Randomly select a master-key $s \in_R \mathbb{Z}_q^*$.
3. Set the system public key by $P_{pub} = sP$.
4. Choose cryptographic hash functions $H_1, H_3, H_5 : \{0,1\}^* \rightarrow \mathbb{G}_1$ and $H_2, H_4 : \{0,1\}^* \rightarrow \mathbb{Z}_q^*$.

PartialKeyGen: Given a user's identity $ID \in \{0,1\}^*$, KGC first computes $Q_{ID} = H_1(ID)$. It then sets this user's partial key $psk_{ID} = sQ_{ID}$ and transmits it to user ID secretly. It is easy to see that psk_{ID} is actually a signature[4] on ID for the key pair (P_{pub}, s), and user ID can check its correctness by checking whether $\hat{e}(psk_{ID}, P) = \hat{e}(Q_{ID}, P_{pub})$.

UserKeyGen: The user ID selects a secret value $x_{ID} \in_R \mathbb{Z}_q^*$ as his secret key usk_{ID}, and computes his public key as $upk_{ID} = x_{ID}P$.

ProxyCertGen: The original signer, A issues a warrant ω which contains all the delegation details, such as $\mathcal{OS}$'s ID, the limitations, the time-window, etc.

On inputs Params, original signer A's identity ID_A, his partial key psk_{ID_A} and user secret key usk_{ID_A},the signer A randomly chooses $r_A \in_R \mathbb{Z}_q^*$, computes

$U_A = r_A P$ and $V_A = h_{A1} \cdot psk_{ID_A} + r_A Q + x_{ID_A} h_{A2}$, where $upk_{ID_A} = x_{ID_A} P$, $h_{A1} = H_2(\omega||ID_A||upk_ID_A||U_A)(a||b$ means string concatenation of a and b), $h_{A2} = H_3(\omega||ID_A||upk_{ID_A})$. Then A broadcasts (ω, U_A, V_A) to all members in $\mathcal{PS}$.

ProxyShadowGen: Each proxy shadow includes 2 parts: one from $\mathcal{OS}$ and the other from $\mathcal{PS}$.

1. A chooses a random polynomial of degree $t-1$ with the secret value r_A as its constant term, i.e., $f(x) = r_A + \sum_{l=1}^{t-1} a_l x^l (mod\ q)$, where $a_l \in Z_q^*$. After that A computes $A_l = a_l V_A$, then publishes $\hat{e}(A_l, P)$ to all members in $\mathcal{PS}$. Finally, A computes $K_i = f(H_4(ID_{B_i}))V_A + psk_{ID_A} + x_{ID_A} h_{A2}$ and sends it through a secure channel to B_i for $1 \leqslant i \leqslant n$;
2. Each proxy signer B_i verifies whether

$$\hat{e}(K_i, P) = \hat{e}(Q_{ID_A}, P_{pub})\hat{e}(U_A, V_A)\hat{e}(upk_{ID_A}, h_{A2}) \prod_{l=1}^{t-1} \hat{e}(A_l, P)^{H_4^l(ID_{B_i})} (1)$$

holds or not. If it holds, B_i keeps K_i as the first part of proxy signing key from original signer.
3. Each B_i $(1 \leqslant i \leqslant n)$ chooses a random number $y_i \in_R Z_q^*$ as her secret value, computes $W_i = y_i P$ and sends W_i to clerk[1]. After receiving all W_i's, clerk computes $W = \sum_{l=1}^n W_i$, and broadcasts W to all members in $\mathcal{PS}$. Then B_i computes $\hat{e}(P, y_i V_A)$ and publishes it to all members in $\mathcal{PS}$.
4. Each B_i chooses a random polynomial of degree $t-1$ with the secret value y_i as its constant term, i.e., $g_i(x) = y_i + \sum_{l=1}^{t-1} b_{i,l} x^l (mod\ q)$, where $b_{i,l} \in Z_q^*$. Then, B_i computes $B_{i,l} = b_{i,l} V_A$, $\hat{e}(P, B_{i,l})$ and publishes $\hat{e}(P, B_{i,l})$ to all members in $\mathcal{PS}$. Then each B_i computes $Y_{j,i} = g_i(H_4(ID_{B_j}))V_A$, and sends it through a secure channel to B_j for $1 \leqslant j \leqslant n, j \neq i$.
5. Each B_j verifies whether

$$\hat{e}(P, Y_{j,i}) = \hat{e}(P, y_i V_A) \prod_{l=1}^{t-1} \hat{e}(B_{i,l}, P)^{H_4^l(ID_{B_j})} \quad (2)$$

holds or not. After receiving all $Y_{j,i}$'s, B_j computes $Y_j = \sum_{i=1}^n Y_{j,i}$ as her second part of proxy signing key.

Sign: Given its proxy signing key, and a message $m \in \{0,1\}^*$, all t members in $\mathcal{RS}$ execute the following procedures to generate their partial signatures:

1. Each C_i $(1 \leqslant i \leqslant t)$ randomly picks $r_i \in_R Z_q^*$, computes $R_i = r_i P$,$T_i = r_i psk_{C_i}$, and broadcasts R_i and T_i to all other signers in $\mathcal{RS}$.
2. After receiving all other $t-1$ R_j's and T_j's, each C_i in $\mathcal{RS}$ computes $R = \sum_{l=1}^t R_i$, $T = \sum_{l=1}^t T_i$, $H = H_5(M||W||R+T)$,$E_i = r_i P_{pub}$ and $S_i = r_i H + \lambda_i(K_i + Y_i)$, where $\lambda_i = \prod_{C_j \in \mathcal{RS}, j \neq i} \frac{0 - H_4(ID_{C_j})}{H_4(ID_{C_i}) - H_4(ID_{C_j})}$, and sends the partial signature (S_i, R_i, T_i, E_i) to clerk[2].

[1] Anyone of the proxy signers can be designated as clerk.
[2] Anyone of the real signers can be designated as clerk.

3. Each C_i computes $\hat{e}(P, \lambda_i(K_i + Y_i))$, and sends them to clerk.

After receiving all partial signatures, the clerk can generate the threshold signature by the following steps:

1. Compute $H = H_5(M||W||\sum_{i=1}^{t} R_i + \sum_{i=1}^{t} T_i)$.
2. Accept if all the following equation holds:

$$\hat{e}(P, S_i) = \hat{e}(R_i, H)\hat{e}(P, \lambda_i K_i)\hat{e}(P, \lambda_i Y_i)) \tag{3}$$

$$\hat{e}(P, E_i + T_i) = \hat{e}(R_i, P_{pub})\hat{e}(E_i, Q_{C_i})) \tag{4}$$

3. Compute $S = \sum_{i=1}^{t} S_i$ and publishes $(\omega, (U_A, V_A),$
   $(S, W, R_1, R_2, ..., R_t, T_1, T_2, ..., T_t))$ as the traceable signature on message M.

Verify: Given a threshold proxy signature $(\omega, (U_A, V_A), (S, W, R_1, R_2, ..., R_t,$
$T_1, T_2, ..., T_t))$ on message M, a verifier does:

1. Verify the warrant ω against the certificate (U_A, V_A) by the following equation:
$$\hat{e}(V_A, P) = \hat{e}(h_{A1}Q_{ID_A}, P_{pub})\hat{e}(U_A, Q)\hat{e}(upk_{ID_A}, h_{A2}) \tag{5}$$
   where $h_{A1} = H_2(\omega||ID_A||upk_{ID_A}||U_A)$, $h_{A2} = H_3(\omega||ID_A||upk_{ID_A})$
2. Accept the signature if and only if the following equation holds:

$$\hat{e}(S, P) = \hat{e}(U_A, V_A)\hat{e}(W, V_A)\hat{e}(Q_{ID_A}, P_{pub})\hat{e}(upk_{ID_A}, h_{A2})\hat{e}(R, H) \tag{6}$$

   where $R = \sum_{i=1}^{t} R_i, T = \sum_{i=1}^{t} T_i, H = H_5(M||W||R + T)$.

Trace: PKG could determine all the real signers by executing the following:

1. Verify the signature is valid.
2. Compute $D_1 = (s^{-1} mod\ q)T_1$.
3. For the pair (R_1, D_1), try all B_j in $\mathcal{PS}$, and call the one that exactly matches the following equation as C_1:

$$\hat{e}(H_1(B_j), R_1) = \hat{e}(D_1, P) \tag{7}$$

   Then C_1 is the one that leaves the trace pair (R_1, T_1).
4. Repeat the previous two steps until to trace out all the signers in the same way, and send $(D_1, D_2, ..., D_t)$ and $(C_1, C_2, ..., C_t)$ to the verifier.

The verifier accepts if $\hat{e}(H_1(C_i), R_i) = \hat{e}(D_i, P)$ holds$(1 \leqslant i \leqslant t)$.

3   Analysis of the Scheme

3.1   Correctness

Let K_i be the first part of proxy signing key from original signer for real signer C_i in $\mathcal{RS}$. Lemma 1 will enable us to establish the correctness of Eq. 6.

Lemma 1. $\sum_{i=1}^{t} \lambda_i K_i = r_A V_A + psk_{ID_A} + x_{ID_A} h_{A2}$, where λ_i and V_A are defined in Section 2.

Proof.

$$\sum_{i=1}^{t} \lambda_i K_i = \sum_{i=1}^{t} \lambda_i (f(H_4(ID_{B_i})) V_A + psk_{ID_A} + x_{ID_A} h_{A2})$$

$$= \sum_{i=1}^{t} \lambda_i f(H_4(ID_{B_i})) V_A + \sum_{i=1}^{t} \lambda_i psk_{ID_A} + \sum_{i=1}^{t} \lambda_i (x_{ID_A} h_{A2})$$

$$= f(0) V_A + psk_{ID_A} + x_{ID_A} h_{A2}$$

$$= r_A V_A + psk_{ID_A} + x_{ID_A} h_{A2}. \qquad \square$$

Similarly, let Y_i be the second part of proxy signing key from $\mathcal{PS}$ for real signer C_i in $\mathcal{RS}$. We can show that:

Lemma 2. $\sum_{i=1}^{t} \lambda_i Y_i = \sum_{j=1}^{n} y_j V_A$, where λ_i and V_A are defined in Section 2.

With Lemma 1 and 2, we then show that Eq. 6 holds for our threshold proxy signature generated by following scheme.

Theorem 1. *If all participants honestly follow the scheme, then the threshold proxy signature can be successfully verified by Eq. 6.*

Proof. Equation 6 can be obtained as follows:

$$\hat{e}(S, P) = \hat{e}(\sum_{i=1}^{t} r_i P, H) \hat{e}(r_A V_A + psk_{ID_A} + x_{ID_A} h_{A2}, P) \hat{e}(\sum_{j=1}^{n} y_j V_A, P)$$

$$= \hat{e}(U_A, V_A) \hat{e}(W, V_A) \hat{e}(Q_{ID_A}, P_{pub}) \hat{e}(upk_{ID_A}, h_{A2}) \hat{e}(R, H). \qquad \square$$

3.2 Security Analysis

Now, we show that our scheme satisfies the following security requirements:

1. **Strong Non-Forgeability:** Our proposed scheme is based on Zhang and Wong's certificateless signature scheme [12]. Therefore, our scheme is secure against forgeability under chosen message attack. Moreover, neither original signer nor third party is able to create a valid signature, since all secret values (y_i' s) from $\mathcal{PS}$ are not available to them. Therefore, our scheme satisfies strong unforgeability [8].
2. **Strong Identifiability:** It contains the warrant ω in a valid proxy signature, so anyone can determine the ID of the corresponding proxy signers from ω.
3. **Strong Non-Deniability:** The valid proxy signature contains ω, which must be verified in the verification phase, it cannot be modified by the proxy signer. Thus once any t or more proxy signers creates a valid proxy signature of an original signer, they cannot repudiate the signature creation.
4. **Prevention of Misuse:** In our threshold proxy signature scheme, using the warrant ω, we had determined the limit of the delegated signing capacity in the warrant ω, so the proxy signers cannot sign some messages that have not been authorized by the original signer.

4   Conclusion

We have presented a traceable certificateless threshold proxy signature scheme from the bilinear maps, which allow an original signer to delegate her signing capability to a group of n proxy signers, and it requires a consonance of at least t proxy signers in order to generate a valid signature. For the privacy protection, all proxy signers remain anonymous but can be traceable by KGC through a tag setting. We note that our scheme may be more efficient than threshold proxy signature schemes in traditional PKC since it avoid the costly computation for the verification of the public key certificates of the signers. And no key escrow in our scheme makes it impossible for the KGC to forge any valid signatures.

Acknowledgment

This work was supported in part by the NSFC under grant No. 60773163, No. 60873238, No. 61003230 and No. 60970135, the National High-Tech Research and Development Plan of China under grant No.2009AA01Z425, as well as China Postdoctoral Science Foundation under Grant No.20100480130.

References

1. Al-Riyami, S., Paterson, K.: Certificateless public key cryptography. In: Laih, C.-S. (ed.) ASIACRYPT 2003. LNCS, vol. 2894, pp. 452–473. Springer, Heidelberg (2003)
2. Alomair, B., Sampigethaya, K., Poovendran, R.: Efficient generic forward-secure signatures and proxy signatures. In: Mjølsnes, S.F., Mauw, S., Katsikas, S.K. (eds.) EuroPKI 2008. LNCS, vol. 5057, pp. 166–181. Springer, Heidelberg (2008)
3. Bao, H., Cao, Z., Wang, S.: Identity-based threshold proxy signature scheme with known signers. In: Cai, J.-Y., Cooper, S.B., Li, A. (eds.) TAMC 2006. LNCS, vol. 3959, pp. 538–546. Springer, Heidelberg (2006)
4. Boneh, D., Lynn, B., Shacham, H.: Short signatures from the weil pairing. Journal of Cryptology 17(4), 297–319 (2004)
5. Kim, S., Park, S., Won, D.: Proxy signatures: Revisited. In: Han, Y., Quing, S. (eds.) ICICS 1997. LNCS, vol. 1334. Springer, Heidelberg (1997)
6. Liu, J., Huang, S.: Identity-based threshold proxy signature from bilinear pairings. Informatica 21(1), 41–56 (2010)
7. Malkin, T., Obana, S., Yung, M.: The hierarchy of key evolving signatures and a characterization of proxy signatures. In: Cachin, C., Camenisch, J.L. (eds.) EUROCRYPT 2004. LNCS, vol. 3027, pp. 306–322. Springer, Heidelberg (2004)
8. Mambo, M., Usuda, K., Okamoto, E.: Proxy signatures for delegating signing operation. In: Proceedings of the 3rd ACM Conference on Computer and Communications Security, pp. 48–57. ACM, New York (1996)
9. Shamir, A.: Identity-based cryptosystems and signature schemes. In: Blakely, G.R., Chaum, D. (eds.) CRYPTO 1984. LNCS, vol. 196, pp. 47–53. Springer, Heidelberg (1985)
10. Xu, J., Zhang, Z., Feng, D.: Id-based threshold proxy signature (2004), http://eprint.iacr.org/2005/250
11. Zhang, K.: Threshold proxy signature schemes. In: Information Security, pp. 282–290 (1998)
12. Zhang, Z., Wong, D., Xu, J., Feng, D.: Certificateless public-key signature: Security model and efficient construction. In: Zhou, J., Yung, M., Bao, F. (eds.) ACNS 2006. LNCS, vol. 3989, pp. 293–308. Springer, Heidelberg (2006)

A Mashup Tool for Cross-Domain Web Applications Using HTML5 Technologies

Akiyoshi Matono, Akihito Nakamura, and Isao Kojima

National Institute of Advanced Industrial Science and Technology (AIST)
1-1-1 Umezono, Tsukuba, Japan
{a.matono,nakamura-akihito}@aist.go.jp, kojima@ni.aist.go.jp

Abstract. Many web applications that do not take reusability and interoperability into account are being published today. However, there are demands that developers provide the ability to collaborate among different web applications. In e-Science, an application uses the results of other application as input data. In this paper, we introduce a mashup tool which can easily create a mashup web application by helping existing cross-domain web applications to collaborate. In our system, we utilize state-of-the-art technologies, such as cross-document messaging and XMLHttpRequest Level 2 of HTML5, and Transforming JSON.

Keywords: mashup, cross-domain, web application, HTML5.

1 Introduction

Today, a huge number of web applications have been published over the Internet. Users want to create new mashup web applications by making those existing web applications collaborate. However, today's web applications are not designed to take reusability and interoperability into account. In e-Science, there are many web applications which have similar components providing the same function developed individually by each project. If existing web applications could seamlessly exchange data, those existing web applications could then be reused to construct new mashup web applications. However, this is difficult because the web applications are often located on different servers, so-called cross-domain servers, and the data structures required by web applications are often different.

In this paper, we provide a mashup tool which makes it possible to construct new mashup web application by allowing cross-domain web applications to collaborate. Our proposed tool provides a sophisticated copy and paste mechanism between distinct web pages which are managed in different domains. Our tool utilizes state-of-the-art web technologies, such as cross-document messaging[3] and XMLHttpRequest Level 2 (XHR2)[4] in HTML5[2], and Transforming JSON (JsonT)[1].

The remainder of this paper is structured as follows. Sec. 2 gives an overview of the architecture, a description of its configuration, and the details of its components. A conclusion is presented in Sec. 3.

2 System Architecture

Fig. 1 illustrates the system architecture of our tool. In this figure, there are three web servers, the first one provides a web page called *mashup engine page* and is used to

X. Du et al. (Eds.): APWeb 2011, LNCS 6612, pp. 382–385, 2011.

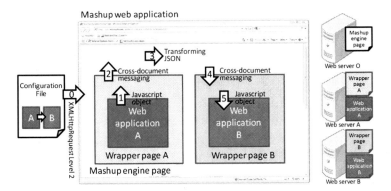

Fig. 1. System architecture

control the entire mashup web application. The other servers each provide a web application as a component of the mashup web application, and also provide *wrapper pages* that are used to intermediate between the mashup engine page and those web applications.

First, the mashup engine reads a configuration file when the mashup engine page is loaded using XHR2, which is provided as a new function of HTML5 (Step 0). The configuration file thus must not exist on the same web server as the mashup engine page. Based on the configuration, the mashup engine opens all the wrapper pages and their corresponding web applications. If a javascript object in web application A is modified, wrapper A sends the data using cross-document messaging, which is also provided in HTML5 (Step 2). At Step 3, the engine transforms the structure of the data using *Transforming JSON (JsonT)*, which can transform the structure of JSON based on configurable rules. Next, in Step 4, the engine sends the transformed data to wrapper B on web server B via cross-document messaging. If wrapper B receives the data, the wrapper puts the data into a variable of web application B as a javascript object.

In this way, the mashup tool can seamlessly exchange data between web applications, even if the web applications are located in different domains, and the data structures required are different. All communication is via the mashup engine page. In this figure, there are only two web applications and data is sent from web application A to web application B, that is, one-way and one-to-one, but our tool can manage any number of web applications, and supports two-way and one-to-many communications, if those definitions are written in the configuration file.

2.1 Configuration

List 1 shows an example of a configuration file. A configuration file is written in the JSON format and is composed of three definitions: application, flow, and JsonT. In lines 1-9, two application definitions are described, and two flow definitions are described in lines 10-16. The JsonT definition is described in each flow definition in lines 13 and 16. In this example, the application definitions mean that this mashup web application is created by the collaboration of two web applications, the first one of which is identified as "appA", located at http://a.example.com/app-a.html, and embedded in an

List 1. An example of a configuration file

```
 1  { "application"  :  {
 2      "appA"  :  {
 3         "url"   :  "http://a.example.com",
 4         "page"  :  "app-a.html",
 5         "view"  :  "iframe1" },
 6      "appB"  :  {
 7         "url"   :  "http://b.test.net",
 8         "page"  :  "app-b.html",
 9         "view"  :  "iframe2" } },
10      "flow"  :  {
11      "appA"  :  [
12        { "destination"  :  "appB",
13          "jsont"        :  { ........ } } ],
14      "appB"  :  [
15        { "destination"  :  "appA",
16          "jsontUrl"     :  "jsont.jt" } ],   } }
```

iframe whose id is "iframe1" on the mashup engine page. The flow definitions define two flows, the first one of which indicates that if an output object in the web application "appA" is modified then the object is to be transformed based on the JsonT definition in line 13 and is to be sent to the web application "appB" as input data. A JsonT definition can be written in the same configuration file directly, or in a different external file as shown as in line 16. A the JsonT definition in an external file is read by the mashup engine using XHR2.

2.2   Wrapper

The HTML body tag of a wrapper page contains only an iframe tag to embed its corresponding web application page, located on the same web server as the wrapper page. A wrapper page has two javascript functions. One is used to monitor a javascript object called "output" on the inner web application page. As soon as the "output" object is modified, the function serializes the object into a string in JSON format using the JSON.stringfy function, and then sends the message string to the mashup engine page using cross-document messaging. The other function is an event handler function called when a message is received by cross-document messaging. After receiving a message, the event handler function checks whether the source of the received message is the mashup engine page or not. If it is from the engine page, then the function deserializes the message into an object using the JSON.parse function, and then puts the object into a variable called "input" on the inner web application page as an input. After a received object is put into the "input" variable, the data can be used freely by the web application with little modification to the application.

2.3   Mashup Engine

The number of iframe tags in the mashup engine page is the same as the number of web applications used in the collaboration. After reading the configuration file, based on the application definitions in it, all wrapper pages and their corresponding web application pages are opened, since they are written in an iframe tag on the wrapper pages.

As soon as the engine page opens the wrapper pages, it sends a connection message to every wrapper page using cross-document messaging, because wrapper pages do not know anything about the engine page and not have the window object of the engine page, although the engine page has the window objects of the wrapper pages. Cross-document messaging uses the window object of its destination page, because the `postMessage` function in the window object is called to send a message. When a wrapper page receives a connection message, the wrapper obtains the source of the message, keeps it as a window object, and then returns an acknowledgment message to the mashup engine page. If the engine page receives acknowledgment messages from all wrappers, the preparation of connection windows is finished.

Then, when the engine page receives a message by cross-document messaging, the engine page checks whether the message is sent from one of the wrapper pages managed. If it is from an allowed wrapper, based on the flow definition, the engine determines the destination of the message. Before sending the message, the message is transformed based on the JsonT definition in the configuration file, if necessary.

3   Conclusion

In this paper, we introduce a mashup tool to construct mashup web applications by utilizing collaboration between existing cross-domain web applications as its components. Our mashup tool uses state-of-the-art mechanisms: cross-document messaging and XMLHttpRequest Level 2 in HTML5, and JsonT, allowing users to easily exchange (and transform) data between multiple existing web applications, even if they are located in different domains.

In the demonstration, we will show two demonstrations which have different aspects. The one is that a sample mashup application which exchanges and transforms sample data in order to prove the feasibility. The other is that a mashup web application which collaborates existing real geographic applications to prove the requirements.

Acknowledgment. This work was partially supported by a Grant-in-Aid for Scientific Research (A) (20240010) from the Ministry of Education, Culture, Sports, Science and Technology.

References

1. Goessner, S.: JsonT - Transforming Json (January 2006),
   http://goessner.net/articles/jsont/
2. Hickson, I., Google Inc.: HTML5: A vocabulary and associated APIs for HTML and XHTML (October 2010), http://www.w3.org/TR/html5/ (w3C Working Draft October 19, 2010)
3. Hickson, I., Google Inc.: HTML5 web messaging (November 2010),
   http://www.w3.org/TR/webmessaging/ (w3C Working Draft November 18, 2010)
4. van Kesteren, A.: XMLHttpRequest Level 2 (September 2010),
   http://www.w3.org/TR/XMLHttpRequest2/ (w3C Working Draft September 07, 2010)

Change Tracer: A Protégé Plug-In for Ontology Recovery and Visualization[*]

Asad Masood Khattak[1], Khalid Latif[2], Zeeshan Pervez[1], Iram Fatima[1],
Sungyoung Lee[1], and Young-Koo Lee[1]

[1] Department of Computer Engineering, Kyung Hee University, Korea
{asad.masood,zeeshan,iram,sylee}@oslab.ac.kr, yklee@khu.ac.kr
[2] School of Electrical Engineering and Computer Science, NUST, Pakistan
khalid.latif@seecs.nust.edu.pk

Abstract. We propose to demonstrate the process of capturing and logging ontology changes and use these logged changes for ontology recovery and visualization. Change Tracer, a Protégé plug-in we have developed for ontology recovery from one state to another (including roll-back and roll-forward) and visualization of ontology changes are structured as RDF graph. Users can also navigate through the history of ontology changes.

1 Demonstration

Knowledge representation and visualization is a collaborative process while dealing with information of high dimensions and complex nature represented in ontology. The goal is to demonstrate our Protégé plug-in i.e., *Change Tracer*, that capture and log changes (happening to these complex ontologies) in a repository i.e., Change History Log (CHL) with conformance to the Change History Ontology (CHO) [1]. On top of these logged changes, applications like ontology change management, ontology recovery, change traceability, and to some extent navigation and visualization of changes and change effects [1, 2] are implemented. The plug-in consist of two main modules;

Recovery module is responsible for rolling back and forwards the applied changes on model in reverse and forward manner for ontology recovery. SPARQL queries are extensively used for the purpose of recovery (see Figure 1-a). For validation of our plug-in (*Change Tracer*), we have checked its change detection accuracy against the *ChangesTab* of Protégé and our plug-in showed good results. Accuracy of high percentage for roll-back and roll-forward algorithm is observed [2].

Visualization module has two sub modules; a) To visualize the ontology and ontology changes in graph like structure. The *TouchGraph* API has been extended for graph drawing. Resources, such as classes, are depicted as nodes and properties as edges. Different resources are represented in different color with respect to one another but in uniform color together. An instance is represented in blue color and expressed in its complete relationships with its associated resources. Numbers of filters are supported in our graph view such as zooming in and out of the graph for detail

---

[*] This research was supported by the MKE(The Ministry of Knowledge Economy), Korea, under IT/SW Creative research program supervised by the NIPA(National IT Industry Promotion Agency)" (NIPA-2010-(C1820-1001-0001)).

X. Du et al. (Eds.): APWeb 2011, LNCS 6612, pp. 386–387, 2011.

view and Fish-eye view effects. A modified version of the *Spring* graph drawing algorithm is implemented in the visualization that ensures esthetically good looking graph structured and well separated nodes. We have also provided a search facility in case if the ontology is too large to find a resource. The graph and each resource in the graph are also drag-able and the graph refreshes itself.

The second sub module; b) Visually navigate through the history of ontology changes with roll-back and roll-forward operations (see Figure 1-b). Appropriate information capturing is very important for accurate recovery. We implemented different listeners (i.e., *KnowledgeBaseListener*, *ClsListener*, *SlotListener*, *FacetListener*, and *InstanceListener*) that actively listen to ontology changes during ontology engineering. For recovery purpose, first the required *ChangeSet* instance is extracted from the log and then all its corresponding changes are extracted. We used the concept of inverse and reverse changes. First, all the changes are converted to their inverse changes (e.g., *RangeAddition* to *RangeDeletion*) and then implemented in reverse sequence of their occurring order on the current version of ontology to get it in previous state. We have provided the playback and play-forward features where not only the ontology but the changes could also be visually navigated and the trend could be analyzed. Starting from the very first version of the ontology, the user can play the ontology changes and visualize as well as visually navigate the ontology and ontology changes. Rest of details on recovery, visualization, and their implementations is available in [2].

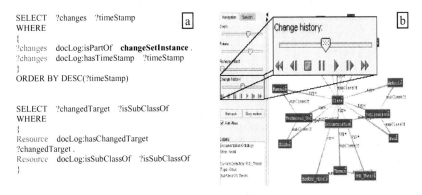

Fig. 1. a) Shows the queries used to extract the changes. b) Visualization of ontology with history playback feature. Users can visually navigate through the history of ontology changes.

The plan is to demonstrate the plug-in features like change capturing, change logging, ontology recovery, ontology visualization, and visual navigation through the history of ontology changes of our developed plug-in.

References

[1] Khattak, A.M., Latif, K., Khan, S., Ahmed, N.: Managing Change History in Web Ontologies. In: International Conference on Semantics, Knowledge and Grid, pp. 347–350 (2008)
[2] Khattak, A.M., Latif, K., Lee, S.Y., Lee, Y.K., Han, M., Kim Il, H.: Change Tracer: Tracking Changes in Web Ontologies. In: 21st IEEE International Conference on Tools with Artificial Intelligence (ICTAI), Newark, USA (November 2009)

Augmenting Traditional ER Method to Support Spatiotemporal Database Applications

Sheng Lin, Peiquan Jin, Lei Zhao, Huaishuai Wang, and Lanlan Zhang

School of Computer Science and Technology,
University of Science and Technology of China, 230027, Hefei, China
jpq@ustc.edu.cn

Abstract. In this paper, we present a new conceptual modeling tool, called STXER (SpatioTemporal eXtended Entity-Relationship), to support spatiotemporal database applications. STXER provides most functions for database conceptual modeling, such as graphical user interface, SQL script generation, XML schema definition, and commercial-DBMS-oriented database creation. Compared with previous tools, STXER is ER-compatible and supports richer spatiotemporal semantics. Thus it can support both non-spatiotemporal and spatiotemporal applications. After an overview of the general features of STXER, we discuss the architecture of STXER. And finally, a case study of STXER's demonstration is presented.

1 Introduction

Previous works in spatiotemporal database were mainly focused on spatiotemporal data models, spatiotemporal indexes, and spatiotemporal query processing. Recently, a few spatiotemporal database prototype systems were proposed [1, 2], in which they proposed to implement a spatiotemporal DBMS based on a commercial object-relational DBMS, such as Oracle, DB2, or Informix. However, few works have been done in spatiotemporal conceptual modeling, which has been an obstacle to build spatiotemporal database applications on the basis of spatiotemporal DBMSs. As traditional conceptual modeling methods such as ER model can not support the modeling of spatiotemporal semantics, it is important to develop a spatiotemporal conceptual modeling tool to support spatiotemporal database applications.

The key challenge to develop a spatiotemporal conceptual modeling tool is how to support various spatiotemporal semantics. The commonly-used ER method fails to represent spatiotemporal entities and relationships. For example, the lifecycle of a geographical object, which refers to its creation, evolution, and elimination, is hard to be modeled according to the ER method. Recently, researchers have proposed some techniques to support spatiotemporal semantics in conceptual modeling tools. Since the ER model was widely used in the conceptual modeling of relational database applications, many people extended the ER model to support spatiotemporal applications [3-7]. For example, N. Tryfona et al [3] presented the STER method for spatiotemporal conceptual modeling. It used some special symbols and syntax to add spatiotemporal characteristics. This method, to our knowledge, is the only one that considered the topological changes into the framework. However, STER does not support modeling of

X. Du et al. (Eds.): APWeb 2011, LNCS 6612, pp. 388–392, 2011.

objects' lifecycle. Besides, STER is very abstractive and no CASE tools have been implemented. The MADS tool [4] is much like the ER model, but it is not ER-compatible and does support rich spatiotemporal characteristics, e.g., the topological change is not supported. The DISTIL tool [5, 6] is ER-compatible but does support lifecycle and topological change.

In this paper, we present a conceptual modeling tool for spatiotemporal database applications, called STXER. The motivation of STXER is to design an effective tool for spatiotemporal applications so that they can create spatiotemporal database schemas in a traditional way used in relational database setting. A preliminary assumption of STEXR is that the fundamental spatiotemporal DBMS is built on a commercial object-relational DBMS such as Oracle and Informix. This assumption is reasonable and practical, as the abstract-type-based approach to implementing a practical spatiotemporal DBMS has been widely studied and promoted since it is proposed by R. H. Güting et al. in 2000 [7]. For example, spatiotemporal extensions on Oracle were proposed by different researchers in recent years [1, 2]. Compared with previous spatiotemporal conceptual modeling methods, STXER has the following unique features:

(1) It is ER-compatible and offers a graphical user interface similar to traditional ER modeling tools, thus it is easy for users to develop spatiotemporal or non-spatiotemporal conceptual models.

(2) It supports richer spatiotemporal semantics than previous methods, so it can support diverse spatiotemporal database applications.

2   Features of STXER

STXER supports four types of spatiotemporal changes, which are described as follows:

(1) **TYPE 1** (*spatial processes*): the spatial attributes of an object change with time, such as spread of fire, flowing of flood, moving of a car, or boundary changes of a land.

(2) **TYPE 2** (*thematic processes*): the thematic attributes of an object change with time, such as changes of soil type and ownership change of a land.

(4) **TYPE 3** (*lifecycle*): changes that result in creation of new objects or deletion of existing objects, such as split of a land or mergence of several lands.

(6) **TYPE 4** (*topological changes*): changes of the topo-logical relationships of two related objects. For example, the topological relationship between a moving car and a road will change with time.

STXER extends traditional ER model in three aspects. First, it adds spatiotemporal types in the type system, so users can use those types to represent spatial or thematic processes (**TYPE** 1 and 2). Second, it adds a new attribute called *LC* to represent the lifecycle of a spatiotemporal object (**TYPE** 3). Third, it introduces a new type of relationship into the ER model, called *transformation*, to represent the topological changes (**TYPE** 4).

STXER contains three components (as shown in Fig.1):

(1) *Graphical User Interface*: This module provides a graphical user interface for users to define an STXER diagram. Fig.2 shows a snapshot of the user interface, which consists of a menu, a toolbox, a model tree view, and a diagram area. We also develop an effective algorithm to arrange the graphical elements in the diagram area.

The most difficult work in the graphical user interface is to adjust the connected lines between entities according to users' operations such as drag and drop of an entity, sizing an entity, or moving a line. We develop a partition-based algorithm in STXER to adjust the layout of lines when the graphical elements of the model are modified. According to this algorithm, the area related with two connected entities is divided into six sub-areas, namely a vertical sub-area, a horizontal sub-area, a south-western sub-area (SW), a north-eastern sub-area (NE), a west-northern sub-area (NW), and a south-eastern sub-area (SE). The vertical and horizontal sub-areas always contain only straight lines, while the SW, NE, NW, and SE sub-areas always contain only polylines with one right angle.

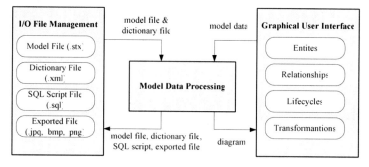

Fig. 1. Architecture of STXER

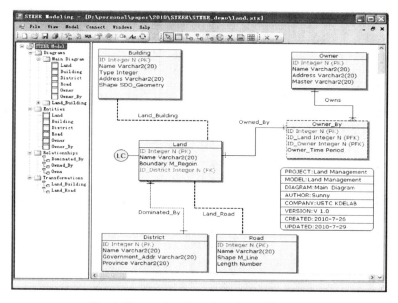

Fig. 2. The graphical user interface of STXER

(2) *Model Data Processing*: In this module, we check the STXER model, and then the legal model can be further processed to generate SQL script or export metadata. During the checking process, we focus on three types of collisions, namely entity collision, relationship collision, and attribute collision.

(3) *I/O File Management*: In this module, we define the input and output format of the STXER model. The STXER model is finally saved as a *.stx* file, and the diagrams can be exported as *.jpg*, *.bmp*, or *.png* files. The metadata of an STXER model is maintained in a XML file. The XML file maintains information about the entities, relationships, transformations as well other elements of an STXER model. The XML file can be further used to generate a SQL script defining a physical database schema. Currently, STXER only allows users to generate the SQL script for Oracle 10g.

3 Demonstration

STXER is implemented in C# language under Windows XP Professional and follow the object-oriented programming standard. Our demonstration will use STXER to create a spatiotemporal conceptual model for a land-use application and then build a spatiotemporal database in Oracle 10g. We will show how to quickly create an STXER diagram using our tool and represent different types of spatiotemporal semantics. In the demonstration, we will first present an overview of STXER and briefly discuss the modeling process of STXER. Then we will choose Oracle 10g as the destination DBMS and connect to the Oracle database server. After that, we will use STXER to create spatiotemporal entities, relationships, lifecycles, and transformations to define an STXER model for a land-use application. We will also show how to check the STXER model and export XML schema and SQL scripts.

Acknowledgements

This work is supported by the National High Technology Research and Development Program ("863" Program) of China (No. 2009AA12Z204), the National Science Foundation of China (no. 60776801), the Open Projects Program of National Laboratory of Pattern Recognition (20090029), the Key Laboratory of Advanced Information Science and Network Technology of Beijing (xdxx1005), and the USTC Youth Innovation Foundation.

References

1. Jin, P., Sun, P.: OSTM: a Spatiotemporal Extension to Oracle. In: Proc. Of NCM 2008, pp. 575–580. IEEE CS Press, Los Alamitos (2008)
2. Pelekis, N., Theodoridis, Y.: An Oracle Data Cartridge for Moving Objects, Technical Report, UNIPI-ISL-TR-2007-04, Dept. of Informatics, Univ. of Piraeus (2007)
3. Tryfona, N., Price, R., Jensen, C.: Conceptual Models for Spatio-temporal Applications. In: Sellis, T.K., Koubarakis, M., Frank, A., Grumbach, S., Güting, R.H., Jensen, C., Lorentzos, N.A., Manolopoulos, Y., Nardelli, E., Pernici, B., Theodoulidis, B., Tryfona, N., Schek, H.-J., Scholl, M.O. (eds.) Spatio-Temporal Databases. LNCS, vol. 2520, pp. 79–116. Springer, Heidelberg (2003)

4. Parent, C., Spaccapietra, S., et al.: Modeling Spatial Data in the MADS Conceptual Model. In: Proc. of SDH (1998)
5. Ram, S., Snodgrass, R., et al.: DISTIL: A Design Support Environment for Conceptual Modeling of Spatio-temporal Requirements. In: Kunii, H.S., Jajodia, S., Sølvberg, A. (eds.) ER 2001. LNCS, vol. 2224, pp. 70–83. Springer, Heidelberg (2001)
6. Khatri, V., Ram, S., Snodgrass, R.: On Augmenting Database Design-support Environments to Capture the Geo-spatiotemporal Data Semantics. Information System 31(2), 98–133 (2006)
7. Güting, R.H., Böhlen, M.H., Erwig, M., et al.: A Foundation for Representing and Querying Moving Objects. ACM Transactions on Database Systems 25(1), 1–42 (2000)

STOC: Extending Oracle to Support Spatiotemporal Data Management

Lei Zhao, Peiquan Jin, Xiaoxiang Zhang, Lanlan Zhang, and Huaishuai Wang

School of Computer Science and Technology,
University of Science and Technology of China, 230027, Hefei, China
jpq@ustc.edu.cn

Abstract. Previous research on spatiotemporal database were mainly focused on modeling, indexing, and query processing, and little work has be done in the implementation of spatiotemporal database management systems. In this paper, we present an extension of Oracle, named STOC (Spatio-Temporal Object Cartridge), to support spatiotemporal data management in a practical way. The extension is developed as a PL/SQL package and can be integrated into Oracle to offer spatiotemporal data types as well as spatiotemporal operations for various applications. Users are allowed to use standard SQL to access spatiotemporal data and functions. After an overview of the general features of STOC, we discuss the architecture and implementation of STOC. And finally, a case study of STOC's demonstration is presented.

1 Introduction

Previous research on spatiotemporal database were mainly focused on spatiotemporal data models [1], spatiotemporal indexes [2], and spatiotemporal query processing [2], whereas little work has been done in the implementation of real spatiotemporal database management systems. Although there are some prototypes proposed in recent years, e.g. BerlinMOD [3], it is still not feasible to use them in commercial applications, because of their incompatibilities with SQL and standard relational DBMS. Some researchers proposed to implement spatiotemporal DBMS based on Oracle [4], however, to our best knowledge, there are no real systems built so far.

Aiming at providing practical support of spatiotemporal data management for various commercial applications, we present a spatiotemporal extension of Oracle in this paper, which is called STOC (SpatioTemporal Object Cartridge). STOC is based on our previous work on spatiotemporal data model [5] and system implementation [6]. Unlike the OSTM extension in [6], STOC is built on Oracle Spatial (while the spatial types in OSTM are defined by ourselves.) and provides more support on spatiotemporal query optimization. STOC is developed using the cartridge technology provided by Oracle, which enables us to add new data types as well as functions and indexing methods into the kernel of Oracle. The unique features of STOC can be summarized as follows:

(1) It is SQL-compatible and built on a widely-used commercial DBMS (*see Section 2.1*), namely Oracle. Thus it can be easily used in real spatiotemporal database

X. Du et al. (Eds.): APWeb 2011, LNCS 6612, pp. 393–397, 2011.
© Springer-Verlag Berlin Heidelberg 2011

applications and provides a practical solution for spatiotemporal data management under current database architecture.

(2) It supports various spatiotemporal data types (*see Section 2.2*), such as *moving number, moving bool, moving string, moving point, moving line,* and *moving region.* Combined with the ten types of spatiotemporal operations (*See Section 2.3*) supported by those new data types, users can represent many types of spatiotemporal data involved in different spatiotemporal applications and query different spatiotemporal scenarios.

2 Implementation of STOC

2.1 Architecture of STOC

The detailed implemental architecture of STOC is shown in Fig.1. The PL/SQL specification provides the signature definition and implementation of all spatiotemporal data types and functions in the STORM model. The STOC cartridge is the component that actually brings spatiotemporal support into Oracle. Once installed, it becomes an integral part of Oracle, and no external modules are necessary. When STOC is installed into Oracle, users can use standard SQL to gain spatiotemporal support from Oracle. No external work imposes on users.

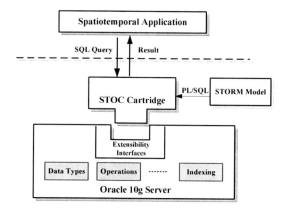

Fig. 1. Architecture of STOC

2.2 The Type System of STOC

STOC extends two categories of new data types into Oracle, namely spatiotemporal data types and temporal data types (as shown in Fig.2). As Oracle has already supported spatial data management from its eighth version, which is known as Oracle Spatial, we build STOC on the basis of Oracle Spatial so as to utilize its mature technologies in spatial data management. The spatiotemporal data types contain moving spatial types and moving base types. The former refers to moving spatial objects in real world, while the latter refers to those numeric, Boolean, or string values changing

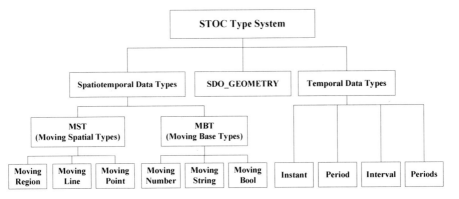

Fig. 2. The type system of STOC

```
CREATE OR REPLACE TYPE M_POINT_U AS OBJECT
  (point mdsys.SDO_GEOMETRY,
   cRX NUMBER,--changeRateofX
   cRY NUMBER,--changeRateofY
   period T_period) ;
CREATE OR REPLACE TYPE M_POINT_U_ARRAY IS VARRAY (102400) OF
  M_POINT_U;
CREATE OR REPLACE TYPE M_POINT AS OBJECT
    (t_type NUMBER,
     srid NUMBER,
   m_point_units M_POINT_U_ARRAY) ;
```

Fig. 3. Defining Moving Point using PL/SQL

with time. All the new data types are implemented by PL/SQL using the CREATE TYPE statement. Fig.3 shows the definition of moving point in STOC.

2.3 The Data Operations in STOC

STOC implements ten types of spatiotemporal operations, which are (1) object data management operation, (2) object attribute operations, (3) temporal dimension project operations, (4) value dimension project operations, (5) temporal selection operations, (6) quantification operations, (7) moving Boolean operations, (8) temporal relation operations, (9) object relation operations, and (10) distance operations. All the operations are implemented by PL/SQL and as member functions of spatiotemporal data types, as shown in Fig.4. For the space limitation, we will not discuss the details about each spatiotemporal operation. However, in the demonstration process, we will show how to use those operations to answer different spatiotemporal queries.

3   Demonstration

Our demonstration will focus on the scenario shown in Fig.4 and use a real dataset describing the movements of the taxi cars in Beijing, China. We will first use STOC

Fig. 4. Scenario of STOC demonstration

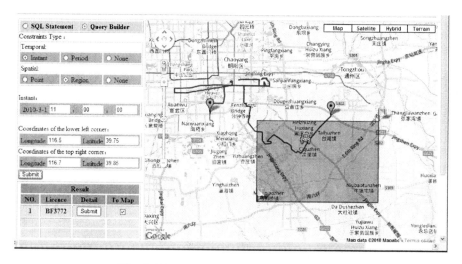

Fig. 5. Client interface of STOC demonstration

to create a spatiotemporal database schema and then transform the original GPS data into the database. After that, we will show how STOC answer different types of spatiotemporal queries. All the queries are conducted through a Web-based client interface (see Fig.5), which uses Google Maps as the spatial reference framework. We provide two methods for users to input a spatiotemporal query, either by entering a SQL statement or by a query builder. The self-inputted SQL statements can represent different kinds of spatiotemporal queries, while the query builder is only able to generate specific queries. The query builder is designed for those who are not experts of Oracle to try the STOC system in an easy way, since it can help them generate spatiotemporal queries.

Acknowledgements

This work is supported by the National High Technology Research and Development Program ("863" Program) of China (No. 2009AA12Z204), the National Science Foundation of China (no. 60776801), the Open Projects Program of National Laboratory of Pattern Recognition (20090029), the Key Laboratory of Advanced Information Science and Network Technology of Beijing (xdxx1005), and the USTC Youth Innovation Foundation.

References

1. Güting, R.H., Böhlen, M.H., Erwig, M., et al.: A Foundation for Representing and Querying Moving Objects. ACM Transactions on Database Systems 25(1), 1–42 (2000)
2. Koubarakis, M., Sellis, T., Frank, A., et al.: Spatio-Temporal Databases: The CHOROCHRONOS Approach. In: Sellis, T.K., Koubarakis, M., Frank, A., Grumbach, S., Güting, R.H., Jensen, C., Lorentzos, N.A., Manolopoulos, Y., Nardelli, E., Pernici, B., Theodoulidis, B., Tryfona, N., Schek, H.-J., Scholl, M.O. (eds.) Spatio-Temporal Databases. LNCS, vol. 2520, Springer, Heidelberg (2003)
3. Düntgen, C., Behr, T., Güting, R.H.: BerlinMOD: A Benchmark for Moving Object Databases. The VLDB Journal 18(6), 1335–1368 (2009)
4. Pelekis, N., Theodoridis, Y.: An Oracle Data Cartridge for Moving Objects, Technical Report, UNIPI-ISL-TR-2007-04, Dept. of Informatics, Univ. of Piraeus (2007)
5. Jin, P., Yue, L., Gong, Y.: Design and Implementation of a Unified Spatiotemporal Data Model. In: Advances in Spatio-Temporal Analysis. Taylor & Francis, Abington (2007)
6. Jin, P., Sun, P.: OSTM: a Spatiotemporal Extension to Oracle. In: Proc. of NCM, pp. 575–580. IEEE CS Press, Gyeongju (2008)

Preface to the 2nd International Workshop on Unstructured Data Management (USDM 2011)

Tengjiao Wang

Key Laboratory of High Confidence Software Technologies (Peking University),
Ministry of Education, China
School of Electronics Engineering and Computer Science,
Peking University Beijing, 100871 China

The management of unstructured data has been recognized as one of the most attracting problems in the information technology industry. With the consistent increase of computing and storage capacities (due to hardware progress) and the emergence of many data-centric applications (e.g. web applications), a huge volume of unstructured data has been generated. Over 80% of world data today is unstructured with self-contained content items. Since most techniques and researches that have proved so successful performing on structured data don't work well when it comes to unstructured data, how to effectively handle and utilize unstructured data becomes a critical issue to these data-centric applications.

The 2nd International Workshop on Unstructured Data Management (USDM 2011) aims at bringing together researchers, developers, and users to discuss and present current technical developments in the area. We have held The 1st Workshop on Unstructured Data Management (USDM 2010) in APWeb 2010 (April 6, 2010, Busan, Korea) successfully. It provided a successful international forum for the dissemination of research results in unstructured data management. In this year, we have received 7 submissions from diverse topics of interest, and 3 of them are selected as accepted papers which resolve some issues in unstructured data querying, retrieval, analysis, and mining, and also in applications of unstructured data management. The Program Committee worked very hard to select these papers through a rigorous review progress and extension discussions, and finally composed a diverse and exciting program for USDM 2011. Here we take a brief look at the techniques involved in this program.

With the popularity and development of information technology, increasingly growth of data on the internet is generated. It brings many challenge topics on analyzing and mining this unstructured data, including querying and searching based on text similarity. During these years, many researches are proposed in leading conference[1,2] to study similarity search. Techniques including Approximate String Joins and Set-Similarity Joins are widely used. In recent years, more and more researches focus on how to process on the explosive growth of data and provide quick response. In this program, authors of **Batch Text Similarity Search with MapReduce** study the problem of batch text similarity search. Specifically, the authors try to solve two problems within the MapReduce environment, i.e. online text similarity search and variable thresholds, and provide real-time online search.

Besides text similarity search, sentiment analysis, which is also referred as sentiment extraction or sentiment classification, has attracted many researchers in recent

X. Du et al. (Eds.): APWeb 2011, LNCS 6612, pp. 398–399, 2011.

years. It aims at determining whether the sentiment expressed in text is positive or negative. Although diverse models, such as generative model[3] that jointly models sentiment words, topic words and sentiment polarity in a sentence as a triple, or global structured model[4] that learns to predict sentiment of different levels of granularity in text, are proposed in droves, sentiment analysis nowadays often maintains a low accuracy. One possible reason is that the rapid growth of the internet makes it too costly or impractical for some models on large scale data. In this program, authors of *An Empirical Study of Massively Parallel Bayesian Networks Learning for Sentiment Extraction from Unstructured Text* propose an algorithm for large scale Bayesian Network structure learning within MapReduce framework. This parallel algorithm could capture the structure of Bayesian Network and obtain a vocabulary for analyzing sentiments.

Nowadays, it is more and more popular to use Resource Description Framework (RDF[5]) in knowledge management. It can maintain data relation, automatically build tables for RDBMS deployment, link across diverse content sets, and etc. There have been many RDF Knowledge Bases, but most of them are with limited sizes. Authors of *ITEM: Extract and Integrate Entities from Tabular Data to RDF Knowledge Base* presents a new system ITEM to enlarge the existing RDF datasets for using them in knowledge management.

The workshop will be a forum not only for presenting new research results but also for the discussions of practical experiences that can help shaping and solving critical problems in unstructured data management. We sincerely hope it provides participants a chance to get more knowledge in this field.

We would like to thank the many people who volunteered their to help making the workshop a success. We thank Steering Chairs Professor Xiaoyong Du and Professor Jianmin Wang for their help and important instructions. We would like to thank the PC members: Professor Hongyan Li, Professor Chaokun Wang and Professor Jiaheng Lu. We thank Zou Miao servicing as USDM workshop local organization chair. Finally, we would also like to thank all the speakers and presenters at the workshop, and all the participates to the workshop, for their engaged and fruitful contributions.

References

1. Berchtold, S., Christian, G., Braunmller, B., Keim, D.A., Kriegel, H.-P.: Fast parallel similarity search in multimedia databases. In: SIGMOD, pp. 1–12 (1997)
2. Dong, X., Halevy, A.Y., Madhavan, J., Nemes, E., Zhang, J.: Similarity search for web services. In: Nascimento, M.A., Ozsu, M.T., Kossmann, D., Miller, R.J., Blakeley, J.A., Schiefer, K.B. (eds.) VLDB, pp. 372–383. Morgan Kaufmann, San Francisco (2004)
3. Eguchi, K., Lavrenko, V.: Sentiment retrieval using generative models. In: Proceedings of the 2006 Conference on Empirical Methods in Natural Language Processing, Sydney, Australia, pp. 345–354 (2006)
4. McDonald, R., Hannan, K., Neylon, T., Wells, M., Reynar, J.: Structured models for fine-to-coarse sentiment analysis. In: Proceedings of the 45th Annual Meeting of the Association of Computational Linguistics, Prague, Czech Republic, pp. 432–439 (2007)
5. Resource Description Framework (RDF): Concepts and Abstract Syntax, http://www.w3.org/TR/rdf-concepts

ITEM: Extract and Integrate Entities from Tabular Data to RDF Knowledge Base

Xiaoyan Guo[1,2], Yueguo Chen[1], Jinchuan Chen[1], and Xiaoyong Du[1,2]

[1] Key Labs of Data Engineering and Knowledge Engineering, Ministry of Education, China
[2] School of Information, Renmin University of China, China
{guoxiaoyan,chenyueguo,jcchen,duyong}@ruc.edu.cn

Abstract. Many RDF Knowledge Bases are created and enlarged by mining and extracting web data. Hence their data sources are limited to social tagging networks, such as Wikipedia, WordNet, IMDB, etc., and their precision is not guaranteed. In this paper, we propose a new system, ITEM, for extracting and integrating entities from tabular data to RDF knowledge base. ITEM can efficiently compute the schema mapping between a table and a KB, and inject novel entities into the KB. Therefore, ITEM can enlarge and improve RDF KB by employing tabular data, which is assumed of high quality. ITEM detects the schema mapping between table and RDF KB only by tuples, rather than the table's schema information. Experimental results show that our system has high precision and good performance.

Keywords: Entity Extraction, Schema Mapping, RDF Knowledge Base.

1 Introduction

Today, it is more and more popular to use RDF [1] in the knowledge management domain [2]. RDF makes it easier to transfer information between different organizations with high agility and flexibility. RDF can maintain all the data's relations, automatically build tables for RDBMS deployment, navigate graphically, support multiple navigation trees, and link across diverse content sets [3].

There have been many *RDF Knowledge Bases* (*RDF KB* for short), such as LUBM, YAGO, SP2Bench, Barton, Freebase, etc. [4, 5, 12]. These datasets are mainly extracted from a very few domains even a single domain. For example, YAGO is mainly created using the Wikipedia and WordNet's data. Furthermore, these RDF datasets are usually constructed by manual work which highly limits their sizes. For those datasets created by automatic tools, it is hard to guarantee precision of the data.

In order to use RDF for knowledge management, we should have large RDF knowledge base in which we can find most of the data we need. Since the existing RDF datasets can't meet our requirements, the problem is then how to enlarge the existing RDF datasets.

An intuitive approach is to integrate all the existing RDF KB into an entire bigger one [6]. But this approach will have some problems. Firstly, the same data may have many copies in different RDF datasets with different accuracy. It is hard to determine

X. Du et al. (Eds.): APWeb 2011, LNCS 6612, pp. 400–411, 2011.
© Springer-Verlag Berlin Heidelberg 2011

which copy should be the correct one. Secondly, it is difficult to design schema mappings between these RDF datasets. Since each RDF dataset is developed independently. And their data models are heterogeneous, such as data semantics, expression of data with same semantic, environments of data sources, etc [7].

Another way is to enlarge a RDF KB by employing some external data sources. With the rapid development of relational database, we can easily find many public relational datasets from the Web. These data have large volume and, since they are created by manual work, are of high accuracy. For example, Google Fusion Tables [10] enables users to collaborate effectively on data management in the cloud, so we can think that the data in Google Fusion Tables are of good quality. Thus the RDF knowledge base can be enlarged with large-volume and high-quality data if we can effectively transfer the relational tables to RDF entities. We will explain this idea with the following example.

Example 1. Assuming that we have a table, as shown in Table 1. We don't have its schema information. But we can discover that each tuple can represent a football player's basic information, including their name, nationality, current club, height, birthday, etc. And we may think the last column may be not useful.

Also suppose we have a RDF KB as shown in Figure 1. We can find that Messi, Ibra. and C.Ronaldo already exist in this RDF KB. But some other players in Table 1 are missing in the RDF KB like Junmin and Kagawa. However, according to the schema of Table 1 and the structure information in Figure 1, we can construct two new entities for these two players, as shown in Figure 2. These two entities can be integrated into the RDF KB, and then the RDF KB will be enlarged.

Table 1. Football players' personal information

Lionel Messi	Argentina	Barcelona	1.69m	24 June 1987	99
Zlatan Ibrahimovic	Sweden	AC Milan	1.95m	3 October 1981	80
Cristinano Ronaldo	Portugal	Real Madrid	1.86m	5 February 1985	90
...					
Hao Junmin	China	Schalke 04	1.78m	24 March 1987	60
Shinji Kagawa	Japan	Dormund	1.73m	17 March 1989	88

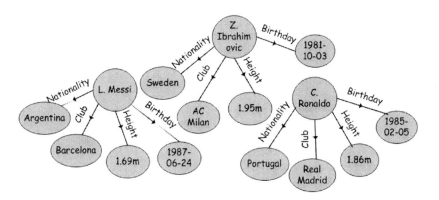

Fig. 1. Entities found in RDF KB

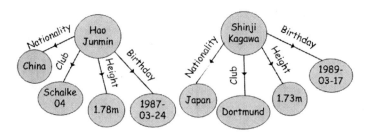

Fig. 2. Entities we can construct from Table 1

To summarize, given a relational table of high quality, we want to enlarge or improve RDF KB by employing these tabular data. In this paper, we build a system, ITEM (Integrate Tabular data and RDF Entities by schema Mapping), which can extract entities from relational tables and integrate them into an RDF KB.

Our contributions in this paper are listed below:

i.   We propose a novel tool, ITEM, for enlarging RDF KB by extracting entities from tables which don't exist in RDF KB.
ii.   We propose an efficient algorithm for matching tuples with entities.
iii.   We efficiently find the optimal schema mapping between a table and an RDF KB by transferring this problem to a known classical problem.
iv.   We examine our proposed method by conducting experiments over real datasets.

The rest of paper is organized as following. In Section 2, we introduce some definitions used in the paper and give the problem definition. In Section 3, we present our system's architecture, and describe the algorithms used in our system. In Section 4, we analyze ITEM and show our system's efficiency and effectiveness. Section 5 analyze related works. Finally, we conclude this paper and discuss some future work in Section 6.

2   Problem Definition

In this section, we will first introduce some definitions used in this paper, and then formally define the problem focused by us.

2.1   Schemas of Table and RDF Knowledge Base

A *relational table*, denoted by T, is a collection of tuples. Each tuple, denoted by t, has many fields. Each field has an attribute associated with it. We use $A = \{A_1, A_2, ..., A_n\}$ to denote the n attributes of T. Also, $t(A_i)$ denotes the value of t's i^{th} attribute.

An *RDF knowledge base*, denoted by D, is a collection of entities. Each entity is a collection of triples. An entity must have a property indicating its name, and may have many other predicates. Each triple is denoted as (name of the entity, predicate, object corresponding to the predicate). We use e to denote an entity and use $P = \{P_1, P_2, ..., P_m\}$ to denote its m predicates, in which P_1 must be its name. Similarly, $e(P_i)$ means the object corresponding to the ith predicate of e.

2.2 Mappings between Table and RDF KB

Definition 1. *A mapping between a tuple and an entity.*
Let us first construct a complete bipartite graph $G(A, P)$. A mapping between t and e, denoted by $M(t, e)$, is a matching in $G(A, P)$ which contains two edges at least, and one of the two edges must be $(t(A_i), e(P_1))$ which connects to the name predicate P_1. Hence, a mapping must contain an edge connecting to the entity's name.

Definition 2. *The optimal mapping between a tuple and an entity.*
Let $sim(t(A_i), e(P_j))$ be the similarity between $t(A_i)$ and $e(P_j)$, we can then define the score of a mapping $M(t, e)$ as follows:

$$score(M) = \sum_{\forall (t(A_i), e(P_j)) \in M} sim(t(A_i), e(P_j))$$

The optimal mapping $M_{opt}(t, e)$ for t and e is the mapping with the largest score among all mappings between t and e. When there are multiple mappings sharing the same largest score, we just randomly choose one as the optimal mapping.

Definition 3. *A match of a tuple t in a KB* is an entity e in the KB such that $M_{opt}(t, e)$ is larger than or equal to the scores of all other mappings between t and any other entities in this KB.
   Formally, e is a match of t in D, *iff*

$$M_{opt}(t, e) = MAX\{M(t, e_i) \mid e_i \in D\}$$

For the case that the score of the mapping between t and its match equals to zero, we say t can't match with this KB.

Definition 4. *A schema mapping between a table T and an RDF KB D*, denoted by $\Gamma(T, D)$, is also a matching in a complete bipartite graph.
Given a tuple $t \in T$, its match e in D, and their optimal mapping $M_{opt}(t, e)$, we can generate a schema mapping Γ, such that $(A_i, P_j) \in \Gamma$ *iff* $(t(A_i), e(P_j)) \in M_{opt}(t, e)$. Here, We call Γ the schema mapping consistent with t, denoted by Γ_t. And this schema mapping also has a $score(\Gamma_t)$ which is equal to $score(M_{opt}(t, e))$.

Definition 5. *Paradoxical schema mappings between a table and an RDF KB.*
Given two schema mappings between T and D: Γ' and Γ''. We say Γ' and Γ'' are paradoxical mappings *iff* $\exists (A_i, P_j) \in \Gamma'$, and $\exists (A_k, P_l) \in \Gamma''$, and either one of the following two conditions is satisfied:

     i. $(A_i = A_k) \wedge (P_j \neq P_l)$

    ii. $(A_i \neq A_k) \wedge (P_j = P_l)$

Thus, if the two schema mappings Γ' and Γ'' are not paradoxical, we say they are *mergeable*. We can merge this two schema mappings into a bigger one as follows:

$$\Gamma = \Gamma' \cup \Gamma'' = \{(A_i, P_j) \mid ((A_i, P_j) \in \Gamma') \vee ((A_i, P_j) \in \Gamma'')\}$$

and $score(\Gamma) = score(\Gamma') + score(\Gamma'')$.

Definition 6. *The optimal schema mapping between a table T and a KB D*.
According to our definitions, each tuple t in T has a consistent schema mapping Γ_t. Let $S_M = \{\Gamma_{t_i} \mid t_i \in T\}$ and S be any subset of S_M such that any two mappings in S are not paradoxical. Given $\Gamma_S = \bigcup_{\Gamma \in S} \Gamma$, then, the optimal schema mapping $\Gamma_{opt}(T, D)$ is the schema mapping such that

$$\forall S \in S_M, score(\Gamma_{opt}(T, D)) \geq score(\Gamma_S).$$

Finally, **the problem focused by us** is:

Given a table T and an RDF KB D, we try to find the optimal schema mapping between T and D, and integrate all tuples in T, which can't find their matches according to this optimal schema mapping, into D.

3 Construct Mappings between Tables and RDF

In this section, we begin with a brief description of the overall system architecture of ITEM. We then detail the algorithms used in the system. We will describe how to compute the similarity of an attribute with an object corresponding to a predicate, how to achieve the entities which are possible matches of a tuple, how to efficiently compute the mapping between a tuple and an entity, and how to efficiently compute the optimal schema mapping between a table and a KB if given each tuple's optimal mapping with the KB.

3.1 System Architecture

Figure 3 shows the main components of ITEM. The system can be split into three main parts. The first part is a crawler, which can crawl over the Web and collect the pages containing relational data. The second part is the main part of the system, which can generate the optimal schema mapping between a table and the KB. The third part will generate many entities according to the schema mapping results and integrate them into the KB.

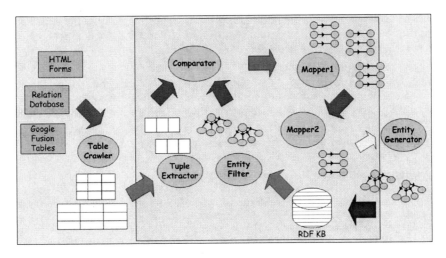

Fig. 3. System Architecture of ITEM

Algorithm 1. *Vector<Entity> ExtractEntitiesFromTable(T, D)*

Input: a table *T*, an RDF KB *D*
Output: all *entities* extracted from *T* and not existing in *D*

1   *create $M_{opt}[T.size()]$ to store T's optimal mappings with D*
2   *for (each tuple t in T) {*
3   *initialize t's optimal mapping $M_{opt}[t]$'s score to zero*
4   *Vector<Entity> vre = findAllRelatedEntities(t);*
5   *for (each entity e in vre) {*
6   *compute similarities between t's attributes and e's objects*
7   *construct a weighted bipartite graph G_1*
8   *invoke KM algorithm to find the matching M*
9   *if (score(M) > score($M_{opt}[t]$))*
10   *$M_{opt}[t] = M$*
11   *}*
12   *}*
13
14   *construct all the schema mapping S[] according to $M_{opt}[]$*
15   *compute all the paradoxical schema mappings in S[]*
16   *construct a graph G_2 according to the paradoxical schema mappings*
17   *invoke MWIS algorithm in G_2 to find the optimal schema mapping S_{opt} in S[]*
18
19   *use S_{opt} to extract all the entities not existing in D, add them to vector<Entity>*
      VE
20
21   *return VE;*

In the second part, the *Tuple Extractor* will extract each tuple from the collected tables, then the *Entity Filter* will extract all the entities in the KB which are similar to this tuple. Given a tuple and an entity, the *Comparator* will compare each pair of attribute and object, and generate a bipartite graph. Then the *Mapper1* will generate the optimal match according to this bipartite graph, and find the optimal mapping between a tuple and the KB. After collecting all the optimal mappings between each tuple and KB, the *Mapper2* will generate the final schema mapping between the table and the KB, and output it. The framework of ITEM is illustrated in Algorithm 1. We will introduce the details in the following sections.

3.2 Compute Similarity of Attributes and Objects

To compute $sim(t(A_i), e(P_j))$, we divide the strings into three categories according to their data formats, i.e. plain name strings, number strings, and time-format strings. And we define a comparator for each category. For plain name strings, we use Jaccard. For number strings, we use the absolute difference. And for time-format strings, we use the time interval between those two times.

If two strings both fall into the same category, we use the comparator of that category. If not, we just regard them as plain name strings. Finally, we normalize each similarity in order to let each value between zero and one.

3.3 Filter Out the Entities can't Be the Match

In order to compute the mapping between a tuple and the KB, we should compute the similarity score of each mapping between each entity in KB and this tuple, and find the one with the largest score. There are a large number of entities in KB, but only a few of them are related to this tuple. Hence, it is a waste of time if we calculate the similarity of every possible mapping pair. We should filter out the non-related or less-related entities to improve efficiency.

In our method, we construct a gram based inverted index for all the entities. We regard each entity as an article, which contains all objects of this entity. Given a keyword, we can then use this inverted index to find the entities containing this keyword.

We regard each attribute of a tuple as a keyword, then perform keywords search using the inverted index, and find all the entities which contain at least one keyword. We consider these entities to be similar or related to the given tuple, and the other entities are irrelevant ones that don't need further work.

3.4 Find the Optimal Mapping of a Tuple and an Entity

For each tuple, we can find all the entities that are similar to it. For each tuple and each entity among them, we can find its mapping, and then find the optimal mapping to these similar entities.

According to Definition 1 and 2, for each tuple and entity, we can construct a weighted bipartite graph, as shown in Figure 4. In this graph, each attribute and each object are regarded as vertices, and we add a weighted edge to each attribute and each object, the weight is the similarity of the attribute and object.

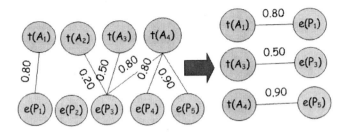

Fig. 4. A weighted bipartite graph generated for a tuple and an entity (The edges which have weight zero are removed)

We now need to find the optimal mapping of this tuple and this entity. In other words, we will find a match in this graph, which has the largest weight sum. So this problem can be converted to a *Maximum Weighted Bipartite Match Problem*, which can be solved in $O(n^3)$ used KM algorithm [8], and with Edmonds and Karp's modifition.

3.5 Find the Optimal Schema Mapping of a Table and KB

We now have all the optimal mappings for all the tuples in the given table. We should merge all these mappings and find the optimal mapping between a table and a KB.

For this objective, we generate a graph as follows. Each mapping is regarded as a vertex which has a weight indicating the score of this mapping computed by *Mapper 1*. Next, for each pair of mappings, if these two mappings are paradoxical, we add an edge between these two vertices. Then we should find a subset of the vertices set, and the sum of these vertices' weights is the largest. This problem can be mapped to the *Maximum Weight Independent Set Problem* [9]. Since this problem is NP-hard, we adopt a greedy algorithm to solve it.

Take an example for more details in Figure 5, we will show how to merge all the schema mappings generated in Sec. 3.4 to find the optimal schema mapping. In this example, supposing we have six schema mappings, we will generate the optimal mapping Γ from them.

Firstly, we generate a graph with six vertices which is M_1 to M_6. For each vertex, we assign a weight to it. The weight is the score of the corresponding schema mapping. For example, $W_1 = score(M_1) = 2.20$ for vertex M_1.

Secondly, we can find that schema mapping M_1 and M_2 are paradoxical because P_1 matches with A_1 in M_1, but matches with A_2 in M_2. Thus we add an edge between M_1 and M_2. Similarly, for each pair of vertices (schema mappings), we add an edge between them *iff* they are paradoxical. In Figure 5, they are (M_1, M_2), (M_2, M_3), (M_5, M_6).

Thirdly, the problem of finding optimal schema mapping is equivalent to finding the maximum weight independent set in the graph generated above. We adopt a greedy algorithm to solve it. We find the vertex which has the largest score, and re-move all the vertices which are adjacent to the vertex. And then repeat this step until all the vertices are consumed. Thus the set $\{M_1, M_3, M_4, M_5\}$ will be the answer, which is shown in the bottom of Figure 5.

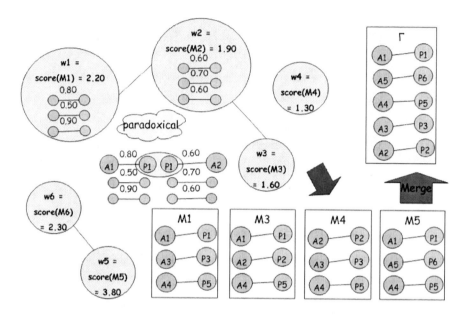

Fig. 5. Merge schema mappings to find optimal schema mapping

Lastly, we will merge all the four schema mappings to one mapping. We extract all the different matches in the four mappings, and put them together to form the optimal schema mapping Γ, which is shown on the upper right part of Figure 5.

4   Experimental Results

4.1   Experiment Setup

All experiments are conducted on a PC with Intel 2.0GHz Core 2 Duo CPU and 2GB memory under Windows Vista Home Basic operating system. All programs are im-plemented in Java.

The tables used in this experiment are crawled from Google Fusion Tables [10]. Google Fusion Tables is a cloud-based service for data management and integration. Google Fusion Tables enables users to upload tabular data files (spreadsheets, CSV, KML). Users can keep the data private, or make it public. [16] Since Google Fusion Tables enables users to collaborate effectively on data management in the cloud, we can assume that the data in Google Fusion Tables are of good quality. The other character of Google Fusion tables is that it contains large quantity of tables, about 200K+ currently.

We download all the public tables of Google Fusion Tables. From them, we randomly select 100 tables in our experiment. On average there are 7.06 attributes, and 131.07 tuples in each of these 100 tables. Among all the attribute values in these 100 tables, about 27.96% are plain strings, 0.66% is date strings, and 71.38% are number strings.

We use YAGO as our RDF Knowledge Base [5]. There are 2,796,962 entities, and 92 different predicates. And the average number of predicates in an entity is 8.46.

4.2 Experiment Results Analysis

Effectiveness:
In order to avoid wrong mappings, in our experiments we only integrate the tables whose optimal schema mapping contains two edges at least, and one of them is related to the name predicate. Then 78 tables can be integrated among all the 100 tables. There are totally 13,107 tuples, and 7,788 of them will be injected into YAGO. Hence, about 4,000 tuples find their match entities in YAGO. By learning their mappings, we transfer all other tuples into entities and insert them into YAGO. Therefore, our method can effectively enlarge RDF KB by automatically mapping existing relational data with RDF KB.

For the sake of evaluating precision, we randomly choose 100 tuples and their match entities, and examine the correctness of the mapping manually. We find that more than 90% matches are correct.

Also, we recheck all the 100 tables, and find 15 new matches that are missed by ITEM. After review them one by one, we find that one major shortcoming of ITEM is that ITEM can't match a predicate with multiple attributes. For example, there are two attributes, *first_name* and *last_name*, in a table. But YAGO uses one predicate to indicate the whole name of a person. We will address this issue in our future work.

Efficiency:
As shown in Sec. 3.3, we construct an inverted list to find the entities which are possible to be the matches of a given tuple. Specifically, we adopt a token-based index,

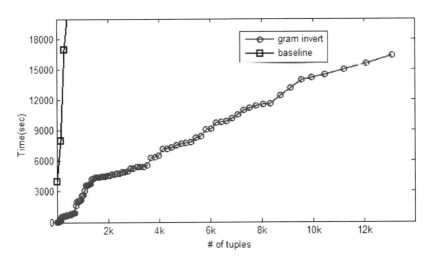

Fig. 6. Time cost (*baseline* vs. *gram_invert*)

i.e. each token of a string will be regarded as a word in this inverted list. We call this method *gram_invert*. In the same time, we also implement a *baseline* method which examines all mappings between a tuple with all entities in KB. The result is shown in Figure 6. Clearly, *gram_invert* performs much better than *baseline*.

5 Related Works

Schema matching is the task of finding semantic correspondences between elements of two schemas. [11] There are many approaches on schema matching. According to a survey [7], automatic schema matching approaches can divide into schema- and instance-level, element- and structure-level, and language- and constraint-based matching approaches. Furthermore, we can also combine many different matchers as a hybrid and composite combination of matchers, and there are many match prototypes such as Cupid, SEMINT, COMA++ [13].

Almost all these methods are schema-level. And the previous instance-level approach doesn't consider the relations of elements of each instance. They don't treat all the elements (attributes) of an instance (tuple) as a whole. Instead, they just regard all the instances of a column as a set, and match columns based on set similarity [14]. Our method considers each tuple as a unit for single matching, and derives spastically matching of the whole table and the KB by aggregating all single matching of tuples.

To our best knowledge, there are few works in discovering entities from relation tables and integrating them with RDF KB entities. Although COMA++ [13] builds mappings between table and RDF, it only uses the schema information to do the mapping rather than considers their instances. And they only transform data between relations and RDF KBs. They therefore don't have the ability to extract entities from relation tables.

The problem is also relevant to the set expansion problem [15]. Given a set of entities in an RDF KB, we want to find even more entities which can be added to this set. But there are some differences between our problem with it. We do not pull some entities to the KB, but initiatively push all the suitable entities into the KB. It makes more sense if we want to incrementally enlarge the knowledge base.

6 Conclusion and Future Work

In this paper, we propose ITEM, a tool to extract and integrate entities from relation tables to RDF Knowledge Base. We first introduce some definitions and formally define the problem. Then we introduce ITEM's architecture and describe the algorithms used in ITEM to show how to efficiently compute the schema mapping between table and RDF knowledge base. Finally, we give some experiments to show the effectiveness and efficiency of ITEM.

In the future, we want to implement our system as a web service and develop a web interface for users to upload their tabular data. Also, we would like to improve our algorithms to obtain better performance and to solve the existing shortcomings like the multi-column matching problem shown in our experiment analysis.

References

1. Resource Description Framework (RDF): Concepts and Abstract Syntax,
   `http://www.w3.org/TR/rdf-concepts`
2. McGlothlin, J.P., Khan, L.R.: RDFKB: efficient support for RDF inference queries and knowledge management. In: Proceedings of IDEAS, pp. 259–266 (2009)
3. Voleti, R., Sperberg, O.R.: Topical Web.: Using RDF for Knowledge Management. Technical Report in XML (2004),
   `http://www.gca.org/xmlusa/2004/slides/sperberg&voleti/`
   `UsingRDFforKnowledgeManagement.ppt`
4. Lehigh University Benchmark (LUBM),
   `http://swat.cse.lehigh.edu/projects/lubm`.
5. Suchanek, F.M., Kasneci, G., Weikum, G.: Yago - A Core of Semantic Knowledge. In: 16th international World Wide Web conference (WWW 2007) (2007)
6. Lenzerini, M.: Data Integration: A Theoretical Perspective. In: ACM Symposium on Principles of Database Systems (PODS), pp. 233–246 (2002)
7. Rahm, E., Bernstein, P.A.: A Survey of Approaches to Automatic Schema Matching. VLDB Journal 10(4) (2001)
8. Munkres, J.: Algorithms for the Assignment and Transportation Problems. Journal of the Society for Industrial and Applied Mathematics 5(1), 32–38 (1957)
9. Maximal Independent Set Problem,
   `http://en.wikipedia.org/wiki/Maximal_independent_set`
10. Google Fusion Tables, `http://www.google.com/fusiontables`.
11. Bergamaschi, S., Castano, S., Vincini, M., Beneventano, D.: Semantic Integration of Heterogeneous Information Sources. Data & Knowledge Engineering 36(3), 215–249 (2001)
12. Abadi, D.J., Marcus, A., Madden, S.R., Hollenbach, K.: Using the Barton Libraries Dataset as an RDF Benchmark. MIT-CSAIL-TR-2007-036. MIT (2007)
13. Aumueller, D., Do, H.-H., Massmann, S., Rahm, E.: Schema and ontology matching with COMA++. In: Proceedings of the ACM SIGMOD (2005)
14. Engmann, D., Massmann, S.: Instance Matching with COMA++. In: BTW Workshop (2007)
15. Wang, R.C., Cohen, W.W.: Language-Independent Set Expansion of Named Entities using the Web. In: ICDM 2007 (2007)
16. Gonzalez, H., Halevy, A., Jensen, C., Langen, A., Madhavan, J., Shapley, R., Shen, W.: Google Fusion Tables: Data Management, Integration and Collaboration in the Cloud. In: SOCC (2010)

Batch Text Similarity Search with MapReduce[⋆]

Rui Li[1,2,3], Li Ju[4], Zhuo Peng[1], Zhiwei Yu[5], and Chaokun Wang[1,2,3]

[1] School of Software, Tsinghua University, Beijing 100084, China
[2] Tsinghua National Laboratory for Information Science and Technology
[3] Key Laboratory for Information System Security, Ministry of Education
[4] Department of Information Engineering, Henan College of Finance and Taxation,
Zhengzhou 450002, China
[5] Department of Computer Science and Technology, Tsinghua University
{r-li09,pengz09,yzw08}@mails.tsinghua.edu.cn, chaokun@tsinghua.edu.cn

Abstract. Batch text similarity search aims to find the similar texts according to users' batch text queries. It is widely used in the real world such as *plagiarism check*, and attracts more and more attention with the emergence of abundant texts on the web. Existing works, such as FuzzyJoin, can neither support the variation of thresholds, nor support the online batch text similarity search. In this paper, a two-stage algorithm is proposed. It can effectively resolve the problem of batch text similarity search based on inverted index structures. Experimental results on real datasets show the efficiency and expansibility of our method.

Keywords: MapReduce, Batch Text Similarity Search.

1 Motivation

Let $X = \{x_1, x_2, \ldots, x_n\}$ be a text database, where x_i is a text record ($1 \leq i \leq n$). Give a collection of query texts $Y = \{y_1, y_2, \ldots, y_m\}$ where y_j is also a text record ($1 \leq j \leq m$). The task of *batch text similarity search* is retrieving all pairs of (x_i, y_j) in which the similarity value between x_i and y_j is no less than a given threshold η_j ($0 < \eta_j \leq 1, 1 \leq j \leq m$).

Text similarity search, a.k.a. text query by example, has been applied widely. A lot of users submit text similarity search tasks every day, e.g. searching for similar webpages, looking for persons having common interests in social networks. There is a wealth of studies on text similarity searching. However, there are still some challenges:

1. The popularity and development of information technology leads to the explosive growth of data on the Internet. For example, the N-gram dataset of Google [1] has 1 billion records.

[⋆] The work is supported by the National Natural Science Foundation of China (No. 60803016), the National HeGaoJi Key Project (No. 2010ZX01042-002-002-01) and Tsinghua National Laboratory for Information Science and Technology (TNLIST) Cross-discipline Foundation.

X. Du et al. (Eds.): APWeb 2011, LNCS 6612, pp. 412–423, 2011.

2. The similarity thresholds change constantly according to users' expectations.
3. Quick response to the users' queries is highly in demand despite huge data in the database.

In this paper, batch text similarity search with MapReduce is proposed to deal with the aforementioned challenges. MapReduce [2] is a new distributed programming model which can process large datasets in parallel fashion. The MapReduce framework may be a simple and effective approach to batch text similarity search due to its automatic load balancing and fault tolerance advantages. Distinguished from FuzzyJoin [3] which is the latest research achievement of set similarity join using MapReduce, batch text similarity join can support both online text similarity search and search with variable threshold effectively.

The main contributions of this work are as follows.

1. We propose the PLT inverted file which contains the text prefix, length, and threshold, in order to support variable threshold.
2. We propose an online batch text similarity searching algorithm Online-Batch-Text-Search (OBTS), and it is able to find out all the query texts whose similarity with the original text database are no less than a given threshold in real-time.
3. Experimental results on real dataset show that OBTS algorithm outperforms previous works.

The rest of the paper is organized as follows. We review the related work on similarity text search in Section 2. In Section 3 we introduce some basic concepts and in Section 4 we give the structure of PLT inverted file . The detailed algorithm is described in Section 5. In Section 6, our algorithm is compared with the existing works and finally we conclude this work in Section 7.

2   Related Work

A large body of work has been devoted to the study of similarity search [4,5,6,7]. The problem of text similarity search has a lot to do with the problem of string joins [8,9] as well as set similarity joins [3,10]. An inverted index based probing method was proposed in [11]. Recently, several filters were proposed to decrease the number of candidate documents needed to be verified, such as prefix filter [12], length filter [10,4], positional filter and suffix filter [13].

Three MapReduce algorithms for text similarity search using vector are introduced in [14]. All the three algorithms can find out the top k similar documents of each query text. However, only the first one, which is in brute force fashion, is suitable for exact similarity search. Furthermore, this algorithm is inefficient because it needs to compute the similarity of every two documents.

A 3-stage approach is proposed in [3] for set-similarity joins. The R-S join case in set-similarity joins is much like similarity search. We can treat the relation R as a query and S as a text database. However, it is worth noting that this approach is inefficient, especially for large-scale datasets, because the 3-stage method have to be run from the beginning to the end when a new query arrives or the similarity threshold is changed.

3   Preliminaries

3.1   Word Frequency Dictionary

We introduce a dictionary containing the total word frequency of each word. A total frequency of a word stands for its occurrences in all records of a text database. The frequencies are ranked in ascending order in our dictionary. For example, there is a text database with just two texts: $t_1 = (a, b, c, d)$ and $t_2 = (c, b, d, e, f)$. A possible form of the word frequency dictionary is (a, e, f, b, c, d).

3.2   Similarity Metrics

There are several similarity functions available, such as Jaccard similarity function, cosine similarity function, and overlap similarity function. Jaccard similarity and cosine similarity for two texts x, y are listed as follows:

$$jaccard(x, y) = \frac{|x \cap y|}{|x \cup y|} \tag{1}$$

$$cosine(x, y) = \frac{\sum_i x_i \bullet y_i}{\sqrt{|x|} \bullet \sqrt{|y|}} \tag{2}$$

In our work, we denote $|x|$ as the length of the text, measured by the number of words. $|x \cap y|$ stands for the common words in x and y while $|x \cup y|$ stands for $|x| + |y| - |x \cap y|$.

There are internal relationships among all sorts of similarity functions [13]. As Jaccard similarity function is widely used in algorithm, we adopt it as our metric. In the example above, the similarity between the two texts t_1 and t_2 is $3/6 = 0.5$.

3.3   Filters

Lemma 1 (prefix filtering principle [12,13]). *Provided that the words in each text are in accordance with the frequency dictionary word order, the p-prefix stands for the p words before x. Let t be the similarity threshold.*
If $jaccard(x, y) > t$, then

$$\{(|x| - \lceil t \bullet |x| \rceil + 1)\text{-}prefix\} \bigcap \{(|y| - \lceil t \bullet |y| \rceil + 1)\text{-}prefix\} \geqslant 1. \tag{3}$$

We can exploit MapReduce framework and Lemma 1 to construct prefix index, which will be discussed in detail in Section 5.

Lemma 2 (length filter principle [10,4]). *For texts x and y, x is longer than y. If $jaccard(x, y) > t$, we have:*

$$|y| \geqslant t \bullet |x| \tag{4}$$

We can construct a length filter by means of Lemma 2. Based on the filter, each text x_i will only have to compare with the text y which satisfies:

$$t \bullet |x_i| \leq |y| \leq \frac{1}{t} \bullet |y| \tag{5}$$

Length filter can further reduce the number of text candidates. The pruning effect is obvious especially when t is close to 1.

4 PLT Inverted File

Normally, there are vast amounts of data in a text database. Pairwise comparison between query texts and all the database texts for similarity search is unrealistic. Normal method of using filters needs to process the database (including computing the prefix of each text and so on) for each search, which is also inefficient. So we consider building an index. For the need of prefix filter and length filter, our index includes the information of prefix, length and similarity threshold, and it is called PLT inverted file.

According to Lemma 1, we find that when given a similarity threshold, the query text's prefix shares at least one common word with the candidate database texts' prefix. So the PLT inverted file needs to contain the prefix information of database texts.

According to Lemma 2, if the length of query text is l, for a given threshold t, query text only needs to match with the database texts whose length are in $[t*l, l/t]$ interval. So the PLT inverted file needs to contain the length information of database texts.

Prefix length has something to do with similarity threshold. The higher the threshold, the longer the prefix length. So we need to calculate the maximum similarity threshold that makes one word appear in the prefix beforehand. Each line in the PLT inverted file is in the form of $<key,\ list\ of\ value>$, where key is a word, and each $value$ contains the following three elements, i.e. id of the text whose prefix contains the key, the length of the text and the maximum similarity threshold that makes the key appear in the prefix.

5 Algorithms

In this section, the algorithm OBTS is presented. OBTS includes two phases. In the first stage, the PLT inverted structure is generated for pruning. In Stage 2, the PLT inverted index structure receives the query text, searches in the PLT inverted file, and matches the query text and the text database. Finally it outputs the text pairs with a threshold no less than the given value.

5.1 Stage 1: Generate a Database Prefix Index

The first stage, generating a database prefix index, is divided into three steps. Step 1 takes charge of ranking of word frequencies. In this step, after the statistics

---

Algorithm 1. *Calculate Overall Word Frequency*

---

1 **procedure** Map(u,x);
2 **for all** token in x **do**
3 | Output (token,1);
4 **end for**

5 **procedure** Combine(x,[1, 1...]);
6 k ← sum of [1,1...];
7 Output(x, k);

8 **procedure** Reduce(x, $[k_1, k_2 ...]$);
9 f ← sum of $[k_1, k_2 ...]$;
10 Output(x, f);

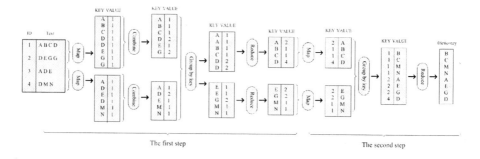

The first step   The second step

Fig. 1. The Generation of Word Frequency Dictionary

for each word frequency is accomplished, word frequencies are sorted in ascending order to generate a word frequency dictionary. Step 2 generates vectors of all texts in the database according to the word frequency dictionary. Each text generates a new vector text in this step. In the final step, the PLT inverted file is generated.

Overall Word Frequency Sort. This step consists of two MapReduce jobs. The first job computes the statistics of word frequency. As shown in Algorithm 1, the input of Map function is the text in database while the output of Map function is the pairs of <*token*, 1>according to each word in the texts. A Combine function is used to aggregate the output in order to reduce the network data transmission from the Map function to the Reduce function. A k is obtained by accumulating of all the *1*s after the same token. Reducer accumulates all the k in the same token, and obtains overall word frequency *f*. The left part of Figure 1 shows the first job. We assume that there are only 4 texts in the database for convenience. Take Text 4 as an example. It contains three words, so Map function will output three pairs: <D, 1>, <M, 1>, <N, 1>. Word D coming from Text 3 and Text 4 are aggregated for they are processed by the same Mapper. Reduce function makes a further aggregation, and all occurrences of the word D in the database are counted.

Algorithm 2. *Word Frequency Sort*

---

1 **procedure** Map(x, f);
2 Output (f, x);

3 **procedure** Reduce$((f, [x_1, x_2 \ldots]))$;
4 **for all** x $\in [x_1, x_2 \ldots]$ **do**
5 $\quad|\quad$ Output $(x_1,$ null);
6 **end for**

The second job is to sort words according to f. As shown in Algorithm 2, the input of Map function is the result of the first job. Map function swaps the positions of x and f in $<x, f>$, and MapReduce framework will automatically sort the pairs according to f. The number of reducer is set to 1 in order to ensure the correctness of sorting results in the distributed environment. As shown in the right part of Figure 1, after grouped by key, all the words have been sorted by f. The output of Reduce function is the word frequency dictionary.

Algorithm 3. *Generate the Vector Text*

---

1 **procedure** Setup( );
2 d $\leftarrow$ LoadDocument (frequency_dictionary);
3 Initialize.Hashtable(H);
4 **for each** word in d **do**
5 $\quad|\quad$ Append(H,$< word, index >$);
6 **end for**

7 **procedure** Map(a, x);
8 v $\leftarrow$ ComputeVector(x, H);
9 Output (a, V);
10 /*Identity Reduce*/

Calculate the Vector Text. Algorithm 3 depicts that Map function receives texts, replaces the word with its position, and then sorts all the position number in the text, finally outputs the pair (a, V) made up by the id and vector of x_i. Take Text 4 in Figure 2 as an example. The positions of words D, M, and N in the dictionary are 8, 3, and 4, respectively. And it is changed to "3, 4, 8" after be sorted. In this stage we use Identity Reduce function to output its input directly.

Generate the PLT Inverted File. As shown in Algorithm 4, the Map function calculates the prefixes of all the vector texts. For each word in the prefix, it generates the pairs which contain the word, text id, length, and threshold value. MapReduce framework groups all the pairs according to word. Reduce function inserts its input into an index, and finally the PLT invert file is generated. In Figure 3, we take document 4 as an example, when given 0.9 as the threshold, the prefix length is 1, so it outputs (3, <4, 3, 0.9>). For threshold 0.8 and 0.7, prefix length remains to be 1. When the threshold is reduced to 0.6, the prefix length increases to 2, and then it outputs (4, <4, 3, 0.6>). Finally we check the threshold value of 0.5 and find that the prefix length is not changed.

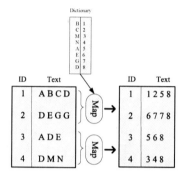

Fig. 2. The Generation of Text Vector

Algorithm 4. *Generate the PLT inverted File*

---

1 **procedure** Map(a, x);
2 p ← ComputePrefix(x, t);
3 **for** each token in p **do**
4 | Output(token,$< id, l, t >$);
5 **end for**

6 **procedure** Reduce(token, $[< id_1, l_1, t_1 >, < id_1, l_1, t_1 > \ldots]$);
7 **for** all $< id, l, t > \in [< id_1, l_1, t_1 >, < id_1, l_1, t_1 > \ldots]$ **do**
8 | list ← Add($< id, l, t >$);
9 **end for**
10 Output (token, list);

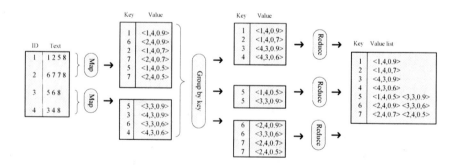

Fig. 3. The Generation of PLT Inverted File

5.2 Stage 2: Query Text Search

The second stage, query text search, is divided into two steps. Firstly we transform the query texts into vector texts, and then we calculate the prefix for each vector text. For each word in the prefix, we search the PLT inverted file, match the texts in database which meet the requirements, and finally output the text pairs with the similarity no less than a given threshold. System continuously detects the coming of new query texts. Once it is detected, the query texts will run the two steps above.

As shown in Algorithm 5, we calculate the prefix for each vector text generating from query text in the Map function. For each word in the prefix, it outputs the pairs consisting of the word (key) and the content of vector text content (value). Reduce function retrieves the PLT inverted file according to the key, returns the information of candidate text, including its id, length, threshold.

Algorithm 5. *Query Text Search*

---

1 **procedure** Map(a, y);
2 p ← ComputePrefix(y, t);
3 **for** each token in p **do**
4 | Output(token, y);
5 **end for**

6 **procedure** Reduce(token, $[y_1, y_2, \ldots]$);
7 **for** all y $\in [y_1, y_2 \ldots]$ **do**
8 | list of x ← Check(y, PLT);
9 | **for** each x in list of x **do**
10 | | sim ← Computesimilarity(x, y);
11 | | **if** sim > t **then**
12 | | | Output (y, $< x, sim >$);
13 | | **end if**
14 | **end for**
15 **end for**

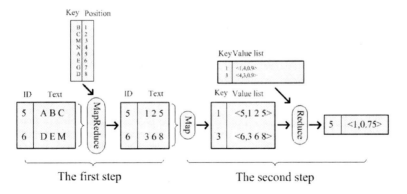

Fig. 4. Query Text Search

After checking whether the length and threshold is desirable, we match the query text with the candidate, and output the text pairs whose similarity is no less than the given threshold. For example, in Figure 4, Let the given threshold be 0.75. For Text 5, we can see the prefix only has a token-1 as the text length is 3. Text 1 is a candidate by searching the PLT inverted file. After the verification, it meets the requirements, with the similarity of 0.75.

6   Experiments

In this section we compare the performance of our algorithm with the fuzzyjoin in [3], whose source code is available at http://asterix.ics.uci.edu/.

6.1   Experimental Setting

Experiments were run on a 5-node cluster. Each node had one Pentium(R) processor E6300 2.8 GHz with two cores, 4 GB of RAM, and 800 GB hard disks. Five nodes were all slave nodes and one of them was master node as well. We installed the Ubuntu 9.10 operating system, Java 1.6.0_12, Hadoop 0.20.1 [15]. In order to maximize the parallelism, the size of virtual memory of each map/reduce task was set to be 2 GB.

6.2   Experimental Dataset

We used MEDLINE as our dataset. It is the U.S. National Library of Medicine's bibliographic database, consisting of more than 11M articles from over 4.8K indexed titles. We only used a part of it. Our dataset contained 48K records and the total size was approximately 290 MB. The average length of each record was 6026 bytes.

6.3   Comparative Experiments

In this subsection, we demonstrate that our algorithm had a better performance by a large number of experimental results.

Influences of similarity threshold. We fixed the number of query texts to 60. Similarity threshold values were changed between 0.5 and 0.9 on the dataset of MEDLINE. The running time consisted of the time cost of two stages: generating the PLT inverted file, and similarity search.

Figure 5 shows the running time for searching MEDLINE containing 10K, 30K, 48K records. As the similarity threshold rose, time cost became lower mainly due to the following two reasons. Firstly, the prefixes of query texts became shorter with the rise of threshold, which made less items need to be searched in the PLT inverted file. Secondly, the rise also made less database texts in each item to be verified. We also noticed that the time cost for generating the PLT inverted file did not increase linearly with the increase of the dataset size. It was because the time cost for launching MapReduce was constant. In order to show worst situations, we used 0.5 as the similarity threshold in experiments below.

Database Size	10K	30K	48K
Time Cost for PLT(s)	257	428	599

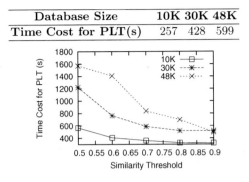

Fig. 5. Influences of similarity threshold

Influences of database size. In this experiment, we also used 60 texts as a query. We selected 5K to 20K texts from the MEDLINE dataset. Figure 6 shows that our algorithm had a better performance. This was mainly due to two reasons. The first was less data was sent through the network in our algorithm, because we used vectors instead of the words. The second was that verification became easier for the use of vectors.

Influences of Query Texts' size. We fixed the number of database texts to 5K, so the time cost of generating the PLT inverted file was constant. Figure 7 shows that as the number of query texts increased, time cost of our algorithm rose slowly. For instance, when we change texts' number from 200 to 300, only 8 extra seconds needed.

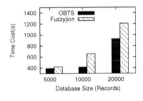

Fig. 6. Influences of database size

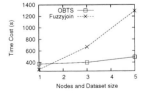

Fig. 7. Influences of Query Texts' size

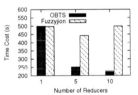

Fig. 8. Influences of Reducer Number

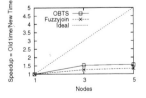

Fig. 9. Speedup

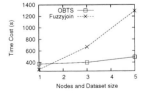

Fig. 10. Scaleup

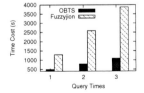

Fig. 11. Online Search

Influences of Reducer Number. In this experiment, we used 500 texts as a query and 2000 texts as a database. Figure 8 shows that the time cost decreased significantly when the number of reducers changed from 1 to 5, and insignificantly when it changed from 5 to 10. As the number of reducers increased, the processing capacity of reducer was improved. But the running time depended on the lowest reducer, not all the reducers finished their jobs at the same time. When the reducer number changed from 1 to 5, the former played a leading role, and when the reducer number changed from 5 to 10, the impact of the latter got larger.

Speedup and scaleup. We calculated the relative speedup and scaleup in order to show the performance of our parallel algorithm.

For the speedup, we fixed the query texts' size as well as the database size and varied the cluster size. Query texts' number was set to 100, and database contained 2K texts. As shown in Figure 9, a dotted straight line shows the ideal speedup. For instance, if we changed cluster nodes from 1 to 3, it should be triple as fast. Both two algorithms did not reach the ideal speedup. Their limited speedup was due to two main reasons. Firstly, network data transmission traffic increased as more nodes added into the cluster. Secondly, the whole task was not equal divided, and the other reducers must wait for the lowest one.

For the scaleup, we increased the dataset (including query text and database) size and the cluster size together by the same factor. Let m be the size of query text and n the size of database, time cost for matching each pair was p. If no filter existed, time cost for matching would be $m \bullet n \bullet p$. If both the query text and the database got t times larger, that would be $(m \bullet t) \bullet (n \bullet t) \bullet p = m \bullet n \bullet p \bullet t^2$. Because of the increase of cluster size, the final result would be $m \bullet n \bullet p \bullet t$. But the start-up time for MapReduce is constant, so the time cost did not increase with t linearly. Figure 10 shows that our algorithm had a better expandability. Because when the cluster size grew larger, OBTS spent the less time for processing the same dataset.

Online Search. OBTS can handle the problem of batch similarity search without processing the database for each query. We used the whole 290MB dataset as the database. Figure 11 shows the advantage of this approach. The advantage is more outstanding at a higher frequency of query due to the existing of PLT inverted file.

7   Conclusions

In this paper, we study the problem of online batch text similarity search using the MapReduce framework. We propose the PLT inverted file in order to avoid processing the database for every query. Based on the PLT inverted file, OBTS algorithm was proposed to support the real-time search. It can also support users to change the similarity threshold each time. We implemented OBTS in Hadoop and compared the performance with fuzzyjoin to show the advantages of OBTS.

References

1. Web-1t: Web 1t 5-gram version 1,
   http://www.ldc.upenn.edu/Catalog/CatalogEntry.jsp?catalogId=LDC2006T13
2. Dean, J., Ghemawat, S.: Mapreduce: Simplified data processing on large clusters.
   In: OSDI (2004)
3. Vernica, R., Carey, M.J., Li, C.: Efficient parallel set-similarity joins using mapreduce. In: SIGMOD (2010)
4. Bayardo, R.J., Ma, Y., Srikant, R.: Scaling up all pairs similarity search. In: WWW,
   pp. 131–140. ACM, New York (2007)
5. Lewis, J., Ossowski, S., Hicks, J., Errami, M., Garner, H.R.: Text similarity: an
   alternative way to search medline. Bioinformatics 22(18) (2006)
6. Berchtold, S., Christian, G., Braunmüller, B., Keim, D.A., Kriegel, H.P.: Fast
   parallel similarity search in multimedia databases. In: SIGMOD, pp. 1–12 (1997)
7. Dong, X., Halevy, A.Y., Madhavan, J., Nemes, E., Zhang, J.: Similarity search for
   web services. In: VLDB, pp. 372–383. Morgan Kaufmann, San Francisco (2004)
8. Gravano, L., Ipeirotis, P.G., Jagadish, H.V., Koudas, N., Muthukrishnan, S., Srivastava, D.: Approximate string joins in a database (almost) for free. In: VLDB,
   pp. 491–500 (2001)
9. Jin, L., Li, C., Mehrotra, S.: Efficient similarity string joins in large data sets.
   In: VLDB (2002)
10. Arasu, A., Ganti, V., Kaushik, R.: Efficient exact set-similarity joins. In: VLDB,
    pp. 918–929. ACM, New York (2006)
11. Sarawagi, S., Kirpal, A.: Efficient set joins on similarity predicates. In: SIGMOD,
    pp. 743–754. ACM, New York (2004)
12. Chaudhuri, S., Ganti, V., Kaushik, R.: A primitive operator for similarity joins in
    data cleaning. In: ICDE, p. 5. IEEE Computer Society, Los Alamitos (2006)
13. Xiao, C., Wang, W., Lin, X., Yu, J.X.: Efficient similarity joins for near duplicate
    detection. In: WWW, pp. 131–140. ACM, New York (2008)
14. Lin, J.: Brute force and indexed approaches to pairwise document similarity
    comparisons with mapreduce. In: SIGIR, pp. 155–162. ACM, New York (2009)
15. Hadoop: Apache Hadoop, http://hadoop.apache.org/

An Empirical Study of Massively Parallel Bayesian Networks Learning for Sentiment Extraction from Unstructured Text

Wei Chen*, Lang Zong, Weijing Huang, Gaoyan Ou,
Yue Wang, and Dongqing Yang

Key Laboratory of High Confidence Software Technologies
(Ministry of Education), School of EECS, Peking University,
Beijing 100871, China
Phone: 86-010-62756382
pekingchenwei@hotmail.com

Abstract. Extracting sentiments from unstructured text has emerged as an important problem in many disciplines, for example, to mine on-line opinions from the Internet. Many algorithms have been applied to solve this problem. Most of them fail to handle the large scale web data. In this paper, we present a parallel algorithm for BN(Bayesian Networks) structure leaning from large-scale dateset by using a MapReduce cluster. Then, we apply this parallel BN learning algorithm to capture the dependencies among words, and, at the same time, finds a vocabulary that is efficient for the purpose of extracting sentiments. The benefits of using MapReduce for BN structure learning are discussed. The performance of using BN to extract sentiments is demonstrated by applying it to real web blog data. Experimental results on the web data set show that our algorithm is able to select a parsimonious feature set with substantially fewer predictor variables than in the full data set and leads to better predictions about sentiment orientations than several usually used methods.

Keywords: Sentiment Analysis, Bayesian Networks, MapReduce, Cloud Computing, Opinion Mining.

1 Introduction

Sentiment analysis of unstructured text has recently received a lot of attention in the information retrieval community. One of the most popular tasks in document-level sentiment analysis is to determine whether the sentiment expressed in a document is positive or negative. Nowadays, sentiment analysis often maintains a low accuracy. One reason is that the rapid growth of the internet makes it easy to collect data on a large scale, often beyond the capacity of individual disks, and too large for processing. Traditional centralized algorithms are often too

---

* Corresponding author.

X. Du et al. (Eds.): APWeb 2011, LNCS 6612, pp. 424–435, 2011.

costly or impractical for large scale data. On the other hand, most algorithms assume that the features used for sentiment analysis are pairwise independent. The goal of this paper is to present a parallel BN learning technique for learning predominant sentiments of on-line texts, available in unstructured format, that:

1. is able to capture dependencies among words and find a minimal vocabulary for categorization purposes, and
2. is able to handle the large scale unstructured web data.

In this paper, we describe our experiences with developing and deploying a BN learning algorithm over large datasets using the MapReduce model of distributed computation. Then, we apply this parallel BN learning algorithm to capture the dependencies among words, and, at the same time, finds a vocabulary that is efficient for the purpose of extracting sentiments. Here, we learn the BN structure from massive real-world online blog data. We show how the parallel BN learning algorithm can be scaled effectively to large datasets. Moreover, BN can allow us to capture semantic relations and dependent patterns among the words, thus approximating the meaning of sentences. Further, performing the sentiment classification task using BN for the sentiment variables has excellent performance.

The rest of the paper is organized as follows. In section 2 we review the related work about sentiment analysis, MapReduce used in machine learning and BN structure learning algorithm TPDA (Three-Phase Dependency Analysis). In section 3 we provide the framework of the distributed BN learning algorithm over large datasets. We demonstrate the performance of our algorithm with experiments on online movie comments in section 4 and conclude in section 5.

2   Related Work

2.1   Sentiment Analysis

The problem of sentiment extraction is also referred to as opinion extraction or semantic classification in the literature. Eguchi and Lavrenko[7] proposed a generative model that jointly models sentiment words, topic words and sentiment polarity in a sentence as a triple. McDonald et al.[10] investigated a global structured model that learns to predict sentiment of different levels of granularity in text. Turney and Littman[16] applied an unsupervised learning algorithm to classify the semantic orientation in the word/phrase level, based on mutual information between document phrases and a small set of positive/negative paradigm words like good and bad. Blitzer et al.[1] focused on domain adaption for sentiment classifiers with respect to different types of products' online reviews. Choi et al.[3] dealt with opinion analysis by combining conditional random fields (CRFs). In the sentence level, a semi-supervised machine learning algorithm was proposed by Pang and Lee[15], which employs a subjectivity detector and minimum cuts in graphs. Another system by Kim and Hovy[8] judges the sentiment of a given sentence by combining the individual word-level sentiment.

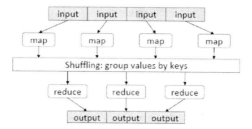

Fig. 1. Illustration of the MapReduce framework: the "mapper" is applied to all input records, which generates results that are aggregated by the "reducer"

2.2   MapReduce Used in Machine Learning

MapReduce is a simple model for distributed computing that abstracts away many of the difficulties in parallelizing data management operations across a cluster of commodity machines. A MapReduce[6] operation takes place in two main stages. In the first stage, the map function is called once for each input record. At each call, it may produce any number of output records. In the second stage, this intermediate output is sorted and grouped by key, and the reduce function is called once for each key. The reduce function is given all associated values for the key and outputs a new list of values (see Fig. 1).

With the wide and growing availability of MapReduce-capable compute infra-structures[6], it is natural to ask whether such infrastructures may be of use in parallelizing common data mining tasks such as BN structure learning. For many data mining operations, MapReduce may offer better scalability with vastly sim-plified deployment in a production setting. Panda et al.[13] described PLANET: a scalable distributed framework for learning tree models over large datasets using the MapReduce model of distributed computation. MRPSO[11] utilizes the Hadoop implementation of MapReduce to parallelize a compute-intensive application, Particle Swarm Optimization. Cohen[5] investigated the feasibility of decomposing useful graph operations into a series of MapReduce processes. Companies such as Facebook and Rackspace[9] use MapReduce to examine log files on a daily basis and generate statistics and on-demand analysis. Chu et al.[4] give an excellent overview of how different popular learning algorithms (e.g., K-means, neural networks, PCA, EM, SVM)can be effectively solved in the MapReduce framework.

2.3   Bayesian Networks

A BN can be formulated as a pair (G, P), where $G = (V, E)$ is a directed acyclic graph (DAG). Here, V is the node set which represents variables in the problem domain $V = \{x_1, x_2, ..., x_n\}$ and E is the arc (directed edge) set which denotes probabilistic relationships among the variables, $E = \{< x_i, x_j > | x_i, x_j \in V\}$, each arc $< x_i, x_j >$ means that x_i implicates x_j. For a node $x \in V$, a parent of x is a node from which there exists an arc to x. $Pa(x_i)$ is the parent set of x_i. The parameter set P describes probability distributions associated with each node, $P = \{p(x_i|Pa(x_i))|x_i \in V\}$, $p(x_i|Pa(x_i))$ is the probability of x_i on $Pa(x_i)$ condition.

Three-Phase Dependency Analysis (TPDA)[2] is one of the widely used BN structure learning algorithm. The BN structure is learned by identifying the conditional independence relationships among the nodes of G. It uses some CI test (mutual information) to find the conditional independence relationships among the nodes of G and use these relationships as constraints to construct a BN.

In [2], a BN is viewed as a network system of information channels, where each node is either active or inactive and the nodes are connected by noisy information channels. The information flow can pass through an active node but not an inactive one. When all the nodes on one undirected path between two nodes are active, we say this path is open. If any one node in the path is inactive, we say the path is closed. When all paths between two nodes are closed given the status of a set of nodes, we say the two nodes are d-separated by the set of nodes. The status of nodes can be changed through the instantiation of a set of nodes. The amount of information flow between two nodes can be measured by using mutual information, when no nodes are instantiated, or conditional mutual information, when some other nodes are instantiated.

In general, the volume of information flow between two variables A and B is measured by mutual information:

$$I(A, B) = \sum_{a,b} P(a, b) log \frac{P(a, b)}{P(a)P(b)} \qquad (1)$$

And the conditional mutual information, with A and B respect to the condition-set C, is formulated as

$$I(A, B|C) = \sum_{a,b,c} P(a, b|c) log \frac{P(a, b|c)}{P(a|c)P(b|c)} \qquad (2)$$

where A, B are two variables and C is a set of variables. Here, we use conditional mutual information as CI tests to measure the average information between two nodes. When $I(A, B|C)$ is smaller than a certain threshold value ϵ , we say that A, B are $d-separated$ by the condition-set C, and they are conditionally independent. A is independent of B whenever $I(A, B) < \epsilon$, for some suitably small threshold $\epsilon > 0$. We will similarly declare conditional independence whenever $I(A, B|C) < \epsilon$.

The three phases of the TPDA algorithm are drafting, thickening and thinning. In "drafting" phase, mutual information of each pair of nodes is computed to produce an initial set of edges based on a simpler test C basically just having sufficient pair-wise mutual information. The draft is a singly-connected graph (a graph without loops). The second "thickening" phase adds edges to the current graph when the pairs of nodes cannot be separated using a set of relevant CI tests. The graph produced by this phase will contain all the edges of the underlying dependency model when the underlying model is DAG-faithful. The third "thinning" phase examines each edge and it will be removed if the two nodes of the edge are found to be conditionally independent.

3   Distributed BN Learning Algorithm with MapReduce

In this section, we discuss the technical details of the major components of BN learning algorithm with MapReduce - the controller that manages the entire BN structure learning process, and the two critical MapReduce that compute the mutual information and conditional mutual information.

Given a set of attributes V and the training dataset D. Here, we assume that the node and arc sets of the BN is small enough to fit in the memory. Thus, the controller can maintain a vector V and a list E stored in a centralized file - BNFile. The controller constructs a BN structure using a set of MapReduce jobs, which compute the mutual information of two nodes and the conditional mutual information of two nodes with respect to their minimum cut set. At any point, the BNFile contains the entire BN structure constructed so far.

3.1   Map Phase

In dataset D, the training data is organized as Table 1. Every line stores all attributes and their values in the form of value-attribute pairs. For example, here v1, v2, v3 and v4 represent four attributes. They take the values High, normal, false, low respectively.

Table 1. Training dataset

line 1 High—v1	Normal—v2	FALSE—v3	Low—v4	......
line 2 High—v1	High—v2	FALSE—v3	Normal—v4	......
line 3 Normal—v1	Normal—v2	FALSE—v3	Low—v4	......
line 4 Normal—v1	Normal—v2	FALSE—v3	Normal—v4	......
... ...				

The training dataset D is partitioned across a set of mappers. Pseudocodes describing the algorithms that are executed by each mapper appear in Algorithm 1 and 2. Algorithm 1 prepare the statistic data for calculating mutual information. Given a line of training dataset, a mapper first emits every (value-attribute, one) as (key, value) output (line 2, Alg. 1). And for every attribute pairs, the mapper emits corresponding (value-attribute pair, one) as (key, value) output (line 5, Alg. 1). For example, a mapper loads the first line of Table 1. It emits following (key, value) lists (see Fig. 2):

(High—v1, 1),   (Normal—v2, 1),
(FALSE—v3, 1),   (Low—v4, 1),
(High—v1+Normal—v2, 1), (High—v1+FALSE—v3, 1),
(High—v1+Low—v4, 1),   (Normal—v2+FALSE—v3, 1),
(Normal—v2+Low—v4, 1),   (FALSE—v3+Low—v4, 1).

Algorithm 1. MR_MI::Map

Require: a line of training dataset L, $L \in D$
1: for $i < L.length$
2:    Output($L[i]$,one);
3: for $i < L.length$
4:    for $j = i + 1$ and $j < L.length$
5:       Output($L[i] + L[j]$,one);

MR_MI::Map

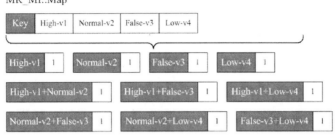

Fig. 2. The map operation of algorithm MR_MI. The mapper emits (value-attribute, one) and (value-attribute pair, one) as (key, value) outputs.

Algorithm 2 prepare the statistic data for calculating conditional mutual information. Each mapper requires training dataset, two attributes and their condition set as input. Given a line of training dataset, a mapper first extracts the corresponding values of the two input attributes and condition set (Line 1, 2, 3, Alg.2). Then the mapper emits (value-attribute and condition-set pair, one) as (key, value) output (line 4, 5, 6, Alg.2). For example, a mapper loads the first line of Table 1, two attribute v1, v2 and condition set (v3, v4) as input. It emits following (key, value) lists (see Fig. 3):
(High—v1+ FALSE—v3+ Low—v4, 1),
(Normal—v2+ FALSE—v3+ Low—v4, 1),
(High—v1+ Normal v2+ FALSE—v3+ Low—v4, 1),
(FALSE—v3+ Low—v4, 1).

3.2   Reduce Phase

The reduce phase, which works on the outputs from the mappers, performs aggregations and computes the sum of the statistic data for calculating mutual information and conditional mutual information.

The pseudocode executed on each reducer is outlined in Algorithm 3. The shuffling model groups values by keys. Each reducer receives (key, a list of all associated values) pairs as inputs. There are three types of keys: value-attribute, value-attribute pair, value-attribute and condition-set pair. The value list contains a list of "one", such as:

(High—v1, 1, 1, 1, 1,......)
(High—v1+Normal—v2, 1, 1, 1, 1,......)
(High—v1+ FALSE—v3+ Low—v4, 1, 1, 1, 1,......)

Reducers process keys in sorted order, aggregate the value lists and emit the sums as outputs (see Fig. 4). The following outputs are illustrated as examples:

(High—v1, 3233)
(High—v1+Normal—v2, 2560)
(High—v1+ FALSE—v3+ Low—v4, 1478)

The controller takes the aggregated values produced by all the reducers and calculate the mutual information for attribute pair and the conditional mutual information for attributes-condition-set pair.

Algorithm 2. MR_CMI::Map

Require: a line of training dataset L, $L \in D$, two attributes v1, v2, condition-set V

1: value1=L.extract(v1);   %extract the value of attribute v1 from L;
2: value2=L.extract(v2);   %extract the value of attribute v2 from L;
3: value3=L.extract(V);   %extract all values of condition-set V from L;
4: Output(value1+value3,one);
5: Output(value2+value3,one);
6: Output(value1+value2+value3,one);
7: Output(value3,one);

MR_CMI::Map

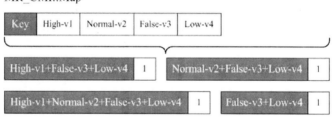

Fig. 3. The map operation of algorithm MR_CMI. The mapper emits (value-attribute and condition-set pair, one) as (key, value) output.

3.3   Controller Design

At the heart of the distributed BN learning algorithm is the "Controller", a single machine that initiates, schedules and controls the entire structure learning process. The controller has access to a computer cluster on which it schedules MapReduce jobs.

The main controller thread (Algorithm 4)schedules the TPDA algorithm[2] until the final BN structure is constructed. While the overall architecture of the controller is fairly straightforward, we would like to highlight a few important design decisions. MR_MI job returns all statistic information for calculating the

Algorithm 3. MR_MI::Reduce
Require: Key, Value_list
1: total=sum(Value_list);
2: Output(total)

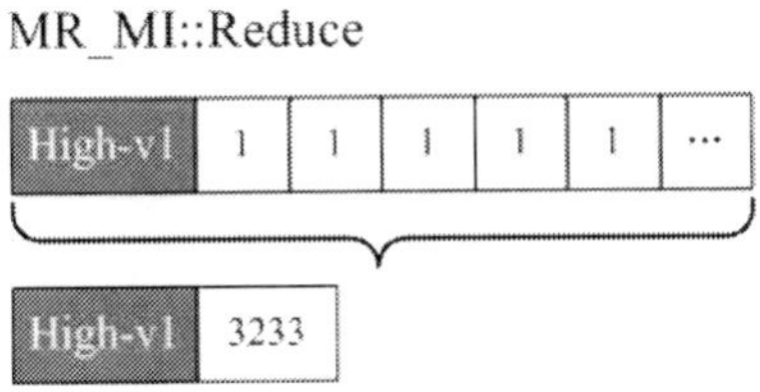

Fig. 4. The reduce operation of algorithm MR_MI. Reducers process keys in sorted order, aggregate the value lists and emit the sums as outputs.

mutual information values of all pairs $< X, Y >$ in V. When the MR_MI job returns, the controller calculate the mutual information with equation (1) using function Calculate_MI(X,Y). MR_CMI job returns all statistic information for calculating the conditional mutual information values of $< X, Y >, D(< X, Y >)$ pair. When the MR_CMI job returns, the controller calculate the conditional mutual information with equation (2) using function Calculate_CMI(X,Y).

4   Empirical Results

4.1   Unstructured Web Data

We tested our method on the data set used in Pang et al[14]. This data set [12] contains approximately 29,000 posts to the rec.arts.movies.reviews newsgroup archived at the Internet Movie Database (IMDb). 10,000 positive posts and 10,000 negative posts are selected. The original posts are available in the form of HTML pages. Some pre-processing was performed to produce the version of the data we used.

In our study, we used words as features. Intuitively the task of sentiment extraction is a hybrid task between authorship attribution and topic categorization; we look for frequent words, possibly not related to the context, that help express lexical patterns, as well as low frequency words which may be specific to few review styles, but very indicative of an opinion. We considered all the words that appeared in more than 8 documents as our input features. A total number of 7,321 words as input features and class feature $Y = \{positive, negative\}$ were selected. In our experiments, each document is represented by a vector, $W = \{w_1, ..., w_{7321}, Y\}$, where each w_i is a binary random variable that takes the value of 1 if the ith word in the vocabulary is present in the document and the value of 0 otherwise.

Algorithm 4. MainControllerThread

Require: $V=\{$all attributes$\}$, $E=\{\}$

Begin [Drafting].

1: NewThread(MR_MI).

2: I(X,Y)=Calculate_MI(X,Y).

3: Let $L = \{X, Y | I(X, Y) > \epsilon), X, Y \in V, X \neq Y\}$

4: Sort L into decreasing order.

5: For each $< X, Y >$ in L:

   If there is no adjacency path between X and Y in current graph E

   add $< X, Y >$ to E and remove $< X, Y >$ from L.

Begin [Thickening]

6: For each $< X, Y >$ in L:

   If EdgeNeeded($(V, E), X, Y$))

   Add $< X, Y >$ to E

Begin [Thinning]

7: For each $< X, Y >$ in E:

   If there are other paths, besides this arc, connecting X and Y,

   $E' = E- < X, Y >$

   if not EdgeNeeded($(V, E), X, Y$))

   $E = E'$

8: For each $< X, Y >$ in E:

   If X has at least three neighbors other than Y, or Y has at least

   three neighbors other than X,

   $E' = E- < X, Y >$

   if not EdgeNeeded($(V, E), X, Y$))

   $E = E'$

9: Return[OrientEdges(V,E)]

Algorithm 5. EdgeNeeded

Require: $< X, Y >$, $V=\{$all attributes$\}$, $E=\{$current edge list$\}$

1: $D(< X, Y >) = MinimumCutSet(< X, Y >, E)$

2: NewThread(MR_CMI($< X, Y >, D(< X, Y >)$)))

3: $I(X, Y | D(< X, Y >))$=Calculate_CMI($< X, Y >, D(< X, Y >)$)))

4: if $I(X, Y | D(< X, Y >)) < \epsilon$

5:    return (false).

6: else return (true).

4.2   Results Analysis of Parallel BN Learning Algorithm

All of our experiments were performed on a MapReduce equipped cluster with 20 machines, where each machine was configured to use 512MB of RAM and 1GB of hard drive space. Running time was measured as the total time between the start of BN learning algorithm and it exiting with the learned structure as output. All the running times have been averaged over multiple runs.

Fig. 5 shows the running time with centralized and distributed algorithms. When the dataset is small, the centralized algorithm needs less running time than distributed algorithm with MapReduce. It is because that MapReduce start up

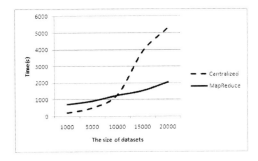

Fig. 5. Running time vs. data size for centralized and distributed algorithms

Table 2. Average performances on the whole feature set

Method	Accuracy	Recall	Features	Size Reduction
TBN	83.03%	54.35%	413	94.36%
OBN	81.86%	58.34%	7321	0
NB	77.77%	57.83%	7321	0
SVM	81.84%	62.62%	7321	0

and tear down costs are primary performance bottlenecks. Therefore, MapReduce does not suit the computation on small dataset. When the dataset grows, the excellent performance of MapReduce becomes explicit.

4.3 Results Analysis of Sentiment Classification

We compare the performance of our parallel BN classifier with naïve Bayes(NB) classifier and support vector machine (SVM) classifier along with a TF-IDF re-weighting of the vectors of word counts. Two BN classifiers are used. One is the original BN(OBN) classifier using 7321 features and Y. The other is the thin BN(TBN) classifier using the features selected from the OBN, including the set of parents of Y, the set of children of Y, the set of parents of parents of Y, the set of children of children of Y and the set of parents of children of Y. Here, we choose 413 features for TBN.

In order to compute unbiased estimates for accuracy and recall we used a ten-fold cross-validation scheme. Table 2 compares the OBN and TBN with the performances of the other classifiers using the whole feature set as input. As we expected, more features did not necessarily lead to better results, as the classifiers were not able to distinguish discriminating words from noise.

Table 3 compares the performance of the OBN and TBN with others classifiers using the same number of features selected by information gain(IG) and principal component analysis(PCA). We notice that feature selection using information gain and PCA criterion does not tell us how many features have to be selected, but rather allows us to rank the features from most to least discriminating instead.

Table 3. Average performances on the same number of features

Method	Accuracy	Recall	Features	Size Reduction
TBN	83.03%	54.35%	413	94.36%
NB+PCA	79.36%	52.35%	413	94.36%
NB+IG	78.61%	55.79%	413	94.36%
SVM+PCA	81.18%	56.84%	413	94.36%
SVM+IG	80.72%	53.14%	413	94.36%

5   Conclusion

In this paper, we present a parallel BN learning technique for learning predominant sentiments of on-line texts, available in unstructured format, that:

1. is able to capture dependencies among words and find a minimal vocabulary for categorization purposes, and
2. is able to handle the large scale unstructured web data.

We describe our experiences with developing and deploying a BN learning algorithm over large datasets using the MapReduce model of distributed computation. Then, we apply this parallel BN learning algorithm to capture the dependencies among words, and, at the same time, finds a vocabulary that is efficient for the purpose of extracting sentiments. Experiment results suggest that our parallel Bayesian network learning algorithm is capable of handling large real-world data sets. Moreover, for problems where the independence assumption is not appropriate, the BN is a better choice and leads to more robust predictions.

Acknowledgment

This work was supported by the National Science & Technology Pillar Program, No. 2009BAK63B08 and China Postdoctoral Science Foundation funded project.

References

1. Blitzer, J., Dredze, M., Pereira, F.: Biographies, bollywood, boom-boxes and blenders: Domain adaptation for sentiment classification. In: Proceedings of the 45th Annual Meeting of the Association of Computational Linguistics, Prague, Czech Republic, pp. 440–447 (2007)
2. Cheng, J., Greiner, R., Kelly, J., Bell, D.: A Learning Bayesian Networks from Data: An Information-Theory Based Approach. The Artificial Intelligence journal 137, 43–90 (2002)
3. Choi, Y., Cardie, C., Riloff, E., Patwardhan, S.: Identifying sources of opinions with conditional random fields and extraction patterns. In: Proceedings of Human Language Technology Conference and Conference on Empirical Methods in Natural Language Processing, Vancouver, British Columbia, Canada, pp. 355–362 (2005)

4. Chu, C.T., Kim, S.K., Lin, Y.A., Yu, Y., Bradski, G., Ng, A.Y., Olukotun, K.: Map-reduce for machine learning on multicore. In: Advances in Neural Information Processing Systems (NIPS 19), pp. 281–288 (2007)
5. Cohen, J.: Graph twiddling in a mapreduce world. Computing in Science and Engineering 11, 29–41 (2009)
6. Dean, J., Ghemawat, S.: MapReduce: Simplified data processing on large clusters. In: Symposium on Operating System Design and Implementation (OSDI), San Francisco, CA, pp. 137–150 (2004)
7. Eguchi, K., Lavrenko, V.: Sentiment retrieval using generative models. In: Proceedings of the 2006 Conference on Empirical Methods in Natural Language Processing, Sydney, Australia, pp. 345–354 (2006)
8. Kim, S.M., Hovy, E.: Determining the sentiment of opinions. In: Proceedings of the 20th international conference on Computational Linguistics, Morristown, NJ, USA, p. 1367 (2004)
9. MapReduce at Rackspace, `http://blog.racklabs.com/?p=66`
10. McDonald, R., Hannan, K., Neylon, T., Wells, M., Reynar, J.: Structured models for fine-to-coarse sentiment analysis. In: Proceedings of the 45th Annual Meeting of the Association of Computational Linguistics, Prague, Czech Republic, pp. 432–439 (2007)
11. McNabb, A.W., Monson, C.K., Seppi, K.D.: Parallel PSO Using MapReduce. In: Proceedings of the Congress on Evolutionary Computation (CEC 2007), pp. 7–14. IEEE Press, Singapore (2007)
12. Movie Review Data, `http://www.cs.cornell.edu/people/pabo/movie-review-data/`
13. Panda, B., Herbach, J.S., Basu, S., Bayardo, R.J.: PLANET: Massively Parallel Learning of Tree Ensembles with MapReduce. In: VLDB, France, pp. 1426–1437 (2009)
14. Pang, B., Lee, L., Vaithyanathan, S.: Thumbs up? Sentiment Classification using Machine Learning Techniques. In: Proceedings of the 2002 Conference on Empirical Methods in Natural Language Processing, pp. 79–86 (2002)
15. Pang, B., Lee, L.: A sentimental education: sentiment analysis using subjectivity summarization based on minimum cuts. In: Proceedings of the 42nd Annual Meeting on Association for Computational Linguistics, Morristown, NJ, USA, p. 271 (2004)
16. Turney, P.D, Littman, M.L.: Unsupervised learning of semantic orientation from a hundred-billion-word corpus. In: Proceedings of CoRR (2002)

Preface of the 2nd International Workshop on XML Data Management

Jiaheng Lu

Key Laboratory of Data Engineering and Knowledge Engineering (DEKE), MOE,
Renmin University of China

It is our great pleasure to welcome you to the 2nd International Workshop on XML Data Management (XMLDM 2011).

XML has gained lot of attention from database and Web researchers who are actively working in one or more of the emerging XML areas. XML data are self-describing, and provide a platform independent means to describe data and therefore, can transport data from one platform to another. XML documents can be mapped to one more of the existing data models such as relational and object-relational, and XML views can be produced dynamically from the pre-existing data models. XML queries can be mapped to the queries of the underlying models and can use their optimization features. XML data integration is useful for E-commerce applications such as comparison-shopping, which requires further study in the domain of data, schema and query based integration. XML change management is another important area that has attracted attention in the context of web warehouse. XML has been in use in upcoming areas such web services, sensor and biological data management. The second International Workshop on XML Data Management focuses on the convergence of database technology with XML technology, and brings together academics, practitioners, users and vendors to discuss the use and synergy between these technologies.

XMLDM attracted 8 submissions from Asia, Europe, and North America. The program committee accepted 4 full papers. These papers cover a variety of topics, including XML views and data mappings, XML query languages and optimization, XML applications in semantic web and so on. We hope that they will serve as a valuable start point for much brilliant thinking in XML data management.

Paper "OSD-DB: A Military Logistics Mobile Database" discusses how to transform and display XML data as HTML tables with multilevel headers, to preserve the original easy-to-read format while having a well defined schema, and describe how it can be used in concrete applications, focusing on a military logistics database. In addition, Rinfret et al. also show how to integrate the XML layer into a more complete application.

Paper "Querying and Reasoning with RDF(S)/OWL in Xquery" investigates how to use the XQuery language for querying and reasoning with RDF(S)/OWL-style ontologies. XQuery library for the Semantic Web and XQuery functions have been proposed by Almendros-Jimenez to handle RDF(S)/OWL triples and encode RDF(S)/OWL.

X. Du et al. (Eds.): APWeb 2011, LNCS 6612, pp. 436–437, 2011.

In paper "A Survey on XML Keyword Search", Tian et al. survey several representative papers with the purpose of extending the general knowledge of the research orientation of the XML keyword search.

In paper "Schema mapping with quality assurance for data integration1", the concept of quality assurance mechanisms was proposed by Bian et al. They discuss that a new model with qualityassurance, and provide a suitable method for this model, and then they propose the strategy of weak branch's convergence on the basis of Schema.

In this workshop, we are very glad to invite Dr. Bogdan Cautis to give a keynote talk. He reports in this talk on recent results on richer classes of XPath rewritings using views. He presents both theoretical and practical results on view-based rewriting using multiple XML views.

Making XMLDM 2011 possible has been a team effort. First of all, we would like to thank the authors and panelists for providing the content of the program. We would like to express our gratitude to the program committee and external reviewers, who worked very hard in reviewing papers and providing suggestions for their improvements. We would thank Steering Chairs Professor Xiaoyong Du and Professor Jianmin Wang for their helps and important instructions. In particular we extend our special thanks to Linlin Zhang for maintaining the XMLDM2011 web site and for his effort in organizing the workshop.

We hope that you will find this program interesting and thought-provoking and that the workshop will provide you with a valuable opportunity to share ideas with other researchers and practitioners from institutions around the world.

XPath Rewriting Using Views:
The More the Merrier

Bogdan Cautis

Telecom ParisTech, Paris, France
cautis@telecom-paristech.fr
http://www.telecom-paristech/~cautis

We report in this talk on recent results on richer classes of XPath rewritings using views.

The problem of equivalently rewriting queries using views is fundamental to several classical data management tasks. Examples include query optimization using a cache of materialized results of previous queries and database security, where a query is answered only if it has a rewriting using the pre-defined security views.

While the rewriting problem has been well studied for the relational data model, its XML counterpart is not yet equally well understood, even for basic XML query languages such as XPath, due to the novel challenges raised by the features of the XML data model.

We have recently witnessed an industrial trend towards enhancing XPath queries with the ability to expose node identifiers and exploit them using intersection of node sets (via identity-based equality). This development enables for the first time multiple-view rewritings obtained by intersecting several materialized view results. We present both theoretical and practical results on view-based rewriting using multiple views. First, we characterize the complexity of the intersection-aware rewriting problem. We then identify tight restrictions (which remain practically interesting) under which sound and complete rewriting can be performed efficiently, i.e. in polynomial time, and beyond which the problem becomes intractable. As an additional contribution, we analyze the complexity of the related problem of deciding if an XPath with intersection can be equivalently rewritten as one without intersection or union.

Then, going beyond the classic setting of answering queries using explicitly enumerated view definitions, we report on results on the problem of querying XML data sources that accept only a limited set of queries. This is motivated by Web data sources which, for reasons such as performance requirements, business model considerations and access restrictions, do not allow clients to ask arbitrary queries. They instead publish as Web Services a set of queries (views) they are willing to answer.

Querying such sources involves finding one or several legal views that can be used to answer the client query. Services can implement very large (potentially infinite) families of XPath queries, and in order to compactly specify such families of queries we adopt a formalism close to context-free grammars (in the same spirit in which a potentially infinite language is finitely specified by a grammar).

X. Du et al. (Eds.): APWeb 2011, LNCS 6612, pp. 438–439, 2011.

We say that query Q is *expressed* by a specification (program) $\mathcal{P}$ if it is equivalent to some expansion of it. Q is *supported* by $\mathcal{P}$ if it has an equivalent rewriting using some finite set of $\mathcal{P}$'s expansions. We present our results on the complexity of expressibility and support and identify large classes of XPath queries for which there are efficient (tractable) algorithms.

This survey mainly summarizes two recent papers, which are joint work Alin Deutsch, Nicola Onose and Vasilis Vassalos.

OSD-DB: A Military Logistics Mobile Database

Denis Rinfret[1,3], Cyrille Chênevert[2,3], and Frédéric Drolet[2,3]

[1] International University
Vietnam National University - Ho Chi Minh City
[2] CFB Kingston, Department of National Defense, Canada
[3] Work done while at the Royal Military College of Canada

Abstract. Government agencies and other organizations often have a huge quantity of data in need of digitization. In some cases, the data are represented on paper as tables with multilevel headers. Storing this data into relational tables probably necessitates a complex redesign of the data, through renaming of columns and/or normalization of the tables. To save this step, an XML data model can be used easily and naturally, but typical non-technical users still prefer to visualize data in the form of tables. We show in this paper how to transform and display XML data as HTML tables with multilevel headers, to preserve the original easy-to-read format while having a well defined schema, and describe how it can be used in concrete applications, focusing on a military logistics database.

Keywords: Usability, XML database, mobile database, government application, presentation model.

1 Introduction

Visualizing data in a tabular form is very convenient and natural for a wide range of people, including non-technical people, especially when multilevel headers are available. On the contrary, visualizing XML data as a tree is not natural and less compact than a table. Additionally, many people, especially non-technical people, find relational tables very limiting because of the necessity of having flat table headers. This is one of the reasons why many people still use spreadsheet software to manage their data [6]. Of course spreadsheets are not sufficient for any significant size projects, but their ease of use is attractive to many people. According to [6],

> We need database systems that reflect the user's model of the data, rather than forcing the data to fit a particular model.

We decided, in this work, to start from a presentation data model (tabular with multilevel headers) that is already working very well, and find a way to make it work well at lower levels.

Many organizations like government agencies still have a lot of data on paper, often represented in tables with multilevel headers. One example is the *Operational Staff Data (OSD)* binder used for logistics by the Department of National

X. Du et al. (Eds.): APWeb 2011, LNCS 6612, pp. 440–449, 2011.
© Springer-Verlag Berlin Heidelberg 2011

Laying Rates

ser	mine_type	by_hand				by_machine				remarks
		surface		buried		surface		buried		
		day	night	day	night	day	night	day	night	
A	20	15	5	3	123	50	40	120	100	Example data
B	25	19	7	5	234	60	50	150	120	Example data

Fig. 1. Mines Laying Rates Table

Armoured Fighting Vehicles

ser	eqpt	abvn	brevity_code	crew	armt	dimensions			wt		MLC	rge		max_speed		fuel		ford_depth	remarks
						l	w	h	empty	loaded		rd	cross_country	rd	cross_country	l2	type		
222	ADATS	ADATS	A12	4	None	8	4	5	3000	3500	11	500	300	40	30	450	O-105	00	N/A

Fig. 2. Example of an Equipment Table

Defence (DND) in Canada. The data in this binder are used for logistics purposes, before, during, and after deployment of military personnel and equipment. Tables are categorized by a wide range of topics, such as simply *Equipment* data (general purpose vehicles, armoured fighting vehicles, ...) with table attributes such as *empty* and *loaded weight*, moving *range* on the *road* or *cross-country*, *maximum speed* on the *road* and *cross-country*, *fuel type* and *capacity* (see Figure 2); and other categories such as "*Weapons*", "*Ammunition and Explosives*", "*Nuclear, Biological and Chemical Agent*" and many others. Another example table containing data about *mines laying rates* is shown in Figure 1. Note that the full schema and data are not shown due to lack of space and confidentiality of some parts of it.

Updating such binders is obviously time consuming and not practical. Some important considerations include not necessitating a database server, which demands more technical expertise to manage, and portability to devices such as Personal Digital Assistants (PDAs), surely influence design choices. Attempts have been made to use software such as MS Access, chosen by non-technical people because of their familiarity with it and its simplicity, have been unsatisfactory. It does satisfy the need to avoid managing a database server but is not well suited to the other needs of the project. Our solution is to use an embedded XML database, such as BerkeleyDB XML [2], as a basis and build a visualization layer on top to represent XML data as HTML tables with multilevel headers. Since the output of our visualization layer is HTML, the results can be viewed in any web browser. The tables can be dumped into files to be viewed statically on any system (including PDAs), or the results of XPath queries or other types of queries can be passed dynamically to the visualization layer to be viewed in a web browser. Since the data changes infrequently, there is no need to keep an on-line read-write version of the database. Updates are usually done in batches by a couple people, and distributed to other people whenever needed.

The rest of this paper in organized in the following way. After a brief overview of related work, common use cases for the system are covered next in Section 3. The XML layer, including the XML schema and style sheet to produce an HTML table from the XML data, are covered in Section 4. Section 5 shows how to integrate the XML layer into a more complete application, and the conclusion follows.

2 Related Work

There is quite a lot of interest in the problem of extracting information from web pages in general, and more particularly extracting data from HTML tables with (possibly) multilevel headers and inserting it into database tables [1, 3, 5, 7], but not the reverse as we do in this paper. We focus on presenting data in an easy-to-read and easy-to-understand way, while they concentrate on extracting information from already published tables (or other structures like lists) and storing it somewhere.

In [9], the authors work on making large HTML tables easily viewable on small screens. Although we don't cover this subject in this paper, being able to collapse columns and/or rows to make large tables readable on small screens would certainly be a useful feature in our project.

The first motivation for tackling this project the way we did is the work of Jagadish et al [6], titled *Making database systems usable*. We thought about the presentation data model first to improve the database usability.

3 Use Cases

An emphasis was put in this project on keeping the current display format (tables with multilevel headers, see Figures 1 and 2). This format works very well for current users of the data, and keeping it reduces the adaptation time to the new system. Certainly the familiarity and simplicity of the display format increases the system's usability. The main goal was to find a way to easily support this format and the following use cases.

As pointed out in the introduction, most users of the data-base are read-only users. Either they are planning some mission or training, or executing it, or debriefing it, using the data without needing to modify it. To make changes, they need to contact the people in charge (the administrators of the database, probably officers located in some logistics department), and get an updated version later on. But this is done infrequently. A static version of the database, demanding less resources, can be kept on a PDA, or smart phone, or vehicle's embedded information system, for easy access and use.

Another goal is to make the system simple enough to not require the administrators to be technical people: keep the simplicity of a spreadsheet but make it more powerful and appropriate for the current project. The full administrator version requires full access (Read/Write) to the schema and the data, and we developed a version for PCs, with search function, queries, ..., and another version for use behind a web server (same search function and queries).

Regular (non-administrator) users can use a read-only version of the full administrator version, a trimmed down version to run on mobile phone or internet tablets (such as the Nokia N800 [8]), or a static HTML version to run on systems with limited resources. Section 5 contains more details on the different versions, and discusses the implementation of the system.

4   XML Layer

The XML layer of the project consists of an XML Schema, many XML documents containing the actual data and an XML Style Sheet that is applied on the XML documents to transform them into HTML tables with multilevel headers.

4.1   XML Schema

The first step is to map the original tables to an XML Schema. Since many tables have the exact same attributes (like *General Purpose Vehicles* and *Armoured Fighting Vehicles* for example), we created XML Schema complex types and used some of them for many tables to avoid redundancy in the schema. Figure 3 shows the type definition for the *Mines Laying Rates* table (with the *by_machine* element removed to save space).

Then follows in the schema all the allowable XML document roots, one for each of the tables. This implies that each table has its own XML document. One big XML document containing everything would be too big and inconvenient, so we choose this approach. Therefore we have a large number of relatively small XML documents, which is not a problem since they are stored in an XML database, and queries and updates are usually easier or at least not harder when using many smaller XML documents compared to one very big document. Refer to Figure 4 and 5 for examples.

To store and manage these XML documents, we could use any XML database or any database system supporting XML (like a relational database with XML extensions), but we choose an embedded XML database, namely BerkeleyDB XML, to satisfy the other requirements of the project. We could even use plain XML text files, without a database, but this would come with a big penalty since we would need to parse the full documents each time we need to access the data and we would not be able to use indexes to help answer queries.

4.2   XSL Style Sheet

We decided to use XML style sheet (XSL) transformations to produce the multilevel header tables from the XML data. The main challenge was to figure out a way to find the correct values for the column span (*colspan*) and row span (*rowspan*) attributes of the HTML table *th* tags, for cells in the HTML table

```
<xs:complexType name="laying_rates_type">
    <xs:all>
        <xs:element name="ser" type="xs:int"/>
        <xs:element name="mine_type"
                    type="xs:string"/>
        <xs:element name="by_hand">
            <xs:complexType>
                <xs:all>
                    <xs:element name="surface">
                        <xs:complexType>
                          <xs:all>
                            <xs:element name="day"
                                type="xs:float"/>
                            <xs:element name="night"
                                type="xs:float"/>
                          </xs:all>
                        </xs:complexType>
                    </xs:element>
                    <xs:element name="buried">
                      <xs:complexType>
                        <xs:all>
                          <xs:element name="day"
                              type="xs:float"/>
                          <xs:element name="night"
                              type="xs:float"/>
                        </xs:all>
                      </xs:complexType>
                    </xs:element>
                </xs:all>
            </xs:complexType>
        </xs:element>
        <xs:element name="remarks"
                    type="xs:string"/>
    </xs:all>
</xs:complexType>
```

Fig. 3. XML schema complex type for the *Mines Laying Rates* table

```
<xs:element name="laying_rates_data">
    <xs:complexType>
        <xs:sequence minOccurs="1"
                     maxOccurs="unbounded">
            <xs:element name="data"
                        type="laying_rates_type"/>
        </xs:sequence>
    </xs:complexType>
</xs:element>
```

Fig. 4. XML schema element for the *Mines Laying Rates* table

```xml
<?xml version="1.0" encoding="UTF-8"?>
<laying_rates_data xmlns:xsi=
    "http://www.w3.org/2001/XMLSchema-instance"
    xsi:noNamespaceSchemaLocation=
        "file:/somewhere/OSD.xsd">
    <data>
        <mine_type>A</mine_type>
        <by_hand>
            <surface>
                <day>20</day>
                <night>15</night>
            </surface>
            <buried>
                <day>5</day>
                <night>3</night>
            </buried>
        </by_hand>
        <ser>123</ser>
        <by_machine>
            <surface>
                <day>50</day>
                <night>40</night>
            </surface>
            <buried>
                <day>120</day>
                <night>100</night>
            </buried>
        </by_machine>
        <remarks>Example data</remarks>
    </data>
</laying_rates_data>
```

Fig. 5. XML document for the *Mines Laying Rates* table

```xml
<xsl:template name="depth">
    <xsl:for-each select="/*/data//*">
        <xsl:sort select="count(ancestor::*)"
                  data-type="number"
                  order="descending"/>
        <xsl:if test="position() = 1">
            <xsl:value-of
                select="count(ancestor::*)+1"/>
        </xsl:if>
    </xsl:for-each>
</xsl:template>
```

Fig. 6. Height of the header template

$treeHeight \leftarrow height(root)$
for all nodes n under the root in breadth-first order **do**
    if n is the first node of its level **then**
        start a new table header row (tag `<tr>`)
    end if
    if n is a leaf **then**
        $rowspan \leftarrow treeHeight - depth(n) + 1$
        $colspan \leftarrow 1$
    else
        $rowspan \leftarrow 1$
        $colspan \leftarrow width(n)$
    end if
    create table cell (tag `<th>`) with $rowspan$ and $colspan$ and the node name as the display text
    if n is the last node of its level **then**
        close the table header row (tag `</tr>`)
    end if
end for

Fig. 7. Algorithm to produce an HTML table with a multilevel header

header. For example, *mine_type* in Figure 1 has a column span of 1 and a row span of 3, and *by_hand* has a column span of 4 and a row span of 1. We wrote an XSL that is generic enough to produce a correct HTML table for any of the tables in our XML schema, for any depth or width of the table header, even if the schema is modified later on, with only a very small modification of the XSL. The only difference between the table transformations is that the main XSL template must match the table name (for example *laying_rates_data*) to get started. In our implementation, we do some string interpolation to insert the table name at the correct place before the actual transformation takes place. The other place we do this is in an *h1* tag containing the table name in the output HTML. We do not include the full XSL here because it would require too much space, but excerpts are included and the algorithm is discussed next.

To be able to get the row span values right, we have to know the height of the tree making up the header. We also need to know the depth of each node as we process them. To get the depth of a node in XSL, we calculate how many ancestors it has. Figure 6 shows the template to compute the height of the header. What we do is sort all the nodes by their number of ancestors in descending order. Then we take the first node with the maximum depth and we return its depth. We need the *+1* because *ancestor::** does not include the root.

Similarly, to get the column span values right, we need to know the width of the tree below each node. So what we really need to count is the number of leaf nodes under the given node. If the node itself is a leaf, then the column span will be 1. If not, we have to recursively count the number of leaves under the node, by repeating the process for each child node of the given node.

The header tree nodes are processed in a breadth-first manner. All the children nodes of the root are processed first, then the children nodes of these nodes, etc...

Figure 7 shows the algorithm. When implementing this algorithm in an XML style sheet, the main difficulty is going through all the nodes in breadth-first order. The best way to do this is to start by matching all the nodes at a depth of 1, i.e. all the nodes under the root. Then match all the nodes one level deeper (depth of 2), then the nodes at depth 3, etc...

5   Implementation

All the programming, except at the XML layer, has been done with Python [4]. The first application developed was a script looping through all the tables, applying the XSL to the XML document containing the data, and dumping the results in HTML files. This *export-to-HTML* program is useful to produce a static HTML version of the database for use on systems with limited resources, or on systems where the other applications, covered next, are not available.

An application with GUI to view and modify the database was developed, in two versions: the first using Python with the GTK library (see Figure 8), aimed at the Linux desktop (but can be ported to MS Windows and Apple Mac OS environments) and the second version to run on the Nokia N800 internet tablet. The implementations are very similar, except that the N800 version doesn't have the possibility to add, delete or modify rows of the tables. There are two main reasons we did that: first, based on the uses cases of the system, PDA or other portable device users should not have the option to modify the data; second, resources are limited and running the full application was very slow.

One big problem on systems with limited resources is applying the XSL to the XML documents to produce the HTML tables, more precisely generating the table headers can be a slow process. Since the schema is unlikely to change often, it is crucial to cache the table headers to avoid generating them all the time. A big performance improvement can be obtained by doing this. On powerful systems, we may save only 0.05 seconds, or even less, depending which table is processed and which XSLT processor is used, but on slower systems it can easily be several seconds lost every time.

We also developed a Web interface to the database. We wrote a stand alone web server with the same functionality as the desktop version. We used the CherryPy Python package [10] to help in this task. The idea was to show yet another way to interact with the database, to show that many interfaces to the same database are possible. We certainly could have used some other technology to do this, for example use any full-fledged web server and run the application behind it, using any programming language. Our web application has two built-in modes, a read-only cached version, and a version with database modification power, aimed at administrators.

Since the core of the system is at the XML layer, implementing a full application is mostly a matter of building a GUI or web interface, except for the caching of table headers as discussed above. If at a later time a new interface is needed, no problem since the data is completely separate from the interface. This will probably be obvious to most computer scientists or engineers, but apparently it

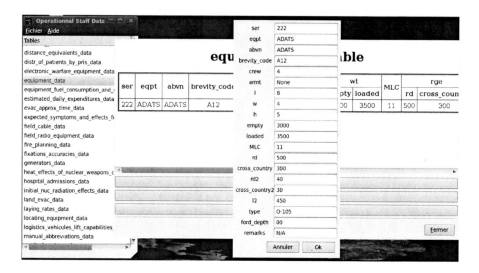

Fig. 8. Application to view and modify the military logistics database

is not so obvious in the military and government world. Many projects involve contractors who put the data into binary files in some undocumented proprietary format, and the next contractor pretty much has to reverse engineer the data files to figure out what's going on before doing anything else. This problem cannot happen in our project.

6   Conclusion

We developed our military logistics database from a usability-first point of view. We kept the paper-based presentation, which is still working very well, and found a good way to support it by using XML and XSL style sheets. Many interfaces are possible and we implemented a few, and we reached the project goals of having a version working on a portable device. For the future, we need to perform a larger field study to get more feedback from users to improve the implementation of the system.

References

[1] Cafarella, M.J., Halevy, A., Khoussainova, N.: Data integration for the relational web. Proc. VLDB Endow. 2, 1090–1101 (2009), ISSN: 2150-8097

[2] Oracle Corporation. Oracle Berkeley DB XML, http://www.oracl.com/us/products/database/berkeley-db/index-066571.html

[3] Crestan, E., Pantel, P.: Web-scale knowledge extraction from semi-structured tables. In: Proceedings of the 19th International Conference on World Wide Web, WWW 2010, Raleigh, North Carolina, USA, pp. 1081–1082 (2010), ISBN: 978-1-60558-799-8

[4] Python Software Foundation. Python Programming Language,
    http://www.python.org

[5] Gatterbauer, W., et al.: Towards domain-independent information extraction from web tables. In: Proceedings of the 16th International Conference on World Wide Web, WWW 2007, Bank, Alberta, Canada, pp. 71–80 (2007), ISBN: 978-1-59593-654-7

[6] Jagadish, H.V., et al.: Making database systems usable. In: Proceedings of the 2007 ACM SIGMOD International Conference on Management of Data, SIGMOD 2007, Beijing, China, pp. 13–24 (2007), ISBN: 978-1-59593-686-8

[7] Lim, S.-J., Ng, Y.-K.: An automated approach for retrieving hierarchical data from html tables. In: CIKM 1999: Proceedings of the Eighth International Conference on Information and knowledge Management, Kansas City, Missouri, United States, pp. 466–474 (1999), ISBN: 1-58113-146-1

[8] Nokia. Nokia N800 Internet Tablet, http://www.nokia-asia.com/find-products/products/nokia-n800-internet-tablet

[9] Tajima, K., Ohnishi, K.: Browsing large html tables on small screens. In: UIST 2008: Proceedings of the 21st annual ACM symposium on User Interface software and Technology, pp. 259–268. ACM, New York (2008), ISBN: 978-1-59593-975-3

[10] CherryPy Team. CherryPy Python Package, http://www.cherrypy.org

Querying and Reasoning
with RDF(S)/OWL in XQuery*

Jesús M. Almendros-Jiménez

Dpto. de Lenguajes y Computación,
Universidad de Almería, Spain
jalmen@ual.es

Abstract. In this paper we investigate how to use the XQuery language for querying and reasoning with RDF(S)/OWL-style ontologies. Our proposal allows the handling of RDF(S)/OWL triples by means of a XQuery library for the Semantic Web, and it encodes RDF(S)/OWL reasoning by means of XQuery functions. We have tested and implemented the approach.

1 Introduction

The *Semantic Web* framework [9,14] proposes that *Web data* represented by *HMTL* and *XML* have to be enriched by means of *meta-data*, in which modeling is mainly achieved by means of the *Resource Description Framework (RDF)* [19] and the *Web Ontology Language (OWL)* [20]. OWL is based on the so-called *Description Logic (DL)* [6], which is a family of logics (i.e. *fragments*) with different expressive power. On the other hand, *XQuery* [11] is a typed functional language devoted to express queries against XML documents. It contains *XPath 2.0* [8] as a sublanguage. *XPath 2.0* supports navigation, selection and extraction of fragments from XML documents. *XQuery* also includes *flowr* expressions (i.e. *for-let-orderby-where-return* expressions) to construct new XML values and to join multiple documents.

In this paper we investigate how to use the XQuery language for querying and reasoning with RDF(S)/OWL-style ontologies. The main features of our proposal can be summarized as follows:

- XQuery has been developed for querying XML documents, however, Web data can be also represented by means of RDF(S) and OWL. Therefore, XQuery should support the simultaneous querying of Web data by both its structure and by its associated meta-data given in form of RDF(S)/OWL-style ontologies. This is an important problem for supporting data discovery tasks against collections of Web data. The proposed approach allows to query XML/RDF(S)/OWL documents and to obtain as output the same kind of documents.

---

* This work has been partially supported by the Spanish MICINN under grant TIN2008-06622-C03-03.

X. Du et al. (Eds.): APWeb 2011, LNCS 6612, pp. 450–459, 2011.

- RDF(S) and OWL querying should be combined with reasoning. RDF(S) and OWL allows to express and infer complex relationships between entities which should be exploited by means of a query language. The proposed approach is able to use semantic information inferred from RDF(S) and OWL resources.
- Finally, we will propose an implementation of the approach. Firstly, RDF(S)-/OWL triples can be handled by means of a XQuery library providing support to the traversal of triples, and the access to triples components. Secondly, RDF(S)/OWL reasoning is *encoded* by means of XQuery functions which facilitates the use of the ontology in queries. In addition, such an encoding allows to reason with the elements of the ontology.

A great effort has been made for defining query languages for RDF(S)/OWL documents (see [7] for a survey about this topic). The proposals mainly fall on extensions of *SQL*-style syntax for handling the *triple-based RDF structure*. In this line the most representative language is *SPARQL* [21]. Some authors [10,15,1] have already investigated how to combine XML and RDF(S)/OWL querying. RDF(S)/OWL can be combined with XML query languages, for instance, by encoding RDF(S)/OWL into XML data and by encoding SPARQL in XQuery. Alternatively, in [13], SPARQL and XQuery are combined in order to provide an unified framework for RDF and XML. We have considered in our approach a different point of view. XQuery can be equipped with a library for the handling of RDF(S)/OWL. And also, the ontologies can be encoded by means of XQuery functions. The advantage of our approach is that XQuery functions are able to reason about the ontology, that is, they are able to infer new information from the **TBox** and **ABox**.

OWL/DL reasoning is a topic of research of increasing interest in the literature. Most of DL reasoners (for instance, *Racer* [16], *FaCT++* [23], *Pellet* [22]) are based on tableaux based decision procedures. We can distinguish two main reasoning mechanism in this context. The first mechanism focuses on checking *consistence* of the ontology. Consistence means that the ontology has a model. The second mechanism focuses on retrieving *logic consequences* from the ontology. Such logic consequence includes the transitive closure of the subclass and subproperty relationships, the membership of individuals to classes, and the relations between individuals, which are not explicitly declared in the ontology.

In our proposal, we have studied a fragment of Description Logic/OWL which can be easily encoded in XQuery. Such an encoding takes Description Logic formulas and encodes them by means of XQuery functions. XQuery is based on the use of *flowr* expressions for querying XML documents. In our case, we use *flowr* expressions to query OWL documents, and it allows to encode OWL ontologies by means of XQuery. Such an encoding allows to *reason* about the elements of the **ABox** of the ontology: we can infer the *membership of individual to classes* and the *roles that two individuals play*. In addition, we are interested in the use of XQuery for querying the meta-data of the **TBox**. In this case, the XQuery library can be used for *retrieving the elements of the* **TBox**.

OWL is a language with different fragments restricted in expressive power (i.e. they consider different subsets of the OWL vocabulary and define restrictions in the use of some of them) in order to retain reasoning capabilities (tractability, scalability, complexity, etc) and decidability. Recently, three fragments of the new OWL 2 have been proposed: OWL EL, OWL RL and OWL QL [20]. In particular, OWL QL is a fragment OWL 2 allowing efficient querying. With respect to the fragment of DL considered in our proposal, we have taken a fragment which is expressive enough that includes the basic elements of OWL. It is a subset of the $\mathcal{ALC}$ fragment with some restrictions incorporating some relations between properties. One key point of our fragment is that the **TBox** has to be acyclic. This condition is required in order to be supported by XQuery. However, we will discuss in the paper the adaptation of our proposal to the recently proposed OWL QL.

We would like to remark that our work continues previous work about XQuery. In [2], we have studied how to define an extension of XQuery for querying RDF(S)/OWL documents. Similarly to the current proposal, such an extension allows to query XML and RDF(S)/OWL resources with XQuery, however, the proposed extension of the quoted papers is based on the implementation of XQuery in a logic language like Prolog (see [5,4,3] for more details).

Finally, we have tested and implemented the proposal of this paper. The XQuery library for handling RDF(S)/OWL together with an example of ontology encoded in XQuery can be downloaded from `http://indalog.ual.es/ XQuerySemanticWeb`.

The structure of the paper is as follows. Section 2 will present the kind of ontologies we consider in our framework. Section 3 will describe the XQuery library for RDF(S)/OWL. Section 4 will show the encoding of ontologies in XQuery. Section 5 will give some examples of ontology querying and finally, Section 6 will conclude and present future work.

2   Ontology Representation

In this section we will define the fragment of OWL of our framework.

Definition 1. *An ontology $\mathcal{O}$ in our framework contains a* **TBox** *including a sequence of definitions of the form:*

$C \sqsubseteq D$ |	(rdfs:subClassof)	$P \equiv Q^-$ |	(owl:inverseOf)
$E \equiv F$ |	(owl:equivalentClass)	$P \equiv P^-$ |	(owl:SymmetricProperty)
$P \sqsubseteq Q$ |	(rdfs:subPropertyOf)	$\top \sqsubseteq \forall P^-.D$ |	(rdfs:domain)
$P \equiv Q$ |	(owl:equivalentProperty)	$\top \sqsubseteq \forall P.D$ |	(rdfs:range)

where C is a class (i.e. concept) description of left-hand side type (denoted by $C \in \mathcal{L}$), of the form: $C ::= C_0 \mid \neg C \mid C_1 \sqcup C_2 \mid C_1 \sqcap C_2 \mid \exists P.C \mid \forall P.C$ and D is a class (i.e. concept) description of right-hand side type (denoted by $D \in \mathcal{R}$), of the form: $D ::= C_0 \mid D_1 \sqcap D_2 \mid \forall P.D$. In all previous cases, C_0 is an atomic class (i.e. class name) and P, Q are property (i.e. role) names. Formulas of equivalence $E \equiv F$ are syntactic sugar for $E \sqsubseteq F$ and $F \sqsubseteq E$. In addition, the **ABox** *contains a sequence of definitions of the form: $P(a,b) \mid C_0(a)$ where P is*

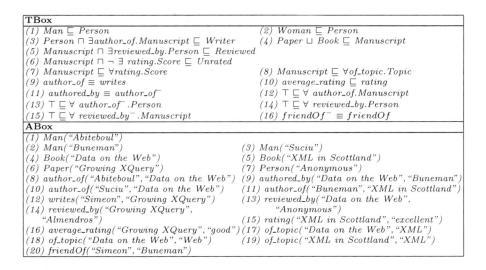

Fig. 1. An Example of Ontology

a property name, C_0 is a class name, and a, b are individual names. *We require that the* **TBox** *does not include concepts and roles that depend themself, that is, the* **TBox** *is acyclic.*

Our fragment of OWL is equipped with the basic elements of OWL: equivalence and inclusion of classes, and equivalence and inclusion of properties. In addition, it allows the specification of inverse and symmetric properties, and the domain and range of properties. We have restricted our approach to *object properties*, it forces to consider properties on *datatypes* as object properties. Let us remark that in right hand-sides of inclusion relations is not allowed to specify either $\neg C$ or $D_1 \sqcup D_2$ or $\exists P.C$. The reason for that is the encoding in XQuery of the ontology. Basically, the encoding allows to query the ontology and to use inclusion and equivalence for reasoning. Finally, we require that the **TBox** is acyclic. The fragment is basically $\mathcal{ALC}$ with restrictions in right hand sides of class axioms, adding some property relations. Let us see now an example of a such a DL ontology (see Figure 1).

The ontology of Figure 1 describes, in the **TBox**, meta-data in which the elements of *Man* and the elements of *Woman* are elements of *Person* (axiom (1) and (2)); and the elements of *Paper* and *Book* are elements of *Manuscript* (axiom (4)). In addition, a *Writer* is a *Person* who is the *author_of* a *Manuscript* (axiom (3)), and the class *Reviewed* contains the elements of *Manuscript* reviewed_by a *Person* (axiom (5)). Moreover, the *Unrated* class contains the elements of *Manuscript* which have not been rated (axiom (6)). The classes *Score* and *Topic* contain, respectively, the values of the properties *rating* and *of_topic* associated to *Manuscript* (axioms (7) and (8)). The property *average_rating* is a subproperty of *rating* (axiom (10)). The property *writes* is equivalent to *author_of*

(axiom (9)), and *authored_by* is the inverse property of *author_of* (axiom (11)).
Finally, the property *author_of*, and conversely, *reviewed_by*, has as domain a *Person* and as range a *Manuscript* (axioms (12)-(15)). Axiom (16) defines *friend_of* as a symmetric relation. The **ABox** describes data about two elements of *Book*: "Data on the Web" and "XML in Scottland" and a *Paper*: "Growing XQuery". It describes the *author_of* and *authored_by* relationships for the elements of *Book* and the *writes* relation for the elements of *Paper*. In addition, the elements of *Book* and *Paper* have been reviewed and rated, and they are described by means of a topic.

3 A Semantic Web Library in XQuery

Now, we will describe the Semantic Web library defined in XQuery. In summary, we have to handle the triples of the **TBox** and **ABox** of the given ontology. For simplicity we will assume that the elements of the **ABox** and the **TBox** has been declared in separated *RDF* items, *one for each formula*. Firstly, in order to handle the **ABox**, we have defined the following XQuery function:

```
declare function sw:abox($doc as node()) as node()*{$doc//owl:Thing[@rdf:about]};
```

which returns the sequence (i.e. by means of the XQuery "node()*" type) of triples of the **ABox** included in the document $doc. In order to access to the components of each triple of the **ABox** we have defined XQuery functions sw:subject, sw:object and sw:property. For instance,

```
declare function sw:subject($triple as node()) as xs:string{ string($triple/@rdf:about)};
```

Now, the **TBox** can be also handled by means of the library. For instance, we can obtain the classes in the **TBox** which are equivalent to a given class as follows:

```
declare function sw:equivalentClass($doc as node(),$class as node()) as node()*{
(for $x in $doc//owl:Class[owl:equivalentClass/@rdf:resource]
        where sw:eq($x/@rdf:about,$class/@rdf:about)
return sw:toClass($x/owl:equivalentClass/@rdf:resource)) union
(for $x in $doc//owl:Class[owl:equivalentClass/@rdf:resource]
        where sw:eq($x/owl:equivalentClass/@rdf:resource,$class/@rdf:about)
return sw:toClass($x/@rdf:about)) union
(for $x in $doc//owl:Class[owl:equivalentClass/owl:intersectionOf]
        where sw:eq($x/@rdf:about,$class/@rdf:about)
return $x/owl:equivalentClass/owl:intersectionOf ) union
(for $x in $doc//owl:Class[owl:equivalentClass/owl:Restriction]
        where sw:eq($x/@rdf:about,$class/@rdf:about)
return $x/owl:equivalentClass/owl:Restriction)
};
```

Let us remark that a class E can be equivalent by means $E \equiv F$, $E \equiv D_1 \sqcap D_2$ and $E \equiv \forall P.D$. The library handles the equivalence formula $E \equiv F$ as not oriented, and then returns E as equivalent to F, and F as equivalent to E. The previous function sw:toClass represents a class name in OWL format. The previous sw:eq is a boolean function to check equality between *RDF* identifiers.

Similarly, we can get the subclasses or superclasses of a given class in the **TBox**, by means of the functions sw:subClassOf and sw:superClassOf, respectively. In the library, there are also XQuery functions for querying about properties. For instance, the following function returns the domain of a property:

```
declare function sw:domain($doc as node(),$property as node()) as node()*{
for $x in $doc//owl:ObjectProperty where sw:eq($property/@rdf:about,$x/@rdf:about)
return sw:toClass($x/rdfs:domain/@rdf:resource)
};
```

There are also functions sw:inverseOf, sw:equivalentProperty, sw:sub-PropertyOf, sw:superPropertyOf and sw:range, and a function sw:Symmetric-Property to retrieve the symmetric properties. Finally, we can query the elements of the ontology like classes, properties and individuals. For instance, the following function returns the classes of the **TBox**:

```
declare function sw:Classes($doc as node()) as node()*{
for $x in $doc//owl:Class[@rdf:about] return sw:toClass($x/@rdf:about)};
```

In the library, there are also functions sw:Properties, sw:Individuals, sw:Unions, sw:Intersections, sw:ClassesofIntersections, sw:Classesof-Unions, sw:Restrictions (i.e. formulas built from $\forall$ and $\exists$). Also, there are also functions sw:oftype, sw:type, sw:ofrole and sw:role, retrieving the individuals of a certain type, the types of an individual, and the same for roles. Finally, given a property filler we can query the related individuals by means of two functions: sw:About and sw:Resource.

Now, using the Semantic Web library we can write queries for retrieving the elements of the ontology, expressing the result in XML format. For instance, we can get the atomic superclasses of a each class:

```
<classes>{
for $x in sw:Classes(doc('example.owl'))
return <class> string($x/@rdf:about) </class>
union
<superclasses>{
for $y in sw:subClassOf(doc('example.owl'),
sw:toClass(string($x/@rdf:about)) where $x/@rdf:about
return string($y/@rdf:about)
} </superclasses>
} </classes>
```

obtaining w.r.t. the running example the XML document:

```
<classes>
       <class>#Man</class>
           <superclasses>#Person</superclasses>
       <class>#Woman</class>
           <superclasses>#Person</superclasses>
....
</classes>
```

In summary, the proposed Semantic library of XQuery allows to query the elements of the **TBox** and the **ABox**. However, for reasoning with the **TBox** and the **ABox**, we have to encode the **TBox** by means of XQuery functions.

4   Encoding of Ontologies by Means of XQuery

Now, we present an example of encoding of an ontology by means of XQuery. The encoding allows to obtain the logic consequences from the **TBox** about the elements **ABox**. For instance, the class *Writer* of the running example can be defined in XQuery as follows:

```
declare function ex:writer($doc as node()) as node()*{
sw:oftype($doc,"Writer") union(
for $x in ex:author_of($doc) where
        (some $z in ex:person($doc) satisfies
        sw:eq(sw:About($z),sw:About($x)))
        and (some $u in ex:manuscript($doc) satisfies
        sw:eq(sw:About($u),sw:Resource($x)))
return sw:toIndividual(sw:About($x)))
};
```

The previous encoding defines the class *Writer* as the union of: the elements of the **ABox** of type *Writer*, and the "*Person*'s which are *author_of* a *Manuscript*", following the **TBox** formula $Person \sqcap \exists author\_of.Manuscript \sqsubseteq Writer$. The elements of the **ABox** of type *Writer* are retrieved by means of a call to the function `sw:oftype` of the Semantic library. The elements of the class *Person* are computed by means of the call to the function `ex:person`, which is defined similarly to `ex:writer`. The same can be said for `ex:author_of` and `ex:Manuscript`. The intersection is encoded by means of the `some-satisfies` construction of XQuery, and the same can be said for the existential quantifier. By calling `ex:writer(doc('example.owl'))`, we would obtain:

```
<owl:Thing rdf:about="#Abiteboul"/>  <owl:Thing rdf:about="#Buneman"/>
<owl:Thing rdf:about="#Suciu"/>      <owl:Thing rdf:about="#Simeon"/>
```

Basically, the idea of the encoding is as follows. Each class C and property P of the ontology defines a function called C and P, respectively. In the running example we would have the functions: `ex:person`, `ex:man`, etc. in the case of classes, and `ex:author_of`, `ex:authored_by`, etc. in the case of properties. Now, the functions for classes are defined as the union of the elements of the **ABox** of the given class, and the elements of subclasses of the given class in the **TBox**. In the case of properties, the functions are defined as the union of: the elements of the **ABox** full-filling the property, and the elements of subproperties of the given property in the **TBox**. In the running example, *author_of* property is encoded as follows:

```
declare function ex:author_of($doc as node()) as node()*{
(sw:ofrole($doc,"author_of")) union
(for $y in sw:ofrole($doc,"writes")
return sw:toFiller(sw:About($y),"author_of",sw:Resource($y))) union
(for $z in sw:ofrole($doc,"authored_by")
return sw:toFiller(sw:Resource($z),"author_of",sw:About($z)))
};
```

Now, by calling `ex:author_of(doc('example.owl'))`, we would obtain:

```
<owl:Thing rdf:about="#Abiteboul">
<author_of rdf:resource="#DataontheWeb"/>
</owl:Thing>
<owl:Thing rdf:about="#Suciu">
<author_of rdf:resource="#DataontheWeb"/>
</owl:Thing>
<owl:Thing rdf:about="#Buneman">
<author_of rdf:resource="#XMLinScottland"/>
</owl:Thing>
```

```
<owl:Thing rdf:about="#Simeon">
<author_of rdfs:resource="#GrowingXQuery"/>
</owl:Thing>
<owl:Thing rdf:about="#Buneman">
<author_of rdfs:resource="#DataontheWeb"/>
</owl:Thing>
```

Let us remark that in both examples, the encoding has used the ontology for reasoning about class assertions and property fillers. In the first case *Abiteboul*, *Suciu*, *Buneman* and *Simeon* are members of the class *Writer* because they are obtained from the **ABox**, and from the **TBox** by means of the reasoning with the formulas $author\_of \equiv writes$ and $authored\_by^- \equiv author\_of$ (i.e. `sw:author_of` returns also elements related by means of *authored_by* and *writes* relationships). The elements of the classes *Paper* and *Book* are elements of *Manuscript* by $Paper \sqcup Book \sqsubseteq Manuscript$.

5   Querying an Ontology by Means of XQuery

Now, we will show some examples of querying and reasoning w.r.t. the running example. Such examples have been tested and can be downloaded from http://indalog.ual.es/XQuerySemanticWeb.

The first example retrieves the *"Authors of a Manuscript"*. Let us remark that the relation *"author_of"* is equivalent a *"writes"* and the inverse of *"authored_by"*. In addition, the class *"Manuscript"* includes the elements of the classes *"Paper* and *"Book"*. The query can be expressed in our proposal as follows:

```
<manuscripts>{
for $x in ex:author_of(doc('example.owl')) return
<item> <author> { sw:subject($x) } </author>
<manuscript> { sw:object($x) } </manuscript> </item>
}</manuscripts>
```

obtaining the answer in XML format as follows:

```
<manuscripts>
   <item> <author>#Abiteboul</author> <manuscript>#DataontheWeb</manuscript> </item>
   <item> <author>#Suciu</author> <manuscript>#DataontheWeb</manuscript></item>
   <item> <author>#Buneman</author> <manuscript>#XMLinScottland</manuscript> </item>
   <item> <author>#Simeon</author> <manuscript>#GrowingXQuery</manuscript> </item>
   <item> <author>#Buneman</author> <manuscript>#DataontheWeb</manuscript> </item>
</manuscripts>
```

The second example retrieves the *"Authors of reviewed manuscripts"* in which the class *Reviewed* includes the manuscripts for which a *"reviewed_by"* relation exists. The query can be expressed in our framework as follows:

```
<manuscripts>{
for $y in ex:author_of(doc('example.owl'))
       where some $x in ex:reviewed(doc('example.owl')) satisfies
       sw:eq(sw:subject($x),sw:object($y))
return <item> <author> { sw:subject($y) } </author>
<manuscript> { sw:object($y) } </manuscript> </item>
}</manuscripts>
```

obtaining as answer:

```
<manuscripts>
    <item><author>#Abiteboul</author> <manuscript>#DataontheWeb</manuscript> </item>
    <item><author>#Suciu</author> <manuscript>#DataontheWeb</manuscript></item>
    <item> <author>#Simeon</author> <manuscript>#GrowingXQuery</manuscript> </item>
    <item> <author>#Buneman</author> <manuscript>#DataontheWeb</manuscript> </item>
</manuscripts>
```

Let us remark that *"XML in Scottland"* is not a reviewed manuscript.

6   Conclusions and Future Work

In this paper we have investigated how to use the XQuery language for querying and reasoning with RDF(S)/OWL-style ontologies. As future work we would like to study how to adapt our proposal to other ontology languages, in particular, how to extend to the recently proposed OWL QL, the W3C proposal for querying OWL ontologies. We have to solve some troubles to cover with OWL QL. Firstly, OWL QL is based on a different fragment of OWL. It is, for instance, restricted to existential quantification of properties, but on the other side it assumes cyclic ontologies. Moreover, it is focused on conjunctive queries. In order to handle conjunctive queries in OWL QL a rewriting query algorithm has been proposed in [12]. In summary, the adaptation of OWL QL to our framework is not a trivial task. Firstly, we have to implement a fix point operator in XQuery in order to cover with cyclic ontologies. In addition, OWL QL assumes the use of conjunctive queries and query rewriting. In this case, we need a formalism for expressing conjunctive queries in XQuery and a mechanism for rewriting. We are investigating how to use SWRL [17] and RIF [18] for expression conjunctive queries and the XQuery itself for query rewriting.

References

1. Akhtar, W., Kopecký, J., Krennwallner, T., Polleres, A.: XSPARQL: Traveling between the XML and RDF worlds – and avoiding the XSLT pilgrimage. In: Bechhofer, S., Hauswirth, M., Hoffmann, J., Koubarakis, M. (eds.) ESWC 2008. LNCS, vol. 5021, pp. 432–447. Springer, Heidelberg (2008)
2. Almendros-Jiménez, J.M.: An RDF Query Language based on Logic Programming. Electronic Notes in Theoretical Computer Science 200(3) (2008)
3. Almendros-Jiménez, J.M.: An Encoding of XQuery in Prolog. In: Bellahsène, Z., Hunt, E., Rys, M., Unland, R. (eds.) XSym 2009. LNCS, vol. 5679, pp. 145–155. Springer, Heidelberg (2009)
4. Almendros-Jiménez, J.M., Becerra-Terón, A., Enciso-Baños, F.J.: Integrating XQuery and Logic Programming. In: Seipel, D., Hanus, M., Wolf, A. (eds.) INAP-WLP 2007. LNCS, vol. 5437, pp. 117–135. Springer, Heidelberg (2009)
5. Almendros-Jiménez, J.M., Becerra-Terón, A., Enciso-Baños, F.J.: Querying XML documents in logic programming. TPLP 8(3), 323–361 (2008)
6. Baader, F., Calvanese, D., McGuinness, D.L., Patel-Schneider, P., Nardi, D.: The Description Logic Handbook: Theory, Implementation, and Applications. Cambridge Univ. Press, Cambridge (2003)

7. Bailey, J., Bry, F., Furche, T., Schaffert, S.: Web and Semantic Web Query Languages: A Survey. In: Eisinger, N., Małuszyński, J. (eds.) Reasoning Web. LNCS, vol. 3564, pp. 35–133. Springer, Heidelberg (2005)

8. Berglund, A., Boag, S., Chamberlin, D., Fernandez, M.F., Kay, M., Robie, J., Siméon, J.: XML path language (XPath) 2.0. In: W3C (2007)

9. Berners-Lee, T., Hendler, J., Lassila, O., et al.: The Semantic Web. Scientific american 284(5), 28–37 (2001)

10. Bikakis, N., Gioldasis, N., Tsinaraki, C., Christodoulakis, S.: Semantic based access over XML data. In: Lytras, M.D., Damiani, E., Carroll, J.M., Tennyson, R.D., Avison, D., Naeve, A., Dale, A., Lefrere, P., Tan, F., Sipior, J., Vossen, G. (eds.) WSKS 2009. LNCS, vol. 5736, pp. 259–267. Springer, Heidelberg (2009)

11. Boag, S., Chamberlin, D., Fernández, M.F., Florescu, D., Robie, J., Siméon, J., Stefanescu, M.: XQuery 1.0: An XML query language. In: W3C (2004)

12. Calvanese, D., De Giacomo, G., Lembo, D., Lenzerini, M., Rosati, R.: Tractable reasoning and efficient query answering in description logics: The DL-Lite family. Journal of automated reasoning 39(3), 385–429 (2007)

13. Droop, M., Flarer, M., Groppe, J., Groppe, S., Linnemann, V., Pinggera, J., Santner, F., Schier, M., Schöpf, F., Staffler, H., et al.: Embedding XPATH Queries into SPARQL Queries. In: Proc. of the 10th International Conference on Enterprise Information Systems (2008)

14. Eiter, T., Ianni, G., Krennwallner, T., Polleres, A.: Rules and ontologies for the semantic web. In: Baroglio, C., Bonatti, P.A., Małuszyński, J., Marchiori, M., Polleres, A., Schaffert, S. (eds.) Reasoning Web. LNCS, vol. 5224, pp. 1–53. Springer, Heidelberg (2008)

15. Groppe, S., Groppe, J., Linnemann, V., Kukulenz, D., Hoeller, N., Reinke, C.: Embedding SPARQL into XQUERY/XSLT. In: Proceedings of the 2008 ACM symposium on Applied computing, SAC 2008, pp. 2271–2278. ACM, New York (2008)

16. Haarslev, V., Möller, R., Wandelt, S.: The revival of structural subsumption in tableau-based description logic reasoners. In: Proceedings of the 2008 International Workshop on Description Logics (DL 2008), CEUR-WS, pp. 701–706 (2008)

17. Horrocks, I., Patel-Schneider, P.F., Boley, H., Tabet, S., Grosof, B., Dean, M.: SWRL: A Semantic Web Rule Language Combining OWL and RuleML. W3C Member Submission (May 21, 2004), http://www.w3.org/Submission/SWRL/

18. Kifer, M.: Rule interchange format: The framework. In: Calvanese, D., Lausen, G. (eds.) RR 2008. LNCS, vol. 5341, pp. 1–11. Springer, Heidelberg (2008)

19. Klyne, G., Carroll, J.J.: Resource Description Framework (RDF): Concepts and Abstract Syntax. Technical report (2004), http://www.w3.org/TR/2004/REC-rdf-concepts-20040210/

20. Motik, B., Patel-Schneider, P.F., Parsia, B., Bock, C., Fokoue, A., Haase, P., Hoekstra, R., Horrocks, I., Ruttenberg, A., Sattler, U., et al.: OWL 2 web ontology language: Structural specification and functional-style syntax. In: W3C (2008)

21. PrudHommeaux, E., Seaborne, A., et al.: SPARQL query language for RDF. In: W3C (2008)

22. Sirin, E., Parsia, B., Grau, B.C., Kalyanpur, A., Katz, Y.: Pellet: A practical OWL-DL reasoner. Web Semantics: Science, Services and Agents on the World Wide Web 5(2), 51–53 (2007)

23. Tsarkov, D., Horrocks, I.: FaCT++ Description Logic Reasoner: System Description. In: Furbach, U., Shankar, N. (eds.) IJCAR 2006. LNCS (LNAI), vol. 4130, pp. 292–297. Springer, Heidelberg (2006)

A Survey on XML Keyword Search

Zongqi Tian, Jiaheng Lu, and Deying Li

School of Information, Renmin University of China, Beijing 100872, China
{tzqruc,jiahenglu}@gmail.com, {deyingli}@ruc.edu.cn

Abstract. Keyword search querying has emerged as one of the most effective paradigms for information discovery, especially over HTML documents in the World Wide Web and much work has been done in this domain. Specifically, With the increasing application of XML in web, the study on XML keyword search has been gaining growing attention of the researchers. Great efforts have also been made in recent years to facilitate XML keyword search to be as convenient as the text document search of web so that the users could conduct query simply by inputting the intended keywords without the necessity of gaining much knowledge about XML schema or XQuery. Despite all the advantages, this query method will certainly give rise to the inaccuracy of the query result, and given that approaches have been proposed by the researchers. This paper survey several representative papers with the purpose of extending the general knowledge of the research orientation of the XML keyword search.

1 Introduction

With the development of information science and information society, an increasing number of information on the internet is stored with XML format. Given the fact that XML is becoming the standard in exchanging and representing data, how to extract accurate information in an effective and efficient way from XML documents has become an important issue arousing much interest of researchers.

A query language for XML, such as XQuery, can be used to extract data from XML documents. XQuery, operating effectively with the structure, can convey complex semantic meaning in the query, and therefore can retrieve precisely the desired results. However, if the user wants to use XQuery to search information efficiently, he needs to possess sufficient knowledge about XQuery and the structure of the XML document. It is, hence, apparent that this method is not a good choice for the beginning users.

Though a variety of research attempts have been made to further explore XQuery, all of their research results fail to settle the problem of complexity of the query syntax.

As the search for technological development, information retrieval(IR)style keyword search on the web has made great achievements. Keyword search is a proven user friendly way of querying HTML documents in the World Wide Web. It allows users to find the information by inputting some keywords. Inspired by

X. Du et al. (Eds.): APWeb 2011, LNCS 6612, pp. 460–471, 2011.

Table 1. Main technique in XML keyword search field recent years

YEAR	Keyword Search Semantics
2003	XKeyword[16], XSEarch[11],XRANK[10]
2004	MLCA[1]
2005	SLCA[2]
2006	Tree Proximity[19]
2007	XSeek[3], SLCA[9], CVLCA[8]
2008	MaxMatch[4], eXtract[15], ELCA[21]
2009	XReal[7], RTF[6], XReD[23]
2010	XBridge[22]

this, it is desirable to support keyword search in XML database. It is a very convenient way to query XML databases since it allows users to pose queries without the knowledge of complex query languages and database schema.

Based on the above situation, XML keyword search in recent years has become a hot research area with a good many papers published and great achievements scored. In this paper, we survey the search semantics and the algorithms presented in several representative papers.With the introduction of these papers, the readers could understand the main research stream of XML keyword search in recent years, as is shown in Table 1.

The research findings of XML keyword search can be generalized from three dimensions:

(1) The efficiency and effectiveness of XML keyword search,
(2) the return results ranked semantics,
(3) the biased snippet and result differentiation of XML keyword search.

We will introduce some papers' main ideas in each dimensions.

2   Efficiency and Effectiveness of XML Keyword Search

The efficiency and effectiveness are both important measures for XML keyword search. Every semantic for XML keyword search should consider these measures. Efficiency often means one method's time cost and space cost. Effectiveness often reflect the meaningfulness of the return result,in the other words, effectiveness can map the results matching degree of user's search intention. Some semantics have been proposed to improve these areas. In this section,we introduce some classical methods based on these semantics.

2.1   XKeyword

XKeyword[16]: Being part of the early research results, XKeyword provides efficient keyword proximity queries for large XML graph databases. This paper adopts the concept that a keyword proximity query is a set of keywords and the results are trees of XML fragments(called *Target Objects*) that contain all

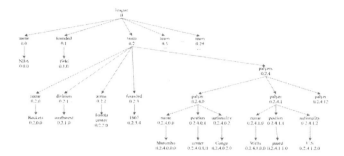

Fig. 1. The XML Tree

the keywords. In order to achieve quick response, XKeyword builds a set of connections,which precompute particular path and tree connections on the *TSS* graph. In this way, the XML data can be stored efficiently to allow the quick discovery of connections among elements that contain keywords; besides, the cost of computing the full presentation graph is very high. Hence, XKsearch adopts an on-demand execution method,where the execution is guided according to the users navigation. Another crucial problem that the result should be meaningful to the user and can not contain too much irrelevant information has been resolved by XKeyword. XKeyword associates a minimal piece of information,called *Target Object*,to each node and displays the target objects instead of the nodes in the results, in this way to make sure the meaningfulness of the result. XKeyword avoids producing duplicate results by employing a smart execution algorithm.

2.2   XSEarch

XSEarch[11]: is a search engine for XML, it developed a syntax for search queries that is suitable for a naive user and facilitates a fine-granularity search. *XSEarch* presents excellent query semantics. It defines three search terms *l:k*, *l:*, *:k*. If an interior node n wants to satisfies one of the search terms, it should be fixed model. Thinking about the effectiveness of the return results, *XSEarch* presents a relation which can be used to determine whether a pair of nodes is meaningfully related. $T_{|n_1,n_2}$ denotes the shortest undirected path between node n_1 and node n_2 consists of the paths from the lowest common ancestor of n_1 and n_2 to n_1 and n_2, then paper[11]defined that n and n' are interconnected if one of the following conditions holds:

1. $T_{|n,n'}$ dose not contain two distinct nodes with the same label or
2. the only two distinct nodes in $T_{|n,n'}$ with the same label are n and n'.

XSEarch also developed a suitable ranking mechanism that takes into account the effectiveness of XML keyword search. we will introduce this ranking sematic in section 2. In order to improve the efficiency of XSEarch, interconnection index and path index are defined.

2.3 SLCA

SLCA[2]: SLCA means *Smallest Lowest Common Ancestor* semantics. A keyword search using the SLCA semantics returns nodes in the XML data that satisfy the following two conditions:(1) the subtrees rooted at the nodes contain all the keywords,and(2)the nodes do not have any proper descendant node that satisfies condition(1).

Example 1: In the Figure 1,we put a query list"Mutombo,position", then the SLCA is "player" with the Dewey id [0.2.4.0]. The answer to the keyword search is the subtree rooted at the "palyer" node with id 0.2.4.0.This result is a better answer than the ssubtrees rooted at "players" or "team".

In the paper[2], two efficient algorithms,*Indexed Lookup Eager* and *Scan Eager* are designed for keyword search in XML documents according to the SLCA semantics. In this paper, XML document is treated as a common tree, and the nodes in the XML tree use *Dewey* numbers as the id, as is showed in Figure 1. The *Indexed Lookup Eager* algorithm is based on four properties of SLCA, one of these properties is:

$$slca(S_1, ..., S_k) = get\_slca(get\_slca(S_1, ..., S_{K-1}), S_K) \ for \ k<2$$

The next algorithm is *Scan Eager* which is a variant of *Index Lookup Eager* Algorithm.When keyword search includes at least one low frequency keyword along with high frequency keywords,then the Scan Eager can be efficiently used.

2.4 CVLCA

CVLCA[8]: In the above, we have introduced three semantics which improved the efficiency and effectiveness of XML keyword search. But it is not difficult to find that they sometimes return meaningless result or incomplete of answers. In order to improve the above situation, *VLCA* and *Compact VLCA* are proposed in the paper[8]. *VLCA* not only eliminates redundant LCAs but also retrieve relevant answers filtered out wrongly,and thus improves both accuracy and completeness of keyword search. The definition of *Homogenous/Heterogenous* and *MDC(Meaningful Dewey Code)* are proposed,they are both important definitions for *CVLCA*. In paper[8] the compact connected trees rooted *CVLCAs* are treated as the answers of keyword queries. Moreover, the paper presents an optimization technique for accelerating the computation of CVLCAs and device an efficient stack-based algorithm to identify the meaningful compact connected trees.

2.5 MaxMatch

MaxMatch[4]: In view of the above-mentioned situation, paper[4] proposes an axiomatic framework that includes two intuitive and non-trivial properties: *monotonicity* and *consistency*, with respect to data and query. Data/Query Monotonicity describes how the number of query results should change upon an update to the data or query. Data/Query Consistency describes how the content of query results should change upon an update to the data or query. These properties are non-trival, non-redundant, and satisfactory. *MaxMatch*, a novel semantics for identifying relevant matches and an efficient algorithm to realize this semantics, have been introduced in paper[4] which satisfies all properties and returns the filtered fragments root at SLCA nodes after pruning the uninteresting nodes. To our knowledge, paper[4] is the first work on reasoning about and evaluating XML keyword search strategies using a formal axiomatic framework.

2.6   RTF

RTF(Relaxed Tightest Fragments): From the section above, we can learn that a contributor-based filter mechanism in the MaxMatch algorithm is proposed. Analysis of paper[6], however, shows that the contributor-based filtering mechanism is not enough to filter out all the uninteresting nodes based on the SLCA. Besides, only focusing the SLCA related fragments is not enough for XML keyword search either. They both suffer *false Positive example* and *Redundancy problem*. Considering these situations, In paper [6] proposes a framework of retrieving meaningful fragments rooted at not only the SLCA nodes but also all of the LCA nodes. The concept of Relaxed Tightest fragment (RTF) is proposed to represent the basic result for the XML keyword search. Then the concept of valid contributor is defined as the following:

(Valid Contributor)*Given an XML tree T and the keyword query $Q = w_1, ..., w_k$, R is a RTF in $T.u,v$ are two nodes in R, and u is the parent of v. The child v is a valid contributor of u if either of the following two conditions holds:*
1. *v is the unique child of u with label $\lambda(v)$;*
2. *v has several siblings $v_1, ..., v_m$ $(m \geq 1)$ with same label as $\lambda(v)$. but the following conditions hold:*
   - *$\nexists v_i, TK_v \subset TK_{v_i}$;*
   - *$\forall v_i \wedge TK_v = TK_{v_i}, TC_v \neq TC_{v_i}$.*

The rule 1 aims to overcome the false positive problem, and rule 2 is used to overcome the redundancy problem. By taking into account the label and the content of a node, the children of a node could be classified as valid contributors and non-contributors. paper [6] also introduces an algorithm based on ValidRTF.

3   Ranking Semantics and Result Type of XML Keyword Search

As was introduced above, XML keyword search provided a simple way to query XML database. However, this kind of simple query format may not be precise and returns too many results, this can bring inconvenience to the users. To address this problem, some methods were proposed in recent years. One proposed way is to first compute the query results and then rank them; another method[14] infers the return node type by analyzing keyword match patters; XReal[7] were proposed recently,which utilized the statics of underlying XML data. In this section we will introduce all of the semantics.

3.1 XRANK

XRANK[10]: XRANK[10] tries to solve the problem of efficiently producing ranked results for keyword search over hyperlinked XML documents.In order to define specific ranking function,the desired properties for ranking functions over hyperlinked XML documents are proposed firstly. There are three desired properties that ranking function should take into account: Result Specificity, Keyword Proximity and Hyperlink Awareness. Result specificity means that the ranking function should rank more specific results higher than less specific results. Some functions respect to different content are well defined in the paper. For example the overall ranking of a result element v_1 for query $Q = k_1, k_2, ..., k_n$ is computed as follows.

$$R(v_1, Q) = \{ \sum_{1 \leq i \leq n} r(v_1, k_i) \} \times p(v_1, k_1, k_2, ...k_n)$$

The overall ranking is the sum of ranks with respect to each query keyword.

Besides, the *ElemRank* which is a measure of the objective importance of an XML element is proposed in the paper[10]. The algorithm for Computing *ElemRank* is as follows:

$$e(v) = \frac{1 - d_1 - d_2 - d_3}{N_d \times N_{de}(v)} + d_1 \sum_{(u,v) \in HE} \frac{e(u)}{N_h(u)} + d_2 \sum_{(u,v) \in CE} \frac{e(u)}{N_c(u)} + d_3 \sum_{(u,v) \in CE^{-1}} e(u)$$

In this function, d_1, d_2, d_3 are the probabilities of navigating through hyperlinks, forward containment edges,and reverse containment edges,respectively. Based on the above definition, two algorithms are designed to rank result for XML keyword search queries: *DIL* query processing algorithm and *RDIL* query processing algorithm. *DIL (Dewey Inverted List)* only stores the IDs of elements that directly contain the keyword, its size is likely to be much smaller the size of the naive inverted list. XRANK system for ranked keyword search over XML documents is designed based on all the content mentioned above.

3.2 XSEEK

XSEEK[3]: Ranking semantics may help users quickly find the information they want to find. But the challenge what appropriate data nodes to be returned has not addressed by the ranking semantics. To solve this problem, paper[3] proposed some definitions ,algorithms and one *XSEEK* system. The first guideline of the *XSEEK* system is to differentiate nodes representing entities from nodes representing attributes,and generate return nodes.when the schema is available, all the node categories are defined as follows:

1. A node represents an *entity* if it corresponds to a *-node in the DTD
2. A node denotes an *attribute* if it dose not correspond to a *-node, and only has one child,which is a value.
3. A node is a *connection* node if it represents neither an entity nor an attribute.

The second guideline of *XSEEK* system is to take keyword macth patterns into consideration,by classfying keywords into two categories:*search predicates* and *return nodes*.The data nodes that match return nodes are output based on their node categories: attributes,entities and connection nodes. On the basis of the above-mentioned semantics, paper[3] presents the algorithms that process keyword search on XML data and achieve the semantics efficiently.

3.3 XReal

XReal[7]: With the increasing attention paid to the study of the effectiveness and ranking semantic of XML keyword search, some problems and approaches are put forward. One of the new proposal is about the effectiveness and ranking mechanism in term of result relevance summarized as three issues in[7]:

> Issue 1:Identify the type of target node(s) that a keyword query intends to search for. *Search for node.*
> Issue 2:Effectively infer the types of condition nodes that a keyword query intends to search via. *Search via nodes.*
> Issue 3:Rank each query result in consideration of the above two issues.

Inspired by the important role of data statics in IR-ranking,[7]maintains and exploits two important basic statics terms, i.e. (1)*XML TF(term frequency)*$f_{a,k}$: The number of occurrences of a keyword k in a given *data node* a in XML data,(2)*XML DF(document frequency)*f_k^T: the number of T-types nodes that contain keyword k in their subtrees in XML data. Then three guidelines of the *search for* node type T are defined: 1. T is related to every query keyword, 2. T should be informative enough, 3. T should not be overwhelming. By incorporating the above guidelines, we define $C_{for}(T,q)$,which is the confidence of node type T to be *search for* node type:

$$C_{for}(T,q) = \log_e(1 + \prod_{k \in q} f_k^T) * r^{depth(T)}$$

The function of the confidence of a node type T to be a desired type to search via is also designed.A novel relevance-oriented ranking scheme called XML *TF*IDF* similarity which can capture the hierarchical structure of XML and resolve Ambiguity1-3 in a heuristic way is introduced, and based on the above definitiona keyword search engine prototype called XReal is implemented.The algorithm of XReal is defined in two steps: first step is *data processing* and *index construction*; then the second step is keyword search and ranking the results.

4   XBridge

XBridge[22]: For common users, each result type implies a possible search intention, so it is desirable to efficiently work out the most relevant result type*(Promising Result Types)* from the retrieved data. The process to determine the promising result type contains three steps:(1) First, computing all query results individually and classifying them, (2)then the score of each type is computed by a proposed ranking function,(3)finally, the type with the highest score is chosen as the promising result type.

Two algorithms are defined for deciding promising type to the query. The main procedure of the algorithms is as follows: (1) first, the corresponding distinct paths of each term in the query should be determined by the given keyword query.(2)then, we can obtain all result types where each type is a distinct label path from the root node to the lowest connected node on its query template.(3) After that, we compute the score for each result by calling a ranking function.(4)finally, the result type with highest score will be considered as the promising type to query. Based on the above contented, the search engine prototype called XBridge is implemented .

5   Snippet and Result Types of XML Keyword Search

From the previous section,we can learn that various ranking schemes have been proposed to assess the relevance of query results so that users can focus on the ones that are deemed to be highly relevant. However,because of the ambiguity of search semantics, it is impossible to design a ranking semantics that always perfectly gauges query result relevance with respect to users' intentions,besides, a user may would like to investigate,compare multiple relevant results for information discovery and decision making. To compensate the inaccuracy of ranking functions and help the user differentiate multiple results, snippet, promising result types,and result differentiation are researched during recent years. In this section,we will introduce these semantics.

5.1   Snippet

Snippet[15]: Most text search engine uses snippets to complement ranking scheme in order to effectively handle user searches. Today, some researches on result exploration of snippets are also under way. Paper[15] is the first paper which tries to carry out research on the area. Four goals are identified for a good result snippet: (1) Self-contained. a query result snippet should be self-contained so that the user can understand it.(2)Distinguishable A snippet should make the corresponding query result distinguishable from the snippets of other query results so that the users can differentiate them with little effort. (3) Representative Snippets. Therefore, the users can grasp the essence of the result from its snippet. (4) Small. A query result snippet should be small so that the user can quickly browse several snippets. The most significant information in the query result that should be selected into a snippet in a snippet information list is identified to meet the first three goals. To satisfy the four requirements, paper[15] designs and implements a novel algorithm to efficiently generate informative yet small snippets.

5.2   Result Differentiation(XRed)

XRed[23]: As we can learn from above, snippets highlight the most dominant features in the results. Without considering the relationships among results, the snippets are not helpful to compare and differentiate multiple results.

In paper[23], some techniques for comparison and differentiation of structured search results are proposed. An algorithm takes as input a set of structured results, and outputs a Differentiation Feature Set(DFS) for each result to highlight their differences within a size bound. Three desiderata for *DFS* are discussed: (1) *limited size, small size of query results* (2)reasonable summary. Otherwise,the differences shown in DFSs do not reflect the actual differences between the corresponding query results.(3)maximal differentiation. For summarizing query results, the DFS should appease two rules: (1) Dominance Ordered (2) Distribution Preserved.

The *DFS* construction problem is proved to be a NP-hard problem. Due to the NP-hardness of the DFS construction problem,two local optimality criteria are proposed: single-swap optimality and multi-swap optimality, then two efficient algorithms for achieving these criteria are designed. XReD system is implemented which satisfies all definitions are mentioned above.

6   Experiments

To verify the effectiveness and efficiency of these approaches we have introduced above, some experiments were designed. all experiments were conducted on a 1.66HZ Intel Core Dou CPU machine with 2GB RAM running Windows XP. The algorithms were implemented in Java and the parsing of the XML files was performed using the SAX API of the Xerces Java Parser.We choose some of algorithms for our experiments. We compare these approaches' efficient, effectiveness. DBLP is used as the test database. The size of the data is 389M.

6.1   Effectiveness

As we said above , effectiveness is the match's degree of users' search intention. To get a fairly objective view of user search intentions in real world, we organize ten persons and tell them some basic information about DBLP XML database. all of the people are asked to write ten to twenty queries and their desired information that can satisfies their search intention. For example, the Fig.2. shows one people's example.

We divide the ten subjects into two groups,whose testing results will be queried with five different methods,and then a comparison will be done to find

Table 2. User Queries and Itention

	Query	Intention
Q_1	xml book	the books about XML
Q_2	xml, twig, paper	the paper about XML twig
Q_3	author jim book	the book wrote by Jim
Q_4	paper 2010 xml	the paper published in the year of 2010
....	......	......

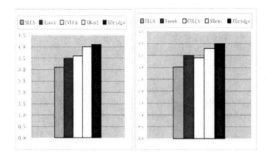

Fig. 2. The Score of Algorithms

the matching degree between the returned results and their search intentions. Each method will be correspondingly graded with points ranging from one to five according to the satisfaction of the subjects. Finally, the average mark of each method as a group is given after the above operation is carried out by the subjects. The results are showed in Figure 2.

From the figure, we can find that the XReal and XBridge do well then the others. XReal decides the *search for node* and *search via node* before calculating the results. XBridge propose a new method of predicting the result type for XML keyword queries. Different semantic of approach cause different effectiveness.

6.2 Efficiency

The efficiency of XML keyword search refers to the time cost of the search step. We evaluate the efficiency of SLCA, XSeek, CVLCA, XReal, XBridge respectively and know that the time costs of various queries vary with the structure of XML database. In this section, we divide the selected fifteen queries into three groups and record the time cost of each query. Next, we compute the average time cost for each corresponding query with a specific method as a group. The following Figure.3 shows the time cost of each method.

According Figure.3, we can see the efficiency of these algorithms. We can infer that the algorithms want to get better effectiveness, it may loose the efficiency of keyword search.

From above, we learn some chosen algorithms' compare. And learn some knowledge about the efficiency and effectiveness of XML keyword search.

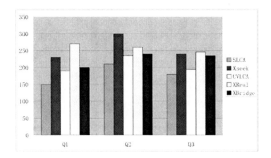

Fig. 3. The Efficiency of Algorithms

7   Conclusion

This paper starts from the introduction of a couple of representative papers and gives a brief explanation of representative semantics and algorithms. It can be concluded that the research of XML keyword search gradually shares many features with the text keyword search, which facilitate the search. As for the future works, existing methods mostly concentrate on retrieval on inverted lists in XML keyword search. Management of efficient indexes is usually not taken into account. There should be efficient mechanism to use B+ tree or other indexing methods to perform search efficiently. Another improvement can be supporting update in XML documents. This is important especially in XML documents which are frequently updated. Finally, another important future work would consider keyword query alternation and suggestion in XML keyword search. An example of this would be modifying " Schwarzeneger California governor " to " Schwarzenegger California governor " to correct the typo. In general a well query-alternation tool is needed for the practical application of XML keyword search platform.

References

1. Li, Y., Yu, C., Jagadish, H.V.: Schema-free xquery. In: VLDB (2004)
2. Xu, Y., Papakonstantinou, Y.: Efficient Keyword Search for Smallest LCAs in XML Databases. In: Proceedings of SIGMOD (2005)
3. Liu, Z., Chen, Y.: Identifying Meaningful Return Information for XML Keyword Search. In: SIGMOD (2007)
4. Liu, Z., Chen, Y.: Reasoning and identifying relevant matches for xml keyword search. In: VLDB (2008)
5. Liu, Z., Chen, Y.: Answering keyword queries on xml using materialized views. In: ICDE (2008)
6. Kong, L., Gilleron, R., Lemay, A.: Retrieving Meaningful Relaxed Tightest Fragments for XML Keyword Search. In: EDBT (2009)
7. Bao, Z., Ling, T.W., Chen, B., Lu, J.: Effective xml keyword search with relevance oriented ranking. In: ICDE (2009)

8. Li, G., Feng, J., Wang, J., Zhou, L.: Effective Keyword Search for Valuable LCAs over XML Documents. In: Proceedings of CIKM (2007)
9. Sun, C., Chan, C.Y., Goenka, A.K.: Multiway slca-based keyword search in xml data. In: WWW, pp. 1043–1052 (2007)
10. Guo, L., Shao, F., Botev, C., Shanmugasundaram, J.: XRANK: Ranked keyword search over XML documents. In: SIGMOD (2003)
11. Cohen, S., Namou, J., Kanza, Y., Sagiv, Y.: XSEarch: A semantic search engine for XML. In: VLDB (2003)
12. Al-Khalifa, S., et al.: Structural joins: A primitive for efficient XML query pattern matching. In: ICDE (2001)
13. Al-Khalifa, S., et al.: Querying structured text in an XML database. In: SIGMOD (2003)
14. Chien, S.-Y., et al.: Efficient structural joins on indexed XML documents. In: VLDB (2002)
15. Huang, Y., Liu, Z., Chen, Y.: Query biased snippet generation in xml search. In: SIGMOD (2008)
16. Hristidis, V., et al.: Keyword proximity search on XML graphs. In: ICDE (2003)
17. Aditya, B., et al.: BANKS: Browsing and keyword searching in relational databases. In: VLDB (2002)
18. Agrawal, S., et al.: DBXplorer: a system for keyword based search over relational databases. In: ICDE (2002)
19. Hristidis, V., Koudas, N., Papakonstantinou, Y., Srivastava, D.: Keyword proximity search in XML trees. In: TKDE (2006)
20. Carmel, D., Maarek, Y.S., Mandelbrod, M., Mass, Y., Soffer, A.: Search xml documents via xml fragments. In: SIGIR (2003)
21. Xu, Y., Papakonstantinou, Y.: Efficient lca based keyword search in xml data. In: EDBT (2008)
22. Li, J., Liu, C.: Suggesttion of Promising Result Types for XML Keyword Search. In: EDBT (2010)
23. Liu, Z., Sun, P., Chen, Y.: Structure Search Result Differentiation. In: VLDB (2010)

Schema Mapping with Quality Assurance for Data Integration[*]

Xu Bian, Hongzhi Wang, and Hong Gao

Department of Computer Science and Engineering, Harbin Institute of Technology,
Harbin 150001, China
bianxu5555@gmail.com, {wangzh,honggao}@hit.edu.cn

Abstract. With the popularity of the internet, more and more data are generated on internet. Because of the usability of Extensible Markup Language(XML for short), more data is organized by XML document format. Because of the flexibility of XML, data organized by XML have a variety of organizational formats which brings a lot of inconvenience to data management. In particular, when the large-scale data operations are performed on XML data, for example data integration, model change, and so on, there are many problems. One of the current implementations is to use Data Exchange to carry out the above operations. The works of predecessors mainly are to analyze the characteristics of Schema Mapping on XML, and institute Data Exchange rules. These rules only consider the data integrity, reliability, but don't consider the quality of the data after conversion. This paper proposes the concept of quality assurance mechanisms. Firstly we discuss that a new model with quality assurance, and provide a suitable method for this model. Then we propose the strategy of weak branch's convergence on the basis of Schema. In the end theoretical analysis and experimental results show that the method is correct and feasible.

Keywords: XML; Schema Mapping; Data Exchange; Quality Assurance.

1 Introduction

Independently developed producers for example Amazon, eBay etc. bring about non-standard data in their respective databases. Consequently, when integrating data from different sources, the lack of a standard data schema is a severe problem. Subsequently Schema-Mapping targeting at the uniformity of data schema is a primary operation. To guarantee the quality of generated data is the main part in data exchange and information integration. We have a very good understanding of mappings between

---

[*] Supported by the This research is partially supported by National Science Foundation of China (No. 61003046), the NSFC-RGC of China (No.60831160525), National Grant of High Technology 863 Program of China (No. 2009AA01Z149), Key Program of the National Natural Science Foundation of China (No. 60933001), National Postdoctoral Foundation of China (No. 20090450126, No. 201003447), Doctoral Fund of Ministry of Education of China(No. 20102302120054), Postdoctoral Foundation of Heilongjiang Province (No. LBH-Z09109), Development Program for Outstanding Young Teachers in Harbin Institute of Technology (No.HITQNJS.2009.052).

X. Du et al. (Eds.): APWeb 2011, LNCS 6612, pp. 472–483, 2011.
© Springer-Verlag Berlin Heidelberg 2011

relational schemas (see, e.g., recent SIGMOD and PODS keynotes on the subject [1, 2]); several advanced prototypes for specifying and managing mappings have been developed and incorporated into commercial systems [3, 4]. There are techniques for using such mappings in data integration and exchange, and tools for handling mappings themselves, for example, for defining various operations on them [2, 5, 6, 7, 8, 9]. But much less is known about mappings between XML schemas.

However, more and more of today's data are represented in non-relational form. In particular, XML is increasingly popular, especially for data published on the Web and data exchanged between organizations. [10] XML data has become the standard of information conversion and integration on the Internet because of the usability of XML. Nevertheless, it's difficult for computer to automatically operate and analyze heterogeneous XML information, so information cannot be extracted and collected effectually. For that reason, Schema-Mapping is required to integrate originally heterogeneous XML information into same Schema. Since XML data is semi-structured, traditional Schema-Mapping operation on the heterogeneous XML data cannot guarantee the quality of generated data. Our work in this paper is committed to achieving Schema-Mapping with Quality Assurance for XML Data Exchange.

The problem we faced has some properties: Heterogeneous Representation (HR) and Inconsistency Hidden (IH) in the semi-structure. 1. HR means that different sources for example Amazon, eBay etc. have different schemas to describe their data. As a semi-structured data description language, XML allows the same data is organized various forms through attributes and elements. 2. IH means that inconsistent data appear in the generated result after the Schema-Mapping operation.

For example, two people, Jack Baker and Jason Brown, are obvious two persons. In the circumstances of writing the full name, they can be distinguished automatically by computer. As a matter of fact, data are not invariably standard as our example. In some case, they both are written as J. B. and to computer, J. B. is one person. Otherwise, Jack Baker is written as Jack B. and computer cannot identify that both names refer to the same person. The former phenomenon is called Homonymy-Name-Problem (HNP), and the latter phenomenon is called Synonym-Name-Problem (SNP).

In general, there are some traditional methods to solve the information integration problem. [11-14] carry out a technique about Schema-Mapping between relational database and XML data. In fact, data integration or data Exchange is an operation of high CPU and Memory cost in database. The predecessor's contribution cannot be simply used in the XML database. Even if the technology of predecessor is applied to implement a system, the system cannot guarantee the quality of data.

In this paper, we are determined to address HNP and SNP to improve data quality. Since entity in the XML data is not correctly identified, HNP and SNP emerge. This will cause that we cannot get complete and accurate information about one entity on the whole. Solving HNP raises the rate of precise (accurate) and solving SNP raises the rate of recall (complete). So we introduce entity identification into Schema-Mapping and bring on Schema-Mapping based on entity.

In this paper, we first summarize the predecessor's works about Schema-Mapping and find the ignored on their methods. Subsequently, we put forward a method with data quality assurance which addresses HNP and SNP based on entity. Afterward a strategy comes up to automatically locate the node with sensitive quality. Ultimately, we prove our algorithm is practicable, feasible and valid through experiment. The rest of the paper is organized as follows. Section 2 presents a motivating example Section 3 describes the model and criterion. Section 4 gives the detailed methods. Section 5 is the implement of automatically positioning with quality problem mechanism. Section 6 uses experiments to verify our methods. Section 7 makes a conclusion. And references are in Section 8.

2   Motivating Example

Let us consider a scenario that a company which has a large number of paper information organized as Schema S carries out data integration that need to update data Schema. We can abstract a simple problem as Figure 1 from Schema-Mapping. For the

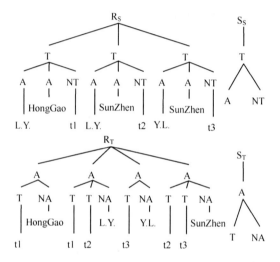

Fig. 1. XML Data Tree and XML Schema Tree

Table 1. The notations used in example

Number	Lable	Description
1	R_S	Root of source data tree
2	R_T	Root of target data tree
3	S_S	Root of source schema tree
4	S_T	Root of target schema tree
5	T	One piece of paper, means Title
6	A	One author, means Author
7	NT	Title of paper, means Name of Title
8	NA	Author name, Name of Author

reason that we mainly focus on XML data which can be represented as node labeled trees [15], the following example will be shown in the form of tree structure. The meaning of label in our example is described in Table 1.

In the source, R_S shows three different paper which can be distinguished by title of paper or paper's authors. R_T shows the author information from original data (after data integration based value). However in the target data, except for the same name (L.Y.), the author of t1 and the author of t2 could not be identified one person, and this is HNP. On the contrary, t2's author L.Y. and t3's author Y.L. are seen as two people even though there is a co-author (SunZhen), and this is SNP. Not only co-author is the description of author, but also tree structure and lable value are the descriptions.

3 Model and Criterion

In this section, we introduce a new model based on entity to solve Schema-Mapping, and propose a criterion to validate that our solution is able to achieve high precision and recall. The model is illustrated as Figure 2 which shows fundamental processes in Schema-Mapping.

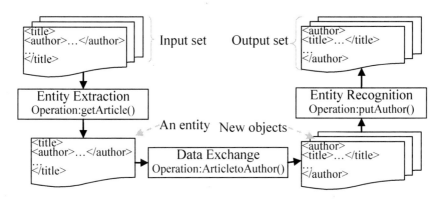

Fig. 2. Data Exchange with Schema-Mapping

We use precision and recall as metrics to evaluation. In order to calculate precision and recall, we generate artificial data. Considering the truth of research team, first of all team data are generated, and every team contains dozens of authors. Then paper data are generated, and every paper is written by several authors from one team. Every person-name might appear in one or more teams, and the aim is to produce HNP. When using person-name in paper, the string of person-name might be mixed into noise information, and the aim is to produce SNP. To evaluate precision, we compare output set with "Gold-Answer" from process of generation. Let O means the set of object-pairs which describe one entity in both output set and Gold-Answer. Let P means the set of object-pairs which describe one entity in output set and different entities in Gold-Answer. Let Q means the set of object-pairs which describe different entities in output set and one entity in Gold-Answer. Then the precision Pre is $\frac{|O|}{|O|+|P|}$ while the

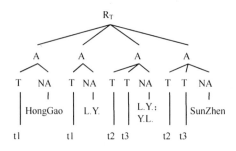

Fig. 3. Gold-Answer

recall Rec is $\frac{|O|}{|O|+|Q|}$. For instance, the output set is R_T in Figure 1, and Gold-Answer is shown in Figure 3. O={<t2,t3>} , P={<t1,t2>} and Q={<t2,t3>}. Pre=0.5 and Rec=0.5.

4   Data Exchange with Schema-Mapping

In this section, the processes of our model are detailed introduced. The Input set is broken down by the Entity Extraction process into separate entity. One or more objects are come into being from every entity by the Data Exchange process. And the Output set is organized by new entities generated from the Entity Recognition process.

4.1   Entity Extraction

The Entity Extraction component is used to extract entities from Input set, and this process ensures that the entity agrees with the source schema. Considering example in Figure 1, R_S is Input set, and every T, son of R_S, is extracted as one entity. The detailed procedures are shown in Operation 1.

Operation 1: getArticle()
1.for each document do
2.   get every article element;
3.   do ArticletoAuthor()
4.end for;

4.2   Data Exchange

The Data Exchange component is used to exchange the old data into new data. And this process acting on old entity retains structure information for produced entities. The detailed operation is various because of different source and target schemas. For our example, the procedures are shown in Operation 2.

Operation 2: ArticletoAuthor()

1.for each article element

2.   get every author element;

3.   restructure author element;

4.   get the structure information of every author;

5.   do putAuthor();

6.end for;

In our example, structure information is CoAuthor, and this is stored in list authors.

4.3 Entity Recognition

The Entity Recognition component is used to distinguish new entity and locate it. And this process aims at solving SNP and HNP through entity recognition. The basic plan is shown in Figure 4. At entrance, edit-distance between strings [16] and string transformations [17] are commendable to find out the hidden objects which are the potential same entity. At exit, the structure information is exploited to eliminate the objects which describe different entity. The procedure is shown in Operation 3.

Operation 3: putAuthor()

1.entity_set;//store all entity

…

2.for each author element do

3.   list=entity_set.find(author.name);//use name value to match entity

4.   for each e in list do

5.   e.match(author);//use structure information

6.   end for;

7.   all entities matched merge;

8.end for;

In the circumstances of data integration and schema mapping, the entity recognition method based on tree edit distance [18] cannot work well. The approach which only considers the simple but significant structure information will obtain a high precise result within reasonable time. In the example, CoAuthor is chosen as structure information.

In operation 3, the entity_set stores all entities and their structure information, dynamic managed in the process of handling object one by one. Similarly the list stores all entities that are candidates of the object, author element. In real application, we have millions of objects that do not necessarily fit in the memory and need to be stored on disk. In putAuthor(), content of entity is stored in entity_set, which can be put in disk. This problem can be solved through adding a mapping structure between entity_set in memory and disk.

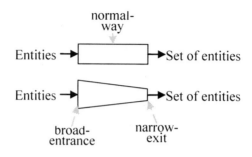

Fig. 4. Basic plan

4.3.1 Broad-Entrance

Broad-entrance is realized by the sentence:

> **list=entity.find(author.name);**

Method find() is defined at entity, and its input is a string of author name while its output is the list of similar entities. The method introduced in [16] and [17] certainly work well in real application. In this paper, our approach below swift obtains list of authors.

In the process of Broad-Entrance, the purpose is to guarantee all entities which are referred by name value are returned. Obviously the string of name is simple but the information hidden carries main ability of distinguish. So the value of name becomes the foundation of Broad-Entrance. Approach in [19], originally proposed in [20], takes as signature all values having a hash value that is divisible by a constant m. When the value of object matches several entities, those entities are returned. This can work well based on clean data where different values mean different entity. Actually there is not clean data in specific applications. One way to deal with this is to replace each string by a set of chunks - e.g. words, q-grams[21] or v-grams[22] occurring in it, and then to compare these sets of chunks. A method for converting strings into sets of chunks is called a chunking strategy. A very simple way for comparing string value based on chunks is to form for each string the set of chunks occurring in any of its values, and then compare these sets. Such an approach has been suggested in [23]. Technically this can be done exactly as before the only difference is that we are now comparing sets of chunks instead of strings containing the original values. While this gives us a rough estimate on whether two strings have something in common, it is not very accurate since information held by structure is not taken into account [24].

By means of chunks, the original value is broken into more short strings. Then the chunks are hashed into different numerical values. The minimum value is chosen as the signature of the set of chunks and the entity. The signature of object obtained as the same method matches entities, then the candidate entities is discovered by filter.

4.3.2 Narrow-Exit

Narrow-exit is realized by the sentence:

> **e.match(author);**

Coauthor is the description of the detected object which means variable author in instance, and stores all coauthors organized as one set. Through comparing the description with existing entities one by one, the truth that the object belongs to a certain entity is discovered. By utilizing Jaccard Measure [25], compare method determines the relation between the object and the entity.

Our approach to schema mapping based on entity is separated into above three operations. To derive the complexity of our method, assume that each of N articles has exactly A authors, and each of M authors has B coauthors. Then the time to process our input set R_S is $O(N \times A \times (M + c \times (A-1) \times B)) = O(N \times A \times M)$. Here c means the number of similar authors. Due to $M >> (A-1) \times B$, the runtime is mainly decided by the scale of article and author.

5   Mechanism of Automatically Positioning with Quality Problem

The model proposed in above section solves the quality problem with given position. In fact, the schema and data are all the information we can get, and the position with quality problem cannot be discovered immediately. The way exploiting programmer's experience can help to manually distinguish the position with quality problem. But it's not a general method. In consequence, the mechanism of automatical positioning with quality problem is proposed in this section.

To realize the auto-mechanism, DTD document, the description of XML data, is detailed analyzed. There are some discoveries:

1. Attribute never needs to be treated as one entity. When the XML data is organized as an XML tree, the attribute can be seen as attribute node.

2. It is probability that the number of certain type node under its father node is only 1 or 0.

A naïve idea is that father node absorbs the single child node as attribute, and a tree which retains main nodes is reorganized. This trick can cut down many weak branches in order to minish the structure of main entity. Based on this idea, the method of weak branch converged is proposed.

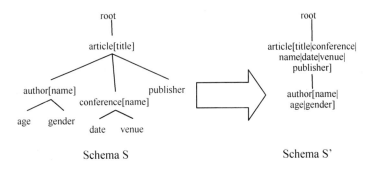

Fig. 5. Weak Branch Converged

5.1 An Example of Weak Branch Converged

The example in motivation is too simple; factually data have more complex structure. Schema S in Figure 5 is article description with more detailed information. Known from common sense, article has one publisher and one conference; author has one age and one gender. Through weak branch converged, S is reformed into S'.

Schema *S* which has 8 nodes without the root contains 2 main nodes: article and author. The pruning effects of strategy causes the space reduction which makes the processes of data exchange and entity recognition work.

5.2 The Strategy of Weak Branch Converged

As the target, a reduced structure is the form which makes the Data Exchange easy-handle. And the strategy is used to transform the complex data schema into a simple schema. Some definitions are given as below:

Central-node: is a relation between two or more nodes which satisfy many-to-many relationship. In Figure 5, article and author are central-node by common sense.

User-node: user might care some specific nodes. **Kernel-node:** the nodes satisfied central-node ship or designated by user.

The reduction process is that **Kernel-nodes** absorb **non-Kernel-nodes**. And target structure is made up of **Kernel-nodes**.

5.3 An Implement of Weak Branch Converged

A common-sense fitted assumption: if A and B have non-kernel-node relationship, they can be detected in local information. This is suitable in mostly circumstances, because information is gathered locally in the process of schema exchange between schema S and schema T. Therefrom three rules for recognition is given as below: 1. If A is B's father in both schema and A:B=1:1, node A absorbs node B and inherits all son nodes under B. 2. If A is B's father in one schema, A:B=1:1 and B is A's father in the other, B:A=1:1, A absorbs node B and inherits all son nodes under B in both schema under the circumstances (e.g. A has no brother node in latter schema). 3. A is B's father and A:B=1:1 in one schema. B is A's father, B:A=1:more, and the type of A is the only type of B's son nodes. Under this circumstances node A absorbs B.

Apply the above rules iteratively until all father-son pairs dissatisfy adsorbility. We use convergence ratio (CR for short) to judge the effect of reduction. CR=U/O, where U means node number of schema T and O means node number of schema S.

6 Experimental Results

Here we present experiment to evaluate our method work well.We ran all of the experiments on an Intel dual core 2.4 Ghz Windows machine with 1G RAM. Our code was implemented by Code::Blocks 8.02.

6.1 Precision and Recall

The result on real data is shown in table 2.

Table 2. Results of different Data Source

Data source	Number of entities	Number of objects	Pre	Rec
DBLP	14	50	97.2	96.5
Book-Order	28	45	98.5	100
CD	17	20	100	100
Course	26	28	100	100

From the results, we can draw some conclusions: 1. Different sources have different situations, and the results of our method are all very perfect. 2. The less objects described the same entity, the better result.

The diagrams which explain trend along with different parameters are illustrated in the following Figure 6. The data is artificial which are mentioned in Section 3.

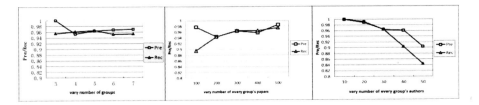

Fig. 6. Precision/recall measures for different parameters

For precision measure, it has nothing with the number of groups and the number of every group's papers, and decreases with the number of every group's authors. For recall measure, there is no explicit relationship between recall and number of groups. It increases with the number of every group's papers and decreases with the number of every group's authors.

6.2 Convergence Ratio

The result on real data is shown in table 3.

Table 3. Result of Weak Branch Converged

source	type	original node number	result node number	ratio
paper	DTD	8	2	0.25
Book-Order	Schema	12	4	0.33
CD	Schema	7	1	0.14
Hardware	DTD	7	2	0.29

It is very common that schema can be converged. And from real data, the converged operation brings about a striking effect (see the small ratio).

7 Conclusion

We have introduced the model and method to solve the data transformation problem with quality assurance. The new model settles HNP and SNP, and the method is based on entity. Meanwhile, a strategy is given to find nodes with sensitive quality. Finally, we prove our technique through experiment.

References

[1] Bernstein, P.: Model management 2.0:manipulating richer mappings. In: SIGMOD 2007, pp. 1–12 (2007)
[2] Kolaitis, P.: Schema mappings, data exchange, and metadata management. In: PODS 2005 (2005)
[3] Miller, R.: The Clio project: managing heterogeneity. SIGMOD Record 30, 78–83 (2001)
[4] Popa, L.: Translating web data. In: VLDB 2002, pp. 598–609 (2002)
[5] Bernstein, P.: Implementing mapping composition. In: VLDB 2006, pp. 55–66 (2006)
[6] Chiticariu, L.: Debugging schema mappings with routes. In: VLDB 2006, pp. 79–90 (2006)
[7] Fagin, R., Kolaitis, P., Popa, L., Tan, W.C.: Composing schema mappings: second-order dependencies to the rescue. ACM TODS 30(4), 994–1055 (2005)
[8] Madhavan, J.: Composing mappings among data sources. In: VLDB 2003, pp. 572–583 (2003)
[9] Nash, A., Bernstein, P., Melnik, S.: Composition of mappings given by embedded dependencies. ACM TODS 32(1), 4 (2007)
[10] Weis, M.: DogmatiX Tracks down Duplicates in XML. In: SIGMOD 2005, pp. 431–442 (2005)
[11] Feng, Y.: Mapping XML DTD to Relational Schema. In: DBTA 2009, pp. 557–560 (2009)
[12] Lu, S., Sun, Y., Atay, M., Fotouhi, F.: A New Inlining Algorithm for Mapping XML DTDs to Relational Schemas. In: ER (Workshops) 2003, pp. 366–377 (2003)
[13] Zhou, R.: Holistic constraint-preserving transformation from relational schema into XML schema. In: Haritsa, J.R., Kotagiri, R., Pudi, V. (eds.) DASFAA 2008. LNCS, vol. 4947, pp. 4–18. Springer, Heidelberg (2008)
[14] Liu, Y., Wang, T., Yang, D., Tang, S.: Propagating Functional Dependencies from Relational Schema to XML Schema Using Path Mapping Rules. In: International Conference on Internet Computing 2007, pp. 294–299 (2007)
[15] Milano, D., Scannapieco, M., Catarci, T.: Structure Aware XML Object Identification. In: CleanDB 2006 (2006)
[16] Ristad, E.S.: Learning String-Edit Distance. IEEE Trans. Pattern Anal. Mach. Intell (PAMI) 20(5), 522–532 (1998)
[17] Arasu, A.: Learning String Transformations From Examples. PVLDB 2(1), 514–525
[18] Lu, C.L., Su, Z.-Y., Tang, C.Y.: A new measure of edit distance between labeled trees. In: Wang, J. (ed.) COCOON 2001. LNCS, vol. 2108, pp. 338–348. Springer, Heidelberg (2001)

[19] Broder, A.: On the resemblance and containment of documents, p. 21. IEEE, Los Alamitos (1997)

[20] Manber, U.: Finding similar files in a large file system. In: USENIX Winter, pp. 1–10 (1994)

[21] Shannon, C.E.: A Mathematical Theory of Communication. CSLI Publications (1948)

[22] Li, C., Wang, B., Yang, X.: Vgram: Improving performance of approximate queries on string collections using variable-length grams. In: VLDB 2007, pp. 303–314 (2007)

[23] Dasu, T., Johnson, T.: Mining database structure; or, how to build a data quality browser. In: SIGMOD 2002, pp. 240–251 (2002)

[24] Köhler, H.: Sampling dirty data for matching attributes. In: SIGMOD 2010, pp. 63–74 (2010)

[25] Hamers, L., Hemeryck, Y., Herweyers, G., Janssen, M., Keters, H., Rousseau, R., Vanhoutte, A.: Similarity measures in scientometric research: The Jaccard index versus Salton's cosine formula. Inf. Process. Manage (IPM) 25(3), 315–318 (1989)

Author Index